BIRKHÄUSER

R.-D. Reiss

M. Thomas

Statistical Analysis of Extreme Values

with Applications to Insurance, Finance, Hydrology and Other Fields

Third Edition

Birkhäuser
Basel · Boston · Berlin

Rolf-Dieter Reiss
Michael Thomas
FB Mathematik
Universität Siegen
Walter-Flex-Str. 3
57068 Siegen
Germany
e-mail: reiss@stat.math.uni-siegen.de
          michael@stat.math.uni-siegen.de

2000 Mathematics Subject Classification 60G70, 62P05, 62P99, 90A09, 62N05, 68N15, 62-07, 62-09, 62F10, 62G07, 62G09, 62E25, 62G30, 62M10

Library of Congress Control Number: 2007924804

Bibliographic information published by Die Deutsche Bibliothek
Die Deutsche Bibliothek lists this publication in the Deutsche Nationalbibliografie;
detailed bibliographic data is available in the Internet at http://dnb.ddb.de

ISBN 978-3-7643-7230-9 Birkhäuser Verlag, Basel - Boston - Berlin

First edition 1997
Second edition 2001

© 2007 Birkhäuser Verlag AG, P.O. Box 133, CH-4010 Basel, Switzerland
Part of Springer Science+Business Media
Printed on acid-free paper produced from chlorine-free pulp. TCF ∞
Cover design: Alexander Faust, Basel, Switzerland
Printed in Germany

ISBN 978-3-7643-7230-9   e-ISBN 978-3-7643-7399-3

9 8 7 6 5 4 3 2 1   www.birkhauser.ch

Preface to the Third Edition

First of all we would like to thank all those who showed a steady, extraordinary interest in this book about extreme value analysis. This gave the motivation, courage and opportunity to provide an update of the book and to work out new topics, which primarily focus on dependencies, conditional analysis based on serial and covariate information, predictions, and the multivariate modeling of extremes.

As mentioned by Chris Heyde[1] while reviewing the second edition: "There is a considerable statistical content in the book, quite apart from its focus on extremes. ... The authors are seeking, quite properly, to embed the analysis of extreme values into the mainstream of applied statistical investigations." This strategy has been continued and strengthened in the new edition.

With each new edition there are complex questions as well as complex solutions. In that context, the cooperation with distinguished experts becomes more and more important.

Parts I–III about the basic extreme value methodology remained unchanged to some larger extent, yet notable are new or extended sections about

- *Testing Extreme Value Conditions with Applications* related to goodness–of–fit tests, co–authored by J. Hüsler and D. Li;

- *The Log–Pareto Model and other Pareto–Extensions* with a view towards super–heavy tailed distributions;

- *An Overview of Reduced–Bias Estimation*, an approach related to the jack-knife method, co–authored by M.I. Gomes;

- *The Spectral Decomposition Methodology* which reduces multivariate questions to univariate ones to some extent,

- *About Tail Independence* with testing tail dependence against certain degrees of tail independence, co–authored by M. Frick.

---

[1]Heyde, C. (2002). Book Review: Statistical Analysis ... Australian & New Zealand J. Statist. 44, 247–248.

Of central importance are the new chapters entitled

- *Extreme Value Statistics of Dependent Random Variables* co–authored by H. Drees;

- *Conditional Extremal Analysis* which provides the necessary technical support for certain applications;

- *Elliptical and Related Distributions* with special emphasis on multivariate Student and sum–stable distributions.

Other new topics are collected within

Part IV: Topics in Hydrology and Environmental Sciences;

Part V: Topics in Finance and Insurance,

Part VI: Topics in Material and Life Sciences.

Within these parts one may find

- a new chapter about *Environmental Sciences*, co–authored by R.W. Katz, with a detailed description of the concepts of cycles, trends and covariates;

- a new section about *Predicting the Serial Conditional VaR*, co–authored by A. Kozek and C.S. Wehn, including remarks about the model validation;

- a new section about *Stereology of Extremes*, co–authored by E. Kaufmann, with remarks about modeling and estimation.

The entire text has been thoroughly updated and rearranged to meet the requirements. The new results and topics are elaborated on about 120 pages.

The book includes the statistical MS Windows application Academic Xtremes 4.1 and StatPascal on CD. The STABLE package for sum–stable distributions is no longer included. The major difference of the academic version compared to the professional one is a restriction of the executable sample sizes. Consequently, not all of the numerical examples in the book can be executed with the academic version.

To keep the book at a reasonable size, the former sections *Implementation in Xtremes* and the separate *Case Studies in Extreme Value Analysis* were omitted. The Appendix about Xtremes and StatPascal of the second edition is considerably shortened. The full description of Xtremes and StatPascal—also including the former sections *Implementation in Xtremes*—may be found in the *Xtremes User Manual* which is enclosed as a pdf–file on the CD.

It is a pleasure to thank Th. Hempfling who showed great efficiency in editing this book.

Siegen, Germany   Rolf–Dieter Reiss
                                                             Michael Thomas

Preface to the Second Edition

With the second edition we continue a project concerning extreme value analysis in combination with the interactive statistical software Xtremes which started in the late 1980's. An early publication was Chapter 6 together with the User's Guide to Xtremes in [16], besides tutorials for the statistical software Xtremes in 1993 and 1995 which had a wider circulation within the extreme value community. These efforts culminated in the first edition of the present book which has found a favorable reception from the side of practitioners and in academic circles.

The new highlights of this extended edition, elaborated on about 160 pages, include

- the statistical modeling of tails in conjunction with the global modeling of distributions with special emphasis laid on heavy–tailed distributions such as sum–stable and Student distributions;

- the Bayesian methodology with applications to regional flood frequency analysis and credibility estimation in reinsurance business;

- von Mises type upper bounds on remainder terms in the exceedance process approximation and a thorough theoretical and practical treatment of the phenomenon of penultimate distributions;

- a section about conditional extremes;

- an extension of the chapter about multivariate extreme value models, especially for the Gumbel–McFadden model with an application to the theory of economic choice behavior;

- a chapter about the bivariate peaks–over–threshold method;

- risk assessment of financial assets and portfolios in the presence of fat and heavy–tailed distributions by means of the Value–at–Risk (VaR);

- VaR under the Black–Scholes pricing and for general derivative contracts;

- sections about corrosion analysis and oldest–old questions.

The "analysis of extreme values must be embedded in various other approaches of main stream statistics" as mentioned in the first edition. In that context, M. Ivette Gomes[2] remarks "its scope is much broader, and I would rather consider it a welcome addition to the reference works in applied statistics ... though there is a unifying basis provided by extreme value theory." For the second edition, it is our declared aim to enforce this characteristic of providing a broad statistical background in the book.

In Part V there is a continuation of the successful program concerning self–contained "Case Studies in Extreme Value Analysis" of other authors. The case studies in the first edition are replaced by new ones with emphasis laid on environmental extreme value statistics. We thank Humberto Vaquera, José Villaseñor, Stuart Coles, Jürg Hüsler, Daniel Dietrich, Dietmar Pfeifer, Pieter van Gelder and Dan Lungu for the new contributions.

It was a pleasure to cooperate with several distinguished experts in various fields, namely, with John Nolan (sum–stable distributions), Edgar Kaufmann (rates of convergence and longevity of humans), Michael Falk (multivariate peaks–over–threshold), Jon Hosking (flood frequency analysis), Michael Radtke (insurance) and Casper de Vries and Silvia Caserta (finance).

The present statistical software environment is much more than an update of Xtremes, Version 2.1. As a consequence of our intention to establish a book about applied statistics—with a unifying basis provided by extreme value statistics—the Xtremes package becomes more and more applicable to various statistical fields. Further introductory remarks may be found in

- Xtremes: Overview and the Hierarchy

- Xtremes and StatPascal Within the Computing Environment RiskTec

after the Prefaces and at the beginning of the Appendix.

We would like to thank colleagues, readers and users of the first edition and Xtremes for their comments, questions and suggestions; among others, Claudio Baraldi, Arthur Böshans, Holger Drees, Harry Harper, Sylvia Haßmann, Claudia Klüppelberg, Elson Lee, Frank Marohn, Alexander McNeil, Richard Smith, Q.J. Wang, Carsten Wehn.

Siegen, Germany   Rolf–Dieter Reiss
                                                                   Michael Thomas

---

[2]Gomes, M.I. (1999). Book Review: Statistical Analysis ... Extremes 2, 111–113.

Preface to the First Edition

This textbook deals with the statistical modeling and analysis of extremes. The restriction of the statistical analysis to this special field is justified by the fact that the extreme part of the sample can be of outstanding importance. It may exhibit a larger risk potential of random scenery such as floods, hurricanes, high concentration of air pollutants, extreme claim sizes, price shocks, incomes, life spans, etc. The fact that the likelihood of a future catastrophe is not negligible may initiate reactions which will help to prevent a greater disaster. Less spectacular yet important, the statistical insight gained from extremes can be decisive in daily business life or for the solution to ecological or technical problems.

Although extreme value analysis has its peculiarities, it cannot be looked at in an isolated manner. Therefore, the analysis of extreme values must be embedded in other various approaches of main stream statistics such as data analysis, nonparametric curve estimation, survival analysis, time series analysis, regression analysis, robust statistics and parametric inference.

The book is divided into

Part I: Modeling and Data Analysis;

Part II: Statistical Inference in Parametric Models;

Part III: Elements of Multivariate Analysis;

Part IV: Topics in Insurance, Finance and Hydrology;

Part V: Case Studies in Extreme Value Analysis,

Appendix: An Introduction to XTREMES.

Whenever problems involving extreme values arise, statisticians in many fields of modern science and in engineering or the insurance industry may profitably employ this textbook and the included software system XTREMES. This book is helpful for various teaching purposes at colleges and universities on the undergraduate and graduate levels. In larger parts of the book, it is merely presumed that the reader has some knowledge of basic statistics. Yet more and more statistical prerequisites are needed in the course of reading this book. Several paragraphs and subsections about statistical concepts are intended to fill gaps or may be regarded as shorter refresher units. Parts I and II (with the exception of Chapter 6) are elementary yet basic for the statistical analysis of extreme values. It is likely that a more profound statistical background is helpful for a thorough understanding of the advanced topics in Chapter 6 and the multivariate analysis in Part III.

Part I sets out the statistical background required for the modeling of extreme values. The basic parametric models are introduced and theoretically justified in Chapter 1. The nonparametric tools introduced in Chapter 2 are most important for our approach to analyzing extreme values. In this context the included statistical software system is helpful to

- get a first insight into the data by means of visualizations;

- employ a data–based parametric modeling and assess the adequacy;

- draw statistical conclusions in a subjective manner;

- carry out the statistical inference in an objective (automatic) manner, and

- control results of parametric inference by nonparametric procedures.

Part II deals with statistical procedures in the parametric extreme value (EV) and generalized Pareto (GP) models. Yet, at the beginning, we start with the statistical inference in normal and Poisson models (Chapter 3) in order to give an outline of our approach to statistical questions within a setting which is familiar to a larger readership. From our viewpoint, the Gaussian model is merely relevant for the center of a distribution and, thus, not for extreme values. Chapters 4 and 5 develop the statistical methodology that is necessary for dealing with extremes. Applied questions are addressed in various examples. These examples also include critical examinations of case studies in publications which are occasionally very ambitious. We will approach extreme value analysis from a practical viewpoint, yet references are given to the theoretical background (as developed in the books [24], [20], [39], [42] and [16]). Applied contributions to extreme value analysis can be found in several journals. It is recommended to have a look at the J. Hydrology, Insurance: Mathematics and Economics, J. Econometrics, J. Royal Statistical Society B and, particularly, at the forthcoming journal Extremes, Statistical Theory and Applications in Science, Engineering and Economics. Other valuable sources for applied work are the recent Gaithersburg proceeding volumes [15] and the hydrology proceedings [13].

Part III contains supplementary material about the analysis of multivariate data and auxiliary results for multivariate distributions applied in Part II. Initially, a textbook for univariate extremes was scheduled. Yet, it is evident that a time–scale must be included in conjunction with time series phenomena, for example, exceedance times and exceedances are jointly visualized in a scatterplot. This was our first step towards multivariate data. Further extensions of our initial framework followed so that, finally, we decided to include some procedures concerning multivariate extremes.

We also want to learn in which manner the methodology provided by the previous parts can be made applicable in certain areas. Part IV deals with important questions in

- insurance (coauthored by Michael Radtke),

- finance (coauthored by Casper G. de Vries),

- hydrology.

We hope that the explanations are also of interest for non–specialists.

Part IV has a certain continuation in Part V which contains several case studies in extreme value analysis. The case studies are written in the form of self–contained articles, which facilitated the inclusion of studies of other authors. One basic requirement for any case study is that the underlying data must be publicly accessible because, otherwise, the hypotheses and conclusions of an analyst cannot be critically examined and further improved by others, which would strongly violate scientific principles. We would like to thank Ana M. Ferreira, Edgar Kaufmann, Cornelia Hillgärtner, Tailen Hsing and Jürg Hüsler for their contributions.

The appendix is a manual for the included statistical software XTREMES. The menu–driven part of XTREMES allows us to surf through data sets and statistical models and reduces the "start up" costs of working in this area. For special problems one may employ the integrated programming language XPL. A short overview of the hierarchy of XTREMES is given after this preface. We believe that an experienced reader can partially handle XTREMES after having read the overview. A further link between the book and the statistical software is provided by sections entitled "Implementation in XTREMES" at the end of the chapters.

We will not make any attempt to give exhaustive references to the extreme value literature. Several footnotes provide hints to papers and books which are important from the historical viewpoint or may be helpful to get a more thorough understanding of special questions. The bibliography merely consists of references to monographs and proceeding volumes that are suggested for further reading or cited several times within the text.

We are grateful to several colleagues for valuable suggestions and stimulating discussions, especially, Sandor Csörgő, Richard A. Davis, Paul Embrechts, Michael Falk, Laurens de Haan, Jürg Hüsler, Edgar Kaufmann, Alex Koning, Ross Leadbetter, Wolfgang Merzenich, Wolfgang Wefelmeyer. We would like to express particular gratitude to the coauthors of Chapters 9 and 10 who had a constructive impact on our work. Very special thanks are due to Sylvia Haßmann for collaboration on the MS–DOS version of XTREMES (documented in [16]) and for assistance in writing a first draft of Section 6.1.

Several generations of students were exposed to the development of this book and the included software system; we acknowledge warmly the assistance of Simon Budig (final version of the UserFormula facility), Andreas Heimel (help system), Jens Olejak (minimum distance estimators), Claudia Schmidt (expert for the Moselle data) and Karsten Tambor (previous version of the multivariate mode of XTREMES). The technical assistance of Sarah Schultz and Maximilian Reiss was very helpful.

Part of the work of the first author was done as a guest professor at the Tinbergen Institute, Rotterdam, and visiting the Center for Stochastic Processes, Chapel Hill. Thanks are due to Laurens de Haan and Ross Leadbetter for their hospitality. The stimulating atmospheres of these institutions had a greater impact on the course of this work. The stay at the Tinbergen Institute also enabled the cooperation with Casper de Vries.

Siegen, Germany   Rolf–Dieter Reiss
                                                                        Michael Thomas

Contents

List of Special Symbols

F	distribution function (df); likewise we use symbols G, W etc. for dfs
$F_{\mu,\sigma}$	df F with added location and scale parameters μ and σ; p. 16
$\overline{F}$	survivor function of df F; pages 11 and 266
f	density of df F with $\int_{\infty}^{x} f(y)\,dy = F(x)$; likewise, we use g, w etc.
F^{-1}	quantile function (qf) pertaining to the df F; p. 40
$F^{[u]}$	exceedance df at u (truncation of df F left of u); p. 12
$F^{(u)}$	excess df at u (residual life df at age u) pertaining to the df F; p. 49
$G_{i,\alpha}$	standard extreme value (EV) df for maximum (ith submodel); p. 15
G_{γ}	standard extreme value (EV) df for maximum (unified model); p. 16
$W_{i,\alpha}$	standard generalized Pareto (GP) df (ith submodel); p. 24
W_{γ}	standard generalized Pareto (GP) df (unified model); p. 25
Φ	standard normal (Gaussian) df; p. 31
F^m	mth power of df F (with $F^m(t) := (F(t))^m$); p. 10
F^{m*}	mth convolution of df F; p. 30
$x_{i:n}$	ith ordered value of data $x_1, \ldots, x_n$ (in non–decreasing order); p. 42
$\widehat{F}_n$	sample df; occasionally written $\widehat{F}_n(\boldsymbol{x}; \cdot)$ to indicate the dependence on $\boldsymbol{x} = (x_1, \ldots, x_n)$; p. 39
$\widehat{F}_n^{-1}$	sample qf; p. 42
$\alpha(F)$	left endpoint of df F; p. 12
$\omega(F)$	right endpoint of df F; p. 11
$I(y \leq x)$	indicator function with $I(y \leq x) = 1$ of $y \leq x$ and 0, otherwise
$E(X)$	expectation of a random variable X
$V(X)$	variance of a random variable X
$\mathrm{Cov(X,Y)}$	covariance of random variables X and Y; p. 71 and 267
$\bar{x}$	sample mean $\frac{1}{n} \sum_{i \leq n} x_i$ (with i running from 1 to n); p. 41
s^2	sample variance $\frac{1}{n-1} \sum_{i \leq n} (x_i - \bar{x})^2$; p. 41
$s_{i,j}$	sample covariance between ith and jth component; p. 71 and 270
$X \stackrel{d}{=} Y$	equality in distribution of X and Y; p. 170

Part I

Modeling and Data Analysis

Chapter 1

Parametric Modeling

Chapter 1 is basic for the understanding of the main subjects treated in this book. It is assumed that the given data are generated according to a random mechanism that can be linked to some parametric statistical model.

In this chapter, the parametric models will be justified by means of mathematical arguments, namely by limit theorems. In this manner, extreme value (EV) and generalized Pareto (GP) models are introduced that are central for the statistical analysis of maxima or minima and of exceedances over a higher or lower threshold, see Sections 1.3 and 1.4. Yet, we do not forget to mention other distributions used in practice for the modeling of extremes. In Section 1.5 we especially deal with fat and heavy–tailed distributions such as, e.g., log–normal and Student distributions.

1.1 Applications of Extreme Value Analysis

At the beginning, the relevance of extreme value analysis to flood frequency analysis, environmental sciences, finance and insurance (these areas are especially dealt with in Chapters 14–17) and other important topics is indicated. We start with flood frequency analysis primarily because of historical reasons.

Flood Frequency Analysis

The ultimate interest of flood frequency analysis is the estimation of the T–year flood discharge (water level), which is the discharge once exceeded on the average in a period of T years. Usually, the time span of 100 years is taken, yet the estimation is carried out on the basis of flood discharges for a shorter period. Consequences of floods exceeding such a level can be disastrous. For example, 100–year flood levels were exceeded by the great American flood of 1993 that caused widespread devastations in the American Midwest (as pointed out in [13], Preface).

Under standard conditions, the T–year level is a higher quantile of the distribution of discharges. Thus, one is primarily interested in a parameter determined by the upper tail of the distribution. Because flood discharges of such a magnitude were rarely observed or are not recorded at all, a parametric statistical modeling is necessary to capture relevant tail probabilities. A careful checking of the validity of the parametric modeling is essential.

If the statistical inference is based on annual maxima of discharges, then our favorite model is the Gumbel or the unified extreme value model, yet we also mention other models employed by hydrologists. Alternatively, the inference is based on a partial duration series, which is the series of exceedances over a certain higher threshold. In that case, our standard model for the flood magnitudes is the generalized Pareto model or certain submodels suggested by the well–known index flood procedure in regional flood frequency analysis. We mention alternative models. A modeling of the exceedance times must be included.

The regional flood frequency provides the possibility to include information from data recorded at nearby gauging stations, respectively, sites having a similar characteristic.

The Actuary's Interest in Extremes

In recent years, actuaries have become more and more aware of the potential risk inherent in very large claim sizes due to catastrophic events. An insurer must compensate the losses—due to the payments for claims of policy holders—by means of an appropriate premium. Thereby, an actuary is first of all interested in estimating the net premium which is the mean of the total claim amount for an individual or a portfolio (collection) of risks.

The total claim amount depends on the size of claims and the frequency of claims within a given period. These two ingredients can be dealt with separately from the statistical viewpoint.

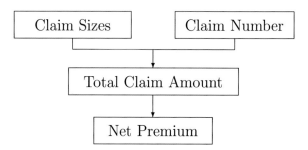

In conjunction with excess–of–loss (XL) reinsurance, when the reinsurer pays for the excess of a higher fixed limit for individual claim sizes, a parametric modeling and estimation of the upper tail of the claim size distribution is of genuine interest to evaluate the net premium.

Our favorite model is again the generalized Pareto model, yet we also study other distributions employed by actuaries (such as Benktander II and truncated Weibull distributions).

For assessing the total premium, the actuary must also consider other parameters of the random losses besides the net premium. We will focus our attention on the interdependence between the ruin probability of a portfolio within a finite time horizon and the choice of an initial capital (reserve).

Extremes in Financial Time Series, Value at Risk

Insurance and financial data can be investigated from the viewpoint of risk analysis (as initiated in the books by H.L. Seal[1] and H. Bühlmann[2]). Therefore, the insight gained from insurance data can also be helpful for the understanding of financial risks (and vice versa).

Financial time series consist of daily or weekly reported speculative prices of assets such as stocks, foreign currencies or commodities such as corn, cocoa, coffee, sugar, etc. Risk management at a commercial bank is interested in guarding against the risk of high losses due to the fall in prices of financial assets held or issued by the bank. It turns out that daily or weekly returns—relative differences of consecutive prices or differences of log–prices (log–returns)—are the appropriate quantities which must be investigated.

Most importantly, there is empirical evidence that distributions of returns can possess fat or heavy tails so that a careful analysis of returns is required. In this context, we deal with the upper tail of loss/profit distributions and, especially, with parameters, which summarize the potential risk to some extent, such as

- the Value–at–Risk (VaR) as the limit which is exceeded by the loss (measured as a positive value) of a given speculative asset or a portfolio with a specified low probability,

- the Capital–at–Risk (CaR) as the amount which may be invested with the possible consequence that the loss exceeds a given limit with a specified low probability.

These two concepts are closely related to each other: we fix either the invested capital (market value) or the limit and, then, compute the other variable.

A special feature of return series is the alternation between periods of tranquility and volatility. Therefore, the VaR, which is essentially the q–quantile of the distribution of log–returns, should vary in time.

The following illustration[3] concerns the log–returns of the Standard & Poors

---

[1]Seal, H.L. (1969). Stochastic Theory of a Risk Business. Wiley, New York.

[2]Bühlmann, H. (1970). Mathematical Methods in Risk Theory. Springer, Berlin.

[3]Illustrations are produced with the *Frame Size (Print/EPS)* facility in the local menus of the plot windows of Xtremes.

500 market index (stored in fm–poors.dat[4]).

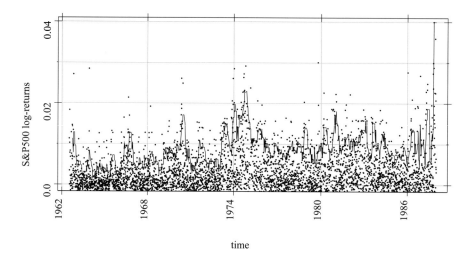

FIG. 1.1. Scatterplot of the log–returns (with changed signs) of the S&P500 index from July 1962 to Dec 1987 and a moving sample 95%–quantile based on the preceding 90 days.

One recognizes the periods of higher volatility which are reflected by periods of higher sample q–quantiles and, thus, a higher VaR.

Material Sciences and Other Important Areas

In most parts of the book, we do not focus our attention on special areas. Such a theoretical approach is justified by the fact that, to some extent, the basic statistical questions have a common feature for all the potential applications. We briefly mention further questions for which extreme value analysis is of interest:

- (Corrosion Analysis.) Pitting corrosion can lead to the failure of metal structures such as tanks and tubes. Extreme value analysis becomes applicable because pits of a larger depth are of primary interest. In analogy to the T–year discharge in hydrology one may estimate the T–unit depths which is the level once exceeded (within a given time span) on the average, if T units or an area of size T are exposed to corrosion. The role of time in other applications (e.g., in hydrology) is now played by the location of the pits, see Section 18.1.

---

[4]Data sets *.dat can be found in the subdirectory `xtremes\dat`, if `xtremes` is taken as the working directory of Xtremes.

- (Telecommunication.) There exist several research papers about teletraffic data in conjunction with a discussion of the concepts of aggregation and self–similarity, see page 169.

- (Strength of Material.) A sheet of metal breaks at its weakest point so that a minimum strength determines the quality of the entire sheet. Related topics are strength of bundles of threads and first failure of equipments with many components, see [20], pages 188–196.

- (Longevity of Human Life.) The insight gained from experiments led to statements such as "the studies are consistently failing to show any evidence for pre–programmed limit to human life span" which found their way to daily news papers[5]. We will study such questions from the viewpoint of extreme value analysis, see Chapter 19.

- (Environmental Sciences.) Higher concentration of certain ecological quantities, like concentration of ozone, acid rain or sulfur dioxide (SO_2) in the air, are of greater interest due to the negative response on humans and, generally, on the biological system, see Chapter 15.

Other important areas of applications include geology, meteorology and seismic risk analysis. It is self–evident that our list is by no means complete.

The T–Year Level and Related Parameters

The T–year level (e.g., for discharges in flood frequency analysis) plays an important role in our analysis. This parameter is dealt with, for example, in Sections 1.2 and 14.2. The same methodology can be applied to the T–unit depths in corrosion analysis, see Section 18.1. In a similar way, we define a T–year initial reserve in insurance (page 430) and a T–day capital in finance (page 390). These parameters are connected to certain small or large quantiles and must be estimated by the methods presented in this book.

1.2 Observing Exceedances and Maxima

It is understood that the reader is familiar with the concepts of a distribution function (df) F, a density f as the derivative of the df, and independent and identically distributed (iid) random variables $X_1, \ldots, X_n$. The df of a real–valued random variable X is given by $F(x) = P\{X \leq x\}$. The expectation and variance of a random variable X are denoted by $E(X)$ and $V(X)$. These quantities are the mean and the variance of the distribution of X.

---

[5]Kolata, G. New views on life spans alter forecasts on elderly. The New York Times, Nov. 16. 1992.

Number of Exceedances over a Threshold

One method of extracting upper extremes from a set of data $x_1, \ldots, x_n$ is to take the exceedances over a predetermined, high threshold u. Exceedances y_j over u (peaks–over–threshold (pot)) are those x_i with $x_i > u$ taken in the original order of their outcome or in any other order. The values $y_j - u$ are the excesses over u.

Subsequently, the number of exceedances over u will be denoted by k or, occasionally, by K to emphasize the randomness of this number. In many cases, the values below u are not recorded or cannot be observed by the statistician. Given random variables $X_1, \ldots, X_n$, we may write $K = \sum_{i \leq n} I(X_i > u)$, where $I(X_i > u)$ is an indicator function with $I(X_i > u) = 1$ if $X_i > u$ holds and zero, otherwise. If the X_i are iid random variables with common df F, then

$$P\{K = k\} = \binom{n}{k} p^k (1-p)^{n-k} =: B_{n,p}\{k\}, \qquad k = 0, \ldots, n, \qquad (1.1)$$

where $B_{n,p}$ is the well–known binomial distribution with parameters n and $p = 1 - F(u)$. The mean number of exceedances over u is

$$\Psi_{n,F}(u) = np = n(1 - F(u)), \qquad (1.2)$$

which defines a decreasing mean value function.

The Poisson Approximation of Binomial Distributions

We give an outline of the well–known fact that the binomial distribution $B_{n,p}$ can be replaced by a Poisson distribution and, thus, the number K of exceedances may be regarded as a Poisson random variable.

Let $X_1, \ldots, X_n$ be iid random variables attaining the values 1 and 0 with probability p and, respectively, $1 - p$. In conjunction with the number of exceedances, these random variables are of the special form $I(X_i \geq u)$. We have

$$P\{X_1 + \cdots + X_n = k\} = B_{n,p}\{k\}, \qquad k = 0, \ldots, n.$$

This formula can be rephrased by saying that success occurs with probability p and failure with probability $1 - p$, then the total number of successes is governed by $B_{n,p}$. Note that np is the expected number of successes, and $np(1 - p)$ is the variance.

Poisson distributions can be fitted to binomial distributions under certain conditions. If $np(n) \to \lambda$ as $n \to \infty$, then

$$B_{n,p(n)}\{k\} \to P_\lambda\{k\}, \qquad n \to \infty, \qquad (1.3)$$

where

$$P_\lambda\{k\} = \frac{\lambda^k}{k!} e^{-\lambda}, \qquad k = 0, 1, 2, 3, \ldots$$

defines the Poisson distribution with parameter λ.

In Fig. 1.2, the binomial and Poisson histograms $B_{n,p}\{k\}$ and $P_{np}\{k\}$, $k = 0, \ldots, 20$, for the parameters $n = 40$ and $p = 0.25$ are displayed, see also the Demos A.1 and A.2.

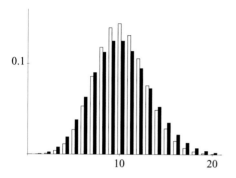

FIG. 1.2. Fitting the Poisson distribution P_{10} (dark) to a binomial distribution $B_{(40, .25)}$ (light).

If X is a Poisson random variable with parameter λ, then $E(X) = \lambda$. In addition, the variance of X is $V(X) = E(X - \lambda)^2 = \lambda$. We see that the expectation and the variance of a Poisson random variable are equal.

It is well known that the inequality

$$|B_{n,p}(A) - P_{np}(A)| \leq p \tag{1.4}$$

holds for each set A of nonnegative integers. Notice that the right–hand side of the inequality does not depend on n. Thus, a Poisson distribution can be fitted to a binomial distribution whenever p is sufficiently small.

The iid condition, under which the Poisson approximation was formulated, can be weakened considerably, yet we do not go into details[6]. In Section 3.4, we also introduce mixtures of Poisson distributions as, e.g., negative binomial distributions.

Maxima

Assume that the given data y_i are maxima, that is,

$$y_i = \max\{x_{i,1}, \ldots, x_{i,m}\}, \qquad i = 1, \ldots, n, \tag{1.5}$$

where the $x_{i,j}$ may not be observable.

If the $x_{i,j}$ in (1.5) can be observed by the statistician, then taking maxima out of blocks is another possibility of extracting upper extreme values from a set of

---

[6]We refer to Haight, F.A. (1967). Handbook of the Poisson Distribution. Wiley, New York; for a more recent monograph see Barbour, A.D., Holst, L. and Janson, S. (1992). Poisson Approximation. Oxford Studies in Probability. Oxford University Press.

data (besides taking exceedances). This method is called annual maxima, blocks or Gumbel method. For example, one takes the maximum values—such as maximum temperatures, water discharges, wind speeds, ozone concentrations, etc.—within a month or a year.

For iid random variables $X_1, \ldots, X_m$ with common df F, one may easily compute the df of maxima; we have

$$P\left\{ \max_{i \leq m} X_i \leq x \right\} = P\{X_1 \leq x, \ldots, X_m \leq x\} = F^m(x). \qquad (1.6)$$

Thus, the y_i's in (1.5) are governed by F^m if the $x_{i,j}$'s are governed by F.

If the iid condition fails, then a df of the form F^m may still be an accurate approximation of the actual df of the maximum. For independent, yet heterogeneous random variables X_j with df F_j, (1.6) holds with F^m replaced by $\prod_{j \leq m} F_j$. A df F^m can be fitted to $\prod_{j \leq m} F_j$ if the deviations of the F_j from each other can be neglected.

Likewise, if there is a slight dependence in the data, a df of the form F^m may still serve as an approximation of the actual df of the maximum. It is plausible to employ this approach in more complex situations as well. In Section 1.3, we go one step further and replace F^m by an extreme value (EV) df G. The EV distributions constitute a parametric model which will be introduced in Section 1.3.

We mention a special EV df, namely the Gumbel df $G_0(x) = \exp(-e^{-x})$, which plays a central role within the extreme value analysis.

EXAMPLE 1.2.1. (Annual Wind–Speed Maxima at Vancouver.) We present annual maxima of daily measurements of wind speeds taken from [21], page 122 (stored in the file em–cwind.dat).

TABLE 1.1. Annual wind–speed maxima in km/h from 1947 to 1984 at Vancouver.

year	speed	year	speed	year	speed	year	speed	year	speed	year	speed	year	speed
47	79.5	53	64.8	59	64.8	65	61.0	71	70.3	77	48.1	83	51.8
48	68.4	54	59.2	60	88.8	66	51.8	72	68.4	78	53.6	84	48.1
49	74.0	55	79.5	61	88.8	67	62.9	73	55.5	79	55.5		
50	59.2	56	62.9	62	75.8	68	64.8	74	64.8	80	62.9		
51	74.0	57	59.2	63	68.4	69	61.0	75	77.7	81	61.0		
52	64.8	58	68.2	64	68.4	70	61.0	76	57.3	82	61.0		

Such data can be visualized by means of nonparametric tools such as, e.g., kernel densities (see Section 2.1). One may try to fit a parametric density—e.g., a Gumbel density with location and scale parameters μ and σ—to the sample density.

In Fig. 1.3, one can recognize a reasonable fit of a Gumbel density to the data (represented by a sample density). For a continuation, see Example 2.1.3.

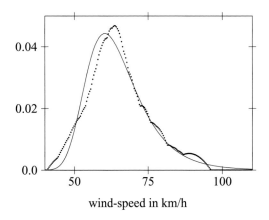

FIG. 1.3. Representing the wind–speed data by a fitted Gumbel density ($\mu = 60.3$ and $\sigma = 8.3$) and a kernel density.

wind-speed in km/h

Minima

Generally, results for minima can be deduced from corresponding results for maxima by writing

$$\min_{i \leq n} X_i = -\max_{i \leq n}(-X_i).$$

In conjunction with minima, it can be useful to present results in terms of the survivor function

$$\bar{F} = 1 - F. \tag{1.7}$$

Utilizing the same arguments as in (1.6), one may compute the survivor function of the minimum of iid random variables X_i with common df F. We have

$$P\left\{\min_{i \leq m} X_i > x\right\} = (1 - F(x))^m = \bar{F}^m(x). \tag{1.8}$$

Therefore, the df of the minimum is

$$P\left\{\min_{i \leq m} X_i \leq x\right\} = 1 - (1 - F(x))^m. \tag{1.9}$$

Actual dfs of minima will be replaced by converse EV dfs in Section 1.3 (see page 22). As an example of a converse EV df, we mention the converse Gumbel df

$$\tilde{G}_0(x) = 1 - G_0(-x) = 1 - \exp(-e^x), \tag{1.10}$$

which is also called Gompertz df.

Magnitudes of Exceedances and Upper Order Statistics

Let the x_i be governed again by a df F and let the threshold u be smaller than the right endpoint $\omega(F) := \sup\{x : F(x) < 1\}$ of the support of F. We speak of a high threshold u, if u is close to the right endpoint $\omega(F)$. In that case, $p = 1 - F(u)$ is small and the number k of exceedances may be regarded as a Poisson random variable. Subsequently, we deal with the magnitudes (sizes) of the exceedances.

Exceedances occur conditioned on the event that an observation is larger than the threshold u. The pertaining conditional df $F^{[u]}$ is called exceedance df at u. If X denotes a random variable with df F, then

$$
\begin{aligned}
F^{[u]}(x) &= P(X \leq x | X > u) \\
&= P\{X \leq x, X > u\}/P\{X > u\} \\
&= \frac{F(x) - F(u)}{1 - F(u)}, \qquad x \geq u.
\end{aligned}
\tag{1.11}
$$

In terms of survivor functions we may write

$$
\overline{F^{[u]}}(x) = \bar{F}(x)/\bar{F}(u), \qquad x \geq u.
\tag{1.12}
$$

One should keep in mind that the left endpoint $\alpha(F^{[u]}) = \inf\{x : F^{[u]}(x) > 0\}$ of $F^{[u]}$ is equal to u. The relationship between exceedances and the exceedance df will be examined more closely in (2.4). Moreover, generalized Pareto (GP) dfs will be fitted to exceedance dfs $F^{[u]}$ in Section 1.4.

Another closely related approach of extracting extreme values from the data is to take the k largest values $x_{n-k+1:n} \leq \cdots \leq x_{n:n}$ of the x_i, where the number k is predetermined. Notice that $x_{n:n}$ is the maximum. Within this approach, the $(k+1)$th largest observation $x_{n-k:n}$ may be regarded as a random threshold.

The T–Year Level

One of the major objectives of extreme value analysis is the estimation of the T–year level. We introduce the T–year level $u(T)$ as the threshold $u(T)$ such that the mean number of exceedances over $u(T)$ within the time span of length T is equal to 1. In this context, it is understood that for each year (or any other period like day, month or season), there is one observation.

Let $X_1, X_2, \ldots, X_T$ be random variables with common df F. Then, $u(T)$ is the solution to the equation

$$
E\left(\sum_{i \leq T} I(X_i \geq u) \right) = 1.
\tag{1.13}
$$

Apparently,

$$
u(T) = F^{-1}(1 - 1/T)
\tag{1.14}
$$

which is the $(1 - 1/T)$–quantile of the df F (also see Section 1.6). We have

$$
P\{X_1 > u(T)\} = 1 - F(u(T)) = \frac{1}{T}
$$

and, thus, the T–year level $u(T)$ is exceeded by the observation in the given year (period) with probability $1/T$.

Exceedance Times and the T–Year Return Level

The primary aim of the following lines is to outline why the T–year level $u(T) = F^{-1}(1 - 1/T)$ is also called T–year return level. In this context, we also get a different interpretation of the T–year level.

Besides the number and sizes of exceedances, one is interested in the times at which the exceedances occur. Given $x_1, \ldots, x_n$, let $x_{i(1)}, \ldots, x_{i(k)}$ be the exceedances over a predetermined threshold u. Let $i(1) \le i(2) \le \cdots \le i(k)$ be the ordered exceedance times.

In conjunction with exceedance times, we are primarily interested in the future occurrence of the next exceedance at a certain higher threshold u for an infinite time horizon. Therefore, one must consider an infinite sequence $X_1, X_2, X_3, \ldots$ of random variables with common df F. The first exceedance time at u is

$$\tau_1 = \min\{m : X_m > u\}, \tag{1.15}$$

whereby it is understood that the threshold u is smaller than the right endpoint $\omega(F)$ of F. We also deal with the random ordered exceedance times

$$\tau_1 < \tau_2 < \tau_3 < \cdots$$

at the threshold u. Occasionally, we also write $\tau_{r,u}$ instead of τ_r for the rth exceedance time to emphasize the dependence on the threshold u. The exceedance times may be defined recursively by

$$\tau_r = \min\{m > \tau_{r-1} : X_m > u\}, \qquad r > 1. \tag{1.16}$$

The sequence of exceedance times has independent, geometrically distributed increments called return periods or interarrival times, see (1.17) and (1.21). Note that

$$\begin{aligned} P\{\tau_1 = k\} &= P\{X_1 \le u, \ldots, X_{k-1} \le u, X_k > u\} \\ &= p(1 - p)^{k-1}, \qquad k = 1, 2, 3, \ldots, \end{aligned} \tag{1.17}$$

with $p = 1 - F(u)$, where the latter equality holds for iid random variables. Thus, the first exceedance time τ_1 at u is distributed according to a geometric distribution with parameter p. Consequently, the mean of the first exceedance time—also called mean return period—at u is $E(\tau_{1,u}) = 1/p$.

A threshold u such that the mean first exceedance time is equal to T is the T–year return level. Thus, the T–year return level is the solution to the equation

$$E(\tau_{1,u}) = \frac{1}{1 - F(u)} = T. \tag{1.18}$$

This equation has the same solution as equation (1.13), namely

$$u(T) = F^{-1}(1 - 1/T). \tag{1.19}$$

This means that the T–year level in (1.14) and the T–year return level coincide under the present conditions.

The first exceedance time $\tau_{1,u(T)}$ at the T–year level $u(T)$ is a geometric random variable with parameter $1/T$. By the definition of $u(T)$, we have $E(\tau_{1,u(T)}) = T$. In addition, the variance is $V(\tau_{1,u(T)}) = (T-1)T$.

A justification for the names "mean return period" and "return level" is provided by the fact that the return periods

$$\tau_1, \tau_2 - \tau_1, \tau_3 - \tau_2, \ldots \tag{1.20}$$

are iid random variables. This yields that the mean return periods $E(\tau_{r+1} - \tau_r)$ are equal to the mean first exceedance time $E(\tau_1)$. Also, the threshold u such that the mean return period is equal to T is the T–year return level in (1.14) and (1.19).

The iid property can be verified easily. For positive integers $k(1), \ldots, k(r)$, we have

$$
\begin{aligned}
&P\{\tau_1 = k(1), \tau_2 - \tau_1 = k(2), \ldots, \tau_r - \tau_{r-1} = k(r)\} \\
={}& P\{\tau_1 = k(1), \tau_2 = k(1) + k(2), \ldots, \tau_r = k(1) + \cdots + k(r)\} \\
={}& P\{X_1 \leq u, \ldots, X_{k(1)-1} \leq u, X_{k(1)} > u, \\
&\qquad\quad X_{k(1)+1} \leq u, \ldots, X_{k(1)+\cdots+k(r)} > u\} \\
={}& \prod_{j \leq r} P\{\tau_1 = k(j)\}.
\end{aligned}
\tag{1.21}
$$

Generally, one may verify that

$$P\{\tau_r = k\} = \binom{k-1}{r-1} p^r (1-p)^{k-r} \tag{1.22}$$

for $k = r, r+1, \ldots$ and, hence, the rth exceedance time is a negative binomial random variable with parameters r and $p = 1 - F(u)$ shifted to the right by the amount r.

1.3   Modeling by Extreme Value Distributions

The actual df of a maximum will be replaced by an extreme value (EV) df. First, we give a list of the standard EV dfs $G_{i,\alpha}$ and G_γ within three submodels and a unified model, where α and γ are certain shape parameters. The pertaining densities are denoted by $g_{i,\alpha}$ and g_γ. One must include location and scale parameters μ and σ to build the full statistical models.

At the end of this section, we will also mention corresponding questions concerning minima.

The α–Parameterization

Here is a list of the three different submodels by writing down the standard EV dfs for the different shape parameters α:

Gumbel (EV0): $\qquad\qquad G_0(x) = \exp(-e^{-x})$, $\qquad$ for all x;

Fréchet (EV1), $\alpha > 0$: $\qquad G_{1,\alpha}(x) = \exp(-x^{-\alpha})$, $\qquad x \geq 0$;

Weibull (EV2), $\alpha < 0$: $\qquad G_{2,\alpha}(x) = \exp(-(-x)^{-\alpha})$, $\qquad x \leq 0$.

The Fréchet df $G_{1,\alpha}$ is equal to zero if $x < 0$; the Weibull df $G_{2,\alpha}$ is equal to one if $x > 0$. We see that each real parameter α determines a standard EV distribution. Notice that $G_{2,-1}$ is the exponential df on the negative half–line. Occasionally, the standard Gumbel df G_0 is also denoted by $G_{0,\alpha}$.

> **Warning!** Our parameterization for Weibull dfs differs from the standard one used in statistical literature, where Weibull dfs with positive shape parameters are taken.

We also note the pertaining densities $g = G'$ of the standard EV dfs:

Gumbel (EV0): $\qquad\qquad g_0(x) = G_0(x)e^{-x}$, $\qquad$ for all x;

Fréchet (EV1), $\alpha > 0$: $\qquad g_{1,\alpha}(x) = \alpha G_{1,\alpha}(x)x^{-(1+\alpha)}$, $\qquad x \geq 0$;

Weibull (EV2), $\alpha < 0$: $\qquad g_{2,\alpha}(x) = |\alpha|G_{2,\alpha}(x)(-x)^{-(1+\alpha)}$, $\qquad x \leq 0$.

EV densities are unimodal. Remember that a distribution and the pertaining density f are called unimodal if the density is non–decreasing left of some point u and non–increasing right of u. Then, u is called a mode. Fréchet densities and the Gumbel density are skewed to the right.

Weibull dfs provide a very rich family of unimodal dfs. Weibull densities

- are skewed to the left if α is larger than -3.6 and have a pole at zero if $\alpha > -1$,

- look symmetrical if α is close to -3.6,

- are skewed to the right—such as Fréchet and Gumbel densities—if α is smaller than -3.6.

Location and Scale Parameters

If a random variable X has the df F, then $\mu + \sigma X$ has the df $F_{\mu,\sigma}(x) = F((x-\mu)/\sigma)$, where μ and $\sigma > 0$ are the location and scale parameters. Full EV models are obtained by adding location and scale parameters μ and $\sigma > 0$. By

$$G_{0,\mu,\sigma}(x) = \exp(-e^{-(x-\mu)/\sigma})$$

and

$$G_{i,\alpha,\mu,\sigma}(x) = G_{i,\alpha}\left(\frac{x-\mu}{\sigma}\right), \qquad i = 1,2,$$

one obtains Gumbel, Fréchet and Weibull dfs with location and scale parameters μ and σ.

We see that the location parameter μ is the left endpoint of the Fréchet df $G_{1,\alpha,\mu,\sigma}$ and the right endpoint of the Weibull df $G_{2,\alpha,\mu,\sigma}$. Furthermore,

$$g_{0,\mu,\sigma}(x) = \frac{1}{\sigma}e^{-(x-\mu)/\sigma}\exp(-e^{-(x-\mu)/\sigma})$$

and

$$g_{i,\alpha,\mu,\sigma}(x) = \frac{1}{\sigma}g_{i,\alpha}\left(\frac{x-\mu}{\sigma}\right), \qquad i = 1,2,$$

are the densities of the Gumbel df $G_{0,\mu,\sigma}$ and the Fréchet and Weibull dfs $G_{i,\alpha,\mu,\sigma}$.

The γ–Parameterization

Up to now, the three different EV models are separated from each other. Yet a visualization of these dfs—or of the pertaining densities—shows that Fréchet and Weibull dfs attain the shape of a Gumbel df when the shape parameter α goes to ∞ and, respectively, to $-\infty$.

By taking the reparameterization $\gamma = 1/\alpha$—due to von Mises[7] (also frequently attributed to Jenkinson[8])—of EV dfs $G_{i,\alpha}$ one obtains a continuous, unified model. In this representation, the Gumbel df again has the parameter γ equal to zero. The standard versions in the γ–parameterization are defined such that

$$G_\gamma(x) \to G_0(x), \qquad \gamma \to 0. \tag{1.23}$$

This is achieved by employing appropriate location and scale parameters in addition to the reparameterization $\gamma = 1/\alpha$. Thus, EV dfs with certain location

---

[7] Mises von, R. (1936). La distribution de la plus grande de n valeurs. Rev. Math. Union Interbalcanique 1, 141–160. Reproduced in Selected Papers of Richard von Mises, Amer. Math. Soc. 2 (1964), 271–294.

[8] Jenkinson, A.F. (1955). The frequency distribution of annual maximum (or minimum) values of meteorological elements. Quart. J. Roy. Meteorol. Soc. 81, 158–171.

and scale parameters unequal to zero and one are taken as the standard EV dfs in the γ–parameterization if $\gamma \neq 0$. We have

$$G_0(x) = \exp(-e^{-x}), \qquad \text{for all } x, \tag{1.24}$$

and

$$G_\gamma(x) = \exp\left(-(1 + \gamma x)^{-1/\gamma}\right), \qquad 1 + \gamma x > 0, \quad \gamma \neq 0. \tag{1.25}$$

By applying the well–known formula $(1 + \gamma x)^{1/\gamma} \to \exp(x)$ as $\gamma \to 0$, one may verify that (1.23) holds. Notice that

- G_0 is the standard Gumbel df;

- G_γ is a Fréchet df if $\gamma > 0$,

- G_γ is a Weibull df if $\gamma < 0$.

The right endpoint of a Weibull df G_γ is equal to $1/|\gamma|$ and the left endpoint of the Fréchet df G_γ is equal to $-1/\gamma$. For $\gamma = 1/\alpha \neq 0$, one obtains the representation

$$G_\gamma = G_{i,\alpha,-\alpha,|\alpha|} \tag{1.26}$$

with $i = 1$ if $\gamma > 0$ and $i = 2$ if $\gamma < 0$. The pertaining densities are

$$g_0(x) = G_0(x)e^{-x}, \qquad \text{for all } x,$$

and

$$g_\gamma(x) = G_\gamma(x)(1 + \gamma x)^{-(1+1/\gamma)}, \qquad 1 + \gamma x > 0, \quad \gamma \neq 0.$$

Also, check that $g_\gamma(x) \to g_0(x)$ as $\gamma \to 0$. Some EV densities around the Gumbel density are displayed in Fig. 1.4.

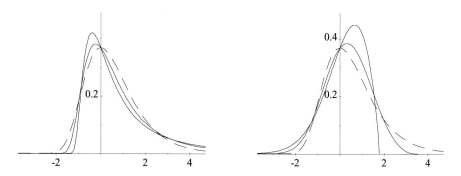

FIG. 1.4. (left.) Gumbel density (dashed) and two Fréchet (solid) densities for parameters $\gamma = .28, .56.$ (right.) Gumbel density (dashed) and two Weibull (solid) densities for parameters $\gamma = -.28, -.56.$

On the right–hand side, we see the Weibull density with shape parameter $\gamma = -.28$ (that is, $\alpha = -3.6$) that looks symmetrical and can hardly be distinguished from a normal (Gaussian) density. Recall that a Weibull density g_γ has the finite right endpoint $1/|\gamma|$. We have $\omega(G_{-.28}) = 3.6$ and $\omega(G_{-.56}) = 1.8$.

Limiting Distributions of Maxima

Recall from (1.6) that $\max_{i \leq n} X_i$ has the df F^n, if $X_1, \ldots, X_n$ are iid random variables with common df F. The choice of Fréchet, Weibull and Gumbel dfs for modeling dfs of maxima becomes plausible from the subsequent remarks:

(**Fisher–Tippett**[9].) If $F^n(b_n + a_n x)$ has a non–degenerate limiting df as $n \to \infty$ for constants b_n and $a_n > 0$, then

$$|F^n(x) - G((x - \mu_n)/\sigma_n))| \to 0, \qquad n \to \infty, \qquad (1.27)$$

for some EV df $G \in \{G_0, G_{1,\alpha}, G_{2,\alpha}\}$ or $G = G_\gamma$ and location and scale parameters μ_n and $\sigma_n > 0$.

If (1.27) holds, then F belongs to the max–domain of attraction of the EV df G; in short, $F \in \mathcal{D}(G)$. For lists of suitable constants b_n and $a_n > 0$, we refer to (1.31) and (2.33), whereby the constants in (2.33) are applicable if a certain von Mises condition is satisfied.

Additionally, convergence also holds in (1.27) for the pertaining densities if F is one of the usual continuous textbook dfs. The illustrations in Fig. 1.5 concern the Gumbel approximation of the maximum of exponential random variables for a sample of size $n = 30$.

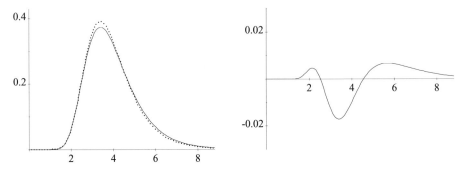

FIG. 1.5. (left.) Density f of df $\left(1 - e^{-x}\right)^{30}$ (solid) and Gumbel density $g_{0,\mu,\sigma}$ (dotted) for parameters $\mu = 3.4$ and $\sigma = .94$. (right.) Plot of the difference $f - g_{0,\mu,\sigma}$.

It can be easily verified that (1.27) holds if, and only if,

$$n\bar{F}(b_n + a_n x) \to -\log G(x), \qquad n \to \infty, \qquad (1.28)$$

for every x in the support of G, where again $\bar{F} = 1 - F$ is the survivor function. According (1.2), this relation concerns the asymptotic behavior of the mean number of exceedances over the threshold $u(n) = \mu_n + \sigma_n x$.

---

[9]Fisher, R.A. and Tippett, L.H.C. (1928). Limiting forms of the frequency distribution of the largest or smallest member of a sample. Proc. Camb. Phil. Soc. 24, 180–190.

More refined necessary and sufficient conditions for (1.27) are due to Gnedenko and de Haan, see, e.g., [16], Theorem 2.1.1 and Section 6.5[10]. A discussion about the accuracy of such approximations may also be found in Section 6.5. If condition (1.27) does not hold or it merely holds with a low rate, then the situation is classified as the "horror case" by some authors. We simply suggest to use a different modeling.

The Max–Stability

EV dfs are characterized by their max–stability. A df F is max–stable if

$$F^n(b_n + a_n x) = F(x) \tag{1.30}$$

for a suitable choice of constants b_n and $a_n > 0$. Thus, the standardized maximum under the df F is distributed according to F.

We give a list of these constants when the max–stable df is one of the standard EV dfs.

$$
\begin{array}{llll}
\text{Gumbel} & G_0: & b_n = \log n, & a_n = 1, \\
\text{Fréchet} & G_{1,\alpha},\ \alpha > 0: & b_n = 0, & a_n = n^{1/\alpha}, \\
\text{Weibull} & G_{2,\alpha},\ \alpha < 0: & b_n = 0, & a_n = n^{1/\alpha}.
\end{array}
\tag{1.31}
$$

For example, $G_{2,-1}^n(x/n) = G_{2,-1}(x)$ for the exponential df $G_{2,-1}$ on the negative half–line. The constants in (1.31) will also be used in Section 6.5 to show the convergence of $F^n(b_n + a_n x)$ to an EV df for a larger class of dfs F.

Moments and Modes of Extreme Value Distributions

The jth moment $E(X^j)$ of a Fréchet or a Weibull random variable X can be written in terms of the gamma function

$$\Gamma(\lambda) = \int_0^\infty x^{\lambda-1} e^{-x}\, dx, \qquad \lambda > 0. \tag{1.32}$$

Recall that $\Gamma(\lambda+1) = \lambda\Gamma(\lambda)$ and $\Gamma(1) = 1$. If X has the df G and density $g = G'$, then the jth moment can be written as

$$m_{j,G} := E(X^j) = \int x^j\, dG(x) = \int x^j g(x)\, dx. \tag{1.33}$$

---

[10]For example, a df F belongs to the max–domain of attraction of the EV df G_γ with $\gamma > 0$ if, and only if, $\omega(F) = \infty$ and

$$\bar{F}(tx)/\bar{F}(t) \to_{t\to\infty} x^{-1/\gamma}, \qquad x > 0. \tag{1.29}$$

The standardizing constants may be chosen as $b_n = 0$ and $a_n = F^{-1}(1 - 1/n)$.

Occasionally, the index G will be omitted. We also write $m_G = m_{1,G}$ for the mean of G. By applying the substitution rule, one obtains

$$m_{j,G_{1,\alpha}} = \Gamma(1 - j/\alpha), \qquad \text{if } \alpha > j,$$

and

$$m_{j,G_{2,\alpha}} = (-1)^j \Gamma(1 - j/\alpha).$$

The jth moment of the Fréchet df $G_{1,\alpha}$ is infinite if $\alpha \le j$. This is due to the fact that the upper tail of the Fréchet density $g_{1,\alpha}(x)$ is approximately equal to $\alpha x^{-(1+\alpha)}$. In that context, one speaks of heavy upper tails of Fréchet distributions.

For the special case $j = 1$, one obtains the mean values of Fréchet and Weibull dfs. We have

$$m_{G_{1,\alpha}} = \Gamma(1 - 1/\alpha), \qquad \text{if } \alpha > 1,$$

and

$$m_{G_{2,\alpha}} = -\Gamma(1 - 1/\alpha).$$

Especially the first moment is infinite whenever $0 < \alpha \le 1$. In contrast to the usual textbook analysis, we are also interested in such distributions.

If the actual distribution is of that type, then a single extreme observation may dominate all the other observations and destroy the performance of a statistic connected to the mean. This is one of the reasons why we are also interested in other functional parameters of GP dfs such as q–quantiles with the median as a special case, cf. Section 1.6.

For centered moments one obtains in analogy to (1.33),

$$E\big((X - E(X))^j\big) = \int (x - m_1)^j \, dG(x) = \int (x - m_1)^j g(x) \, dx,$$

where $m_1 = m_{1,G}$. For $j = 2$ one obtains the variance.

The variances $\mathrm{var}_{G_{i,\alpha}}$ of Fréchet and Weibull dfs can be deduced easily, because $\mathrm{var} = m_2 - m_1^2$. We have

$$\mathrm{var}_{G_{i,\alpha}} = \Gamma(1 - 2/\alpha) - \Gamma^2(1 - 1/\alpha), \qquad \text{if } 1/\alpha < 1/2.$$

In the γ–parameterization, the mean and variance of an EV df is given by

$$m_{G_\gamma} = (\Gamma(1 - \gamma) - 1)/\gamma, \qquad \text{if } \gamma < 1,$$

and

$$\mathrm{var}_{G_\gamma} = \big(\Gamma(1 - 2\gamma) - \Gamma^2(1 - \gamma)\big)/\gamma^2, \qquad \text{if } \gamma < 1/2.$$

The case $\gamma = 0$ is included in the latter formulas by considering limits with γ tending to zero. We have

$$m_{G_0} = \lim_{\gamma \to 0} m_{G_\gamma} = \int_0^\infty (-\log x) e^{-x} \, dx = \lambda, \qquad (1.34)$$

where $\lambda = 0.577216\ldots$ is Euler's constant. Moreover,

$$\text{var}_{G_0} = \lim_{\gamma \to 0} \text{var}_{G_\gamma} = \pi^2/6, \tag{1.35}$$

and

$$m_{2,G_0} = \lambda^2 + \pi^2/6.$$

Normalizing centered moments by the standard deviation $\text{var}^{1/2}$ one obtains

$$E\left(\frac{X - EX}{\text{var}_G^{1/2}}\right)^j = \frac{E(X - EX)^j}{\text{var}_G^{j/2}}. \tag{1.36}$$

For $j = 1, 2$, the normalized, centered moments are equal to zero and one, respectively.

For $j = 3$, one obtains the skewness coefficient of G

$$\text{skew}_G = \frac{E(X - EX)^3}{\text{var}_G^{3/2}}, \tag{1.37}$$

if $\alpha > 3$ or $\gamma < 1/3$, and X is a random variable with df G. Note that skew $= (m_3 - 3m_1 m_2 + 2m_1^3)/\sigma^3$. The 4th normalized, centered moment is the kurtosis, cf. page 31. Check that the normalized, centered moments are independent of location and scale parameters.

In Figure 1.6, the skewness coefficient of EV dfs G_γ is plotted against γ. The skewness coefficient is equal to zero at $\gamma_0 = -.2776\ldots$ in accordance with the remark on page 15 that the density of the Weibull df with shape parameter $\gamma = -.28$ (or $\alpha = -3.6$) looks symmetrical. We have $\text{skew}_{G_\gamma} > 0$, if $\gamma > \gamma_0$, and $\text{skew}_{G_\gamma} < 0$, if $\gamma < \gamma_0$. Furthermore, $\text{skew}_{G_0} = 1.1395\ldots$.

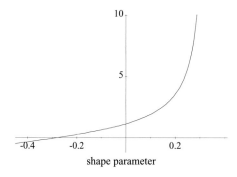

FIG. 1.6. Plotting the skewness coefficients skew_{G_γ} against $\gamma < 1/3$.

Finally, we deal with the unique modes, cf. page 15, of EV dfs G which will be denoted by mod_G. We have

$$\text{mod}_{G_{1,\alpha}} = (\alpha/(1+\alpha))^{1/\alpha}$$

and, thus, $\mathrm{mod}_{G_{1,\alpha}} \to 0$ as $\alpha \to 0$ and $\mathrm{mod}_{G_{1,\alpha}} \to 1$ as $\alpha \to \infty$. In addition, $\mathrm{mod}_{G_{2,\alpha}} = 0$, if $-1 \leq \alpha < 0$, and

$$\mathrm{mod}_{G_{2,\alpha}} = -(\alpha/(1+\alpha))^{1/\alpha}, \qquad \alpha < -1.$$

Thus, $\mathrm{mod}_{G_{2,\alpha}} \to 0$ as $\alpha \to -1$ and $\mathrm{mod}_{G_{2,\alpha}} \to -1$ as $\alpha \to -\infty$. In the γ–parameterization there is

$$\mathrm{mod}_{G_\gamma} = ((1+\gamma)^{-\gamma} - 1)/\gamma, \qquad \gamma \neq 0,$$

and the mode of the standard Gumbel distribution G_0 is zero.

Limiting Distributions of Minima

There is a one–to–one relationship between limiting dfs of maxima and minima. If

$$P\Big\{\max_{i \leq n}(-X_i) \leq b_n + a_n z\Big\} \to G(z), \qquad n \to \infty,$$

then

$$P\Big\{\min_{i \leq n} X_i \leq d_n + c_n z\Big\} \to 1 - G(-z), \qquad n \to \infty,$$

with $c_n = a_n$ and $d_n = -b_n$ (and vice versa). This yields that the limiting dfs of sample minima are the converse Gumbel, Fréchet and Weibull dfs

$$\widetilde{G}_{i,\alpha}(x) = 1 - G_{i,\alpha}(-x) \tag{1.38}$$

in the α–parameterization, and

$$\widetilde{G}_\gamma(x) = 1 - G_\gamma(-x) \tag{1.39}$$

in the γ–parameterization. Moreover, if $G_{i,\alpha,\mu,\sigma}$ or $G_{\gamma,\mu,\sigma}$ is the df of $-Y_i$, then $\widetilde{G}_{i,\alpha,-\mu,\sigma}$ or $\widetilde{G}_{\gamma,-\mu,\sigma}$ is the df of Y_i.

   The limiting dfs $\widetilde{G}$ of minima are also called extreme value dfs or, if necessary, converse extreme value dfs. Here are three special cases:

- the converse Gumbel df $\widetilde{G}_0$ is also called Gompertz df; this is the df that satisfies the famous Gompertz law, see page 54;

- $\widetilde{G}_{2,-1}$ is the exponential df on the positive half–line,

- $\widetilde{G}_{2,-2}$ is the Rayleigh df that is also of interest in other statistical applications. Note the following relationship: if the areas of random circles are exponentially distributed, then the diameters have a Rayleigh df.

The Min–Stability

The limiting dfs of minima—these are the converse EV dfs—are characterized by the min–stability. A df F is min–stable if

$$P\left\{\min_{i \leq n} X_i \leq d_n + c_n x\right\} = 1 - (1 - F(d_n + c_n x))^n = F(x) \qquad (1.40)$$

for a certain choice of constants d_n and $c_n > 0$, where $X_1, \ldots, X_n$ are iid random variables with common df F.

The min–stability can be expressed in terms of the survivor function $\bar{F} = 1 - F$ of F. We have

$$P\left\{\min_{i \leq n} X_i > d_n + c_n x\right\} = \bar{F}^n(d_n + c_n x) = \bar{F}(x). \qquad (1.41)$$

Some Historical Remarks

We compare the symbols employed for EV dfs in this textbook with others also used in relevant statistical publications:

Gumbel	EV0	G_0	Type I	Λ
Fréchet	EV1	$G_{1,\alpha}$	Type II	Φ_α
Weibull	EV2	$G_{2,\alpha}$	Type III	Ψ_α

In this book—as in many statistical textbooks—the normal (Gaussian) df is denoted by Φ and, therefore, the same symbol cannot be used for the Fréchet df. By combining the numbering of EV dfs in [20] with the letter G used in [39] (presumably, in honor of B.V. Gnedenko), one obtains the symbol $G_{i,\alpha}$ for EV dfs also taken in [42] and [16]. In contrast to the latter representation, the Gumbel df also gets the parameter $\alpha = 0$ in the present α–parameterization. The books [20] by J. Galambos and [39] by M.R. Leadbetter, G. Lindgren and H. Rootzén, together with the booklet [25] by L. de Haan, laid the probabilistic foundations of modern extreme value theory. We refer to [20] and [42] for a more detailed account of the history.

1.4 Modeling by Generalized Pareto Distributions

The standard generalized Pareto (GP) dfs $W_{i,\alpha}$ and W_γ are the adequate parametric dfs for exceedances, cf. also (1.11). The pertaining densities are denoted by $w_{i,\alpha}$ and w_γ. There is the simple analytical relationship

$$W(x) = 1 + \log G(x), \qquad \text{if} \quad \log G(x) > -1,$$

between GP dfs W and EV dfs G.

The α–Parameterization

First we introduce a representation for GP dfs within three submodels corresponding to that for EV dfs.

Exponential (GP0): $W_0(x) = 1 - e^{-x}$, $x \geq 0$,

Pareto (GP1), $\alpha > 0$: $W_{1,\alpha}(x) = 1 - x^{-\alpha}$, $x \geq 1$,

Beta (GP2), $\alpha < 0$: $W_{2,\alpha}(x) = 1 - (-x)^{-\alpha}$, $-1 \leq x \leq 0$.

The exponential df W_0 is equal to zero for $x < 0$; the Pareto dfs $W_{1,\alpha}$ are
equal to zero for $x < 1$; the beta dfs $W_{2,\alpha}$ are equal to zero for $x < -1$ and equal
to 1 for $x > 0$.

Note that $W_{2,-1}$ is the uniform df on the interval $[-1, 0]$. One should be
aware that the dfs $W_{2,\alpha}$ constitute a subclass of the usual family of beta dfs.
Subsequently, when we speak of beta dfs only dfs $W_{2,\alpha}$ are addressed (except of
those in Section 4.3, where we mention the full beta family).

> **Warning!** Our parameterization for beta dfs differs from the stan-
> dard one used in the statistical literature, where beta dfs with positive
> shape parameters are taken.

Once more, we also note the pertaining densities $w = W'$:

Exponential (GP0): $w_0(x) = e^{-x}$, $x \geq 0$,

Pareto (GP1), $\alpha > 0$: $w_{1,\alpha}(x) = \alpha x^{-(1+\alpha)}$, $x \geq 1$,

Beta (GP2), $\alpha < 0$: $w_{2,\alpha}(x) = |\alpha|(-x)^{-(1+\alpha)}$, $-1 \leq x < 0$.

The Pareto and exponential densities are decreasing on their supports. This
property is shared by beta densities with shape parameter $\alpha < -1$. For $\alpha = -1$,
one gets the uniform density on $[-1, 0]$ as mentioned above. Finally, the beta
densities with shape parameter $\alpha > -1$ are increasing, having a pole at zero.

We see that a GP density $w_{i,\alpha}$ possesses an upper tail similar (asymptotically
equivalent) to that of an EV density $g_{i,\alpha}$. Roughly speaking, the EV density is a
GP density tied down near the left endpoint.

Again, one must add location and scale parameters μ and $\sigma > 0$ in order to
obtain the full statistical families of GP dfs. Notice that the left endpoint of the
Pareto df $W_{1,\alpha,\mu,\sigma}(x) = W_{1,\alpha}((x - \mu)/\sigma)$ is equal to $\mu + \sigma$.

The γ–Parameterization

The unified GP model is obtained by applying a representation that corresponds to that for EV dfs in Section 1.3. The standard versions are defined such that

$$W_\gamma(x) \to W_0(x), \qquad \gamma \to 0. \tag{1.42}$$

By putting $\gamma = 1/\alpha$ and choosing location and scale parameters as in Section 1.3, the GP dfs in the γ–parameterization are $W_0(x) = 1 - e^{-x}$ for $x > 0$, and

$$W_\gamma(x) = 1 - (1 + \gamma x)^{-1/\gamma} \quad \text{for} \quad \begin{cases} 0 < x, & \gamma > 0; \\ & \text{if} \\ 0 < x < 1/|\gamma|, & \gamma < 0. \end{cases}$$

It is straightforward to verify that (1.42) holds for this choice of standard GP distributions. Note that W_0 is the standard exponential df again. Corresponding to (1.26), we have for $\gamma = 1/\alpha \neq 0$,

$$W_\gamma = W_{i,\alpha,-\alpha,|\alpha|}$$

with $i = 1$ if $\gamma > 0$ and $i = 2$ if $\gamma < 0$. The left endpoint of W_γ is equal to zero for all γ. Therefore, the γ–parameterization has the additional advantage that the location parameter is always the left endpoint of the support of the distribution.

The pertaining densities $w_\gamma = W_\gamma'$ are $w_0(x) = e^{-x}$, $x > 0$, if $\gamma = 0$, and

$$w_\gamma(x) = (1 + \gamma x)^{-(1+1/\gamma)} \quad \text{for} \quad \begin{cases} 0 \leq x, & \gamma > 0; \\ & \text{if} \\ 0 \leq x < 1/|\gamma|, & \gamma < 0. \end{cases}$$

The convergence $w_\gamma(x) \to w_0(x)$ as $\gamma \to 0$ holds again. In Fig. 1.7, Pareto and beta densities around the exponential density are displayed.

One can recognize that the left endpoint of a standard GP distribution W_γ is equal to zero; the right endpoint is finite in the case of beta (GP2) distributions.

The POT–Stability

GP dfs are the only continuous dfs F such that for a certain choice of constants b_u and a_u,

$$F^{[u]}(b_u + a_u x) = F(x),$$

where $F^{[u]}(x) = (F(x) - F(u))/(1 - F(u))$ is again the exceedance df at u pertaining to F as introduced in (1.11). This property is the pot–stability of GP dfs. The following three examples are of particular interest:

- for exponential dfs with left endpoint equal to zero,

$$W_{0,0,\sigma}^{[u]} = W_{0,u,\sigma}; \tag{1.43}$$

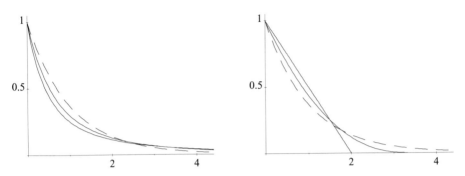

FIG. 1.7. (left.) Exponential (dashed) and Pareto (solid) densities for parameters $\gamma = .5, 1.0$. (right.) Exponential (dashed) and beta (solid) densities for parameters $\gamma = -.3, -.5$.

- for Pareto dfs $W_{1,\alpha,\mu,\sigma}$ in the α–parameterization with $\mu + \sigma < u$,

$$W_{1,\alpha,\mu,\sigma}^{[u]} = W_{1,\alpha,\mu,u-\mu}, \tag{1.44}$$

- for GP dfs $W_{\gamma,\mu,\sigma}$ in the γ–parameterization with $\mu < u$ and $\sigma + \gamma(u-\mu) > 0$,

$$W_{\gamma,\mu,\sigma}^{[u]} = W_{\gamma,u,\sigma+\gamma(u-\mu)}, \tag{1.45}$$

whereby (1.43) is a special case.

According to these relations the truncated version of a GP distribution remains in the same model, a property that exhibits the outstanding importance of GP distributions for our purposes. A beta df, truncated left of 1, is displayed in Fig. 1.8.

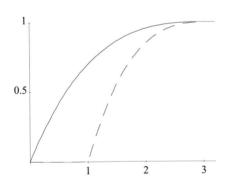

FIG. 1.8. Beta df $W_{-0.3}$ (solid) and a pertaining exceedance df $W_{-0.3}^{[1]} = W_{-0.3,\,1,\,0.7}$ (dashed).

In Fig. 1.8, one recognizes that the truncated beta df has a left endpoint equal to the truncation point 1 and the same right endpoint $1/|\gamma| = 10/3$ as the original beta df.

Limiting Distributions of Exceedances

The parametric modeling of exceedance dfs $F^{[u]}$ by GP dfs is based on a limit theorem again. Recall that our intention is the modeling of exceedances dfs at high thresholds and, hence, one should consider thresholds tending to the right endpoint of the actual df F.

(**Balkema–de Haan–Pickands**[11].) If $F^{[u]}(b_u + a_u x)$ has a continuous[12] limiting df as u goes to the right endpoint $\omega(F)$ of F, then

$$|F^{[u]}(x) - W_{\gamma,u,\sigma_u}(x)| \to 0, \qquad u \to \omega(F), \tag{1.46}$$

for some GP df with shape, location and scale parameters γ, u and $\sigma_u > 0$.

Note that the exceedance df $F^{[u]}$ and the approximating GP df possess the same left endpoint u. If (1.46) holds, then F belongs to the pot–domain of attraction of the GP df W_γ. It is evident that this limit theorem can be formulated likewise in terms of GP dfs in the α–parameterization. It is evident that (1.46) is closely related to (1.28).

One can easily prove that every EV df G_γ belongs to the pot–domain of attraction of W_γ. For this particular case, we also compute a rate of convergence. By employing a Taylor expansion of the exponential function around zero, one should first verify that the relation

$$\bar{G}_\gamma(x) = \overline{W}_\gamma(x)\Big(1 + O\big(\overline{W}_\gamma(x)\big)\Big) \tag{1.47}$$

holds for the survivor functions $\bar{G}_\gamma$ and $\overline{W}_\gamma$. Then, deduce

$$|G_\gamma^{[u]}(x) - W_\gamma^{[u]}(x)| = O\big(\overline{W}_\gamma(u)\big). \tag{1.48}$$

Now, (1.46) follows from (1.45), whereby $\sigma_u = 1 + \gamma u$. Moreover, if one of the usual textbook dfs satisfies (1.46), then the convergence also holds for the pertaining densities, also see Section 6.5.

Fig. 1.9 provides two illustrations in terms of densities: The first example treats the standard Weibull and beta densities g_γ and w_γ with shape parameter $\gamma = -.3$. The second example concerns the standard Cauchy density $f(x) = 1/(\pi(1 + x^2))$ and a Pareto density with shape parameter $\alpha = 1$.

In Fig. 1.9, one can see that the GP (beta and Pareto) densities fit to the upper tail of the Weibull and Cauchy densities. Elsewhere, the densities are substantially different. The preceding Pareto density with $\mu = -.25$ replaced by $\mu = 0$ also fits to the upper tail of the Cauchy density.

---

[11] Balkema, A.A. and de Haan, L. (1974). Residual life time at great age. Ann. Probab. 2, 792–804. Parallel work was done by J. Pickands (1975). Statistical inference using extreme value order statistics. Ann. Statist. 3, 119–131.

[12] An example of a discrete limiting df is, e.g., the geometric df $H_p(k) = 1 - (1 - p)^k$, $k \geq 1$, mentioned on page 13.

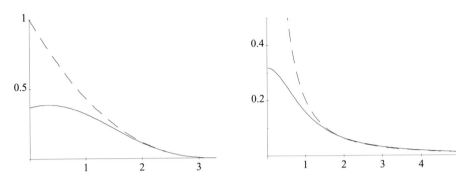

FIG. 1.9. (left.) Weibull density g_γ (solid) and beta density w_γ (dashed) with shape parameter $\gamma = -.3$. (right.) Standard Cauchy density (solid) and Pareto density $w_{1,\alpha,\mu,\sigma}$ (dashed) with $\alpha = 1$, $\mu = -.25$ and $\sigma = 1/\pi$.

Fitting a Generalized Pareto Distribution to the Upper Tail

As suggested by the limiting relation (1.46), we assume that a GP df $W_{\gamma,u,\sigma}$ can be fitted to the exceedance df $F^{[u]}$ for certain parameters γ and σ (in short: $F^{[u]} \approx W_{\gamma,u,\sigma}$). This implies that a certain GP df can be fitted to the original df F. More precisely, if $F(u)$ is known, we deduce from a relation $F^{[u]} \approx W_{\gamma,u,\sigma}$ that

$$F(x) \approx W_{\gamma,\tilde{\mu}(x),\tilde{\sigma}}, \quad x \geq u, \tag{1.49}$$

for certain parameters $\tilde{\mu}$ and $\tilde{\sigma}$. According to the definition (1.11) of exceedance dfs we have

$$
\begin{aligned}
F(x) &= (1 - F(u))F^{[u]}(x) + F(u) \\
&\approx (1 - F(u))W_{\gamma,\mu,\sigma}(x) + F(u) \\
&= W_{\gamma,\tilde{\mu},\tilde{\sigma}}(x), \quad x \geq u,
\end{aligned}
\tag{1.50}
$$

where the latter equality holds for the parameters

$$\tilde{\sigma} = \sigma / \left(1 + \gamma W_\gamma^{-1}(F(u))\right) \quad \text{and} \quad \tilde{\mu} = u - \tilde{\sigma} W_\gamma^{-1}(F(u)). \tag{1.51}$$

There is a unique relation between the two pairs of parameters μ, σ and $\tilde{\mu}$, $\tilde{\sigma}$ determined by the equations

$$W_{\gamma,\tilde{\mu},\tilde{\sigma}}^{[u]} = W_{\gamma,u,\sigma} \quad \text{and} \quad W_{\gamma,\tilde{\mu},\tilde{\sigma}}(u) = F(u). \tag{1.52}$$

In terms of survivor functions, (1.50) can be written $\bar{F}(x) = \bar{F}(u)\overline{F^{[u]}}(x) \approx \bar{F}(u)\overline{W}_{\gamma,u,\sigma}(x) = \overline{W}_{\gamma,\tilde{\mu},\tilde{\sigma}}(x)$, $x \geq u$.

Related conclusions hold in the α–parametrization: if the Pareto df $W_{1,\alpha,\mu,\sigma}$ with left endpoint $\mu + \sigma = u$ fits to the exceedance df $F^{[u]}$, then the Pareto df

$W_{1,\alpha,\tilde{\mu},\tilde{\sigma}}$ determined by $W^{[u]}_{1,\alpha,\tilde{\mu},\tilde{\sigma}} = W_{1,\alpha,\mu\sigma}$ and $W_{1,\alpha,\tilde{\mu},\tilde{\sigma}}(u) = F(u)$ fits to the upper tail of the df F. We have $\tilde{\mu} = \mu$ and $\tilde{\sigma} = \sigma(1 - F(u))^{1/\alpha}$.

Note that the first line in (1.50) can be extended to the representation

$$F(x) = F(u)P(X \le x | X \le u) + (1 - F(u))P(X \le x | X > u), \quad x \text{ real}, \quad (1.53)$$

where X is a random variable with df F.

For a continuation of this topic we refer to page 58, where relations between GP and sample dfs are investigated.

Maxima of Generalized Pareto Random Variables

It is a simple exercise to deduce from (1.47) that $W_0^n(x + \log n)$ and $W_{i,\alpha}^n(a_n x)$—with a_n as in (1.31)—are EV dfs G_0 and $G_{i,\alpha}$ in the limit as $n \to \infty$. The rate of convergence is $O(n^{-1})$.

There is another interesting non–asymptotic relationship between maxima of GP random variables and EV distributions. Generally, let $X_1, X_2, X_3, \ldots$ be non–negative iid random variables with common df F which has the left endpoint zero. Let N be a Poisson random variable with parameter $\lambda > 0$ which is independent of the X_i. Then, with $F^0(x) = 1$ we have

$$
\begin{aligned}
P\left\{ \max_{i \le N} X_i \le x \right\} &= \sum_{n=0}^{\infty} P\left\{ \max_{i \le N} X_i \le x, N = n \right\} \\
&= \sum_{n=0}^{\infty} P\{N = n\} F^n(x) \\
&= \exp(-\lambda(1 - F(x))) \quad (1.54)
\end{aligned}
$$

for $x \ge 0$ and zero otherwise. Notice that there is a jump at zero. Plugging in a GP df one obtains a df that equals an EV df right of zero. If λ is sufficiently large, then the jump is negligible and the df in (1.54) can be replaced by a GP df.

Moments of Generalized Pareto Distributions

The jth moment $m_{j,W}$ of a Pareto and beta dfs W can be computed in a straightforward way. Recollect that

$$m_{j,W} = E(X^j) = \int x^j \, dW(x) = \int x^j w(x) \, dx,$$

where $w = W'$ is the density of W and X is a random variable distributed according to W. We have

$$m_{j,W_{1,\alpha}} = \alpha/(\alpha - j), \quad \text{if } \alpha > j,$$

and

$$m_{j,W_{2,\alpha}} = (-1)^j \alpha/(\alpha - j).$$

The jth moment of a Pareto df $W_{1,\alpha}$ is infinite if $\alpha \leq j$. For the special case of $j = 1$, one obtains the mean values $m_W = m_{1,W}$ of Pareto and beta dfs W. For $i = 1, 2$ we have

$$m_{W_{i,\alpha}} = |\alpha|/(\alpha - 1), \tag{1.55}$$

if $1/\alpha < 1$. If $1/\alpha < 1/2$, the variances are

$$\mathrm{var}_{W_{i,\alpha}} = \alpha/((\alpha - 1)^2(\alpha - 2)). \tag{1.56}$$

The mean values and variances of the GP dfs W_γ are

$$m_{W_\gamma} = 1/(1 - \gamma), \qquad \text{if } \gamma < 1, \tag{1.57}$$

and

$$\mathrm{var}_{W_\gamma} = 1/((1 - \gamma)^2(1 - 2\gamma)), \qquad \text{if } \gamma < 1/2.$$

In the special case of $\gamma = 0$, one gets the well–known result that the mean value and variance of the standard exponential distribution are both equal to 1.

1.5 Heavy and Fat–Tailed Distributions

There are several different concepts of fat or heavy–tailedness of a distribution. We say that a df F has a heavy upper tail if a jth moment $\int_0^\infty x^j \, dF(x)$—evaluated over the positive half–line—is equal to infinity for some positive integer j. Therefore, all Pareto dfs are heavy–tailed. In addition, all the dfs in the pot–domain of attraction of a Pareto df are heavy–tailed. Likewise one may speak of a heavy lower tail.

Fat–tailedness may be introduced by comparing the kurtosis to that of the normal distribution.

The Normal (Gaussian) Model

Let $\Phi(x) = \int_{-\infty}^x \varphi(y) \, dy$ denote the standard normal df, and

$$\varphi(y) = \exp(-y^2/2)/\sqrt{2\pi}$$

the pertaining normal density.

We introduce the mth convolution of a df F, which is given by

$$F^{m*}(x) = P\{X_1 + \cdots + X_m \leq x\},$$

where $X_1, \ldots, X_m$ are iid random variables with common df F. A parametric modeling of convolutions by means of normal dfs is adequate, because under the conditions of the central limit theorem, we know that

$$\sup_x |F^{m*}(x) - \Phi((x - \mu_m)/\sigma_m)| \to 0, \qquad m \to \infty, \tag{1.58}$$

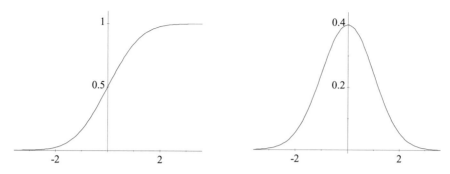

FIG. 1.10. (left.) Standard normal df Φ. (right.) Standard normal density φ.

where μ_m and $\sigma_m > 0$ are location and scale parameters.

It is noteworthy that the normal df Φ is sum–stable because $\Phi^{m*}(m^{1/2}x) = \Phi(x)$. The Cauchy df F is another sum–stable df with $F^{m*}(mx) = F(x)$. This topic will be further pursued in Section 6.4 within the framework of sum–stable distributions.

The Kurtosis and a Concept of a Fat–Tailedness

The kurtosis of a df F is another moment ratio which is defined by

$$\int (x - m_F)^4 \, dF(x) \Big/ \mathrm{var}_F^2, \tag{1.59}$$

where m_F and var_F are the mean and the variance of F. If the density has a high central peak and long tails, then the kurtosis is typically large. Notice that the kurtosis is independent of location and scale parameters. A df with kurtosis larger than 3, which is the kurtosis of normal dfs, is fat–tailed or leptokurtic. We also speak of fat upper and lower tails in this context.

It is easily proven that the mixture of two Gaussian dfs—see (1.61) and Example 2.3.2—with equal location parameters and unequal scale parameters is leptokurtic.

Log–Normal Distributions

The property of fat–tailedness is also discussed in conjunction with the log–normal distribution which has a short (finite) lower tail, yet one can speak of a fat upper tail.

If one has a strong preference for the normal model, yet the positive data $x_1, \ldots, x_n$ indicate a fat or heavy upper tail of the underlying distribution, then one may try to fit a normal df $\Phi_{\mu,\sigma}$ to the transformed data $T(x_1), \ldots, T(x_n)$ of a reduced magnitude, cf. also page 37. The first choice of such a transformation is $T(x) = \log x$.

If a normal df $\Phi_{\mu,\sigma}$ can be fitted to $\log(x_1), \ldots, \log(x_n)$, then the log–normal df

$$F_{(\mu,\sigma)}(x) = \Phi_{\mu,\sigma}(\log(x)), \qquad x > 0, \tag{1.60}$$

can be fitted to the original data $x_1, \ldots, x_n$. Conversely, exp–transformed normal data are log–normal which justifies to say that the log–normal df has a fatter upper tail than the normal one.

We have $F_{(0,\sigma)}(1 + \sigma x) \to \Phi(x)$ as $\sigma \to 0$. Therefore, instead of employing a log–transformation to the data, one may also try to fit a log–normal df to the original data.

The log–normal df is not heavy–tailed in our terminology, cf. page 30, because all moments are finite. Note that normal as well as log–normal dfs belong to the pot–domain of attraction of the exponential df , cf. [20] on pages 67–68.

Notice that $F_{(\mu,\sigma)}(x) = \Phi_{0,\sigma}(\log(x/\exp(\mu)))$ and, hence, $\exp(\mu)$ is a scale parameter of the log–normal df. Log–normal distributions are skewed to the right.

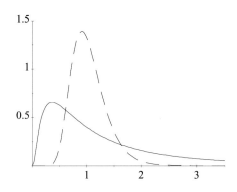

FIG. 1.11. Plots of log–normal densities $f_{(0,1)}$ (solid) and $f_{(0,.3)}$ (dashed).

The preceding illustration concerns the standard log–normal density

$$f_{(0,\sigma)}(x) = \frac{1}{\sqrt{2\pi}\sigma x} \exp\big(-(\log(x))^2/2\sigma^2\big), \qquad x > 0,$$

with the shape parameter σ.

Mixtures of Gaussian and Heavy–Tailed Distributions

To enlarge a given model one may take mixtures

$$F = (1 - d)\, F_0 + d\, F_1, \tag{1.61}$$

where F_0 and F_1 are taken from an initial parametric model and the mixing parameter d ranges between 0 and 1. An important example is a mixture of two normal distributions.

Actual dfs are mixtures, if data are generated according to random mechanism with dfs F_0 and F_1 with a probability $1 - d$ and, respectively, d. A more detailed interpretation of mixtures is given in Section 8.1 in conjunction with conditional distributions.

Subsequently, we deal with mixtures $(1 - d)\Phi + d\,F$, where F is another df that may be regarded as a contamination of the ideal normal model and $0 \le d \le 1$. We consider two special cases of a densities f which may serve as a contamination of the normal density.

- (Student Distributions.) For $\alpha > 0$, we introduce densities

$$f_\alpha(x) = c(\alpha)\left(1 + \frac{x^2}{\alpha}\right)^{-(1+\alpha)/2} \tag{1.62}$$

  where $c(\alpha) = \Gamma((1+\alpha)/2)/(\Gamma(\alpha/2)\Gamma(1/2)\alpha^{1/2})$. For positive integers α this is the Student distribution with α degrees of freedom. The variance is $\alpha/(\alpha-2)$ for $\alpha > 2$. In (6.16), such distributions are introduced as mixtures of normal distributions.

- (Generalized Cauchy Distributions.) The density of the standard generalized Cauchy df with shape parameter $\alpha > 0$ is[13]

$$f_\alpha(x) = c(\alpha)/\left(1 + |x|^{1+\alpha}\right), \tag{1.63}$$

  where $c(\alpha) = ((1 + \alpha)/(2\pi))\sin(\pi/(1 + \alpha))$.

In both cases one obtains the standard Cauchy density, cf. page 27, for $\alpha = 1$. In addition, these densities are unimodal with mode equal to zero and possess heavy tails like the Pareto dfs with shape parameter α.

The mixture density $(1 - d)\varphi(x) + d\,f_\alpha(x)$, with d being close to zero, has the shape of a normal density in the center, yet possesses heavier tails. It is advisable to apply trimmed estimators of μ and σ to reach an estimate of the central normal part of the distribution.

The following illustrations concern mixtures between the standard normal distribution and Cauchy and Weibull distributions.

In Fig. 1.12 (left), one can clearly observe the heavier tails of the mixture, although the fraction of the Cauchy distribution is only ten percent. The center of the mixture can hardly be distinguished from a normal density.

Remarks About Robust Statistics

The contamination of a normal distribution by a heavy–tailed distribution (such as the Cauchy distribution) is one of the favorite topics in robust statistics, because

---

[13]Drees, H. and Reiss, R.–D. (1992). Tail behavior in Wicksell's corpuscle problem. In: Probability Theory and Applications, J. Galambos and I. Kátai (eds.), 205–220, Kluwer, Dortrecht.

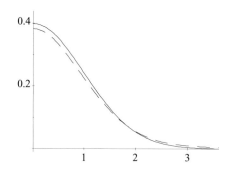 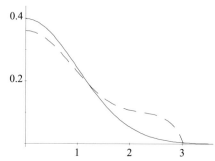

FIG. 1.12. Normal density φ (solid) and mixture densities $.9\varphi + .1f_1$ (left) and $0.9\varphi + 0.1g_{2, -1.6, 3, 1}$ (right) with Cauchy and Weibull components.

observations under the heavy tail—frequently regarded as outliers—destroy the performance of classical statistical procedures such as the sample mean.

A primary aim of robust statistics may be formulated in the following manner (cf. [27]): "Describe the structure best fitting to the bulk of the data: we tentatively assume a parametric model ... taking explicitly into account that the model may be distorted." For this (central) bulk of the data the normal model is the dominating parametric model.

A Converse View Towards Robust Statistics

In the preceding lines, a fat or heavy–tailed distribution was regarded as a contamination of the normal df. A converse view can be expressed in the following manner[14]:

"In many applications extreme data do not fit to the ideology of normal samples and, therefore, extremes are omitted from the data set or one uses statistical procedures upon which extremes have a bounded influence. The converse attitude is to regard extremes as the important part of the data set, yet one must still follow certain principles of robust statistics and check the validity of the statistical model selection."

The decisive step was the fitting of certain dfs—such as GP dfs—to extreme data. One must be aware that

- the actual df deviates from any of the chosen parametric ones, and

- there can be gross errors in the measurements which may heavily influence the performance of statistical procedures.

[14] Reiss, R.–D. (1989). Robust statistics: A converse view. 23rd Semester on Robustness and Nonparametric Statistics. Stefan Banach Center, Warsaw, Abstracts of Lectures, Part II, 183–184.

The difficulties arising from the fact that the actual distribution deviates from any EV or GP distribution is the central topic in most research papers on extreme value theory. This question is briefly touched upon in this book:

- we already mentioned limit theorems for maxima and exceedance dfs (cf. Sections 1.3 and 1.4);

- a von Mises condition will be introduced in Section 2.1 in conjunction with reciprocal hazard functions, and

- the concept of a δ–neighborhood of a GP distribution and related weaker conditions are studied in Section 6.5.

Less is known about robustifying statistical procedures in the extreme value setting against gross errors. The explanations in the Sections 3.1 and 5.1 about M–estimates should be regarded as a first step.

1.6 Quantiles, Transformations and Simulations

The concept of q–quantiles and quantile functions (qfs) is introduced in greater generality and in a more rigorous manner compared to our remarks about quantiles on page 12. We also deal with technical questions that can be best described by means of quantile functions.

Quantiles

In conjunction with the T–year level, we already introduced the q–quantile of a df F as the value z such that $F(z) = q$. Thus, the q–quantile z is a value along the measurement scale with the property that the fraction q of the distribution is left of z.

In Fig. 1.13, this will be illustrated for the Gumbel df $G_0(x) = \exp(-e^{-x})$, where $z = -\log(-\log(q))$.

If the q–quantile is not unique, one may take the smallest z such that $F(z) = q$. For discrete dfs and, generally, for discontinuous dfs, it may happen that there is no value z such that $F(z) = q$. Then, the q–quantile is the smallest value x such that $F(x) \geq q$. The $1/2$–quantile is the median of the df F.

Quantile Functions

If the df F is continuous and strictly increasing on its support $(\alpha(F), \omega(F))$, then the quantile function (qf) F^{-1} is the usual inverse of F.

Usually, the quantile function (qf) F^{-1} is a "generalized inverse"

$$F^{-1}(q) := \inf\{x : F(x) \geq q\}, \qquad 0 < q < 1. \tag{1.64}$$

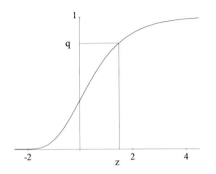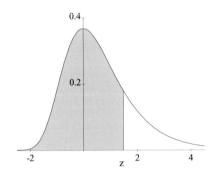

FIG. 1.13. (left.) The Gumbel df G_0 evaluated at the q–quantile z. (right.) The area under the Gumbel density $g_0(x) = e^{-x} \exp(-e^{-x})$ left of the q–quantile z is equal to q.

Notice that $F^{-1}(q)$ is the q–quantile of F as introduced in the preceding lines. If $F_{\mu,\sigma}$ is a df with location and scale parameters μ and σ, then

$$F_{\mu,\sigma}^{-1} = \mu + \sigma F^{-1}, \tag{1.65}$$

where $F = F_{0,1}$ is the standard df.

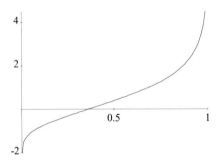

FIG. 1.14. Plot of the Gumbel qf $G_0^{-1}(q) = -\log(-\log(q))$.

We state the explicit form of EV and GP qfs in both parameterizations starting with EV qfs in their α–parameterization.

Gumbel (EV0): $G_0^{-1}(q) = -\log(-\log(q));$

Fréchet (EV1), $\alpha > 0$: $G_{1,\alpha}^{-1}(q) = (-\log(q))^{-1/\alpha};$

Weibull (EV2), $\alpha < 0$: $G_{2,\alpha}^{-1}(q) = -(-\log(q))^{-1/\alpha}.$

Next, we present EV qfs in their γ–representation:

EV, $\gamma \neq 0$: $\qquad\qquad G_\gamma^{-1}(q) = ((-\log(q))^{-\gamma} - 1)/\gamma,$

and for $\gamma = 0$, we have the Gumbel qf G_0^{-1} again.

We continue with GP qfs in the α–parameterization:

Exponential (GP0): $\qquad\quad W_0^{-1}(q) = -\log(1-q);$

Pareto (GP1), $\alpha > 0$: $\qquad W_{1,\alpha}^{-1}(q) = (1-q)^{-1/\alpha};$

Beta (GP2), $\alpha < 0$: $\qquad\; W_{2,\alpha}^{-1}(q) = -(1-q)^{-1/\alpha}.$

Finally, there is the γ–representation of GP dfs:

GP, $\gamma \neq 0$: $\qquad\qquad\qquad W_\gamma^{-1}(q) = ((1-q)^{-\gamma} - 1)/\gamma,$

and again $W_0^{-1}(q) = -\log(1-q)$.

Recall that the medians of these distributions are obtained by plugging in the value $q = 1/2$. Verify, by means of

$$(x^y - 1)/y \to \log x, \qquad y \to 0, \qquad\qquad (1.66)$$

that EV and GP qfs in the γ–parameterization again built a continuous family.

Transformations

If a statistician is preoccupied with certain parametric dfs, say F_ϑ, yet none of these dfs fits to the data $x_1, \ldots, x_n$, one may deal with the transformed data $T(x_1), \ldots, T(x_n)$. If F_ϑ fits to $T(x_1), \ldots, T(x_n)$ for some parameter ϑ, then the df $F_\vartheta(T)$ fits to the original data $x_1, \ldots, x_n$. For example, the transformation $T(x) = \log(x)$ reduces the magnitude of positive data. The df $F_\vartheta(\log(x))$ is called the log–F_ϑ df. We may say that the log–df has a fatter upper tail than the original df in view of the exp–transformation of the data.

- (Exponential Model as Log–Gompertz Model.) If X is an exponential random variable with scale parameter σ, then $\log X$ is a Gompertz random variable with location parameter $\mu = \log \sigma$.

- (Pareto Model as Log–Exponential Model.) If X is a Pareto random variable with shape and scale parameters α and σ, then $\log X$ is an exponential random variable with location and scale parameters $\log \sigma$ and $1/\alpha$.

- (Fréchet Model as Log–Gumbel Model.) A corresponding relationship holds between a Fréchet random variable with shape and scale parameters α and σ and a Gumbel random variable with location and scale parameters $\log \sigma$ and $1/\alpha$.

- (Pareto and Beta Model.) If X is once again a Pareto random variable with shape and scale parameters α and σ, then $-1/X$ is a beta random variable with shape and scale parameters $-\alpha$ and $1/\sigma$.

- (Fréchet and Weibull Model.) A corresponding relationship holds between a Fréchet random variable with shape and scale parameters α and σ and a Weibull random variable with shape and scale parameters $-\alpha$ and $1/\sigma$.

Such transformations are suggested by the quantile and probability transformations:

1. **Quantile Transformation:** if U is a $(0,1)$–uniformly distributed random variable, then $F^{-1}(U)$ has the df F,

2. **Probability Transformation:** conversely, if X is a random variable with continuous df F, then $F(X)$ is $(0,1)$–uniformly distributed.

We mention further examples of transformed random variables: if X is a converse exponential random variable, then

$$T(X) = G_{2,\alpha}^{-1}(G_{2,-1}(X)) = -(-X)^{-1/\alpha} \qquad (1.67)$$

is a Weibull random variable with shape parameter α. Likewise,

$$T(X) = -(-X)^{-1/\alpha} \qquad (1.68)$$

is a beta random variable with shape parameter α if X is $(-1,0)$–uniformly distributed.

Generation of Data

According to our preceding remarks, one can easily generate data under EV and GP dfs: if $u_1, \ldots, u_n$ are governed by the uniform df on the interval $(0,1)$, the transformed values $F^{-1}(u_1), \ldots, F^{-1}(u_n)$ are governed by the df F. For example, the quantile transformation technique can be applied to a Cauchy random variable which possesses the qf $F_1^{-1}(q) = \tan(\pi(q - 0.5))$.

If location and scale parameters μ and σ are included, then the transformation can be carried out in two steps because $F_{\mu,\sigma}^{-1}(u_i) = \mu + \sigma F^{-1}(u_i)$.

We mention two other simulation techniques[15]:

- the normal qf is not feasible in an analytical form and the quantile transformation is not applicable. In this case, the polar method can be utilized,

- the interpretation of mixture distributions within a two–step experiment stimulates a technique to carry out simulations, cf. page 33 and Section 8.1.

---

[15] For details see, e.g., Devroye, L. (1986). Non–Uniform Random Variate Generation. Springer, New York.

Chapter 2

Diagnostic Tools

In this chapter, we catch a glimpse of the real world in the condensed form of data. Our primary aim is to fit extreme value (EV) and generalized Pareto (GP) distributions, which were introduced in the foregoing chapter by means of limit theorems, to the data.

For that purpose, fairly simple nonparametric techniques for visualizing data are introduced. Those tools are especially selected which are of high relevance to extreme values. For example, sample excess and sample reciprocal hazard functions should be close to a straight line if a GP hypothesis is valid. Of course, Q–Q plots are also on the agenda. We discuss Q–Q plots for location and scale parameter families as well as for EV and GP models. Finally, certain tools are introduced which are relevant in conjunction with time series phenomena.

2.1 Visualization of Data

First, we describe visualization techniques such as the sample df, the sample qf, histograms and kernel densities. In contrast to Chapter 1, we do not assume that the data are maxima, exceedances or sums.

The Sample Distribution Function

The sample df $\widehat{F}_n(x)$ at x for a series of univariate data $x_1, \ldots, x_n$ is the relative number of the x_i that are smaller or equal to x. Thus,

$$\widehat{F}_n(x) := \frac{1}{n} \sum_{i \leq n} I(x_i \leq x), \tag{2.1}$$

where the indicator function is defined by $I(y \leq x) = 1$ if $y \leq x$ and 0, elsewhere; furthermore, the summation runs over $i = 1, \ldots, n$. Sample dfs are particularly useful for representing samples of a smaller size.

The data $x_1, \ldots, x_n$ ordered from the smallest to the largest are denoted by

$$x_{1:n} \leq \cdots \leq x_{n:n}.$$

We have $\widehat{F}_n(x_{i:n}) = i/n$ if $x_{i:n}$ is not a multiple point. Notice that $\widehat{F}_n$ is constant between consecutive ordered values. The ordered values can be recaptured from the sample df and, thus, there is a one–to–one correspondence between the sample df and the ordered values. One may take a linear interpolation between consecutive points as well (also see pages 42 to 47 for other modifications).

Occasionally, we write $\widehat{F}_n(\boldsymbol{x}; x)$ in place of $\widehat{F}_n(x)$ to indicate the dependence on the vector of data $\boldsymbol{x} = (x_1, \ldots, x_n)$. We will primarily deal with situations where each of the x_i is generated under a common df F and the sample df $\widehat{F}_n$ is approximately equal to F. This relationship will be briefly written

$$\boxed{\widehat{F}_n(x) \approx F(x).} \tag{2.2}$$

In view of (2.2), we say that $\widehat{F}_n(x)$ is an estimate of $F(x)$.

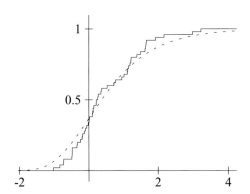

FIG. 2.1. Gumbel df (dotted) and sample df of a Gumbel data set with 50 points.

To a larger extent, our statistical arguments are based on the relationship (2.2) between the sample df $\widehat{F}_n$ and the underlying df F so that, apparently, the statistical prerequisites for understanding this textbook are of an elementary nature.

The Sample Exceedance Distribution Function

Let $\widehat{F}_n(\boldsymbol{x}; \cdot)$ be the sample df based on the data x_i as in the preceding lines. Recall that the exceedances y_i over the threshold u are those x_i with $x_i > u$ taken in the original order of their outcome. The sample df $\widehat{F}_k(\boldsymbol{y}; \cdot)$ based on the vector

$\boldsymbol{y} = (y_1, \ldots, y_k)$ of exceedances is the sample exceedance df. We have

$$\widehat{F}_k(\boldsymbol{y}; \cdot) = \left(\widehat{F}_n(\boldsymbol{x}; \cdot)\right)^{[u]}. \qquad (2.3)$$

Thus, the sample exceedance df is equal to the exceedance df of the sample df.

From the fact that the sample df $\widehat{F}_n(\boldsymbol{x}; \cdot)$ is an estimate of the underlying df F, as pointed out in (2.2), one may conclude that

$$\widehat{F}_k(\boldsymbol{y}; \cdot) \approx F^{[u]}, \qquad x \geq u; \qquad (2.4)$$

that is, the sample exceedance df is an estimate of the exceedance df.

Sample Moments

Now we introduce sample versions of the mean, variance and skewness coefficient. These sample functionals can be deduced from the sample moments

$$m_{j,n} = \frac{1}{n} \sum_{i \leq n} x_i^j. \qquad (2.5)$$

Subsequently, assume that $x_1, \ldots, x_n$ are governed by the df F. The representation $m_{j,n} = \int x^j \, dF_n(x)$ and (2.2) make it plausible that the jth sample moment $m_{j,n}$ is an estimate of the jth moment $m_j = \int x^j \, dF(x)$ of F if this moment exists.

We also write

$$\bar{x}_n = \frac{1}{n} \sum_{i \leq n} x_i$$

for the sample mean. The sample variance is

$$s_n^2 = \frac{1}{n-1} \sum_{i \leq n} (x_i - \bar{x}_n)^2. \qquad (2.6)$$

A factor $1/(n-1)$ is employed instead of $1/n$ in the definition of s_n^2 to obtain an unbiased estimator (cf. page 89) of the variance of F. Note that $s_n^2 = m_{2,n} - m_{1,n}^2$ if the factor $1/n$ is taken. We also write $\bar{x}$ and s^2 instead of $\bar{x}_n$ and s_n^2.

The sample skewness coefficient can be written

$$\frac{1}{n} \sum_{i \leq n} \left(\frac{x_i - \bar{x}_n}{s_n}\right)^3 = (m_{3,n} - 3m_{1,n}m_{2,n} + 2m_{1,n}^3)/s_n^3. \qquad (2.7)$$

The sample skewness coefficient is also a natural estimate of the skewness coefficient skew_F of F.

Sample Quantiles, Sample Quantile Functions

We believe that qfs and densities are more useful than dfs for the visual discrimination between distributions. In the subsequent lines we present the sample versions of qfs and densities.

The sample qf $\widehat{F}_n^{-1}$ may be introduced by

$$\widehat{F}_n^{-1}\left(\frac{i}{n+1}\right) := x_{i:n}, \tag{2.8}$$

where $x_{1:n} \leq \cdots \leq x_{n:n}$ are the data arranged in increasing order. In statistical publications, it is also suggested to employ plotting positions $(i-0.5)/n$ in place of $i/(n+1)$ in (2.8). In addition, let $\widehat{F}_n^{-1}$ be constant between consecutive points $i/(n+1)$. Likewise, one may employ a linear interpolation. Note that the sample qf remains constant in case of multiple points.

The sample qf $\widehat{F}_n^{-1}$ is the qf—in the sense of (1.64)—of the sample df $\widehat{F}_n$ if the sample qf is taken left–continuous. Also, $\widehat{F}_n^{-1}(q)$ is the sample q–quantile. Usually, special constructions are employed for the sample median.

Conclude from the basic relationship (2.2) for dfs that

$$x_{i:n} = \widehat{F}_n^{-1}\left(\frac{i}{n+1}\right) \approx F^{-1}\left(\frac{i}{n+1}\right), \tag{2.9}$$

where F^{-1} is the qf pertaining to the df F, cf. (1.64). We also have $x_{[nq]:n} \approx F^{-1}(q)$.

In conjunction with grouped data and the kernel method, we will also present smoothed versions of the sample qf.

Linearly Interpolated Sample Distributions

If the underlying df is continuous, then it is plausible to estimate this df by means of a continuous sample df. Such a df can be constructed by linearly interpolating the sample df $\widehat{F}_n$ in (2.1) over bins $(t_j, t_{j+1}]$, where the $t_j < t_{j+1}$ constitute a grid on the real line. One gets the continuous sample df

$$
\begin{aligned}
F_n(x) &= \widehat{F}_n(t_j) + \frac{x - t_j}{t_{j+1} - t_j}\left(\widehat{F}_n(t_{j+1}) - \widehat{F}_n(t_j)\right) \\
&= \widehat{F}_n(t_j) + \frac{(x - t_j)n_j}{t_{j+1} - t_j}, \qquad \text{for } t_j < x \leq t_{j+1}, \tag{2.10}
\end{aligned}
$$

where n_j is the frequency of data $x_1, \ldots, x_n$ in the bin $(t_j, t_{j+1}]$. Thus, the sample df F_n only depends on the data in a grouped form.

EXAMPLE 2.1.1. (Sample Distribution and Quantile Functions of Grouped Fire Claim Data.) The sample df and sample qf of grouped claim data are displayed. From the original data set of UK fire claim data, observed during a four year period, we just take the 387 claim sizes over £51,200 (these data are taken from [10] and stored in the file ig–fire1.dat).

TABLE 2.1. Number $n(j)$ of claim sizes within class limits in £1000

lower class limit	51.2	72.41	102.4	250	500	750	1000	2000	3000
number of claims $n(j)$	108	88	117	47	12	4	8	3	0

The sample df and sample qf pertaining to the histogram are plotted in Fig. 2.2.

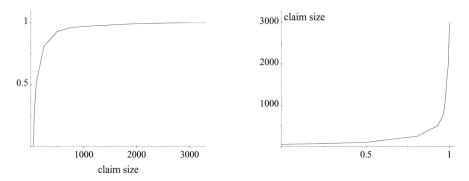

FIG. 2.2. Sample df (left) and sample qf (right) for fire claim data over £51,200.

The first overall impression is that the sample qf is a reflection of the sample df at the diagonal (as it should be because the sample qf is a generalized inverse of the sample df). It is evident that the sample df and qf provide the same information about the data although the viewpoints are different.

The sample df based on grouped data is piecewise continuously differentiable and, therefore, it has a density in the form of a histogram. We will also deal with histograms for discrete data and with kernel densities for continuous data. The latter concept can be regarded as a modification of histograms for grouped data.

Histograms for Grouped Data

Again let n_j be the frequency of data in the bin $(t_j, t_{j+1}]$. Taking the derivative of the preceding sample df F_n based on grouped data as given in (2.10), one gets the probability density

$$f_n(x) = \frac{n(j)}{n(t_{j+1} - t_j)}, \qquad t_j < x \le t_{j+1}. \tag{2.11}$$

It is very natural to visualize frequencies by means of such a histogram. It is apparent that this histogram is an appropriate estimate of the density f of F. The histogram may also be addressed as sample density.

Practitioners use histograms because of their simplicity in representing data, even if the data are given in a continuous form. One disadvantage of a histogram is that one must choose the location of the grid. Later we will introduce an alternative method for representing data, namely by means of a kernel density.

Histograms for Discrete Data

In the case of integer–valued data, a sample histogram is given by

$$p_n(j) = n(j)/n,$$

where $n(j)$ is the number of data $x_1, \ldots, x_n$ equal to the integer j. In analogy to (2.2), we have

$$\boxed{p_n(j) \approx P\{j\},} \qquad (2.12)$$

where P is the underlying discrete distribution (under which the x_i were generated). Note that the discrete values x_i—ordered according to their magnitudes—can be recaptured from the histogram. Histograms for binomial and Poisson distributions are given in Fig. 1.2, also see Fig. 3.2.

Kernel Densities

Starting with continuous data $x_1, \ldots, x_n$, the histogram for grouped data may be constructed in the following manner. Replace each point x_i in the bin $(t_j, t_{j+1}]$ by the constant function

$$g(x, x_i) = \frac{1}{n(t_{j+1} - t_j)}, \qquad t_j < x \leq t_{j+1}, \qquad (2.13)$$

with weight $1/n$. In summing up the single terms $g(x, x_i)$, one gets the histogram for grouped data in the representation $f_n(x) = \sum_{i \leq n} g(x, x_i)$. If continuous data are given, then the choice of the grid is crucial for the performance of the histogram.

We represent an alternative construction of a sample density. In contrast to (2.13), replace x_i by the function

$$g_b(x, x_i) = \frac{1}{nb} k\left(\frac{x - x_i}{b}\right),$$

where k is a function (kernel) such that $\int k(y)\, dy = 1$ and $b > 0$ is a chosen bandwidth. If $k \geq 0$, then $k((x - x_i)/b)/b$ may be regarded as a probability density with location and scale parameters x_i and $b > 0$. The function $g_b(\cdot, x_i)$ again possesses the weight $1/n$.

In summing up the single terms, one gets the kernel density

$$f_{n,b}(x) := \sum_{i \leq n} g_b(x, x_i) = \frac{1}{nb} \sum_{i \leq n} k\left(\frac{x - x_i}{b}\right) \tag{2.14}$$

which is a probability density if $k \geq 0$.

In subsequent illustrations, the Epanechnikov kernel

$$k(x) = \frac{3}{4}(1 - x^2)I(-1 \leq x \leq 1)$$

is taken which is the optimal kernel under a certain criterion. A kernel related directly to the terms $g(x, x_i)$ in the histogram is the "naive" one

$$k(x) = 0.5 \times I(-1 \leq x < 1).$$

Other interesting choices of kernels are

$$k(x) = \frac{1}{8}(9 - 15x^2)I(-1 \leq x \leq 1) \tag{2.15}$$

or

$$k(x) = \frac{15}{32}(3 - 10z^2 + 7z^4)I(-1 \leq x \leq 1), \tag{2.16}$$

because these kernels satisfy the additional condition $\int x^2 k(x)\, dx = 0$.

In analogy to the choice of the grid for the histogram—particularly of the bin–width—the choice of an appropriate bandwidth b is crucial for the performance of the kernel density.

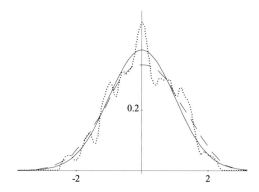

FIG. 2.3. Plots of normal density (solid) and kernel densities for the bandwidths $b = 1.1$ (dashed) and $b = 0.3$ (dotted) based on 50 Gaussian data.

If the bandwidth b is small, which is related to a small scale parameter, then one can still recognize terms $g_b(x, x_i)$ representing the single data. If b is large, then an oversmoothing of the data may prevent the detection of certain clues in the data.

An Automatic Bandwidth Selection

An automatic bandwidth selection is provided by cross–validation (see, e.g., the review article by Marron[1] or [50]). For finite sample sizes, the automatic choice of the bandwidth must be regarded as a first crude choice of the bandwidth. It is useful to vary the bandwidth around the automatically selected parameter; e.g., decrease the bandwidth until the graph of the kernel density becomes bumpy.

EXAMPLE 2.1.2. (TV Watching in Hours per Week.) We analyze 135 data of TV watching in hours per week stored in the file su–tvcon.dat. The bulk of the data is below 20 hours. The lower 100 observations range from 1.68 to 19.5. A list of the 35 data exceeding 20 hours is given.

TABLE 2.2. Hours per week over 20.

20	20.5	23	24	26	27.5	28.5	31.5	45
20	22	23	24.75	26	27.5	29	33	49
20	22	23	25	27	28	29.5	37	63
20.5	22	23.9	25	27	28	30	40	

These data are represented by a kernel density in Fig. 2.4. The automatic bandwidth selection by means of cross–validation leads to the choice $b = 12.5$.

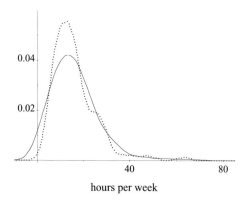

FIG. 2.4. Two kernel densities for bandwidths $b = 12.5$ (solid) and $b = 5$ (dotted) for TV data.

The cross–validation with the selected bandwidth $b = 12.5$ seems to oversmooth the data. The bandwidth $b = 5$ is small enough so that a more detailed structure becomes visible in the upper part of the distribution which might indicate a subpopulation. Yet, this bandwidth is sufficiently large so that the kernel density does not become bumpy.

---

[1]Marron, J.S. (1988). Automatic smoothing parameter selection: A survey. Empirical Economics 13, 187–208.

It is likely that in the classical statistical analysis—from a methodological viewpoint one may specify this as the work done before R.A. Fisher[2]—certain observations, such as the data point 63, are regarded as outliers and are omitted from the sample.

Kernel Distribution and Quantile Functions

By taking the df pertaining to the kernel density $f_{n,b}$, one obtains a competitor $\widehat{F}_{n,b}$ of the sample df $\widehat{F}_n$. We have

$$\widehat{F}_{n,b}(x) := \frac{1}{n} \sum_{i \leq n} K\left(\frac{x - x_i}{b}\right), \tag{2.17}$$

where

$$K(x) = \int_{-\infty}^{x} k(y)\, dy.$$

By taking the qf pertaining to $\widehat{F}_{n,b}$, one obtains a competitor of the sample qf.

Another version of the sample qf is constructed by directly smoothing the sample qf by a kernel k. Let

$$\widehat{F}_{n,b}^{-1}(q) = \frac{1}{b} \int_0^1 k\left(\frac{q - y}{b}\right) \widehat{F}_n^{-1}(y)\, dy.$$

One must apply a variable bandwidth selection around the boundary points.

Kernel Densities With Bounded Support

If it is known that none of the observations is below or, respectively, above a specific threshold—e.g., life spans are non–negative or exceedances over a certain threshold t exceed t—then the foregoing smoothing of data should not result in shifting weight below or above such thresholds (also compare the kernel densities in Fig. 2.4 and Fig. 2.10).

One may reflect the sample at the minimum and/or maximum value and apply the preceding method or, alternatively, take bandwidths that vary with the location.

If one realizes that there is a mode at a boundary point—as in the case of the exponential density at zero—then one should employ less smoothing around this point.

---

[2]who discussed the problem of outliers : "... the rejection of observations is too crude to be defended: and unless there are other reasons for rejection than the mere divergences from the majority, it would be more philosophical to accept these extremes, not as gross errors, but as indications that the distribution of errors is not normal" on page 322, respectively, page 289 in Fisher, R.A. (1922). On the mathematical foundation of theoretical statistics. Phil. Trans. Roy. Soc. A 222, 309–368, and Collected Papers of R.A. Fisher, Vol. I, J.H. Bennett, ed., pp. 274–335. University of Adelaide, 1971.

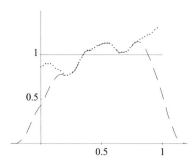 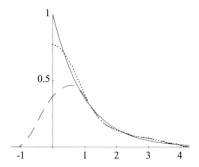

FIG. 2.5. (left.) $[0, 1]$–uniform (solid), kernel (dashed) and bounded kernel density (dotted) for $b = .2$ based on 400 uniform data. (right.) Exponential (solid), kernel (dashed) and left bounded kernel density (dotted) for $b = 1.1$ based on 50 exponential data.

Critical Remarks About Sample Distribution Functions

Visualizing data by the sample df yields a severe regularization of the data in so far that

- there is a severe averaging, and

- one gets a monotone function approaching one (zero) in the upper (lower) tail.

This makes the sample df particularly applicable for small sample sizes. Yet for moderate and large sample sizes, the sample qf and, to some extent, sample densities such as the histogram and the kernel densities are more useful. This discussion will be continued in Section 2.4.

The Scatterplot

Points $(i, x(i))$ or, generally, $(t(i), x(i))$ for $1 \leq i \leq n$ are plotted, thus resulting in a scatterplot. It is evident that such a scatterplot is also useful for plotting a function. The scatterplot is an indispensable tool for visualizing time series phenomena (see Section 2.5).

EXAMPLE 2.1.3. (Continuation of Example 1.2.1 about Annual Wind–Speed Maxima at Vancouver.) In Fig. 2.6 the annual wind–speed maximum is plotted against the year of occurrence.

The scatterplot exhibits a certain decreasing tendency in the data which is captured by a least squares line, cf. Section 2.5, given by $s(x) = 938.49 - 0.44x$. Such a decrease in the annual wind–speed maxima may be due to changes in the climate or an urbanization near the gauging station.

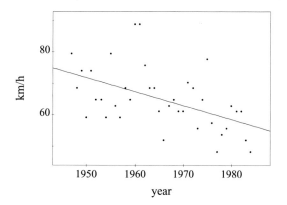

FIG. 2.6. Annual wind–speed maxima measured in km/h plotted against years.

This part of the book only concerns univariate extremes. Whenever bivariate data are treated, then the first component is regarded as a time–scale. In extreme value theory, it is customary to replace the time i by i/n in order to obtain an elegant formulation of asymptotic results. We are particularly interested in scatterplots of points where the second component $x(i)$ exceeds a selected threshold u. Then, the values i/n locate the exceedance times, cf. page 13.

For a continuation of this topic we refer to Sections 2.5 to 2.7.

2.2 Excess and Hazard Functions

Mean and median excess functions are important to extreme value analysis from a diagnostic viewpoint, because excess functions of GP dfs are straight lines. The pertaining sample versions provide further useful techniques for visualizing data.

Another diagnostic tool, taken from survival analysis, is the hazard function. Yet, we are primarily interested in the reciprocal hazard function which is a straight line for GP dfs.

Excess Distribution Functions

As introduced in Section 1.2, the excesses

$$y'_i := y_i - u$$

over the threshold u are a variant of the exceedances y_i. The excesses are the exceedances shifted to the left by the amount u. The pertaining excess df $F^{(u)}$ at u is

$$F^{(u)}(x) = F^{[u]}(x+u) = \frac{F(x+u) - F(u)}{1 - F(u)}, \qquad x \geq 0, \qquad (2.18)$$

where $F^{[u]}$ is the exceedance df. Notice that the left endpoint of $F^{(u)}$ is equal to zero.

The excess df $F^{(u)}$ can be introduced as a conditional df: if X is a random variable with df F, then

$$F^{(u)}(x) = P(X - u \le x | X > u) \qquad (2.19)$$

is the conditional df of $X - u$ given $X > u$.

The excess df $F^{(u)}$ is alternatively called residual life df at age u, where $F^{(u)}(x)$ is the probability that the remaining life time is smaller or equal to x given survival at age u.

In this section, we are especially concerned with functional parameters of the excess df $F^{(u)}$ as a function in u. For example, the mean excess function describes the expected remaining life given survival at age u.

Mean Excess Functions

The mean excess function e_F of a df F (respectively, of a random variable X) is given by the conditional expectation of $X - u$ given $X > u$. We have

$$e_F(u) = E(X - u | X > u) = \int x \, dF^{(u)}(x), \qquad u < \omega(F). \qquad (2.20)$$

It is evident that $e_F(u)$ is the mean of the excess df at u. The mean excess function e_F is also called the mean residual life function, see, e.g., [28]. Note that

$$e_{F_{\mu,\sigma}}(u) = \sigma e_F \left((u - \mu)/\sigma \right), \qquad (2.21)$$

where μ and σ are the location and scale parameters.

In conjunction with the visualization of data, we are expressly interested in dfs, where the mean excess function is a straight line. It is well known that the GP dfs are the only dfs where this goal is achieved.

We also deal with converse Weibull dfs

$$\widetilde{G}_{2,\alpha}(x) = 1 - \exp(-x^{-\alpha}), \qquad x > 0,$$

with parameter $\alpha < 0$ (introduced in (1.38)). Note that $\widetilde{G}_{2,-1} = W_0$ is the exponential df on the positive half–line.

Here is a list of mean excess functions for GP and converse Weibull dfs.

Exponential (GP0): $e_{W_0}(u) = 1, \qquad u > 0,$

Pareto (GP1), $\alpha > 1$: $e_{W_{1,\alpha}}(u) = u/(\alpha - 1), \qquad u > 1,$

Beta (GP2), $\alpha < 0$: $e_{W_{2,\alpha}}(u) = u/(\alpha - 1), \qquad -1 \le u \le 0,$

GP: $e_{W_\gamma}(u) = \frac{1 + \gamma u}{1 - \gamma}$ for $\begin{cases} 0 < u, & 0 \le \gamma < 1, \\ 0 < u < -1/\gamma, & \gamma < 0, \end{cases}$ if

Converse Weibull, $\alpha < 0$: $e_{\widetilde{G}_{2,\alpha}}(u) = |\alpha|^{-1}u^{1+\alpha}(1 + O(u^{\alpha}))$.

We see that the mean excess function $e_{W_{\gamma,\mu,\sigma}}$ has the slope $\beta_1 = \gamma/(1-\gamma)$ and the intercept $\beta_0 = (1 - \gamma\mu)/(1-\gamma)$. Therefore, the slopes of the mean excess functions of GP dfs $W_{\gamma,\mu,\sigma}$ are increasing in γ.

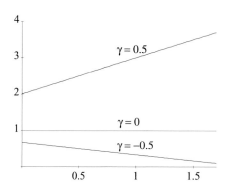

FIG. 2.7. Mean excess functions of GP dfs for parameters $\gamma = -0.5, 0, 0.5$.

If the excess df $F^{(s)}$ is close to a GP df W, then it is obvious from (2.20) that $e_F(u)$, $u \geq s$, is close to the corresponding straight line determined by W. Mean excess functions do not exist for Pareto dfs with a shape parameter $\alpha \leq 1$.

We also refer to (5.34), where Benktander II dfs are constructed with mean excess functions equal to $|\alpha|^{-1}u^{1+\alpha}$.

We now mention another representation of the mean excess function. First notice that

$$e_F(u) = \frac{E((X-u)I(X>u))}{P\{X>u\}}, \qquad u < \omega(F).$$

Secondly, one gets

$$E((X-u)I(X>u)) = \int I(u<x)(x-u)\,dF(x)$$
$$= \int_u^\infty (1 - F(x))\,dx, \qquad (2.22)$$

where the second equation can be verified by writing

$$x - u = \int I(u \leq y)I(y<x)\,dy, \qquad x \geq u,$$

and by interchanging the order of integration (applying Fubini's theorem). Therefore, the mean excess function can also be written as

$$e_F(u) = \frac{\int_u^\infty (1 - F(x))\,dx}{1 - F(u)}, \qquad u < \omega(F). \qquad (2.23)$$

The expectations

$$E(\max\{X - u, 0\}) = \int_u^\infty (1 - F(x))\, dx \qquad (2.24)$$

are also called "tail probabilities".

Sample Mean Excess Functions

Let e_F again be the mean excess function of the df F. By plugging in the sample df $\widehat{F}_n$ based on $x_1, \ldots, x_n$, one obtains by

$$e_n(u) := e_{\widehat{F}_n}(u) = \frac{\sum_{i \le n}(x_i - u)\, I(x_i > u)}{\sum_{i \le n} I(x_i > u)}, \qquad x_{1:n} < u < x_{n:n}, \qquad (2.25)$$

an estimate of e_F. Note that $e_n(u)$ is the sample mean of the excesses over the threshold u.

Taking the mean or the median of the excesses over u entails a certain smoothing (regularization) of the data, yet one should be aware that excess functions are neither increasing or decreasing nor do they approach a constant at the upper end of the range.

Sample mean excess functions will be one of our basic tools to verify the validity of a GP hypothesis in the upper tail of a distribution. The applicability of this approach is confirmed by the relationship to the reciprocal hazard function, cf. (2.27), in conjunction with the von Mises condition (2.32).

Trimmed Mean Excess Functions

Because the mean excess function does not exist for Pareto dfs with shape parameter $\alpha < 1$ and due to the fact that e_n is an inaccurate estimate of e_F if $\alpha > 1$ is close to 1 (for a discussion see [16], page 150), a trimmed version

$$e_{F,p}(u) = \frac{1}{p} \int_{-\infty}^{(F^{(u)})^{-1}(p)} x\, dF^{(u)}(x)$$

is of interest, where $0 < p < 1$. There is no trimming if $p = 1$. Trimmed mean excess functions of GP dfs again form a straight line[3]. In the Pareto case we have

$$e_{W_{1,\alpha},p}(u) = \left(\frac{1}{p} \int_{1-p}^1 y^{-1/\alpha} dy - 1 \right) u, \qquad u > 1.$$

Moreover, $e_{W_{2,\alpha},p}$ is strictly decreasing and $e_{W_{0,1},p}$ is a constant. A sample version is also obtained by plugging in the sample df.

---

[3]Drees, H. and Reiss, R.–D. (1996). Residual life functionals at great age. Commun. Statist.–Theory Meth. 25, 823–835.

Median Excess Functions

The median excess function is defined by the medians of excess dfs. We have

$$m_F(u) := (F^{(u)})^{-1}(1/2).$$

The median excess function has properties corresponding to those of the trimmed mean excess function. For standard Pareto dfs $W_{1,\alpha}$, we have

$$m_{W_{1,\alpha}}(u) = (2^{1/\alpha} - 1)u, \qquad u > 1.$$

Once again, a sample version is obtained by plugging in the sample df. A formula corresponding to (2.21) also holds for trimmed mean and median excess functions.

Hazard Functions

The hazard function h_F of a df F with density f is

$$h_F(t) = f(t)/(1 - F(t)), \qquad t < \omega(F).$$

Note that h_F is the derivative of the cumulative hazard function

$$H_F(t) = -\log(1 - F(t)), \qquad t < \omega(F).$$

The value $h_F(t)$ is also called hazard rate or mortality rate at age t. It is the right–hand derivative taken at zero of the residual life df $F^{(t)}$ at age t. We have

$$F^{(t)}(x) \approx h_F(t)x$$

for small x. Recall that $F^{(t)}(x)$ is the probability that the remaining life time is less than the instant x given survival at age t. Therefore, one gets the interpretation that the mortality rate is approximately equal to the probability that the remaining life is less than 1 given survival at age t. Check that

$$h_{F_{\mu,\sigma}}(t) = h_F\left((t - \mu)/\sigma\right)/\sigma,$$

where μ and σ denote the location and scale parameters.

We include a list of the hazard functions of GP, converse Weibull and converse Gumbel dfs.

Exponential (GP0): $\qquad h_{W_0}(t) = 1, \qquad t > 0;$

Pareto (GP1), $\alpha > 1$: $\qquad h_{W_{1,\alpha}}(t) = \alpha/t, \qquad t > 1;$

Beta (GP2), $\alpha < 0$: $\qquad h_{W_{2,\alpha}}(t) = \alpha/t, \qquad -1 \le t \le 0.$

In the γ–representation, we have:

GP: $\qquad h_{W_\gamma}(t) = \frac{1}{1+\gamma t}$ for $\begin{cases} 0 < t, & 0 \le \gamma, \\ & \text{if} \\ 0 < t < 1/|\gamma|, & \gamma < 0; \end{cases}$

Converse Weibull, $\alpha < 0$: $\qquad h_{\widetilde{G}_{2,\alpha}}(t) = |\alpha| t^{-(1+\alpha)}, \qquad t > 0;$

Converse Gumbel: $\qquad h_{\widetilde{G}_0}(t) = e^t.$

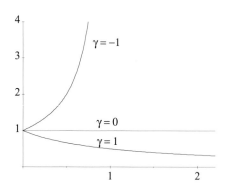

$\gamma = -1$

$\gamma = 0$

$\gamma = 1$

FIG. 2.8. Hazard functions of GP dfs for $\gamma = -1, 0, 1$.

The hazard function of the converse Weibull df $\widetilde{G}_{2,\alpha}$ is a straight, increasing line if $\alpha = -2$. Recall that $\widetilde{G}_{2,-2}$ is the Rayleigh df. Moreover, one can easily check that the converse Gumbel dfs $\widetilde{G}_{0,\mu,\sigma}$ are the only ones that satisfy the famous Gompertz law postulating a mortality rate of the form[4]

$$h(x) = ae^{bx}.$$

Hence $\widetilde{G}_0$ is also called Gompertz df. Because

$$1 - F(t) = \exp\left(-\int_{\alpha(F)}^t h_F(x)\, dx\right), \qquad \alpha(F) < t < \omega(F), \qquad (2.26)$$

we see that the survivor function and, thus, also the df F can be regained from the hazard function.

The hazard function and, therefore, the survivor function can be written in terms of the mean excess function. By multiplying both sides in (2.23) with $1 - F(u)$ and taking derivatives, one obtains

$$h_F(t) = \frac{1 + e_F'(t)}{e_F(t)}, \qquad \alpha(F) < t < \omega(F). \qquad (2.27)$$

---

[4]Gompertz, B. (1825). On the nature of the function expressive of the law of human mortality etc. Phil. Trans. Roy. Soc. 115, 513–585.

Combining this with (2.26), we have

$$1 - F(t) = \exp\left(-\int_{\alpha(F)}^{t} \frac{1 + e'_F(x)}{e_F(x)} \, dx\right), \qquad \alpha(F) < t < \omega(F). \tag{2.28}$$

Sample Hazard Functions

The sample hazard function

$$h_{n,b}(t) = \frac{f_{n,b}(t)}{1 - \widehat{F}_{n,b}(t)}, \qquad t < x_{n:n},$$

is an estimator of the hazard function h_F, where $f_{n,b}$ is the kernel density in (2.14), and $\widehat{F}_{n,b}$ is the kernel estimator of the df in (2.17). Note that $\widehat{F}_{n,b}$ may be replaced by the sample df $\widehat{F}_n$. The quality of the sample hazard function as an estimate of the hazard function heavily depends on the choice of the bandwidth b.

It can be advantageous to consider left or right–bounded versions of the sample hazard function as it was done for the kernel density itself.

For grouped data, again with frequencies $n(j)$ in cells $[t_j, t_{j+1})$, the sample hazard function is defined by means of the histogram. Consequently,

$$h_n(t) = \frac{n(j)}{(t_{j+1} - t_j) \sum_{i \geq j} n(i)}, \qquad t_j \leq t < t_{j+1}.$$

Another version of the sample hazard function may be obtained by taking the derivative of a smoothed sample cumulative hazard function $\log(1 - \widehat{F}_n)$.

Reciprocal Hazard Functions, a Von Mises Condition

The reciprocal $1/h_W$ of the hazard function of a GP df W is a straight line which is clearly of interest for visual investigations. Moreover, this observation leads to a condition (due to von Mises) that guarantees that a df belongs to the max and the pot–domain of an EV and GP distribution. Thus, the reciprocal hazard function is also of theoretical interest.

Reciprocal hazard and mean excess functions are related to each other. We have

$$e_{W_{i,\alpha}} = 1/((\alpha - 1)\alpha h_{W_{i,\alpha}}), \tag{2.29}$$

if $i = 1$ and $\alpha > 1$ or $i = 2$, and

$$e_{W_\gamma} = 1/((1 - \gamma)h_{W_\gamma}), \tag{2.30}$$

if $\gamma < 1$. Note that $e_{W_0} = h_{W_0} = 1$.

The reciprocal hazard function was also mentioned due to technical reasons. Observe that the reciprocal hazard function of a GP df $W_{\gamma,\mu,\sigma}$ satisfies

$$\frac{1}{h_{W_{\gamma,\mu,\sigma}}(t)} = \frac{1 - W_{\gamma,\mu,\sigma}(t)}{w_{\gamma,\mu,\sigma}(t)} = \sigma + \gamma(t - \mu).$$

Therefore, the first derivative of the reciprocal hazard function is equal to γ. We have

$$\left(\frac{1 - W_{\gamma,\mu,\sigma}}{w_{\gamma,\mu,\sigma}} \right)' = \gamma \qquad (2.31)$$

on the support of $W_{\gamma,\mu,\sigma}$. From (2.26) deduce that GP dfs are the only dfs which possess this property.

If this condition is approximately satisfied for large t by some df F, then F belongs to the max and pot–domain of attraction of the EV and GP df with the given parameter γ. More precisely, the

von Mises condition: $\qquad \lim_{x \to \omega(F)} \left(\frac{1 - F}{f} \right)'(x) = \gamma \qquad (2.32)$

is sufficient for the relations in (1.27) and (1.46), namely

- G_γ is the limiting df of $F^n(b_n + a_n x)$ as $n \to \infty$, and

- W_γ is the limiting df of $F^{[t]}(b_t + a_t x)$ as $t \to \omega(F)$

for certain normalizing constants.

Check, for example, that the normal df Φ satisfies condition (2.32) for $\gamma = 0$ and, hence, the Gumbel df is the limiting df of maxima of normal random variables.

Under condition (2.32), one may take the following constants a_n and b_n if the standard EV dfs in the α–parameterization are taken as limiting dfs.

$$
\begin{aligned}
&G_0: &b_n &= F^{-1}(1 - 1/n), &a_n &= 1/(nf(b_n)); \\
&G_{1,\alpha}: &b_n &= 0, &a_n &= F^{-1}(1 - 1/n), \\
&G_{2,\alpha}: &b_n &= \omega(F), &a_n &= \omega(F) - F^{-1}(1 - 1/n).
\end{aligned} \qquad (2.33)
$$

For iid random variables $X_1, \ldots, X_n$ with common df F, the expected number of exceedances over $F^{-1}(1 - 1/n)$ is equal to one. For GP dfs $F = W_{i,\alpha}$, one obtains in (2.33) the standardizing constants under which the max–stability of the standard EV dfs $G_{i,\alpha}$ holds, cf. also (1.31).

2.3   Fitting Parametric Distributions to Data

In this section we visualize data by parametric dfs or densities. Thereby, one may also visually control the validity of a parametric model. This idea will be exemplified for EV, GP and, in addition, for Poisson and normal distributions.

Fitting an Extreme Value Distribution to Maxima

Let $\boldsymbol{x} = (x_1, \ldots, x_n)$ be the vector of maxima x_i of blocks of size m as in (1.5). Recall from (2.2) that the sample df $\widehat{F}_n(\boldsymbol{x}; \cdot)$ based on $\boldsymbol{x}$ is approximately equal to the underlying df F^m if the number n of maxima is sufficiently large.

Combining (2.2) and (1.27) one gets

$$\widehat{F}_n(\boldsymbol{x}; \cdot) \approx F^m \approx G_{\gamma, \mu_m, \sigma_m},$$

if n and m are sufficiently large. Therefore, an EV df can be fitted to the sample df based on maxima.

It is likely that one of our basic assumptions is violated if none of the EV dfs fits to the sample df. Correspondingly, fit an EV density to the histogram or kernel density f_n based on the x_i. An efficient, interactive software package is needed to carry out a visual selection of the parameters.

Of course, the selection of the parameters is done in some subjective manner, yet one may follow certain principles.

- If the selection is based on the sample df, one may choose the parametric df by minimizing the maximum deviation between the curves.

- If the selection is based on a kernel density, then we suggest to single out a parametric density so that the area between both curves is minimized; this corresponds to minimizing the L_1–distance.

- There is a strong dependence on the smoothing parameter in the latter case if the selection is based on the maximum deviation between the parametric and the sample curve.

Fitting a Generalized Pareto Distribution to Exceedances

Let $\widehat{F}_k(\boldsymbol{y}; \cdot)$ be the sample exceedance df, cf. page 41, based on the exceedances $y_1, \ldots, y_k$ over the threshold u. Combining (2.4) and (1.46) one gets

$$\widehat{F}_k(\boldsymbol{y}; \cdot) \approx F^{[u]} \approx W_{\gamma, u, \sigma}, \tag{2.34}$$

if k and u are sufficiently large. Therefore, a GP df can be fitted to the sample exceedance df. Once again, it is likely that one of our basic assumptions is violated if none of the GP dfs fits to the sample exceedance df.

If excesses $y_i' = y_i - u$ are taken in place of the exceedances y_i, then a GP df with location parameter (left endpoint) equal to zero must be fitted to the data. Likewise, one may fit a Pareto df $W_{1, \alpha, \mu, \sigma}$ with left endpoint $\mu + \sigma = u$ to the sample exceedance df $\widehat{F}_k(\boldsymbol{y}; \cdot)$.

Fitting a Generalized Pareto Distribution to the Original Data

In the preceding lines, a GP df was fitted to the sample exceedance df. By changing the location and scale parameters a GP df can be fitted to the upper tail of the original sample df and, therefore, to the original data.

This is an application of (1.49) with $F^{[u]}$ and F replaced by the sample exceedance df $\widehat{F}_k(\boldsymbol{y}; \cdot)$ and and sample df $\widehat{F}_n(\boldsymbol{x}; \cdot)$. If (2.34) is valid, then (1.49) yields

$$W_{\gamma,\tilde{\mu},\tilde{\sigma}}(x) \approx \widehat{F}_n(\boldsymbol{x}; x), \qquad x \geq u, \tag{2.35}$$

where

$$\tilde{\sigma} = \sigma(k/n)^{\gamma} \tag{2.36}$$

and

$$\tilde{\mu} = u - \sigma\big(1 - (k/n)^{\gamma}\big)/\gamma. \tag{2.37}$$

Notice that $\widehat{F}_n(\boldsymbol{x}; u) = 1 - k/n$ and k is the number of exceedances.

There is a unique relation between the two pairs of parameters μ, σ and $\tilde{\mu}$, $\tilde{\sigma}$ determined by the equations $W^{[u]}_{\gamma,\tilde{\mu},\tilde{\sigma}} = W_{\gamma,u,\sigma}$ and $W_{\gamma,\tilde{\mu},\tilde{\sigma}}(u) = 1 - k/n$.

Fig. 2.9 illustrates the two different approaches of fitting a density to the exceedances and to the original data. The illustration on the right–hand side essentially shows a magnification of the curves on the left–hand side right of u.

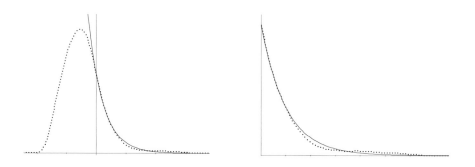

FIG. 2.9. (left.) Fitting a GP density (solid) to the upper tail of a kernel density (dotted) based on the original x_i. (right.) Fitting a GP density (solid) to a kernel density (dotted) based on exceedances y_i over u.

Let $W_{1,\alpha,\mu,\sigma}$ be a Pareto df with $\mu + \sigma = u$ which is the left endpoint. If this Pareto df fits to the sample exceedance df $\widehat{F}_k(\boldsymbol{y}; \cdot)$, then $W_{1,\alpha,\tilde{\mu},\tilde{\sigma}}$, determined by $W^{[u]}_{1,\alpha,\tilde{\mu},\tilde{\sigma}} = W_{1,\alpha,\mu,\sigma}$ and $W_{1,\alpha,\tilde{\mu},\tilde{\sigma}}(u) = 1 - k/n$, fits to the original sample df $\widehat{F}_n(\boldsymbol{x}; \cdot)$. Moreover, $\tilde{\mu} = \mu$ and $\tilde{\sigma} = \sigma(k/n)^{1/\alpha}$.

Fitting a Poisson Distribution to Discrete Data

Let $x_1, \ldots, x_n$ be governed by a Poisson distribution P_λ. Remember from (2.12) that $p_n(j) = n(j)/n \approx P_\lambda\{j\}$, where $n(j)$ is again the number of data being equal to j. Therefore, a Poisson hypothesis should be rejected if none of the Poisson histograms $P_\lambda\{j\}$ visually fits to the sample histogram $p_n(j)$.

Local Fitting of a Normal Distribution

The preceding arguments can also be applied in the Gaussian case, if (1.27) is replaced by (1.58). Select location and scale parameters μ and σ so that $\Phi_{\mu,\sigma} \approx \widehat{F}_n(\boldsymbol{y}; \cdot)$ in order to get a visualization of sums $y_1, \ldots, y_n$ via a normal df.

The procedure of fitting a GP df to upper extremes may be regarded as a local statistical modeling. Such a local approach can be employed in the Gaussian setting as well. For example, fit the main component of the normal mixture in the subsequent Fig. 2.10 to the central observations (visualized by a kernel density).

To some extent, the local fitting of parametric distributions to data corresponds to nonparametric density estimation, where the estimate of the density at a fixed point x is just based on data around x.

Mixtures of Normal Distributions

If a parametric model as dealt with above is untenable, then a modification of this approach, such as dealing with a mixture, can be beneficial.

EXAMPLE 2.3.1. (Mixture of Normal Distributions.) We continue the analysis of 135 data of TV watching in hours per week, cf. Example 2.1.2. From the form of the kernel density (Epanechnikov kernel and bandwidth $b = 5$) in Fig. 2.4, we already know that a normal density cannot be fitted to the data. From Fig. 2.10, we see that a better fit is obtained by a normal mixture. This modeling is still unsatisfactory in the upper tail of the distribution.

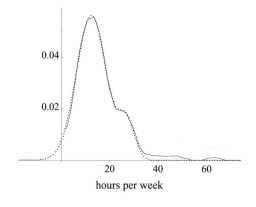

FIG. 2.10. Normal mixture (dotted) $0.875\varphi_1 + 0.125\varphi_2$, where

$$\varphi_1(x) = \varphi((x - 12.4)/6.2)/6.2,$$

$$\varphi_2(x) = \varphi((x - 27)/3.4)/3.4,$$

and a left–bounded kernel density.

If the means of the two normal components are slightly closer together, then the heterogeneity of the data becomes less visible. Then, a representation of the data by means of a right–skewed distribution, such as the Gumbel distribution (cf. Fig. 1.3), would also be acceptable.

Conversely, if a Gumbel modeling for certain data seems to be adequate, then a mixture distribution cannot be excluded. It depends on the posed question or the additional information about the data as to which type of modeling (in other words, hypothesis formulation) is preferable.

It is likely that mixtures of normal distributions having identical means are of higher interest. Recall from page 31 that such distributions are fat–tailed, that is, the kurtosis is lager than the kurtosis 3 of a normal distribution.

EXAMPLE 2.3.2. (Fitting a Mixture of Two Normal Distributions to Financial Data.) Parametric densities are fitted to a kernel density—with bandwidth $b = 0.002$—based on the centered financial data which are displayed in Fig. 1.1.

A normal mixture—a maximum likelihood estimate—with mixing parameter $d = 0.362$ and scale parameters $\sigma_1 = 0.00285$ and $\sigma_2 = 0.00898$ is hardly distinguishable from the kernel density (both represented by a solid line).

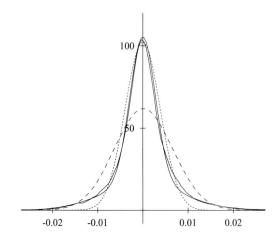

FIG. 2.11. Fitting normal (dotted, dashed) and a mixture of two normal densities (solid) to financial data.

Two normal densities with $\sigma = 0.0039$ (dotted) and $\sigma = 0.0065$ (dashed) are included. The first one fits well to the kernel density in the center, yet there is a significant deviation in the tails. The second one—determined by the sample standard deviation—strongly deviates from the kernel density in the center.

In Chapter 16, page 379, we also a apply maximum likelihood estimators in statistical models of Student and sum–stable distributions, cf. Sections 6.3 and 6.4, to these data.

The Art of Statistical Modeling

One major goal of this book is to select distributions which simultaneously fit to
the central as well as to the extreme data. The appropriate insight can be gained
by using the available tools such as Q–Q plots (subsequent Section 2.4) and sample
mean excess functions (Section 2.2).

A parametric modeling can be useful in reducing the variance of an estimation
procedure. If the parametric model is incorrect, then one must take a bias into
account. An optimal choice of the parametric model is achieved if there is a certain
balance between the variance and the bias. In Section 6.6 much effort is invested
to reduce the bias of estimators within the extreme value setting by introducing
higher order conditions.

We believe that visual procedures are preferable to automatic ones in many
situations. In that context, we also cite an interesting argument (translated from
German) in Pruscha[5], page 62: "For larger sample sizes n, visual diagnostic tools
can be preferable to goodness–of–fit tests. A parametric hypothesis will be rejected
for larger n, even if the deviation of this hypothesis is negligible (from a practical
viewpoint) due to the high power of test procedures."

2.4   Q–Q and P–P Plots

Q–Q plots are usually defined for location and scale parameter families and, therefore, we first review the main ideas for such models. An extension to EV and GP
models is obtained by employing estimators of the shape parameter.

Q–Q Plots in Location and Scale Parameter Families

Assume that the data $x_1, \ldots, x_n$ are governed by a df

$$F_{\mu,\sigma}(x) = F((x - \mu)/\sigma)$$

with location and scale parameters μ and $\sigma > 0$. Thus, $F = F_{0,1}$ is the standard
version. Values $\widehat{F}_n^{-1}(q)$ of the sample qf will be plotted against $F^{-1}(q)$. More
precisely, one plots the points

$$\left(F^{-1}(q_i), \, \widehat{F}_n^{-1}(q_i)\right), \qquad i = 1, \ldots, n,$$

where $q_i = i/(n+1)$. Notice that location and scale parameters need not be selected
in advance when a Q–Q plot is applied to the data. Because $\widehat{F}_n^{-1}(q_i) = x_{i:n}$, the
relationship (2.9) between the sample qf and the underlying qf yields

$$\widehat{F}_n^{-1}(q_i) \approx F_{\mu,\sigma}^{-1}(q_i) = \mu + \sigma F^{-1}(q_i), \tag{2.38}$$

---

[5] Pruscha, H. (1989). Angewandte Methoden der Mathematischen Statistik. Teubner,
Stuttgart.

and, hence, the Q–Q plot of points

$$\left(F^{-1}(q_i),\, x_{i:n}\right), \qquad i = 1, \ldots, n, \tag{2.39}$$

is close to the graph $(x, \mu + \sigma x)$. The Q–Q plot can be visualized by a scatterplot, whereby a linear interpolation may be added. Apparently, the intercept and the slope of the Q–Q plot provide estimates of μ and σ. Another frequently taken choice of q_i is $(i - 0.5)/n$.

The selected location/scale parameter family is untenable if the deviation of the Q–Q plot from a straight line is too strong.

EXAMPLE 2.4.1. (Continuation of Example 2.1.2 about TV Data.) A greater problem in Example 2.1.2 was the selection of an appropriate bandwidth parameter. In conjunction with Q–Q plots, one must not choose a bandwidth, but a parametric model. Subsequently, we selected the normal location and scale parameter family. A straight line will be fitted to that part of the Q–Q plot which represents the bulk of the data.

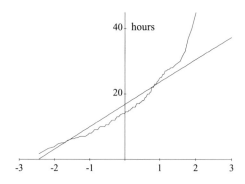

FIG. 2.12. Normal Q–Q plot based on 135 TV data and a straight line visually fitted to the Q–Q plot.

The normal Q–Q plot is sufficiently close to a straight line below the quantity of 18 hours, yet there is a stronger deviation above this threshold.

Of course, one may also check whether the data were generated under a specific df F. This is just the case when the actual df is the standard df in the preceding considerations. Then, the Q–Q plot must be close to the main diagonal in the plane. The disadvantage of Q–Q plots is that the shape of the selected parametric distribution is no longer visible.

Q–Q Plots in Extreme Value and Generalized Pareto Models

In EV and GP models, one must keep in mind that there is an additional parameter, namely the shape parameter, besides the location and scale parameters. We suggest applying a Q–Q plot with the unknown shape parameter having been replaced by an estimate. If there is a stronger deviation of the Q–Q plot from a straight line, then either the estimate of the shape parameter is inaccurate or the model selection is untenable.

P–P Plots

We introduce the P–P plot in conjunction with a location and scale parameter family of dfs $F_{\mu,\sigma}$. The P–P plot is given by

$$\Big(q_i,\; F\big((x_{i:n} - \mu_n)/\sigma_n\big)\Big), \qquad i = 1, \ldots, n,$$

where μ_n and σ_n are estimates of the location and scale parameters (such estimates will be presented in the next section).

Because

$$F\big((x_{i:n} - \mu_n)/\sigma_n\big)\big) = F_{\mu_n,\sigma_n}\big(\widehat{F}_n^{-1}(q_i)\big),$$

a strong deviation of the P–P plot from the main diagonal in the unit square indicates that the given model is incorrect (or the estimates of the location and scale parameters are inaccurate). The values of the P–P plot will be close to one (or zero) and, thus, close to the diagonal in the upper (or lower) tail, even if the choice of the model is wrong.

Further Remarks About Regularization and Smoothing

Let us continue our permanent discussion about the usefulness of sample dfs, qfs, densities, etc. for the visualization of data. A remark made about regularization in Section 2.1 is also relevant to Q–Q and P–P plots.

By applying the sample qf, sample excess functions, the sample hazard function or the Q–Q plot, one is able to extract the information inherent in the data in a suitably way by achieving a compromise between the following two requirements:

- the random fluctuation of the data must be reduced, and

- special features and clues contained in the data must be exhibited.

We believe that this goal is not achieved in the same manner by dfs and related tools such as the P–P plot because there is an oversmoothing particularly in the upper and lower tails of the distribution.

Exaggerating a bit, one may say that one should apply the sample df F_n (or, likewise, the survivor function $1 - F_n$) and the P–P plot if one wants to justify a hypothesis visually. The other tools are preferable whenever a critical attitude towards the modeling is adopted.

2.5   Trends, Seasonality and Autocorrelation

This section concerns several aspects of exploratory time series analysis. Many observations recorded at specified times exhibit a dependence on time. This dependence may be caused, for example, by a certain tendency in the climate, an increasing population, inflation, or seasonal effects.

   We collect and discuss some statistical procedures known from regression and time series analysis for measuring and removing a trend or a seasonal component in a series of data. The autocorrelation function is also on the agenda. After having removed a trend or a seasonal component from the data, one obtains residuals which may be dealt with by the tools provided in foregoing sections.

The Linear Least Squares Method

A trend is a long–term change of a series of data. First let a linear tendency be visible, cf. Example 2.1.3, that will be captured and removed by a least squares line. Thus, a straight line is fitted to the points $(t_1, y_1), \ldots, (t_n, y_n)$ in a scatterplot by applying the least squares method. The t_i need not be integer–valued or equidistant.

   The least squares line

$$s(t; \beta_0, \beta_1) = \beta_0 + \beta_1 t$$

with regression slope β_1 and intercept β_0 is chosen such that the cumulated squared distances

$$\sum_{i \le n} (y_i - \beta_0 - \beta_1 t_i)^2 \tag{2.40}$$

between the values y_i and $s(t_i; \beta_0, \beta_1)$ at t_i are minimal. The well–known solutions are the estimates

$$\beta_{1,n} = \frac{\sum_{i \le n} (y_i - \bar{y})(t_i - \bar{t})}{\sum_{i \le n} (t_i - \bar{t})^2}$$

of the regression slope, and

$$\beta_{0,n} = \bar{y} - \beta_{1,n} \bar{t}$$

of the intercept, where $\bar{y}$ and $\bar{t}$ are again the averages of the y_i and t_i.

   One gets a decomposition

$$y_i = s(t_i; \beta_{0,n}, \beta_{1,n}) + x_i \tag{2.41}$$

where $s(t_i; \beta_{0,n}, \beta_{1,n}) = \beta_{0,n} + \beta_{1,n} t_i$ represents a linear part in the data and the x_i are the residuals which fluctuate irregularly around zero.

Nonlinear (Polynomial) Least Squares Methods

If one recognizes a nonlinear trend in the scatterplot of points (t_i, y_i), one may employ parametric trend functions $s(t; \beta_0, \ldots, \beta_p)$, where the parameters $\beta_0, \ldots, \beta_p$ are selected such that

$$\sum_{i \leq n} \left(y_i - s(t_i; \beta_0, \ldots, \beta_p) \right)^2 \tag{2.42}$$

is minimal. There is a greater variety of parametric trend functions. For example, the linear approach can be extended to polynomials

$$s(t; \beta_0, \ldots, \beta_p) = \beta_0 + \sum_{j \leq p} \beta_j t^j$$

of degree p. For $p = 0$ one gets the sample mean which determines a straight line. Explicit solutions $\beta_{0,n}, \ldots, \beta_{p,n}$ to the least squares minimization for polynomials can be obtained within the bounds of multiple, linear regression.

Parametric Regression for a Fixed Design

We reformulate the preceding considerations within a stochastic framework. Let

$$Y_i = s(t_i; \beta_0, \ldots, \beta_p) + \varepsilon_i,$$

where the random variable Y_i is observable at time (at the position) t_i, and ε_i is a random residual with expectation $E\varepsilon_i = 0$. Thus, Y_i is a random variable with expectation

$$EY_i = s(t_i; \beta_0, \ldots, \beta_p).$$

Notice that the y_i and x_i in (2.41) may be regarded as realizations of Y_i and ε_i.

The least squares solutions $\beta_{j,n}$—within the polynomial framework—provide unbiased estimators of the unknown parameters β_j and, therefore,

$$m_n(i) = s(t_i; \beta_{0,n}, \ldots, \beta_{p,n}) \tag{2.43}$$

is an unbiased estimator of the expectation $E(Y_i)$.

It suggests itself also to employ certain averages to estimate such expectations. Below, we deal with moving averages of the Y_j pertaining to adjacent points t_j of t_i. These moving averages provide nonparametric estimates of a parametric or nonparametric trend function.

Parametric regression for a random design will be studied in Section 8.1.

Moving Averages, Nonparametric Estimation of a Trend

To eliminate—or, at least, to reduce—the irregular fluctuation of measurements y_i in a nonparametric manner one may average those y_j pertaining to adjacent points t_j of t_i. These averages capture a trend in the y_i.

For example, take the Nadaraya–Watson moving average

$$m_n(i) = \frac{1}{K_n} \sum_{j \le n} I(t_i - b \le t_j \le t_i + b) y_j \tag{2.44}$$

at the position t_i, where

$$K_n = \sum_{j \le n} I(t_i - b \le t_j \le t_i + b)$$

is the number of points t_j in the interval of length $2b$ around t_i. The value $b > 0$ is called a bandwidth.

Generally, averages can be expressed by

$$m_n(i) = \sum_{j \le n} k\left(\frac{t_j - t_i}{b}\right) y_j \bigg/ \sum_{j \le n} k\left(\frac{t_j - t_i}{b}\right) \tag{2.45}$$

at the positions t_i, where k is a kernel such that $\int k(t)\,dt = 1$ and $b > 0$ is a bandwidth (cf. also page 44, where, e.g., the Epanechnikov kernel is introduced). Usually, k is taken symmetrical around zero. In (2.44) there is the uniform kernel

$$k(t) = 0.5 \times I(-1 \le t \le 1).$$

Again one gets a decomposition

$$y_i = m_n(i) + x_i,$$

where the x_i are the residuals which fluctuate irregularly around zero.

Special kernels (e.g., truncated versions) must be taken at corner points. We suggest to employ truncated versions of the given kernel.

Subsequently, we assume that the positions t_i are arranged in increasing order and, in addition, the t_i are equidistant, with $t_i = i$ or $t_i = i/n$ as special cases. Then, (2.45) can be written

$$m_n(i) = \sum_j a_j y_{i+j}, \tag{2.46}$$

for certain weights a_j satisfying $a_j = a_{-j}$ and $\sum_j a_j = 1$. In the case of the uniform kernel $k(t) = 0.5 \times I(-1 \le t \le 1)$, one obtains the moving average

$$m_n(i) = \frac{1}{2v + 1} \sum_{|j| \le v} y_{i+j}. \tag{2.47}$$

A slightly modified version is

$$m_n(i) = \frac{1}{2v}\left(\frac{1}{2} y_{i-v} + \sum_{|j| \le v-1} y_{i+j} + \frac{1}{2} y_{i+v}\right). \tag{2.48}$$

There is a trade–off between the two requirements that

- the irregular fluctuation should be reduced,

- the long term variation in the data such as a quadratic or cubic tendency should not be distorted by oversmoothing the data.

A balance between these two requirements can be gained by an appropriate choice of the bandwidth b or of the numbers v in (2.47) and (2.48).

In addition, it can be useful to employ kernels as in (2.15) or (2.16) with

$$\int k(t)t^2 \, dt = 0$$

in order to preserve a quadratic or cubic tendency in the data. This corresponds to the condition $\sum_j a_j j^2 = 0$ for the weights a_j in (2.47), if $x(i) - x(i-1) = 1$. An example is provided by Spencer's 15 point moving average, where the weights are

$$(a_0, a_1, \ldots, a_7) = (74, 67, 46, 21, 3, -5, -6, -3)/320.$$

In Fig. 2.13 (right), Spencer's moving average is applied to 200 iid standard normal data.

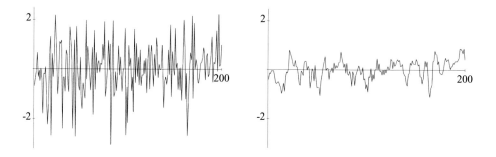

FIG. 2.13. (left.) Scatterplot of 200 standard normal data. (right.) Spencer's 15 point moving average of these data.

The strong fluctuation in the normal data cannot be smoothed appropriately by Spencer's moving average. Apparently, the choice of the number of points is more important than the selection of the kernel.

EXAMPLE 2.5.1. (Maximum Daily Temperature at Death Valley.) We consider the maximum daily temperatures at Furnace Creek, Death Valley National Park, from Aug. 1994 to Aug. 1995, by which the measurements for Dec. 1994 were not available.

For the missing data, we filled in values deduced from a quadratic least squares procedure applied to points around the gap. In Fig. 2.14 (left), the maximum daily temperature is plotted against the day of the measurement.

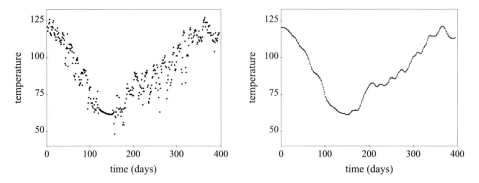

FIG. 2.14. (left.) Scatterplot of maximum daily temperature measured in Fahrenheit. (right.) Moving average (41 points) employing the Epanechnikov kernel.

Relative to the given time span of 13 months, we observe a long term variation (a trend) in the data that can be removed by means of a moving average. This data set is stored in the file et–deval.dat.

Given a time span of several years, the variation in the preceding temperature data must be interpreted as a seasonal effect (dealt with below). Thus, the time horizon greatly influences the methodology of handling a series of data.

Our general references for this section are the mathematically oriented book [5] and [26] for economic time series, yet the application of kernels and the question of an appropriate choice of the number of points are not dealt with in these books.

Modifications of Moving Averages

In certain applications, it is desirable to employ modified versions of the averages as introduced in the preceding lines. Other characteristics of the sample, such as medians and quantiles, are also also of interest. We give some details.

- (One–Sided Moving Averages.) This is the construction employed for the ordinary moving average at the upper corner point.

- (Moving Averages With a Random Bandwidth Chosen by Nearest Neighbors.) If there is not a grid of equidistant points, as, e.g., a full grid of integers, then it can be advisable to employ a nearest neighbor method. A random bandwidth b is determined in the following manner. Given a non-negative integer r, let b be the minimum of the distances between the fixed point t_i, where the moving average is evaluated, and its rth upper and, respectively, rth lower neighbor t_j. Near to the upper or lower corner point, one merely evaluates the distance to the rth lower or, respectively, the rth upper neighbor. Thus, one is averaging over $r+1$ to $2r+1$ values y_j.

- (Local Weighted Regression.) As before choose neighboring t_j of a fixed value t_i and carry out a weighted local least squares (lowess) procedure based on these neighbors, cf. the book [8] by Cleveland.

- (Moving Medians.) Alternatively, one may use moving medians of the form $\left(t_i, \operatorname{med}_{|j|\leq k} y_{i+j}\right)$ to reduce the fluctuation of a time series. This can be necessary when the observations come from a heavy–tailed distribution.

- (Moving Quantiles.) Later we will be particularly interested in moving higher q–quantiles in conjunction with series of log–returns of financial data (cf. Fig. 1.1 and Chapter 16).

In the latter context, we also use the parametric approach for estimating a higher quantile.

The Seasonal Component

If the moving average exhibits a variation that is annual in period—in other words, seasonal—then a refined decomposition of the measurements y_i is suggested. For simplicity, let $t_i = i$ for $i = 1, \ldots, n$, where $n = lp$ and l, p are the number and length of periods.

Now we also single out a periodic component s_n, $i = 1, \ldots, p$, satisfying

$$s_n(i + jp) = s_n(i), \qquad j = 0, \ldots, l - 1; \; i = 1, \ldots, p, \tag{2.49}$$

and $\sum_{i\leq p} s_n(i) = 0$. Thus, we have a decomposition in mind

$$y_i = m_n(i) + s_n(i) + x_i, \tag{2.50}$$

where the $m_n(i)$ represent the smooth trend components and the x_i are the residuals. This is done in three steps.

- (Preliminary Determination of a Trend Component.) By applying moving averages as introduced in (2.47) or (2.48) with $p = 2v+1$ or $p = 2v$, one gets a preliminary trend component $\widetilde{m}_n(i)$ that is not affected by any periodic component.

- (Determination of a Period (Cycle).) From the residuals $y_i - \widetilde{m}_n(i)$, single out a periodic component determined by

$$\tilde{s}_n(i) = \frac{1}{l} \sum_{j=0}^{l-1} \left(y_{i+jp} - \widetilde{m}_n(i + jp)\right), \qquad i = 1, \ldots, p,$$

or

$$s_n(i) = \tilde{s}_n(i) - \frac{1}{p} \sum_{j\leq p} \tilde{s}_n(j), \qquad i = 1, \ldots, p,$$

whereby the second version is preferable because it satisfies the additional requirement $\sum_{i \leq p} s_n(i) = 0$.

If the number of periods is small, yet one can postulate a smooth periodic component, then it is plausible to apply the preceding operations to a slightly smoothed version of the $y_i - \widetilde{m}_n(i)$.

- (Final Determination of the Trend Component.) Finally, compute a moving average or a parametric trend function $m_n(i)$ based on the deseasonalized data $y_i - s_n(i)$. One gets the residuals

$$x_i = y_i - s_n(i) - m_n(i) \tag{2.51}$$

by combining these steps,

If the detrended and deseasonalized data $x_1, \ldots, x_n$ are realizations of random variables $X_1, \ldots, X_n$ with expectation $EX_i = 0$, and $X_1, \ldots, X_n$ are uncorrelated or independent, then standard statistical procedures become applicable.

EXAMPLE 2.5.2. (Water Levels of the Moselle River.) We examine the water levels (in meters) of the Moselle River measured in Zeltingen from Nov. 1964 to Dec. 1977 and from Jan. 1981 to Jan. 1996 (stored in the file ht–mosel.dat). The measurements from the years 1978–1980 are missing. Since 1988, the measurements are daily maxima. Before 1988, there was one measurement each day.

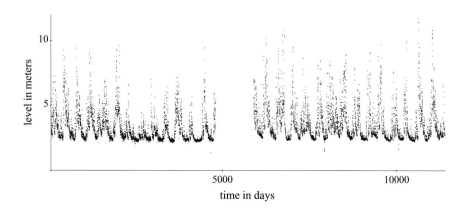

FIG. 2.15. Scatterplot of the Moselle River levels from Nov. 1964 to Jan. 1996, with a gap of the years 1978–1980 due to missing data.

Of course, the missing data from the years 1978 to 1980 caused an additional problem. To simplify the matter, this gap was filled by corresponding neighboring measurements. The gap from Jan. 1978 to June 1979 was filled with the values from Jan. 1976 to June 1977, and, likewise, the gap from July 1979 to Dec. 1980 was filled with the values from July 1981 to Dec. 1982 (this completed data set is stored in the file ht–mofil.dat).

In Fig. 2.16, we see the estimated seasonal component with and without smoothing. The smoothing reduces the irregular fluctuation to some extent.

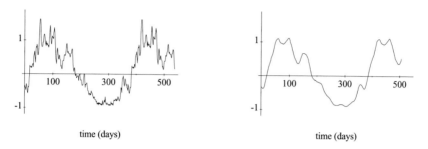

time (days)   time (days)

FIG. 2.16. (left.) Seasonal component without smoothing. (right.) Seasonal component with an additional 25 points smoothing.

The seasonal component attains its maximum values within the period from the end of December to mid–February. There is another remarkable peak in April.

The Moselle River data will be analyzed more intensively in Chapter 14. One must cope with the facts that these data are serial correlated and seasonally varying in the variance (heterosketastic) and in higher moments.

Serial Analysis of Stationary Data: the Autocovariance Function

For random variables X, Y with EX^2, $EY^2 < \infty$ the covariance is

$$\mathrm{Cov}(X, Y) = E\big((X - EX)(Y - EY)\big). \tag{2.52}$$

Loosely speaking, there is a tendency that the random variables X and Y simultaneously exceed or fall below their expectations EX and EY, if there is a positive covariance.

The random variables X and Y are uncorrelated if $\mathrm{Cov}(X, Y) = 0$. Recall that independent random variables are uncorrelated, yet the converse conclusion is not valid.

The pertaining sample covariance is

$$s_{x,y,n} = \frac{1}{n-1} \sum_{i \leq n} (x_i - \bar{x})(y_i - \bar{y}), \tag{2.53}$$

where $\bar{x}$ and $\bar{y}$ are the sample means of the data $x_1, \ldots, x_n$ and $y_1, \ldots, y_n$.

Subsequently, we study the serial dependence structure of a detrended and deseasonalized time series by means of the autocovariance function. A sequence

$X_1, \ldots, X_n$ is stationary if $EX_i = EX_1$, $EX_i^2 < \infty$ and the covariances

$$\mathrm{Cov}(X_i, X_{i+h}) = \mathrm{Cov}(X_1, X_{1+h}) =: r(h), \qquad i + h \leq n, \qquad (2.54)$$

merely depend on the time lag h. One also speaks of a weakly or covariance stationary series. The function r, with the time lag h as a variable, is the autocovariance function. It is particularly assumed that the expectations and variances are constant. The autocovariance function $r(h)$ can be estimated by the sample version

$$\hat{r}_n(h) = \frac{1}{n} \sum_{i \leq n-h} (x_i - \bar{x})(x_{i+h} - \bar{x}), \qquad (2.55)$$

where $\bar{x} = \sum_{i \leq n} x_i / n$ is the sample mean. The estimation is accurate if $n - h$ is sufficiently large. Notice that $\hat{r}_n(0)$ is the sample variance with the factor $1/(n-1)$ replaced by $1/n$ to reduce the random fluctuation for larger lags h.

The ratio

$$\rho(h) = r(h)/r(0) \qquad (2.56)$$

is the autocorrelation function. The sample autocorrelation function ρ is

$$\hat{\rho}_n(h) = \hat{r}_n(h)/\hat{r}_n(0). \qquad (2.57)$$

In Fig. 2.17, the sample autocorrelations are represented by bars.

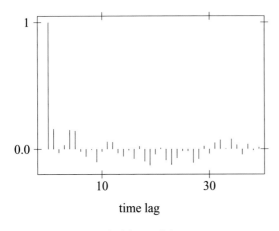

FIG. 2.17. Sample autocorrelations of 200 iid standard exponential data for lags $h = 0, \ldots, 40$.

Notice that $\hat{\rho}_n(0) = \rho(0) = 1$ and the autocorrelation function attains values between -1 and 1 under the stationarity condition.

In the case of detrended and deseasonalized data, we may assume that the expectation is equal to zero. An estimate of the autocovariance function r is

$$\hat{r}_n(h) = \frac{1}{n} \sum_{i \leq n-h} x_i x_{i+h} \ .$$

If $\hat{r}_n(h)$ is sufficiently close to zero for $h \geq q$, it is legitimate to assume that the residuals $x_1, x_{1+q}, x_{1+2q}, \ldots$ are realizations of uncorrelated or independent random variables. Then, standard statistical procedures for uncorrelated or independent random variables become applicable to the subsequence.

In conjunction with a non–trivial autocorrelation function, one speaks of a time series. As a theoretical example of such a series, we mention a Gaussian AR(1) time series $\{X_i\}$.

EXAMPLE 2.5.3. (Gaussian AR(1) Series.) For $0 \leq d \leq 1$, let

$$
\begin{aligned}
X_1 &= Y_1, \\
X_k &= d\,X_{k-1} + (1 - d^2)^{1/2}\,Y_k, \qquad k > 1,
\end{aligned}
\tag{2.58}
$$

where the Y_i are iid standard Gaussian random variables (normal random variables are often called Gaussian in the time series context).

This is the usual construction when defining standard Gaussian random variables X_1 and X_2 with correlation d. Verify that $\{X_k\}$ is a stationary sequence of standard Gaussian random variables with autocorrelation function

$$
\rho(h) = E(X_1\,X_{1+h}) = d^h, \qquad h \geq 0.
$$

Thus, we have a geometrically decreasing autocorrelation function. Notice that independence holds for $d = 0$ and total dependence for $d = 1$.

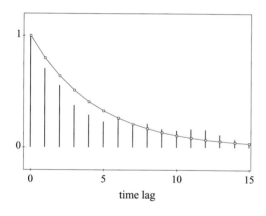

FIG. 2.18. Theoretical autocorrelations of Gaussian AR(1) series with $d = 0.8$ (solid line) and sample autocorrelations of 200 Gaussian AR(1) data under the parameter $d = 0.8$ (bars) for time lags $h = 0, \ldots, 15$.

Next we want to illustrate the sample behavior of AR(1) series for uncorrelated (independent) and strongly correlated random variables. In Fig. 2.19, Gaussian AR(1) data series of size 200 are plotted which were generated under the correlation parameters $d = 0$ and $d = 0.95$.

The data randomly fluctuate around the x–axis on the left–hand side, yet seem to exhibit a certain trend on the right–hand side although there is a stationary series.

The discussion about AR(1) series and related time series will be continued in the Sections 6.2, 16.7 and 16.8.

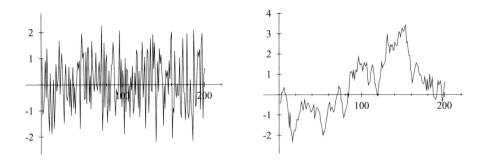

FIG. 2.19. Gaussian AR(1) data generated under $d = 0$ (left) and $d = 0.95$ (right).

2.6   The Tail Dependence Parameter

The following remarks should be regarded as a preliminary technical introduction to the concept of tail dependence and tail independence which concerns a certain property of the bivariate survivor function in the upper tail.

Later on, namely in Section 12.1, tail independence is interpreted as the property of the upper tail of a bivariate distribution which entails that the componentwise taken maxima are asymptotically independent. In Section 13.3 there is also a detailed discussion of other tail independence parameters which measure the rate at which the tail independence is attained.

At the beginning we introduce a certain tail dependence parameter by means of the bivariate survivor function. The definition of a auto–tail–dependence function ist added which is related to the autocovariance function. In addition, certain sample versions of the tail dependence parameter and of the auto–tail–dependence function are suggested.

An Introduction to Tail Dependence

Let X and Y be random variables with the joint df F and univariate marginal dfs F_X and F_Y. The dependence in the upper tail region of the distribution may be expressed by the conditional probability

$$P(Y > y | X > x) = \frac{P\{X > x, Y > y\}}{P\{X > x\}} \tag{2.59}$$

of $Y > y$ given $X > x$. Such conditional probabilities were studied by Sibuya[6] and other authors in conjunction with the asymptotic independence of componentwise taken maxima, see (12.8).

---

[6]Sibuya, M. (1960). Bivariate extreme statistics. Ann. Inst. Math. Statist. 11, 195–210.

The conditional probability in (2.59) is independent of the marginal dfs F_X and F_Y when x and y are replaced by the quantiles $F_X^{-1}(u)$ and $F_Y^{-1}(v)$.

In the sequel, let $u = v$. The tail dependence parameter $\chi(u)$ of X and Y at the level u is

$$
\begin{aligned}
\chi(u) &= P\big(Y > F_Y^{-1}(u)\,\big|\,X > F_X^{-1}(u)\big) \\
&= P(V > u\,|\,U > u),
\end{aligned} \tag{2.60}
$$

where

$$(U, V) = (F_X(X), F_Y(Y))$$

is the pertaining copula random vector with $(0, 1)$–uniformly distributed marginal random variables U and V, if F_X and F_Y are continuous (according to the probability transformation).

It is always understood that u is close to 1, that is, we are dealing with probabilities in the upper tail region of the joint distribution of X and Y or, respectively, of U and V. The term

$$\chi = \lim_{u \to 1} \chi(u) \tag{2.61}$$

is addressed as tail dependence parameter. We have tail independence, if $\chi = 0$, and total tail dependence if $\chi = 1$.

We list some properties of the tail dependence parameters:

- $\chi(u)$ and χ are symmetric in X and Y;

- $\chi(u)$ and χ range between zero and one;

- if X and Y are stochastically independent (in the usual sense), then $\chi(u) = 1 - u$ and $\chi = 0$; therefore, independence implies tail independence,

- if $X = Y$, then $\chi(u) = 1$ and $\chi = 1$.

However, one should be aware that tail independence does not imply independence. Let

$$W(x, y) = 1 + x + y$$

for $x, y \le 0$ and $x + y \ge -1$. Then, $\chi(u) = 0$ if $u \ge 1/2$ and, therefore, also $\chi = 0$. Thus, we have tail independence. Yet, W is the joint df of the totally dependent rvs Z and $-(1 + Z)$ where Z is on $(-1, 0)$–uniformly distributed.

The sample version pertaining to $\chi(u)$ and χ, based on data (x_i, y_i) under the df F, is

$$\chi_n(u) = \frac{1}{n(1 - u)} \sum_{i \le n} I(x_i > x_{[nu]:n},\; y_i > y_{[nu]:n}). \tag{2.62}$$

For a continuation of this topic we refer to the Chapters 12 and 13 which concern multivariate extreme value and multivariate generalized Pareto models. For such multivariate distributions, the dependence structure is of central importance (besides of the univariate marginals). We particularly refer to Section 12.1, where we describe the relationship of the tail dependence parameter χ to the Pickands dependence function D, and to Section 13.3 for the definition of other tail dependence parameters $\bar{\chi}$ and β which determine the rate at which the tail independence is attained.

The Auto–Tail–Dependence Function

Next, let $X_1, \ldots, X_n$ be a series of identically distributed random variables with common df F. Assume, in addition, that the series has stationary dependencies in the upper tail in the sense that for $i \leq n - h$,

$$P\big(X_{i+h} > F^{-1}(u)\big|X_i > F^{-1}(u)\big) = P\big(X_{1+h} > F^{-1}(u)\big|X_1 > F^{-1}(u)\big)$$
$$=: \rho(u, h) \tag{2.63}$$

which defines the auto–tail–dependence function at the level u. Likewise one defines an auto–tail–dependence function $\rho(h)$ by

$$\rho(h) = \lim_{u \to 1} \rho(u, h) \tag{2.64}$$

as the limit of $\rho(u, h)$ for $u \to 1$.

The auto–tail–dependence functions $\rho(u, h)$ at the level u can be estimated by the sample version

$$\rho_n(u, h) = \frac{1}{n(1-u)} \sum_{i \leq n} I\big(\min(x_i, x_{i+h}) > x_{[nu]:n}\big) \tag{2.65}$$

based on data $x_1, \ldots, x_n$.

It is clear that $\rho_n(u, h)$ also serves as an estimate of $\rho(h)$, where a sufficiently large u must be selected by the statistician. The level (threshold) u should be sufficiently large to reduce the bias, and small enough to reduce the variance of the estimator.

Another auto–tail–dependence function $\bar{\rho}$ will be introduced in Section 13.3 in conjunction with the tail dependence parameter $\bar{\chi}$ which measures the degree of tail independence.

2.7　Clustering of Exceedances

In conjunction with extreme values, it has been implicitly assumed in foregoing sections that the data $x_1, \ldots, x_n$ are generated independently from each other or, at least, the dependence between the data is negligible.

In Section 2.6 we already started with a preliminary discussion about dependence concepts in conjuction with tail dependence parameters. In that context, another feature will be captured by the phenomenon that stronger dependence may lead to a clustering of extreme values. The clustering of data is also related to another important parameter in extreme value theory, namely the extremal index θ (as indicated in this section, see below, and further mentioned in Section 6.2).

Building Clusters by Runs, the Mean Cluster Size

Given $x_1, \ldots, x_n$ let $x_{i(1)}, \ldots, x_{i(k)}$ again denote the exceedances over a predetermined threshold u. For some choice of a positive integer r, called the run length, define clusters of exceedance times $i(j)$ in the following manner.

Any run of at least r consecutive observations x_i below the threshold u separates two clusters. Hence, there is a minimal gap of length r between two consecutive clusters of exceedance times.

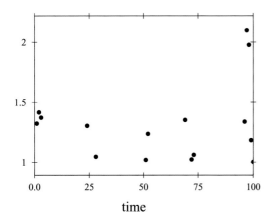

FIG. 2.20. Six clusters above $u = 1$ with respect to a run length $r = 3$ of a Gaussian AR(1) series with $d = 0.8$ with cluster sizes between 1 and 5.

We introduce the mean cluster size that characterizes the clusters to some extent. Let $n(u, r)$ denote the number of clusters over u. The mean cluster size, relative to u and the run length r, is

$$\mathrm{mcsize}(u, r) = k/n(u, r). \qquad (2.66)$$

The mean cluster size is a useful statistic for describing extreme data besides an estimate of the tail index.

The Blocks Method

Clusters of exceedance times may also be built by the blocks method (also called Gumbel method) as mentioned in Section 1.2. Each block containing at least one exceedance time is treated as a cluster (cf. [16], pages 242–243, and the literature cited therein for more details).

Declustering

To obtain data that correspond more closely to a model of iid random variables, one may reduce the sample of exceedances to that of the cluster maxima. This topic will be further pursued within the framework of Poisson counting processes, see Section 9.2.

The Cluster Size Distribution

The following discussion merely concerns the run length definition of clusters.

Let $|\mu|$ denote the size of a cluster μ of exceedance times for the given exceedances $x_{i(1)}, \ldots, x_{i(k)}$ over u according to the run length r. The relative number of clusters with size j defines the (dicrete) cluster size distribution $P_{u,r}$ on the set $1, \ldots, k$. We have

$$P_{u,r}(\{j\}) := \frac{|\{\mu : |\mu| = j\}|}{n(u,r)}, \qquad 1 \leq j \leq k. \qquad (2.67)$$

The mean of the cluster size distribution $P_{u,r}$ is the mean cluster size mcsize(u,r) as introduced in (2.66) before.

The illustration in Fig. 2.21 concerns the exceedances over a threshold $u = 1$ of 4000 Gaussian AR(1) data generated under the correlation coefficient $d = 0.8$. The exceedances as well as the exceedance times are plotted. One clearly recognizes the clustering in the exceedances $x_{i(j)}$ as well as in the exceedance times.

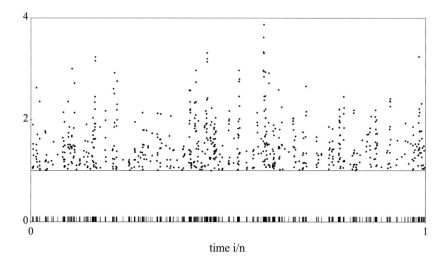

time i/n

FIG. 2.21. Scatterplot of exceedances over the threshold $u = 1$ of 4000 Gaussian AR(1) data with correlation coefficient $d = 0.8$.

We remark that the number of exceedances over 1 (respectively, 2, 3) in Fig. 2.21 is 707 (respectively, 102, 12).

Next, the run length r is chosen equal to 1. In Fig. 2.22 (left), the cluster size distribution for $u = 1$ and the (reciprocal) mean cluster sizes for varying $u \geq 1$ are plotted.

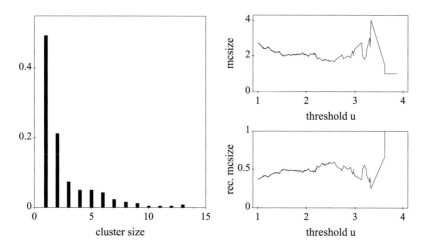

FIG. 2.22. Cluster size distribution for $u = 1$ and $r = 1$ (left). Mean cluster sizes (right, top) and reciprocal mean cluster sizes for $u \geq 1$ and $r = 1$ (right, bottom).

The reciprocal mean cluster sizes $1/\mathrm{mcsize}(u, r)$ will also be written $\theta(u, r)$. The reciprocal mean cluster sizes $\theta(u, r)$ are related to the extremal index, denoted by θ, which will be discussed in Section 6.2. In Fig. 2.23, $\theta(\cdot, r)$ is plotted against the threshold u for run lengths $r = 1, 2, 5, 10$.

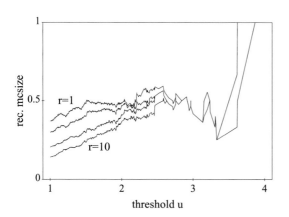

FIG. 2.23. Reciprocal mean cluster sizes for $u \geq 1$ and $r = 1, 2, 5, 10$.

We also include plots of the reciprocal mean cluster sizes $\theta(\cdot, r)$ of Gaussian

AR(1) data for the sample size $n = 4000$, corresponding to those in Fig. 2.23, for $d = 0.4$ and $d = 0.6$.

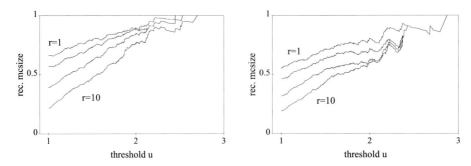

FIG. 2.24. Reciprocal mean cluster sizes $\theta(u, r)$ for $d = 0.4$ and $d = 0.6$ and $u \geq 1$, $r = 1, 2, 5, 10$

For smaller d, that is the situation closer to independence, the reciprocal mcsizes are closer to 1. Likewise, one may deal with mean cluster sizes and cluster size distributions depending on the number k of exceedances as a parameter.

Part II

Statistical Inference in Parametric Models

Chapter 3

An Introduction to Parametric Inference

In the preceding chapters, we emphasized the visual viewpoint of representing data and fitting parametric distributions to the data. This is the exploratory approach to analyzing data. In the present chapter, we add some parametric estimation and test procedures which have been partially deduced from the visual ones.

This chapter gives us the opportunity to reinforce previous knowledge as well as to fill gaps. For example, we give a detailed description of the parametric bootstrap in conjunction with confidence intervals in Section 3.2. The p–value is employed instead of a fixed significance level in testing problems.

We introduce some classical estimation procedures in the exponential, Gaussian and Poisson models in order to give an outline of our approach within a familiar setting. This is done in the Sections 3.1, 3.4 and 3.5. Recall that exponential distributions belong to the family of generalized Pareto (GP) distributions. Gaussian distributions are on the agenda because we want to compare classical statistical procedures in Gaussian models with those influenced by extreme value analysis.

The Sections 3.2 and 3.3 are devoted to confidence intervals and test procedures for parametric models. Poisson distributions are of interest in extreme value analysis because the number of exceedances above a higher threshold can be modeled by such distributions.

Bayesian analysis gains more and more importance in our investigations. In Section 3.5, we give an introduction to the Bayesian estimation principle within a decision theoretic framework and collect some relevant examples of Bayesian estimators in continuous and discrete models.

3.1 Estimation in Exponential and Gaussian Models

In the preceding chapter, a normal df was visually fitted to the sample df. Subsequently, automatic procedures are provided using estimators of the location and scale parameters. We present some prominent estimators of the mean μ and the standard deviation σ within the normal (Gaussian) model $\{\Phi_{\mu,\sigma} : \mu \text{ real}, \sigma > 0\}$. Keep in mind that the normal dfs and densities will not be represented by the variance σ^2, but by the standard deviation (scale parameter) σ.

   We start with a likelihood–based estimator, namely, the maximum likelihood estimator in exponential and Gaussian models. Other likelihood–based estimators are the Bayesian estimators.

Maximum Likelihood Estimation in the Exponential Model

We compute the maximum likelihood estimate (MLE) for the model of exponential densities $g_\vartheta(x) = \vartheta \exp(-\vartheta x)$, $x > 0$, where $\vartheta > 0$ is the unknown reciprocal scale parameter.

   Remember that the joint density of iid exponential variables $X_1, \ldots, X_n$ with parameter ϑ is

$$g(\boldsymbol{x}|\vartheta) = \prod_{i \le n} g_\vartheta(x_i), \qquad \mathbf{x} = (x_1, \ldots, x_n),$$

cf. also (10.10) in the multivariate part of this book. The MLE $\hat\vartheta_n$ maximizes the likelihood function

$$L(\vartheta) = \prod_{i \le n} g_\vartheta(x_i)$$

for the given sample $\boldsymbol{x} = (x_1, \ldots, x_n)$.

   Now compute the MLE by taking the derivative of the log–likelihood function $\log L(\vartheta)$ and solving the likelihood equation

$$(\log L)'(\vartheta) = 0.$$

The solution is the reciprocal sample mean $\hat\vartheta_n = 1/\bar{x} = n \Big/ \sum_{i \le n} x_i$.

   Likewise, the sample mean is the MLE of the scale parameter in the exponential model.

Maximum Likelihood Estimation in the Gaussian Model

The joint density of iid normal random variables $X_1, \ldots, X_n$ with mean and standard deviation (location and scale parameters) μ and σ is

$$\varphi(\boldsymbol{x}|\mu, \sigma) = \prod_{i \le n} \varphi_{\mu,\sigma}(x_i), \qquad \mathbf{x} = (x_1, \ldots, x_n),$$

where the density of $\Phi_{\mu,\sigma}$ is denoted by $\varphi_{\mu,\sigma}$ again.

The MLEs of μ and σ in the normal model are the sample mean and the sample standard deviation

$$\hat{\mu}_n = \bar{x} \quad \text{and} \quad \hat{\sigma}_n = \left(\frac{1}{n} \sum_{i \leq n} (x_i - \hat{\mu}_n)^2 \right)^{1/2}. \tag{3.1}$$

The estimate $(\hat{\mu}_n, \hat{\sigma}_n)$ maximizes the likelihood function $L(\mu, \sigma) = \varphi(\boldsymbol{x}|\mu, \sigma)$ for the given sample $x_1, \ldots, x_n$. The values $\hat{\mu}_n$ and $\hat{\sigma}_n$ may be computed as the the solutions to the likelihood equations

$$\frac{\partial}{\partial \mu} \log L(\mu, \sigma) = 0 \quad \text{and} \quad \frac{\partial}{\partial \sigma} \log L(\mu, \sigma) = 0$$

obtained by the partial derivatives of the log–likelihood function.

EXAMPLE 3.1.1. (Michelson's Determination of the Velocity of Light in the Air from 1879.) Using a refinement of Foucault's method, Michelson obtained $n = 100$ measurements of the velocity of light in the air[1]. The values in Table 3.1 plus 299,000 in km/sec are Michelson's determinations of the light speed in the air (stored in the order of the outcome in the file nu–miche.dat).

TABLE 3.1. Michelson's 1879 measurements of the velocity of light in the air.

620	760	800	810	840	850	870	880	930	960
650	760	800	810	840	850	870	880	930	960
720	760	800	810	840	850	870	890	940	970
720	760	800	810	840	850	880	890	940	980
720	770	800	810	840	850	880	880	940	980
740	780	810	820	840	860	880	900	950	980
740	780	810	820	840	860	880	900	950	1000
740	790	810	830	850	860	880	910	950	1000
750	790	810	830	850	870	880	910	960	1000
760	790	810	840	850	870	880	920	960	1070

The Q–Q plot in Fig. 3.1 confirms a normal modeling for the measurements. The MLEs are $\hat{\mu} = 852.4$ and $\hat{\sigma} = 78.6$. In addition, the plots of the estimated parametric density and a kernel density are sufficiently close to one another. Thus, 299,841.7 km/sec is a parametric estimate of the light speed in the air.

The universally–accepted light speed in a vacuum is about 299,792.5 km/sec. To obtain the light speed in the air, this value must be multiplied by a correction factor which depends on atmospheric humidity, pressure and temperature. By employing the correction factor taken by Michelson, one arrives at the value 299,734.5; other reasonable choices of

---

[1]See also Andrews, D.F. and Herzberg, A.H. (1985). Data. Springer, New York.

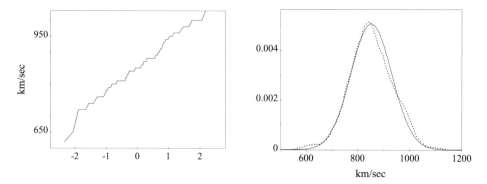

FIG. 3.1. (left.) Normal Q–Q plot for Michelson's data. (right.) Kernel density (dotted) with $b = 70$ and normal density (solid) with parameters given by the MLE.

the correction factor would lead to a similar conclusion. Thus, due to a systematic error in the experiment, the correct normal modeling of Michelson's data and the skillfully defined MLE are merely of limited relevance to evaluating the target parameter, namely the "true" light speed. For a continuation, see Example 3.2.1, where a confidence interval is presented.

It is likely that some of the readers are disappointed or even frustrated by the foregoing example, yet it was deliberately chosen to enforce a critical attitude and to exhibit certain limitations of statistical inference.

The Moment Estimation Method

The MLEs $\hat{\mu}_n$ and $\hat{\sigma}_n$ in the normal model may be classified as moment estimates, because $\hat{\mu}_n$ and $\hat{\sigma}_n^2$ are the sample mean and sample variance, and μ and σ^2 are the mean and the variance of the normal df $\Phi_{\mu,\sigma}$.

Generally, moment estimates are obtained by equating the sample moments with the pertaining moments of parametric distributions. A similar method is introduced in Section 14.5, where ordinary moments are replaced by L–moments.

The Quantile Estimation Method, L–Statistics

Recall from (2.9) that $x_{[nq]:n}$ is an estimate of the q–quantile. In a location and scale parameter family, we have $x_{[nq]:n} \approx \mu + \sigma F^{-1}(q)$. Such a relation can be employed to estimate μ and σ (and further parameters if necessary). In the normal case, the median $x_{[n/2]:n}$ is an estimate of μ. By taking differences, one eliminates the location parameter and finds an estimate of σ. For example, the interquartile range leads to the well–known robust estimate

$$(x_{[3n/4]:n} - x_{[n/4]:n})/(2\Phi^{-1}(3/4))$$

of σ in the normal location and scale parameter family. Estimators of this type may be classified as quick or systematic estimators.

EXAMPLE 3.1.2. (Fitting a Normal Distribution When Fat Tails are Correct.) To data $x_1, \ldots, x_n$, which are governed by a symmetric distribution F with fat or heavy tails, one may fit a normal df with location parameter $\mu = 0$ and scale parameter given by the sample standard deviation, or quick estimators such as

$$\hat{\sigma}_n = x_{[nq]:n} / \Phi^{-1}(q) \tag{3.2}$$

based on the x_i, or, under a condition of symmetry imposed on F,

$$\tilde{\sigma}_n = y_{[nq]:n} / \Phi^{-1}((1+q)/2) \tag{3.3}$$

based on $y_i = |x_i|$.

By the definition of $\hat{\sigma}_n$, the normal df $\Phi_{0,\hat{\sigma}_n}$ has the q–quantile $x_{[nq]:n}$ and, therefore, the weight of $\Phi_{0,\hat{\sigma}_n}$ and the number of data beyond the point $x_{[nq]:n}$ correspond to each other. Typically, $\hat{\sigma}_n > s_n$ and $\tilde{\sigma}_n > s_n$ and the normal dfs pertaining to the quick estimates are more appropriate as estimators of the tails of F.

More generally, one may deal with L–statistics (linear combination of order statistics) of the form $\sum_{i \leq n} c_{i,n} x_{i:n}$ which provide a rich class of statistics for estimating parameters (see, e.g., [48]). Special L–statistics will by utilized in conjunction with the L–moment estimation method which will be introduced in Section 14.5.

The Least Squares Estimation Method

Location and scale parameters μ and σ may be estimated by a least squares method.

The points of a normal Q–Q plot in (2.39) are close to the straight line $(x, \sigma x + \mu)$ and, therefore, also close to the least squares line $(x, ax + b)$ as defined by (2.40). Consequently, the two straight lines $(x, \sigma x + \mu)$ and $(x, ax + b)$ are close together. This yields that the intercept b and the slope a of the least squares line provide plausible estimates of the location and scale parameters μ and σ.

It is advisable to use a trimmed version of the least squares procedure. Thus, first omit a certain number of upper and lower extremes from the sample, and then apply the least squares method. This least squares method for estimating location and scale parameters can be utilized for any df F in place of the normal df Φ.

The Minimum Distance Method

By visually fitting a normal df or density to the sample df or, respectively, to a kernel density or histogram, one is essentially applying a minimum distance (MD) method.

Let d be a distance on the family of dfs. Then, $(\hat{\mu}_n, \hat{\sigma}_n)$ is an MDE[2], if

$$\mathrm{d}(\widehat{F}_n, \Phi_{\hat{\mu}_n, \hat{\sigma}_n}) = \inf_{\mu, \sigma} \mathrm{d}(\widehat{F}_n, \Phi_{\mu, \sigma}),$$

where $\widehat{F}_n$ again denotes the sample df. The distance may also be based on a distance between normal densities $\varphi_{\mu, \sigma}$ and sample densities f_n. One must apply the Hellinger distance

$$H(\varphi_{\mu, \sigma}, f_n) = \left(\int \left(\varphi_{\mu, \sigma}^{1/2}(x) - f_n^{1/2}(x) \right)^2 dx \right)^{1/2} \tag{3.4}$$

to obtain asymptotically efficient estimators[3] (a property which is shared by the MLE). The Hellinger or L_2 distances also possess computational advantages, because, then, the distances $\mathrm{d}(\varphi_{\mu, \sigma}, f_n)$ are differentiable in μ and σ.

The M–Estimation Method

In view of an application in Section 5.1 we deal with M–estimators in scale parameter models $\{F_\sigma : \sigma > 0\}$ with special emphasis laid on the exponential scale parameter model. Let f be the density of $F = F_1$. The MLE of the scale parameter is a special M–estimator as the solution to the likelihood equation

$$\sum_{i \leq n} \psi_{\mathrm{ML}}(x_i/\sigma) = 0,$$

where the M–function is $\psi_{\mathrm{ML}}(x) = -x f'(x)/f(x) - 1$.

In the exponential model, where $f(x) = e^{-x}$, $x \geq 0$, we have

$$\psi_{\mathrm{ML}}^*(x) = x - 1, \tag{3.5}$$

and the MLE of the scale parameter is the sample mean.

Generally, M–estimates of σ are solutions to M–equations

$$\sum_{i \leq n} \psi(x_i/\sigma) = 0, \tag{3.6}$$

where ψ is the M–function with $\int \psi(x) f(x) dx = 0$. In addition,

---

[2] The L_2 distance between normal densities and histograms was used in Brown, L.D. and Gene Hwang, J.T. (1993). How to approximate a histogram by a normal density. The American Statistician 47, 251–255.

[3] Beran, R.J. (1977). Minimum Hellinger distance estimates for parametric models. Ann. Statist. 5, 445–463.

- boundedness of ψ is required to achieve robustness against gross errors,

- ψ should be close to ψ_{ML} in order to gain efficiency of the estimator.

An estimator satisfying both requirements is obtained by truncating ψ_{ML} (see [27], page 122). Such M–functions go back to P.J. Huber [30]).

For the exponential model, we use M–functions of the form

$$\psi_b^*(x) = -\exp(-x/b) + b/(1+b). \tag{3.7}$$

Check that ψ_b^* is bounded and $b\psi_b^*(x) \to \psi_{\mathrm{ML}}^*(x)$ as $b \to \infty$ (apply (1.66)). Because ψ_{ML}^* is differentiable, one can apply the Newton–Raphson iteration procedure to solve the M–equation (3.6).

Covering Probabilities

The accuracy of an estimator can also be measured directly by the df of the estimator: the probability that the absolute deviation of the sample mean $\bar{X}$ from the true mean m_F is smaller or equal to some $t > 0$ has the representation

$$P\{|\bar{X} - m_F| \leq t\} = P\{\bar{X} \leq m_F + t\} - P\{\bar{X} < m_F - t\}. \tag{3.8}$$

If $F = \Phi_{\mu,\sigma}$ and the X_i are independent, this yields

$$P\{|\bar{X} - \mu| \leq t\sigma/n^{1/2}\} = 2\Phi(t) - 1. \tag{3.9}$$

In greater generality, we have

$$P\{|\bar{X} - m_F| \leq t s_F/n^{1/2}\} \approx 2\Phi(t) - 1, \tag{3.10}$$

for every df F which possesses a finite standard deviation s_F.

These two formulas lead to exact and asymptotic confidence intervals for the mean of a distribution (cf. Section 3.2).

The Bias and the Mean Squared Error (MSE) of an Estimator

Let us deal with the special case of estimating the location parameter μ in the normal model and, generally, with estimating the mean $m_F = \int x \, dF(x)$ of a df F. The natural estimator of m_F is the sample mean $\bar{X} = \frac{1}{n}\sum_{i \leq n} X_i$, where the X_i are identically distributed with common df F. We have $E\bar{X} = m_F$. Hence, the bias $E\bar{X} - m_F$ is equal to zero, and $\bar{X}$ is an unbiased estimator of m_F. For such an unbiased estimator, the variance $E(\bar{X} - m_F)^2$ is an appropriate measure for the accuracy.

Generally, the performance of an estimator $\hat{\vartheta}_n$ of a parameter ϑ can be measured by the mean squared error (MSE) which is the expected squared deviation of the estimator from the target parameter. We have

$$\begin{aligned} E(\hat{\vartheta}_n - \vartheta)^2 &= E\big(\hat{\vartheta}_n - E(\hat{\vartheta}_n)\big)^2 + \big(E(\hat{\vartheta}_n) - \vartheta\big)^2 \\ &=: V(\hat{\vartheta}_n) + \mathrm{Bias}^2(\hat{\vartheta}_n). \end{aligned} \tag{3.11}$$

For unbiased estimators, the MSE is the variance. The bias that occurs when the model is incorrect is essential for the performance of an estimator.

The MSE plays a central role in the definition of the Bayes estimator which will be dealt with in the subsequent lines.

3.2 Confidence Intervals

Intervals based on the data are constructed so that the true parameter falls into such an interval in $(1-\alpha) \times 100\%$ of the trials. One obtains 95% or 99% confidence intervals, if $\alpha = 0.05$ or $\alpha = 0.01$.

We mostly deal with confidence intervals based on estimators. Such a confidence interval also provides a measure for the accuracy of the estimator. First, we will employ (3.9) and (3.10) to construct confidence intervals for the mean of a distribution.

Confidence Intervals for the Mean

First assume that the data $x_1, \ldots, x_n$ are governed by a normal df Φ_{μ,σ_0} with μ unknown and σ_0 fixed. Let $u(\alpha) = \Phi^{-1}(1 - \alpha)$ denote the $(1 - \alpha)$–quantile of Φ. (3.9) yields that, with probability $1 - \alpha$, a sample $x_1, \ldots, x_n$ is drawn such that the interval

$$[\bar{x} - \sigma_0 u(\alpha/2)/n^{1/2}, \; \bar{x} + \sigma_0 u(\alpha/2)/n^{1/2}] \tag{3.12}$$

covers μ. Thus, we gain a $(1 - \alpha)$ confidence interval for the mean.

If σ_0 in (3.12) is unknown, then the interval bounds are also unknown to the statistician. Yet, if σ_0 is replaced by the sample standard deviation s_n, one gets an interval which is merely based on the data. This leads to a confidence interval that approximately attains the level $1 - \alpha$.

To obtain a confidence interval such that the level $1 - \alpha$ is attained exactly, one must adopt quantiles of the t–df with $n - 1$ degrees of freedom, see (1.62).

EXAMPLE 3.2.1. (Continuation of Example 3.1.1 about Michelson's 1879 Determination of the Velocity of Light in the Air.) The confidence interval of level .99 for the unknown mean value μ^* is $[\bar{x} - 2.58 s_n n^{-1/2}, \; \bar{x} + 2.58 s_n n^{-1/2}] = [832.1, \, 872.7]$. Thus, when Michelson's experiment is repeated for several times, then μ^* would fall into such intervals in 99 % of the trials. Whereas the value of 299,841.7 km/sec determined by Michelson might be an acceptable estimate of the light–speed in the air—there is a deviation of 107.2 km/sec—the measuring of the accuracy of the estimate by means of the confidence interval must be regarded as a failure.

From (3.10), we know that an asymptotic confidence interval for the mean is still obtained by the preceding construction if the normal distribution is replaced by any df which has a finite variance. This construction can be extended to other functional parameters.

Asymptotic Confidence Intervals for Functional Parameters

We deal with the fairly general case of two–sided confidence intervals for a functional parameter $T(\vartheta)$ of a parametric df F_ϑ with $\vartheta \in \Theta$. We assume that a consistent estimator $\hat\vartheta_n$ of the parameter ϑ is available.

In analogy to (3.10), assume that $\widehat{T}_n(\boldsymbol{X})$ is an estimator of $T(\vartheta)$ such that

$$P\{|\widehat{T}_n(\boldsymbol{X}) - T(\vartheta)| \le ts(\vartheta)/n^{1/2}\} \approx 2\Phi(t) - 1, \tag{3.13}$$

where $\boldsymbol{X} = (X_1, \ldots, X_n)$ is a vector of n iid random variables X_i with common df F_ϑ, and $s(\vartheta)$ are normalizing constants varying continuously in ϑ.

Let $\boldsymbol{x}$ be the vector of data x_i governed by F_ϑ. If $\hat\vartheta_n(\boldsymbol{x}) \approx \vartheta$, then $s(\hat\vartheta_n(\boldsymbol{x})) \approx s(\vartheta)$. Therefore, corresponding to (3.12), the vectors $\boldsymbol{x} = (x_1, \ldots, x_n)$ are drawn, with probability approximately equal to $1 - \alpha$, such that the intervals

$$\left[\widehat{T}_n(\boldsymbol{x}) - \frac{s(\hat\vartheta_n(\boldsymbol{x}))u(\alpha/2)}{n^{1/2}}, \widehat{T}_n(\boldsymbol{x}) + \frac{s(\hat\vartheta_n(\boldsymbol{x}))u(\alpha/2)}{n^{1/2}}\right] \tag{3.14}$$

cover the functional parameter $T(\vartheta)$.

For the preceding construction, one needs a precise knowledge of the asymptotic behavior of the estimator of the functional parameter $T(\vartheta)$. In particular, the normalizing constants $s(\vartheta)$ must be known to the statistician. Otherwise, confidence intervals may be constructed by means of the bootstrap[4] approach.

Parametric Bootstrap Confidence Intervals

We will briefly explain the bootstrap approach within a parametric framework (also see [49] and the review paper by Manteiga and Sánchez[5]). Let $\widehat{T}_n(\boldsymbol{x})$ be an estimate of the functional parameter $T(\vartheta)$ based on $\boldsymbol{x} = (x_1, \ldots, x_n)$. Assume that $\hat\vartheta_n$ is a consistent estimator of ϑ. Usually, we have $\widehat{T}_n(\boldsymbol{x}) = T(\hat\vartheta_n(\boldsymbol{x}))$

The parametric bootstrap df based on the vector $\boldsymbol{x}$ of data is given by

$$\widehat{B}_n(\boldsymbol{x};t) = P\{|\widehat{T}_n(\boldsymbol{Z}) - T(\hat\vartheta_n(\boldsymbol{x}))| \le t\}, \tag{3.15}$$

where $\boldsymbol{Z} = (Z_1, \ldots, Z_n)$ is a vector of iid random variables with common df $F_{\hat\vartheta_n(\boldsymbol{x})}$. Notice that $\widehat{B}_n(\boldsymbol{x}; \cdot)$ is known to the statistician (at least, theoretically). Denote the $(1 - \alpha)$–quantile of the bootstrap df $\widehat{B}_n(\boldsymbol{x}; \cdot)$ by

$$c_{n,\alpha}(\boldsymbol{x}) = \widehat{B}_n^{-1}(\boldsymbol{x}; 1 - \alpha). \tag{3.16}$$

---

[4]A term coined in Efron, B. (1979). Bootstrap methods: another look at the jackknife. Ann. Statist. 7, 1–26.

[5]Manteiga, W.G. and Sánchez, J.M.P. (1994). The bootstrap—a review. Comp. Statist. 9, 165–205

Under certain regularity conditions, the intervals in (3.14) can be replaced by the bootstrap confidence intervals

$$\left[\widehat{T}_n(\boldsymbol{x}) - c_{n,\alpha}(\boldsymbol{x}), \widehat{T}_n(\boldsymbol{x}) + c_{n,\alpha}(\boldsymbol{x})\right]. \tag{3.17}$$

A justification of the bootstrap approach will be provided at the end of this section. First, the value $c_{n,\alpha}(\boldsymbol{x})$ will be computed by employing the Monte Carlo method.

Simulating the Bootstrap Confidence Bounds

Generate data $z_{1,j}, \ldots, z_{n,j}$, $j = 1, \ldots, N$, according to the df $F_{\hat{\vartheta}_n(\boldsymbol{x})}$. Notice that the bootstrap sample

$$b_j = |\widehat{T}_n(z_{1,j}, \ldots, z_{n,j}) - T(\hat{\vartheta}_n(\boldsymbol{x}))|, \qquad j = 1, \ldots, N, \tag{3.18}$$

is governed by the bootstrap df $\widehat{B}_n(\boldsymbol{x}; \cdot)$. According to (2.9), the bootstrap sample $(1 - \alpha)$–quantile $b_{[(1-\alpha)N]+1:N}$ satisfies the relation

$$b_{[(1-\alpha)N]+1:N} \approx \widehat{B}_n^{-1}(\boldsymbol{x}; 1 - \alpha) \tag{3.19}$$

which yields that $\tilde{c}_{n,\alpha}(\boldsymbol{x}) = b_{[(1-\alpha)N]+1:N}$ may be taken in (3.17) in place of $c_{n,\alpha}(\boldsymbol{x})$.

In order to obtain a sufficiently accurate estimate of the bootstrap $(1 - \alpha)$–quantile in (3.19) for typical values $\alpha = .01$ and $\alpha = .05$, we suggest taking $N = \max(n, 2000)$ as the sample size for the simulation.

A Justification of the Bootstrap Approach

One must verify that

$$P\{|\widehat{T}_n(\boldsymbol{X}) - T(\vartheta_0)| \le c_{n,\alpha}(\boldsymbol{X})\} \approx 1 - \alpha \tag{3.20}$$

where $\boldsymbol{X} = (X_1, \ldots, X_n)$ is a vector of iid random variables with common df F_{ϑ_0}. If (3.13) holds uniformly in a neighborhood of ϑ_0 and $s(\hat{\vartheta}_n(\boldsymbol{x})) \approx s(\vartheta_0)$, then the bootstrap df is an estimate of the centered distribution of the estimator $\widehat{T}_n(\boldsymbol{X})$, that is,

$$\widehat{B}_n(\boldsymbol{x}; t) \approx P\{|\widehat{T}_n(\boldsymbol{X}) - T(\vartheta_0)| \le t\}. \tag{3.21}$$

Moreover, from (3.13) we know that

$$P\{|\widehat{T}_n(\boldsymbol{X}) - T(\vartheta_0)| \le d_{n,\alpha}\} \approx 1 - \alpha, \tag{3.22}$$

where $d_{n,\alpha} = n^{1/2}\Phi^{-1}(1 - \alpha/2)/s(\vartheta_0)$. Now, deduce from (3.21) and (3.22) that $d_{n,\alpha}/c_{n,\alpha}(\boldsymbol{x}) \approx 1$ which implies that $d_{n,\alpha}$ can be replaced by $c_{n,\alpha}(\boldsymbol{X})$ in (3.22). Thus, (3.20) holds true.

Extensions

In many applications, a framework more general than that of replicates $x_1, \ldots, x_n$ is required. For example, one may deal with measurements x_i at time t_i governed by a df $F_{\beta_0 + \beta_1 t_i}$; the joint experiment is determined by the parameter vector (β_0, β_1).

Likewise, one may deal with one–sided bootstrap confidence intervals. For that purpose take a bootstrap sample with $\widehat{T}_n(z_{1,j}, \ldots, z_{n,j}) - T(\hat{\vartheta}_n(\boldsymbol{x}))$ instead of b_j in (3.18).

3.3 Test Procedures and p–Values

In this section we briefly discuss critical regions of a significance level α and their representation by means of a p–value. In statistical software packages, the output of a test usually consists of the p–value, whereby it is understood that the user has some significance level in mind. The major advantage of the p–value is that the significance level must not be specified in advance.

Test Statistics and Critical Values

Let $\{F_\vartheta\}$ be a family of dfs, where ϑ varies over a parameter space Θ. Let Θ_0 and Θ_1 be the null hypothesis and the alternative, where Θ_0 and Θ_1 constitute a partition of Θ. A decision for or against the null hypothesis may be based on a critical region C. If $\boldsymbol{x} = (x_1, \ldots, x_n)$ belongs to the critical region C, then the null hypothesis is rejected.

Throughout, we consider critical regions of the special form

$$C_\alpha = \{\boldsymbol{x} : T(\boldsymbol{x}) \geq G^{-1}(1 - \alpha)\} \tag{3.23}$$

of a significance level α, where T is a test statistic and G is a df. The usual significance levels are $\alpha = .05$ or $\alpha = .01$. Thus, the null hypothesis is rejected whenever $T(\boldsymbol{x})$ exceeds the critical value $G^{-1}(1 - \alpha)$.

Let $\boldsymbol{X} = (X_1, \ldots, X_k)$ be a vector of iid random variables with common df F_ϑ, where ϑ belongs to the null hypothesis. Usually, G is the exact or asymptotic df of the test statistic $T(\boldsymbol{X})$ for a certain parameter ϑ in the null hypothesis. We have

$$P\{T(\boldsymbol{X}) \geq G^{-1}(1 - \alpha)\} \leq \alpha. \tag{3.24}$$

The null hypothesis is rejected, although it is true with a probability bounded by α.

The p–Value

We introduce the p–value in conjunction with critical regions as given in (3.23). Let T be the test statistic and let G be the exact or approximate df of $T(\boldsymbol{X})$ under some appropriate parameter in the null hypothesis. Notice that

$$\{x : T(x) \geq G^{-1}(1 - \alpha)\} \quad = \quad \{x : G(T(x)) \geq 1 - \alpha\}$$
$$= \quad \{x : p(x) \leq \alpha\},$$

where
$$p(x) = 1 - G(T(x)). \tag{3.25}$$

Now the decision will be based on the p–value $p(x)$. If one has a significance level α in mind, then the null hypothesis is rejected whenever $p(x) \leq \alpha$. The p–value is also called the sample significance level, because $p(x)$ is the smallest significance level α such that the null hypothesis is rejected[6].

According to the probability transformation (cf. page 38), the random variable $p(X)$ is uniformly distributed on $[0, 1]$ if G is the distribution of $T(X)$ under the null hypothesis.

Evaluation of p–Values for the Normal Model

We mention three simple, well–known testing problems in the normal model and specify the p–values.

- (One–Sided Test of the Mean with Known Variance.) The testing of the null hypothesis $H_0 : \mu \leq \mu_0$ against the alternative $H_1 : \mu > \mu_0$, with $\sigma = \sigma_0$ known, is based on the critical region

$$C_\alpha = \left\{ x : \frac{\bar{x} - \mu_0}{\sigma_0/n^{1/2}} \geq \Phi^{-1}(1 - \alpha) \right\},$$

  where $\bar{x}$ is the sample mean again. Then, the p–value is

$$p(x) = 1 - \Phi\left(\frac{\bar{x} - \mu_0}{\sigma_0/n^{1/2}}\right).$$

- (One–Sided t–Test of the Mean with Unknown Variance.) When testing the same hypotheses as before with σ being unknown, let

$$C_\alpha = \left\{ x : \frac{\bar{x} - \mu_0}{s/n^{1/2}} \geq t_{n-1}^{-1}(1 - \alpha) \right\},$$

  where t_{n-1} is the Student df with $n - 1$ degrees of freedom, and s^2 is the sample variance again. The explicit form of the Student density may be found in (1.62) and (6.14). The p–value is

$$p(x) = 1 - t_{n-1}\left(\frac{\bar{x} - \mu_0}{s/n^{1/2}}\right).$$

---

[6]See, e.g., Rice, J.A. (1988). Mathematical Statistics and Data Analysis. Wadsworth & Brooks, Pacific Grove.

- (One–Sided Testing of the Mean with Asymptotic p–Value.) Because the Student distribution is asymptotically normal as $n \to \infty$, we know that t_{n-1} can be replaced by the standard normal df Φ for larger sample sizes n. The p–value is

$$p(\boldsymbol{x}) = 1 - \Phi\left(\frac{\bar{x} - \mu_0}{s/n^{1/2}}\right).$$

This p–value can be generally employed for testing the mean of a distribution under the conditions of the central limit theorem.

We started with a likelihood ratio (LR) test statistic for simple hypotheses and replaced the unknown standard deviation by the corresponding sample coefficient. The latter test statistic is based on the first and second sample moments.

Asymptotic p–Values for the Multinomial Model, χ^2 and Likelihood Ratio Statistics

We mention the χ^2 and LR–statistics for the multinomial model and specify the pertaining p–values. Let $X_1, \ldots, X_n$ be iid random variables and let $B_0, \ldots, B_m$ be a partition of the real line (or, generally, of the Euclidean d–space). Then, the joint distribution of the random numbers

$$N_i = \sum_{j \le n} I(X_j \in B_i), \qquad i = 0, \ldots, m,$$

is a multinomial distribution $M_{n,p}$ with parameter vector $\boldsymbol{p} = (p_0, \ldots, p_m)$, where $p_i = P\{X_1 \in B_i\}$. For $\boldsymbol{n} = (n_0, \ldots, n_m)$ with $\sum_{j=0}^{m} n_j = n$, we have

$$
\begin{aligned}
M_{n,p}(\{\boldsymbol{n}\}) &= P\{N_0 = n_0, \ldots, N_m = n_m\} \\
&= \frac{n!}{n_0! \cdots n_m!} \prod_{j=0}^{m} p_j^{n_j}.
\end{aligned}
$$

Notice that the parameter vector belongs to the space

$$K_m = \left\{ \boldsymbol{p} : p_j \ge 0, \sum_{j=0}^{m} p_j = 1 \right\}.$$

We are testing a composite null hypothesis K_0 which is a subspace of K_m.

- (The χ^2–Statistic.) The χ^2–statistic is

$$T_{\chi^2}(\boldsymbol{n}) = \sum_{j=0}^{m} \frac{(n_j/n - \hat{p}_{j,0}(\boldsymbol{n}))^2}{\hat{p}_{j,0}(\boldsymbol{n})/n}, \qquad (3.26)$$

where $\hat{\boldsymbol{p}}_0(\boldsymbol{n}) = (\hat{p}_{0,0}(\boldsymbol{n}), \ldots, \hat{p}_{m,0}(\boldsymbol{n}))$ is an MLE for K_0. Under the null hypothesis, the asymptotic df of the χ^2–statistic is a χ^2–df χ_k^2 with k degrees of freedom (see Section 4.3), where k is the difference of the dimension m of K_m and the dimension of the null hypothesis K_0. Therefore, the (asymptotic) p–value is

$$p_{\chi^2}(\boldsymbol{n}) = 1 - \chi_k^2(T_{\chi^2}(\boldsymbol{n})).$$

If the null hypothesis is simple, say, $K_0 = \{\boldsymbol{p}_0\}$, then $\hat{\boldsymbol{p}}_0 = \boldsymbol{p}_0$ and $k = m$. The χ^2–statistic in (3.26) is the Pearson χ^2–statistic.

- (The Likelihood Ratio (LR) Statistic.) The LR statistic for the preceding testing problem is

$$
\begin{aligned}
T_{\mathrm{LR}}(\boldsymbol{n}) &= 2\log \frac{\sup_{\boldsymbol{p} \in K_m} M_{n,\boldsymbol{p}}(\{\boldsymbol{n}\})}{\sup_{\boldsymbol{p} \in K_0} M_{n,\boldsymbol{p}}(\{\boldsymbol{n}\})} \\
&= 2n \sum_{j=0}^{m} \frac{n_j}{n} \log \frac{n_j}{n\hat{p}_{j,0}(\boldsymbol{n})},
\end{aligned}
\tag{3.27}
$$

where $\hat{\boldsymbol{p}}(\boldsymbol{n}) = (n_0/n, \dots, n_m/n)$ is the MLE for the full parameter space K_m and $\hat{\boldsymbol{p}}_0(\boldsymbol{n}) = (\hat{p}_{0,0}(\boldsymbol{n}), \dots, \hat{p}_{m,0}(\boldsymbol{n}))$ is the MLE again for K_0. Under the null hypothesis, the LR statistic has the same asymptotic df as the preceding χ^2–statistic. Therefore, the p–value is

$$p_{\mathrm{LR}}(\boldsymbol{n}) = 1 - \chi_k^2(T_{\mathrm{LR}}(\boldsymbol{n})).$$

Examples of χ^2 and LR–tests are given in the subsequent section, where a goodness–of–fit test is employed to the Poisson model.

3.4 Inference in Poisson and Mixed Poisson Models

In this section, estimating and testing within the Poisson model and the negative binomial model will be studied, whereby the latter model consists of mixed Poisson distributions. The Poisson model is relevant to extreme value analysis—as pointed out in Section 1.2—because the number of exceedances may be regarded as a Poisson random variable.

Estimation in the Poisson Model

Let $x_1, \dots, x_n$ be nonnegative integers governed by a Poisson distribution P_λ with unknown parameter $\lambda > 0$. Recall that λ is the mean and variance of P_λ.

The sample mean $\hat{\lambda}_n = \bar{x}$ is the natural estimate of the unknown parameter λ. Note that the mean $\hat{\lambda}_n$ is also the MLE for the Poisson model $\{P_\lambda : \lambda > 0\}$.

Statistics for such discrete data can be expressed in terms of the number $n(j)$ of data x_i equal to j. For example, we may write

$$\hat{\lambda}_n = \sum_j jn(j) \Big/ \Big(\sum_j n(j) \Big)$$

for the sample mean.

There are three different approaches for estimating a Poisson distribution:

- (Nonparametric Approach.) Take the sample histogram $p_n(j) = n(j)/n$ as an estimate of the Poisson histogram $P_\lambda\{j\}$ (see page 59);

- (Parametric Approach: Poisson Model) Compute the MLE $\hat{\lambda}_n$ in the Poisson model and display $P_{\hat{\lambda}_n}\{j\}$;

- (Parametric Approach: Normal Model.) Compute the MLE $(\hat{\mu}_n, \hat{\sigma}_n)$ in the normal model (or any related estimator) based on $x_1, \ldots, x_n$ and take the normal density $\varphi_{\hat{\mu}_n, \hat{\sigma}_n}$ as an estimate of the Poisson histogram $P_\lambda\{j\}$, where this procedure is merely accurate if λ is sufficiently large.

Another estimate of the parameter λ is the sample variance s_n^2, due to the fact that the mean and the variance of a Poisson distribution are equal. One obtains the representation

$$s_n^2 = \sum_j j^2 n(j) \Big/ \Big(\sum_j n(j) \Big) - \hat{\lambda}_n^2.$$

A strong deviation of the sample mean from the sample variance indicates that the Poisson assumption is violated.

Goodness–of–Fit Test for the Poisson Model

The null hypothesis—that the data are generated under a Poisson distribution P_λ with unknown parameter λ—is tested against any other distribution on the nonnegative integers. The likelihood ratio and $\hat{\chi}^2$–test in the multinomial model will be made applicable by grouping the data.

Under the null hypothesis, let $X_1, \ldots, X_n$ be iid random variables with common Poisson distribution P_λ where λ is unknown. Let $B_j = \{j\}$ for $j = 0, \ldots, m-1$ and $B_m = \{m, m+1, m+2, \ldots\}$. Thus, we are observing $\boldsymbol{n} = (n_0, \ldots, n_m)$, where n_j is the frequency of the observations x_i in the cell B_j.

The null hypothesis is

$$K_0 = \{(p_0(\lambda), \ldots, p_m(\lambda)) : \lambda > 0\},$$

where

$$p_j(\lambda) = P_\lambda\{j\} = \frac{\lambda^j}{j!} e^{-\lambda}, \qquad j = 0, \ldots, m-1,$$

and

$$p_m(\lambda) = 1 - \sum_{j=0}^{m-1} p_j(\lambda).$$

The null hypothesis K_0 is a parameter space of dimension 1 and, therefore, the limiting df for the likelihood ratio (LR) and χ^2–statistics is the χ^2–df with $k = m - 1$ degrees of freedom. To make (3.27) and (3.26) applicable one must compute the MLE

$$\boldsymbol{p}(\hat{\lambda}) = (p_0(\hat{\lambda}), \ldots, p_m(\hat{\lambda}))$$

for K_0. For that purpose, one must find the solution to the likelihood equation

$$\frac{\partial}{\partial \lambda} \log M_{n,\boldsymbol{p}(\lambda)}\{\boldsymbol{n}\} = 0$$

which is equivalent to the equation

$$\lambda = \frac{1}{n} \left(\sum_{j=0}^{m-1} jn_j + \frac{n_m}{p_m(\lambda)} \sum_{j=m}^{\infty} jP_\lambda\{j\} \right).$$

If $n_m = 0$, then $\hat{\lambda}(\boldsymbol{n})$ is the sample mean which is the MLE in the Poisson model. The likelihood ratio (LR) and the χ^2–statistics for the present problem are

$$T_{\mathrm{LR}}(\boldsymbol{n}) = 2n \sum_{j=0}^{m} \frac{n_j}{n} \log \frac{n_j}{np_j(\hat{\lambda}(\boldsymbol{n}))} \qquad (3.28)$$

and

$$\hat{\chi}^2(\boldsymbol{n}) = \sum_{j=0}^{m} \frac{(n_j - np_j(\hat{\lambda}(\boldsymbol{n})))^2}{np_j(\hat{\lambda}(\boldsymbol{n}))}. \qquad (3.29)$$

The pertaining p–values for testing the Poisson hypothesis are

$$p_{\mathrm{LR}}(\boldsymbol{n}) = 1 - \chi^2_{m-1}(T_{\mathrm{LR}}(\boldsymbol{n}))$$

and

$$p_{\chi^2}(\boldsymbol{n}) = 1 - \chi^2_{m-1}(\hat{\chi}^2(\boldsymbol{n})).$$

Mixtures of Poisson Distributions, Negative Binomial Distributions

Remember that the mean and variance of a Poisson distribution are equal. The Poisson modeling is untenable if the sample mean and sample variance deviate significantly from each other. Therefore, we also deal with mixed Poisson distributions Q given by

$$Q\{k\} = \int_0^\infty P_\lambda\{k\}f(\lambda)\,d\lambda, \qquad k = 0, 1, 2, \ldots, \qquad (3.30)$$

where f is a mixing density. Subsequently, the mixing will be done with respect to a gamma density.

When the mixing is carried out with respect to a gamma density (see (3.42)) with shape parameter $r > 0$ and reciprocal scale parameter d, one finds the negative binomial distribution

$$B^-_{r,p}\{k\} = \frac{\Gamma(r+k)}{\Gamma(r)\Gamma(k+1)}p^r(1-p)^k, \qquad k = 0, 1, 2, \ldots, \qquad (3.31)$$

with parameters $r > 0$ and $p = d/(1+d)$. Notice that $B^-_{1,p}$ is a geometric distribution (also see page 13).

The mean and the variance of the negative binomial distribution $B^-_{r,p}$ are $r(1-p)/p$ and $r(1-p)/p^2$. We see that the variance is larger than the mean (a property which is shared by all mixed Poisson distributions that differ from a Poisson distribution).

If $r(1 - p(r)) \to \lambda$ as $r \to \infty$, then the mean and variance of the negative binomial distribution tend to λ. Moreover, one ends up with a Poisson distribution with parameter λ in the limit. We have

$$B^-_{r,p(r)}\{k\} \to P_\lambda\{k\}, \qquad k = 0, 1, 2, \ldots .$$

In similarity to the Poisson approximation of binomial distributions, a much stronger inequality holds. We have

$$|B^-_{r,p}(A) - P_{r(1-p)/p}(A)| \leq \frac{1-p}{\sqrt{2p}} \qquad (3.32)$$

for each set A of nonnegative integers[7]. Note that the negative binomial and the approximating Poisson distribution in (3.32) have identical mean values. Thus, a modeling by means of a Poisson as well as a negative binomial distribution is justified, if p is sufficiently close to 1.

Estimation in the Negative Binomial Model

From the mean and variance of negative binomial distributions, one may easily deduce moment estimators for that model. We also mention MLEs.

- Moment Estimator: Given a sample of nonnegative integers x_i, the moment estimates $\hat{r}_n$ and $\hat{p}_n$ of the parameters r and p are the solutions to the equations

$$r(1-p)/p = \bar{x} \quad \text{and} \quad r(1-p)/p^2 = s^2,$$

  where $\bar{x}$ and s^2 again denote the sample mean and sample variance. We have

$$\hat{p}_n = \bar{x}/s^2 \quad \text{and} \quad \hat{r}_n = \bar{x}^2/(s^2 - \bar{x}).$$

---

[7]Matsunawa, T. (1982). Some strong ϵ–equivalence of random variables. Ann. Inst. Math. Statist. 34, 209–224.

If $s^2 \leq \bar{x}$, take some value for $\hat{p}_n$ close to 1 and $\hat{r}_n = \bar{x}\hat{p}_n/(1 - \hat{p}_n)$; another plausible choice would be a Poisson distribution with parameter $\bar{x}$.

• Maximum Likelihood Estimator: One must compute a solution to the likelihood equations

$$\frac{\partial}{\partial r} \log L(r, p) = 0 \quad \text{and} \quad \frac{\partial}{\partial p} \log L(r, p) = 0, \tag{3.33}$$

where

$$L(r, p) = \prod_{i \leq n} B_{r,p}^-\{x_i\}$$

denotes the likelihood function. Deduce from the 2nd equation in (3.33) that

$$p = r/(r + \bar{x}).$$

By adopting the formula $\Gamma(r + 1) = r\Gamma(r)$ and inserting the value for p in the 1st equation (3.33), one obtains

$$\frac{1}{n} \sum_{i \leq n} \sum_{j \leq x_i} \frac{1}{r + j - 1} = \log(1 + \bar{x}/r) \tag{3.34}$$

with $\sum_{j \leq 0} = 0$. This equation must be solved numerically, whereby the moment estimate for r may serve as an initial value of an iteration procedure. For that purpose, it is advisable to write the left–hand side of (3.34) as

$$\sum_{j \leq \max\{x_i\}} \left(\sum_{i \leq n} I(x_i \geq j)\right) \Big/ (r + j - 1).$$

In the subsequent example, Poisson and negative binomial distributions are fitted to a data set of numbers of car accidents.

EXAMPLE 3.4.1. (Number of Car Accidents.) We deal with the classical problem of modeling the number of car accidents caused by a single driver within a given period. The present data set records the number of accidents over the period from Nov. 1959 to Feb. 1968 for 7842 drivers in the state of California (stored in the file id–cars2.dat).

TABLE 3.2. Frequencies $n(j)$ of number of accidents equal to j.

j	0	1	2	3	4	5	6	7	8	9	10	11
$n(j)$	5147	1859	595	167	54	14	5	0	0	0	0	1

Since the sample mean and sample variance are 0.49 and 0.68, one hesitates to fit a Poisson distribution to the data. This critical attitude is confirmed by plots of sample and

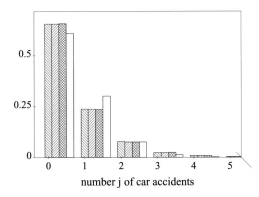

number j of car accidents

FIG. 3.2. From left to right: Sample histogram; MLE and Moment histograms in negative binomial model; MLE histogram in Poisson model.

Poisson histograms (cf. Figure 3.2) and a χ^2–goodness–of–fit test. The usual suggestion is to fit a negative binomial distribution to such numbers.

The sample histogram and the negative binomial histograms pertaining to the MLE ($r = 1.34$, $p = 0.73$) and Moment estimates ($r = 1.31$, $p = 0.73$) cannot be distinguished visually (as far as the height of the bars is concerned); there is a remarkable deviation of the Poisson histogram from the other histograms.

The usual explanation for the failure of fitting a Poisson distribution to data—such as those in Example 3.4.1—is that the number of accidents of a single driver may fit to a Poisson distribution P_λ, yet the parameter λ varies from one driver to another. Therefore, one deals with the following two–step stochastic experiment:

1st Step.   A driver with an individual accident characteristic λ is randomly drawn according to a certain mixing distribution.

2nd Step.   The final outcome (namely the number of claims for this driver) is generated under the Poisson distribution P_λ.

For the parametric modeling, one may take the mixture of Poisson distributions P_λ with respect to a gamma distribution with shape and reciprocal scale parameters r and d, which is a negative binomial distribution with parameters r and $p = d/(1 + d)$, cf. (3.31).

The data in Example 3.4.1 can be regarded as the outcome of 7842 independent repetitions of such a two–step stochastic experiment.

Bayesian inference in Poisson models will be dealt with at the end of the subsequent section.

3.5   The Bayesian Estimation Principle

Bayesian estimation is another likelihood–based method besides the maximum likelihood (ML) method. In addition to the statistical model, the statistician must specify a prior distribution.

In this section, we introduce Bayes estimators within a decision–theoretical framework. We estimate a real–valued functional parameter $T(\boldsymbol{\vartheta})$, where

$$\boldsymbol{\vartheta} = (\vartheta_1, \dots, \vartheta_m)$$

is a parameter vector. This includes, e.g., the estimation

- of the jth component ϑ_j of $\boldsymbol{\vartheta}$ if $T(\boldsymbol{\vartheta}) = \vartheta_j$,

- of the mean $T(\boldsymbol{\vartheta}) = \int x \, dF_{\boldsymbol{\vartheta}}(x)$ of the underlying df $F_{\boldsymbol{\vartheta}}$ represented by $\boldsymbol{\vartheta}$.

The latter problem will be studied in Section 17.6 in conjunction with estimating the net premium of an individual risk.

The Prior Density, Minimizing the Bayes Risk

Let $\widehat{T}(X)$ be an estimator of the functional parameter $T(\boldsymbol{\vartheta})$, where X is a random variable having a distribution represented by $\boldsymbol{\vartheta}$. For example, $X = (X_1, \dots, X_n)$ is a sample of size n, $T(\boldsymbol{\vartheta})$ is the mean of the common distribution and $\widehat{T}(X)$ is the sample mean $\overline{X}$. The performance of $\widehat{T}(X)$ as an estimator of $T(\boldsymbol{\vartheta})$ can be measured by the mean squared error (MSE)

$$E\left(\left(\widehat{T}(X) - T(\boldsymbol{\vartheta})\right)^2 \big| \boldsymbol{\vartheta}\right) := E\left(\widehat{T}(X) - T(\boldsymbol{\vartheta})\right)^2,$$

cf. also page (3.11), where the left–hand side emphasizes the fact that the expectation is taken under the parameter $\boldsymbol{\vartheta}$.

The performance of the estimator can be made independent of a special parameter vector $\boldsymbol{\vartheta}$ by means of a "prior" probability density $p(\boldsymbol{\vartheta})$ which may be regarded as a weight function. Some prior knowledge about the parameter of interest is included in the statistical modeling by means of the prior density $p(\boldsymbol{\vartheta})$.

The Bayes risk of the estimator $\widehat{T}$ with respect to the prior $p(\boldsymbol{\vartheta})$ is the integrated MSE

$$R(p, \widehat{T}) = \int E\left(\left(\widehat{T}(X) - T(\boldsymbol{\vartheta})\right)^2 \big| \boldsymbol{\vartheta}\right) p(\boldsymbol{\vartheta}) \, d\vartheta_1 \cdots d\vartheta_m. \qquad (3.35)$$

An estimator $\widehat{T}$, which minimizes the Bayes risk, is called Bayes estimator.

Computing the Bayes Estimator, the Posterior Density

We introduce the posterior density, which is determined by the prior density $p(\vartheta)$ and the likelihood function, and deduce an explicit representation of the Bayes estimator by means of the posterior density.

Let $\boldsymbol{X} = (X_1, \ldots, X_n)$ be a vector of iid random variables with common df $F_{\boldsymbol{\vartheta}}$ and density $f_{\boldsymbol{\vartheta}}$, where again $\boldsymbol{\vartheta} = (\vartheta_1, \ldots, \vartheta_m)$ is the parameter vector. As in the discussion about the MLE on page 84, let

$$L(\boldsymbol{x}|\boldsymbol{\vartheta}) = \prod_{i \leq n} f_{\boldsymbol{\vartheta}}(x_i) \tag{3.36}$$

be the likelihood function given the sample vector $\boldsymbol{x} = (x_1, \ldots, x_n)$. Using the likelihood function, one gets the representation

$$E\left((\widehat{T}(X) - T(\boldsymbol{\vartheta}))^2 | \boldsymbol{\vartheta}\right) = \int (\widehat{T}(\boldsymbol{x}) - T(\boldsymbol{\vartheta}))^2 L(\boldsymbol{x}|\boldsymbol{\vartheta}) \, dx_1 \cdots dx_n \tag{3.37}$$

of the MSE.

We verify that the Bayes estimate can be written

$$T^*(\boldsymbol{x}) = \int T(\boldsymbol{\vartheta}) p(\boldsymbol{\vartheta}|\boldsymbol{x}) \, d\vartheta_1 \cdots d\vartheta_m, \tag{3.38}$$

where the function

$$p(\boldsymbol{\vartheta}|\boldsymbol{x}) = \frac{L(\boldsymbol{x}|\boldsymbol{\vartheta}) p(\boldsymbol{\vartheta})}{\int L(\boldsymbol{x}|\boldsymbol{\vartheta}) p(\boldsymbol{\vartheta}) \, d\vartheta_1 \cdots d\vartheta_m} \tag{3.39}$$

is the "posterior" density for a given sample vector $\boldsymbol{x}$, whenever the denominator is larger than zero. If ϑ is a one-dimensional parameter and $T(\vartheta) = \vartheta$, then the Bayes estimate is the mean $\int \vartheta p(\vartheta|\boldsymbol{x}) \, d\vartheta$ of the posterior distribution according to (3.38).

Thus, by means of the prior density $p(\boldsymbol{\vartheta})$ and the likelihood function $L(\boldsymbol{x}|\boldsymbol{\vartheta})$ one gets the posterior density $p(\boldsymbol{\vartheta}|\boldsymbol{x})$. Notice that the posterior density in (3.39) and the Bayes estimate can be computed whenever $L(\boldsymbol{x}|\boldsymbol{\vartheta}) p(\boldsymbol{\vartheta})$, as a function in $\boldsymbol{\vartheta}$, is known up to a constant. Writing $g(\boldsymbol{\vartheta}) \propto f(\boldsymbol{\vartheta})$, when functions g and f are proportional, we have

$$p(\boldsymbol{\vartheta}|\boldsymbol{x}) \propto L(\boldsymbol{x}|\boldsymbol{\vartheta}) p(\boldsymbol{\vartheta}). \tag{3.40}$$

To prove the representation (3.38) of the Bayes estimate, combine (3.35) and (3.37) and interchange the order of the integration. The Bayes risk with respect to the prior $p(\boldsymbol{\vartheta})$ can be written

$$
\begin{aligned}
R(p, \widehat{T}) &= \int (\widehat{T}(\boldsymbol{x}) - T(\boldsymbol{\vartheta}))^2 L(\boldsymbol{x}|\boldsymbol{\vartheta}) p(\boldsymbol{\vartheta}) \, dx_1 \cdots dx_n \, d\vartheta_1 \cdots d\vartheta_m \tag{3.41} \\
&= \int \left(\int (\widehat{T}(\boldsymbol{x}) - T(\boldsymbol{\vartheta}))^2 p(\boldsymbol{\vartheta}|\boldsymbol{x}) \, d\vartheta_1 \cdots d\vartheta_m \right) f(\boldsymbol{x}) \, dx_1 \cdots dx_n,
\end{aligned}
$$

where $f(\boldsymbol{x}) = \int L(\boldsymbol{x}|\boldsymbol{\vartheta})p(\boldsymbol{\vartheta})\,d\vartheta_1 \cdots d\vartheta_m$. To get the Bayes estimate, compute for every $\boldsymbol{x}$ the value z which minimizes the integral $\int \left(z - T(\boldsymbol{\vartheta})\right)^2 p(\boldsymbol{\vartheta}|\boldsymbol{x})\,d\vartheta_1 \cdots d\vartheta_m$. This value z can be computed by showing that the derivative

$$
\begin{aligned}
\frac{\partial}{\partial z} \int \left(z - T(\boldsymbol{\vartheta})\right)^2 & p(\boldsymbol{\vartheta}|\boldsymbol{x})\,d\vartheta_1 \cdots d\vartheta_m \\
&= \int 2\big(z - T(\boldsymbol{\vartheta})\big)p(\boldsymbol{\vartheta}|\boldsymbol{x})\,d\vartheta_1 \cdots d\vartheta_m \\
&= 2\big(z - T^*(\boldsymbol{x})\big)
\end{aligned}
$$

is equal to zero, if z is equal to the value $T^*(\boldsymbol{x})$ in (3.38).

In those cases where the posterior density $p(\boldsymbol{\vartheta}|\boldsymbol{x})$ is of the same type as the prior density $p(\boldsymbol{\vartheta})$ one speaks of a conjugate prior. An example of a conjugate prior will be given for the exponential model in (3.44).

For a different view toward the Bayesian principle we refer to Section 8.4. The results of Section 3.5 are revisited within a Poisson process setting in Sections 9.3 and 9.5. The Sections 14.5 and 17.6 concern applications of Bayesian estimators in regional flood frequency analysis and reinsurance business.

The Gamma and Reciprocal Gamma Distributions

Our prior densities will be usually specified by means of gamma densities or modifications of gamma densities. The gamma density with shape parameter $s > 0$ and reciprocal scale parameter $d > 0$ is given by

$$
h_{s,d}(\vartheta) = \frac{d}{\Gamma(s)}(d\vartheta)^{s-1}\exp(-d\vartheta), \qquad \vartheta > 0. \tag{3.42}
$$

It is clear from the integral representation of the gamma function in (1.32) that $h_{s,d}$ is a probability density. The gamma distribution in (3.42) has the mean s/d and variance s/d^2. It is the type III distribution in the Pearson system. We also remark that $h_{1,d}$ is an exponential density.

We also introduce the reciprocal gamma distribution. If a random variable X has the gamma density $h_{a,b}$, then $1/X$ has an reciprocal gamma density

$$
\widetilde{h}_{a,b}(x) = \frac{1}{b\Gamma(a)}(x/b)^{-(1+a)}\exp(-b/x), \qquad x > 0. \tag{3.43}
$$

A prominent example of a reciprocal gamma density is the sum–stable Lévy density with index $a = 1/2$, also see (6.18). The mean, variance and coefficient of variation are $b/(a-1)$, $b^2/\left((a-1)^2(a-2)\right)$ and $(a-2)^{-1/2}$, if $a > 1$ and $a > 2$, respectively. The mode is equal to $b/(a+1)$.

A First Example: Bayesian Estimators in the Exponential Model

We compute a Bayes estimator for the model of exponential densities

$$f_\vartheta(x) = \vartheta \exp(-\vartheta x), \qquad x > 0,$$

where $\vartheta > 0$ is the unknown reciprocal scale parameter. The gamma density

$$p(\vartheta) = h_{s,d}(\vartheta), \tag{3.44}$$

with parameters $s, d > 0$, is taken as a prior.

Writing the likelihood function as $L(\boldsymbol{x}|\vartheta) = \vartheta^n \exp\left(-\vartheta \sum_{i \leq n} x_i\right)$ and representing the posterior density by

$$p(\vartheta|\boldsymbol{x}) \propto L(\boldsymbol{x}|\vartheta) h_{s,d}(\vartheta) \propto h_{s',d'}(\vartheta),$$

with $s' = s + n$ and $d' = d + n\bar{x}$ ($\bar{x}$ denoting the sample mean), one obtains the representation

$$p(\vartheta|\boldsymbol{x}) = h_{s',d'}(\vartheta) \tag{3.45}$$

of the posterior density which is again a gamma density. We see that gamma densities are conjugate priors for the exponential model.

According to (3.38), the Bayes estimator ϑ_n^* of ϑ is the mean of the posterior distribution. We have

$$\vartheta_n^*(\boldsymbol{x}) = \int \vartheta p(\vartheta|\boldsymbol{x}) \, d\vartheta = \frac{s+n}{d+n\bar{x}}. \tag{3.46}$$

Notice that the Bayes estimates $\vartheta_n^*(\boldsymbol{x})$ approach the reciprocal sample mean $1/\bar{x}$, which is the MLE in the exponential model, with a rate of order $O(1/n)$. The estimates can be written as the linear combination $\alpha s/d + (1 - \alpha)/\bar{x}$ with $\alpha = 1/(1 + n\bar{x}/d)$.

Bayesian estimation in Pareto models and Poisson process models with Pareto marks is dealt with on the pages 129–133 and 254–245.

Bayesian Estimation in the Poisson Model

We also deal with the Bayes estimation in discrete models. For that purpose, the likelihood function for continuous data, as utilized on page 102, must be replaced by the likelihood function for discrete data. This will be illustrated for the Poisson model with repect to two different parameterizations.

- Bayes Estimator: we compute an explicit representation of the Bayes estimator within the Poisson model. The likelihood function for a sample vector $\boldsymbol{k} = (k_1, \ldots, k_n)$ of size n within this discrete model is given by $L(\boldsymbol{k}|\lambda) = \prod_{i \leq n} P_\lambda\{k_i\}$. The MSE of an estimator $\hat{\lambda}_n$ of λ can be written

$$E\big((\hat{\lambda}_n - \lambda)^2|\lambda\big) = \sum(\hat{\lambda}_n(\boldsymbol{k}) - \lambda)^2 L(\boldsymbol{k}|\lambda),$$

where the summation runs over all vectors $\boldsymbol{k} = (k_1, \ldots, k_n)$ of nonnegative integers. In the subsequent lines, we just follow the arguments concerning the Bayes risk and Bayes estimators as outlined at the beginning of this section (or apply the general arguments in Section 8.4). The gamma density

$$p(\lambda) = h_{r,c}(\lambda), \tag{3.47}$$

with parameters r and c, cf. (3.42), is chosen as a prior. Verify that the posterior density $p(\lambda|\boldsymbol{k})$ satisfies $p(\lambda|\boldsymbol{k}) \propto L(\boldsymbol{k}|\lambda)h_{r,c}(\lambda) \propto h_{r',c'}(\lambda)$ with $r' = r + \sum_{i \leq n} k_i$ and $c' = c + n$. Therefore, the posterior density

$$p(\lambda|\boldsymbol{k}) = h_{r',c'}(\lambda) \tag{3.48}$$

is also a gamma density. We see that gamma densities are conjugate priors for the Poisson model.

Because the Bayes estimate of λ is the mean of the posterior density, we have

$$\lambda_n^*(\boldsymbol{k}) = \int \lambda p(\lambda|\boldsymbol{k}) \, d\lambda = \frac{r + n\bar{k}}{c + n}, \tag{3.49}$$

where $\bar{k}$ is the sample mean.

- Bayes Estimator in a Modified Poisson Model: for later purposes (cf. Section 9.3), we shortly mention the model of reparameterized Poisson distributions $P_{\lambda T}$ for parameters $\lambda > 0$, where $T > 0$ is some fixed value. Moreover, let $n = 1$. The likelihood function is

$$L(k|\lambda) = \left((\lambda T)^k/k!\right)\exp(-\lambda T) \tag{3.50}$$

for every observed nonnegative integer k. Choosing again a gamma density

$$p_0(\lambda) = h_{r,c}(\lambda) \tag{3.51}$$

as a prior, one obtains the posterior density

$$p_0(\lambda|k) = h_{r',c'}(\lambda), \tag{3.52}$$

which is the gamma density with parameters $r' = r + k$ and $c' = c + T$. Therefore, the Bayes estimate is

$$\lambda_k^* = \int \lambda p_0(\lambda|k) \, d\lambda = \frac{r + k}{c + T} \tag{3.53}$$

for all nonnegative integers k.

Chapter 4

Extreme Value Models

This chapter is devoted to statistical procedures in parametric extreme value (EV) models which are especially designed for maxima. It is worth recalling that minima can be dealt with by changing the sign of the data.

In Section 4.1, we deal with estimators within the Gumbel (EV0), Fréchet (EV1), Weibull (EV2) and unified extreme value (EV) models. Q–Q plots based on estimators of the shape parameter will be employed to check the validity of such models. Tests are addressed in Section 4.2. Extensions of extreme value models and certain alternative models for the modeling of maxima are briefly discussed in Section 4.3.

4.1   Estimation in Extremes Value Models

Consider data $x_1, \ldots, x_n$ generated under a df F^m as in Section 1.2. Thus, each x_i is the maximum of m values that are governed by the df F. Think back to Section 2.1, where we found that the df F^m and the pertaining density or qf can be estimated by the corresponding sample versions, namely the sample df $\widehat{F}_n$, the histogram or kernel density f_n or the sample qf $\widehat{F}_n^{-1}$.

Remember that some extreme value (EV) df G can be fitted to the df F^m of the maximum if the arguments in Section 1.2 are valid. This includes that, for a while, the $x_1, \ldots, x_n$ may be regarded as observations governed by G. Therefore, statistical procedures within a parametric model can be applied.

The Gumbel (EV0) Model

This is the traditional model in extreme value analysis which had about the same status as the normal model in other applications (with just the same advantages and negative consequences). The major advantage of the Gumbel model is that the distribution can be specified by location and scale parameters as in the Gaussian case. Thus, the Gumbel model is an ordinary location and scale parameter model.

One of our declared aims is to overcome the potential dislike for an additional parameter—besides the location and scale parameters—and to convince the reader that an extension of the Gumbel (EV0) model to the EV model can be worthwhile.

Recall that the standard Gumbel df is given by

$$G_0(x) = \exp\left(-e^{-x}\right).$$

By adding a location and scale parameters μ and σ, one gets the Gumbel (EV0) model

EV0: $\{G_{0,\mu,\sigma} : \mu \text{ real}, \ \sigma > 0\}.$

We mention the following estimators for μ and σ.

- MLE(EV0): the MLEs μ_n and σ_n of the location and scale parameters must be evaluated numerically. First, compute σ_n as the solution to the equation

$$\sigma - n^{-1}\sum_{i\leq n} x_i + \left(\sum_{i\leq n} x_i \exp(-x_i/\sigma)\right)\Big/\left(\sum_{i\leq n}\exp(-x_i/\sigma)\right) = 0. \qquad (4.1)$$

  The least squares estimate for σ can be taken as an initial value of the iteration procedure. Then,

$$\mu_n = -\sigma_n \log\left(n^{-1}\sum_{i\leq n}\exp(-x_i/\sigma_n)\right). \qquad (4.2)$$

- Moment(EV0): estimators of μ and σ are deduced from the sample mean and variance. From (1.34) and (1.35), conclude that $\mu + \sigma\lambda$ and $\sigma^2\pi^2/6$ are the mean and the variance of $G_{0,\mu,\sigma}$, where λ again denotes Euler's constant. Therefore,

$$\sigma_n = 6^{1/2} s_n/\pi$$

  and

$$\mu_n = \bar{x} - \sigma_n\lambda$$

  are the moment estimates of the location and scale parameters μ and σ in the Gumbel model, where $\bar{x}$ and s_n^2 are the sample mean and variance again.

If G_{0,μ_n,σ_n} strongly deviates from $\widehat{F}_n$, then the modeling of F^m by Gumbel dfs $G_{0,\mu,\sigma}$ is critical (or the estimates are inaccurate)! Likewise, one may compare the estimated density g_{0,μ_n,σ_n} with the kernel density $f_{n,b}$ for some choice of a bandwidth b.

EXAMPLE 4.1.1. (Annual Floods of the Feather River.) We partially repeat the analysis of Benjamin and Cornell[1] and Pericchi and Rodriguez–Iturbe[2] about the annual floods of the Feather River in Oroville, California, from 1902 to 1960. The following flood data are stored in the file ht–feath.dat.

TABLE 4.1. Annual maximum discharges of Feather River in ft^3/sec from 1902 to 1960.

year	flood	year	flood	year	flood	year	flood	year	flood
1907	230,000	1943	108,000	1911	75,400	1916	42,400	1934	20,300
1956	203,000	1958	102,000	1919	65,900	1924	42,400	1937	19,200
1928	185,000	1903	102,000	1925	64,300	1902	42,000	1913	16,800
1938	185,000	1927	94,000	1921	62,300	1948	36,700	1949	16,800
1940	152,000	1951	92,100	1945	60,100	1922	36,400	1912	16,400
1909	140,000	1936	85,400	1952	59,200	1959	34,500	1908	16,300
1960	135,000	1941	84,200	1935	58,600	1910	31,000	1929	14,000
1906	128,000	1957	83,100	1926	55,700	1918	28,200	1955	13,000
1914	122,000	1915	81,400	1954	54,800	1944	24,900	1931	11,600
1904	118,000	1905	81,000	1946	54,400	1920	23,400	1933	8,860
1953	113,000	1917	80,400	1950	46,400	1932	22,600	1939	8,080
1942	110,000	1930	80,100	1947	45,600	1923	22,400		

By applying the MLE(EV0), one obtains $\mu_n = 47,309$ and $\sigma_n = 37,309$. The pertaining Gumbel distribution can be well fitted to the upper part of the flood data (according to the Q–Q plot, etc.). Therefore, this parametric approach may be employed to estimate T–year flood levels (cf. (1.19)): the 50 and 100–year flood levels are $u(50) = 192,887$ and $u(100) = 218,937$ ft^3/sec. Notice that the estimated 100–year flood level was exceeded in the year 1907. The estimated Gumbel distribution has a remarkable 2.8% of its weight on the negative half–line. Yet, this does not heavily influence the estimation of parameters of the upper tail like T–year flood levels (also see Example 5.2.1).

Pericchi and Rodriguez–Iturbe also select gamma (Pearson type III), log–gamma (log–Pearson type III) and log–normal models (cf. (4.6), (4.12) and (1.60)), yet there seems to be no clear conclusion which of these models is most preferable.

---

[1] Benjamin, J.R. and Cornell, C.A. (1970). Probability, Statistics and Decisions for Civil Engineers. McGraw–Hill, New York.

[2] Pericchi, L.R. and Rodriguez–Iturbe, I. (1985). On the statistical analysis of floods. In: A Celebration of Statistics. The ISI Centenary Volume, A.C. Atkinson and S.E. Fienberg (eds.), 511–541.

Moreover, these authors suggest to use excess dfs and hazard functions in flood studies. In this context, these functions are called residual flood df and flood rate.

The following Fréchet and Weibull models can be deduced from the Gumbel model by means of the transformations $T(x) = \log(x)$ and $T(x) = -1/x$ (cf. page 37). Thus, results and procedures (such as the MLE) for the Gumbel model can be made applicable to these models (and vice versa). Finally, we deal with estimators in the unified extreme value (EV) model.

The Fréchet (EV1) Model

We deal with another submodel of extreme value dfs. Recollect that the standard Fréchet df with shape parameter $\alpha > 0$ is given by $G_{1,\alpha}(x) = \exp(-x^{-\alpha})$ for $x > 0$. By adding a scale parameter σ, one gets the EV1 model

$$\text{EV1:} \qquad \{G_{1,\alpha,0,\sigma} : \alpha, \sigma > 0\}.$$

Keep in mind that the left endpoint of $G_{1,\alpha,0,\sigma}$ is always equal to zero. If all potential measurements are positive—e.g., due to physical reasons—and, in addition, a left endpoint equal to zero is acceptable, then it can be worthwhile to check the validity of the EV1 model.

When applying the maximum likelihood principle for this model, one should make sure that the left endpoint of the actual df is larger than zero (at least this should hold true for the data).

The Weibull (EV2) Model

The standard Weibull df with shape parameter $\alpha < 0$ is given by $G_{2,\alpha}(x) = \exp(-(-x)^{-\alpha})$ for $x \leq 0$. Recall from Section 1.3 that our parameterization for Weibull dfs differs from the standard one used in the statistical literature, where a positive shape parameter is taken.

In the following model, a scale parameter is included:

$$\text{EV2:} \qquad \{G_{2,\alpha,0,\sigma} : \alpha < 0, \sigma > 0\}.$$

Notice that the upper endpoint of each of the dfs is equal to zero.

For many data which may be regarded as minima, as, e.g., data concerning the strength of material, a converse Weibull df $\widetilde{G}_{2,\alpha,0,\sigma}(x) = 1 - G_{2,\alpha,0,\sigma}(-x)$ is a plausible candidate of a distribution, also see (1.38) and (1.39).

The Unified Extreme Value Model

The unified EV model—in the γ–parameterization—is the most important one for modeling maxima. By including scale and location parameters σ and μ, we obtain the model

$$\text{EV:} \qquad \{G_{\gamma,\mu,\sigma} : \mu, \gamma \text{ real, } \sigma > 0\},$$

where G_γ are the standard versions as defined on page 16.

Occasionally, the extreme value (EV) model is called the generalized extreme value (abbreviated GEV) model in statistical references. The latter notion can be a bit confusing because this model only consists of extreme value distributions.

Here is a list of estimation procedures.

- MLE(EV): The MLE in the EV model must be numerically evaluated as a solution to the likelihood equations.

  The MLE determines a local maximum of the likelihood function if the iterated values remain in the region $\gamma > -1$. If γ varies in the region below -1, then neither a global nor a local maximum of the likelihood function exists, see R.L. Smith[3].

- MDE(EV): Let d be a distance on the family of dfs. Then, $(\gamma_n, \mu_n, \sigma_n)$ is an MDE (minimum distance estimator), if

$$\mathrm{d}(\widehat{F}_n, G_{\gamma_n, \mu_n, \sigma_n}) = \inf_{\gamma, \mu, \sigma} \ \mathrm{d}(\widehat{F}_n, G_{\gamma, \mu, \sigma}).$$

  Likewise, an MDE may be based on distances between densities. This estimation principle is related to the visualization technique as dealt with in preceding sections.

- LRSE(EV): This is a class of estimators that are linear combinations of ratios of spacings (RS's)

$$\hat{r} = \frac{x_{[nq_2]:n} - x_{[nq_1]:n}}{x_{[nq_1]:n} - y_{[nq_0]:n}},$$

  where $q_0 < q_1 < q_2$. Note that this statistic is independent of the location and scale parameters in distribution (in other words, $\hat{r}$ is invariant under affine transformations of the data). As a consequence of (2.38),

$$\hat{r} \approx \frac{G_\gamma^{-1}(q_2) - G_\gamma^{-1}(q_1)}{G_\gamma^{-1}(q_1) - G_\gamma^{-1}(q_0)} = \left(\frac{-\log q_2}{-\log q_0} \right)^{-\gamma/2},$$

  if q_0, q_1, q_2 satisfy the equation $(-\log q_1)^2 = (-\log q_2)(-\log q_0)$. In this manner, one obtains the estimate $\gamma_n = 2\log(\hat{r})/\log(\log(q_0)/\log(q_1))$ of γ. Such an estimate was suggested in [31]—attributed to S.D. Dubey—for estimating the shape parameter of Weibull distributions.

  If $q_0 = q$, $q_1 = q^a$, $q_2 = q^{a^2}$ for some $0 < q, a < 1$, then

$$\gamma_n = \log(\hat{r})/\log(1/a). \tag{4.3}$$

---

[3]Smith, R.L. (1985). Maximum likelihood estimation in a class of nonregular cases. Biometrika 72, 67–90.

The efficiency can be improved by taking a linear combination of such RSE's where each of the single terms is defined for $q = i/n$.

The least squares method may be applied to obtain the additionally required estimators of the location and scale parameters μ and σ.

The L–moment estimators within EV models will be dealt with in Section 14.5 in the context of flood frequency analysis.

Simulating the Mean Squared Error (MSE)

Given an estimator γ_n and a df H, generate a set of estimates, say, $\gamma_{n,1}, \ldots, \gamma_{n,N}$, where N is the selected number of simulations. Our standard number of simulations is $N = 4000$.

We simulate the MSE of estimators based on Gumbel data, maxima of $m = 30$ exponential data, and maxima of $m = 30, 100$ standard normal data (indicated by the dfs G_0, W^{30} and Φ^m) for sample sizes $n = 20, 100, 500$.

TABLE 4.2. MSE of estimators γ_n of the shape parameter $\gamma = 0$ for block sizes $m = 30, 100$.

	MSE					
	G_0			W_0^{30}		
sample size n	20	100	500	20	100	500
MLE(EV)	0.057	0.006	0.001	0.058	0.006	0.001
LRSE(EV)	0.165	0.022	0.004	0.167	0.023	0.004

	Φ^{30}			Φ^{100}		
sample size n	20	100	500	20	100	500
MLE(EV)	0.086	0.024	0.017	0.081	0.019	0.012
LRSE(EV)	0.156	0.041	0.023	0.163	0.034	0.016

For the sample size $n = 500$, the MSE is close to the squared bias. Based on the $\gamma_{n,i}$ one may also simulate the df or density of the estimator.

EV Q–Q Plots

Recall from page 62 that the unknown shape parameter γ must be replaced by an estimator γ_n when a Q–Q plot is employed in the EV model. If there is a stronger deviation of the EV Q–Q plot from a straight line, then either the estimator of the shape parameter is inaccurate (perhaps try another one), or the EV model is untenable.

EXAMPLE 4.1.2. (De Bilt September Data.) The de Bilt (September) data consist of 133 monthly maxima of temperature in de Bilt measured during the years 1849–1981. Thus, the background of the data suggests to use the EV model. These data are stored in the file em–dbilt.dat jointly with the maxima of other months and the annual maxima.

The EV Q–Q plot (cf. Fig. 4.1 on the left–hand side) seems to confirm the EV assumption. The maximum temperature within the whole period, measured in the year 1949, is equal to 34.2 degrees Celsius. The values of the MLE in the EV model are $\gamma = -0.19$, $\mu = 24.39$ and $\sigma = 2.76$; the right endpoint of the estimated Weibull distribution is equal to 38.76. In Fig. 4.1, on the right–hand side, the kernel density (with Epanechnikov kernel and a bandwidth equal to 1.8) and the estimated Weibull density are displayed.

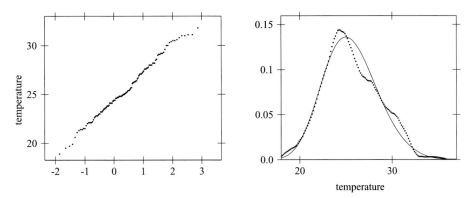

FIG. 4.1. (left.) EV Q–Q plot of the de Bilt (September) data with MLE(EV) $\gamma = -0.19$. (right.) A kernel density (dotted) and the Weibull density (solid) pertaining to the MLE(EV).

When a least squares line and a two–sided moving average are displayed in a scatterplot window, there is no significant trend visible in the data. There is some evidence of periods, yet we believe that this is negligible.

Notice that the estimated Weibull density in the preceding example is skewed to the right. We remark that the shapes of the estimated Weibull distributions vary from month to month. For example, the estimated Weibull density for the month May is skewed to the left[4].

Estimation of Parameters When a Trend Must be Dealt With

Next, we admit a certain trend in the data y_i that are measurements at time t_i. Assume that y_i is governed by an EV df $G_{\gamma,\mu(t_i;\beta_0,...,\beta_p),\sigma}$, where the location parameter depends on the time t_i and further parameters $\beta_0, ..., \beta_p$. For example,

---

[4] Also see [42] and Buishand, T.A. (1989). Statistics of extremes in climatology. Statist. Neerlandica 43, 1–30.

the model is specified by the four parameters γ, β_0, β_1 and σ, if $\mu(t; \beta_0, \beta_1) = \beta_0 + \beta_1 t$. The estimation in such a complex parametric model can be carried out by using the ML or MD estimation methods. A particular challenge is to theoretically evaluate the properties of such estimators and to implement such estimation procedures in a statistical software package.

Subsequently, we handle this question in a more pragmatic way and suggest an estimation method that can be carried out in three steps:

- estimate the trend in the data by the least squares method or in some non-parametric manner (without taking the EV model into account);

- employ the preceding parametric estimators in the EV model to the residuals,

- deduce estimators of γ, $\mu(t_i; \beta_0, \ldots, \beta_p)$ and σ.

This approach will be exemplified in conjunction with the least squares method as introduced in Section 2.5. Let $m_n(i)$ be the least squares estimate based on $y_1, \ldots, y_n$ of the mean of $G_{\gamma, \mu(t_i; \beta_0, \ldots, \beta_p), \sigma}$ as introduced in (2.43), where we assume that $\gamma < 1$ so that the mean exists. Next, assume that the residuals $x_i = y_i - m_n(i)$ are governed by an EV df $G_{\gamma, \mu, \sigma}$, where μ is independent of i. Let $\hat{\gamma}_n$, $\hat{\mu}_n$ and $\hat{\sigma}_n$ be estimators of γ, μ and σ based on the residuals $x_1, \ldots, x_n$. We propose $\hat{\gamma}_n$, $\hat{\mu}_n + m_n(i)$ and $\hat{\sigma}_n$ as estimates of γ, $\mu(u_i; \beta_0, \ldots, \beta_p)$ and σ in the original model.

EXAMPLE 4.1.3. (Iceland Storm Data[5].) We report preliminary investigations concerning storm data for the years between 1912 and 1992, with the measurements from the years 1915 and 1939 missing (stored in em–storm.dat). The annual maxima resemble those of the storm data at Vancouver given in Table 1.1.

TABLE 4.3. Iceland storm data.

year	1912	1913	1914	1916	$\cdots$	1989	1990	1991	1992
annual maxima	38	69	47	53	$\cdots$	60	68	89	63

In Fig. 4.2, the annual maxima are plotted against the years. One can recognize a slight, positive trend. The least squares line $f(x) = 48.05 + 0.12(x - 1911)$ is added. A quadratic least squares line leads to a similar trend function so that the hypothesis of a linear trend seems to be justified.

Firstly, by applying the MLE(EV) to the annual maxima in Table 4.3, we get the parameters $\alpha = -3.60$, $\mu = 98.53$ and $\sigma = 50.59$ (taken in the α–parameterization). Thus, the estimated Weibull distribution is nearly symmetric around 53. Recall that $\mu = 98.53$ is also the upper endpoint of the distribution.

---

[5]cf. also Stoyan, D., Stoyan, H. and Jansen, U. (1997). Umweltstatistik. Teubner, Stuttgart.

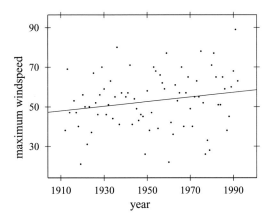

FIG. 4.2. Scatterplot of storm data and linear least squares line.

Secondly, we take the slight, positive trend in the data into account. The MLEs for the residuals (detrended data) are $\alpha = -3.09$, $\mu = 38.85$ and $\sigma = 43.43$. Thus, the estimated Weibull distribution is now slightly skewed to the left. By adding the values of the least squares line $48.05 + 0.12(x - 1911)$ for each of the years x, one can estimate the location parameter of the Weibull distribution for the year x by

$$\mu(x) = 86.90 + 0.12(x - 1911).$$

The estimates for α and σ remain unchanged. Moreover, the estimated upper endpoints are equal to 86.90 and 96.62 for the years 1912 and 1992.

We estimate the T–year thresholds with and without a linear trend hypothesis. In the latter case, equation (9.5) must be applied to $F_k = G_{\alpha,\mu(k),\sigma}$ with $\alpha = -3.09$, $\mu(k) = 97.1 + 0.12k$ and $\sigma = 43.43$, when the first year is 1997. Upper bootstrap confidence bounds are added.

TABLE 4.4. Estimated T–year thresholds (initial year 1997) with 95% upper bootstrap confidence bounds in brackets.

	T–year thresholds $u(T)$		
T	50	100	200
without trend	81.41 (85.8)	84.42 (89.8)	86.9 (103.5)
with trend	89.02 (89.7)	95.47 (109.7)	107.17 (134.7)

As mentioned at the beginning, these investigations were of a preliminary nature. Although the analysis was carried out skillfully, according to the present state–of–the–art, the results are partly of a limited relevance due to the fact that the measurements do not directly concern wind speeds, but a certain index given by

$$\frac{\text{number of stations measuring at least 9 on the Beaufort scale}}{\text{total number of stations}} \times 100 \; .$$

Thus, the values in the 2nd column of Table 4.4 concern the annual maximum of such daily indices which are necessarily below 100. This implies that the estimated 200–year threshold and some of the confidence bounds in Table 4.4 are nonsense (sorry!).

If the estimation is carried out in the Weibull (EV2) model (without trend) with a fixed upper endpoint equal to 100, then the estimated parameters are $\alpha = -3.72$, $\mu = 100$ and $\sigma = 52.13$. Again, T–year thresholds and pertaining upper bootstrap confidence bounds may be calculated. Directions of the winds are also included in this data set.

This example shows that one has to work closely with people in applications in order to carry out a correct analysis.

In conjunction with wind speeds it is of interest to include the direction of the wind speeds in the analysis for design assessment of large buildings and other structures[6]. We also refer to the case studies by Zwiers[7] and Ross[8]; we believe that the analysis of these wind–speed data would deserve particular attention, yet the full data set seems to be unavailable to the scientific community.

The question of estimating a trend in the scale parameter of generalized Pareto (GP) dfs is dealt with in Section 5.1, page 139.

Asymptotic Distributions of Estimators and Confidence Intervals in Extreme Value Models

We specify the asymptotic normal distributions of the MLE(EV0) and deduce the asymptotic normality of the pertaining estimators of the T–year threshold by means of (11.6). Detailed explanations about multivariate normal distributions are postponed until Section 11.1. Asymptotic confidence intervals can easily be established by replacing unknown parameters in covering probabilities by estimators, cf. Section 3.2.

The MLEs μ_n and σ_n of the location and scale parameters μ and σ in the Gumbel EV0 model are jointly asymptotically normal with mean vector (μ, σ) and covariance matrix Σ/n, where

$$\Sigma = \frac{6\sigma^2}{\pi^2} \begin{pmatrix} \frac{\pi^2}{6} + (1-\lambda)^2 & (1-\lambda) \\ (1-\lambda) & 1 \end{pmatrix}$$

and λ is Euler's constant. Let $g_T(\mu, \sigma) = G_{0,\mu,\sigma}^{-1}(1 - 1/T) = \mu + \sigma c(T)$ denote the

---

[6]Coles, S.G. and Walshaw, D. (1994). Directional modeling of extreme wind speeds. Appl. Statist. 43, 139–157.

[7]Zwiers, F.W. (1995). An extreme–value analysis of wind speeds at five Canadian locations. In [21], pp. 124–134.

[8]Ross, W.H. (1995). A peaks–over–threshold analysis of extreme wind speeds. In [21], pp. 135–142.

T–year threshold, where $c(T) = -\log(-\log(1 - 1/T))$. Because

$$D = \left(\frac{\partial g_T}{\partial \mu}(\mu, \sigma), \frac{\partial g_T}{\partial \sigma}(\mu, \sigma) \right) = (1, c(T)),$$

we know that the pertaining estimator $g_T(\mu_n, \sigma_n)$ is asymptotically normal with mean $g_T(\mu, \sigma)$ and variance $D\Sigma D^t/n$, where

$$D\Sigma D^t = \sigma^2 \left(1 + 6(1 - \lambda + c(T))^2/\pi^2 \right). \tag{4.4}$$

This is [24], 6.2.7, formula (2), in a paragraph written by B.F. Kimball.

For a discussion of the asymptotic normality of the MLE(EV) we refer to the article by Smith cited on page 111 or [31], Chapters 20 and 21.

Minima

As mentioned before, results for minima can be deduced from those for maxima. If $x_1, \ldots, x_n$ may be regarded as minima, then the $y_i = -x_i$ may be regarded as maxima. If the EV df $G_{i,\alpha,\mu,\sigma}$ or $G_{\gamma,\mu,\sigma}$ can be fitted to the y_i, then the converse EV df $\widetilde{G}_{i,\alpha,-\mu,\sigma}$ or $\widetilde{G}_{\gamma,-\mu,\sigma}$, see (1.38) and (1.39), can be fitted to the original data x_i.

EXAMPLE 4.1.4. (Tensile Strength of Sheet Steel.) We partly repeat the statistical analysis in the book by D. Pfeifer[9] concerning sheet steel with a cross–section of 20×0.7 mm^2. The tensile strength is measured as the elongation in % in a test according to DIN 50145.

TABLE 4.5. Tensile strength of sheet steel (elongation in %).

35.9	37.3	38.7	39.4	40.5	41.2	41.5	42.3	42.8	45.0
36.5	37.3	38.8	39.9	40.8	41.2	41.6	42.4	42.9	46.7
36.9	37.8	38.9	40.3	40.8	41.2	41.7	42.5	43.0	
37.2	37.8	38.9	40.4	41.1	41.3	41.9	42.6	43.7	
37.2	38.1	39.0	40.5	41.1	41.5	42.2	42.7	44.3	
37.3	38.3	39.3	40.5	41.1	41.5	42.2	42.7	44.9	

From P–P plots, Pfeifer deduced a converse Weibull df with parameters $\alpha = -3.34$, $\mu = 33.11$ and $\sigma = 8.38$. Notice that $\mu = 33.11$ is also the left endpoint of the distribution. The MLE(EV) gives the parameters $\alpha = -2.87$, $\mu = 34.53$ and $\sigma = 6.84$, which are comparable to the preceding ones.

One may argue that there is no plausible explanation for a left endpoint not equal to zero. Therefore, we also applied the MLE in the EV2 model, where the left endpoint is fixed and equal to zero. The estimates are $\alpha = -18.1$, $\mu = 0$ and $\sigma = 41.7$, and there is also a reasonable fit to the data. The data are stored in gu–steel.dat.

---

[9]Pfeifer, D. (1989). Einführung in die Extremwertstatistik. Teubner, Stuttgart.

Estimation of Actual Functional Parameters

What do we estimate if the actual distribution is unequal yet close to an EV distribution? Let us return to our initial model, where the $x_1, \ldots, x_n$ are generated under F^m. Now, there is no one–to–one relationship between parameters and distributions and, therefore, parameters lose their operational meaning. There is a close relationship between our question and a corresponding one in the book [27] by Hampel et al., page 408: "What can actually be estimated?" There is a plain answer in our case: by estimating the parameters γ, μ and σ, we also estimate $G_{\gamma,\mu,\sigma}$ and, therefore, the actual df F^m of the maximum, respectively, functional parameters of F^m.

4.2   Testing within Extreme Value Models

We may continue the discussion of the previous lines by asking: what are we testing? We are primarily interested in the question whether the actual distribution of the maximum is sufficiently close to an EV distribution so that the actual distribution can be replaced by an ideal one to facilitate the statistical inference. Therefore, the test should not be too stringent for larger sample sizes because, otherwise, small deviations from an EV df would be detected.

Of course, one may also test the domain of attraction as it was done, e.g., in the paper by Castillo et al.[10] Accepting the null–hypothesis does not necessarily mean that the actual distribution is closer to one of the specific EV distributions in the null–hypothesis.

Likelihood Ratio (LR) Test for the Gumbel (EV0) Model

Within the EV model, we are testing

$$\mathrm{H}_0 : \ \gamma = 0 \quad \text{against} \quad \mathrm{H}_1 : \ \gamma \neq 0 \tag{4.5}$$

with unknown location and scale parameters. Thus, the Gumbel distributions are tested against other EV distributions for a given vector $\boldsymbol{x} = (x_1, \ldots, x_n)$ of data. The likelihood ratio (LR) statistic is

$$T_{\mathrm{LR}}(\boldsymbol{x}) = 2 \log \frac{\prod_{i \leq n} g_{\hat{\gamma},\hat{\mu},\hat{\sigma}}(x_i)}{\prod_{i \leq n} g_{0,\tilde{\mu},\tilde{\sigma}}(x_i)}$$

with $(\hat{\gamma}, \hat{\mu}, \hat{\sigma})$ and $(\tilde{\mu}, \tilde{\sigma})$ denoting the MLEs in the EV and EV0 models. Because the parameter sets have dimensions 3 and 2, we know that the LR–statistic is

---

[10]Castillo, E., Galambos, J. and Sarabia, J.M. (1989). The selection of the domain of attraction of extreme value distribution from the set of data. In [14], pp. 181–190.

asymptotically distributed according to the χ^2–df χ_1^2 with 1 degree of freedom under the null–hypothesis. Consequently, the p–value is

$$p_{\mathrm{LR}}(\boldsymbol{x}) = 1 - \chi_1^2(T_{\mathrm{LR}}(\boldsymbol{x})).$$

The significance level is attained with a higher accuracy by employing the Bartlett correction[11] when the LR–statistic T_{LR} is replaced by $T_{\mathrm{LR}}/(1 + 2.8/n)$. In this case the p–value is

$$p(\boldsymbol{x}) = 1 - \chi_1^2\big(T_{\mathrm{LR}}(\boldsymbol{x})/(1 + 2.8/n)\big).$$

There is a general device to replace an LR–statistic T_{LR} by $T_{\mathrm{LR}}/(1+b/n)$ to achieve a higher accuracy in the χ^2–approximation[12].

A Test–Statistic Suggested by LAN–Theory

Marohn[13] applies the LAN–approach to deduce a test statistic for the testing problem (4.5). Let

$$
\begin{aligned}
v_{1,\mu,\sigma}(x) &= -(x-\mu)/\sigma + ((x-\mu)/\sigma)^2 \left(1 - \exp\left(-(x-\mu)/\sigma\right)\right)/2, \\
v_{2,\mu,\sigma}(x) &= -1/\sigma + ((x-\mu)/\sigma)\left(1 - \exp\left(-(x-\mu)/\sigma\right)\right)/\sigma, \\
v_{3,\mu,\sigma}(x) &= (1 + \exp\left(-(x-\mu)/\sigma\right))/\sigma \ .
\end{aligned}
$$

The test–statistic can be approximately represented by

$$T_{n,\mu,\sigma}(\boldsymbol{x}) = \Big| 1.6449 \sum_{i\leq n} v_{1,\mu,\sigma}(x_i) - \sigma \times 0.5066 \sum_{i\leq n} v_{2,\mu,\sigma}(x_i)$$

$$- \sigma \times 0.8916 \sum_{i\leq n} v_{3,\mu,\sigma}(x_i) \Big| \Big/ (n^{1/2} 3.451).$$

The p–value is

$$p(\boldsymbol{x}) = 2\Phi(T_{n,\hat{\mu}(\boldsymbol{x}),\hat{\sigma}(\boldsymbol{x})}(\boldsymbol{x})/0.69) - 1,$$

where $\hat{\mu}(\boldsymbol{x})$ and $\hat{\sigma}(\boldsymbol{x})$ are the MLEs of μ and σ.

Other Test Procedures

In those cases where the null–hypothesis is a location and scale parameter family it is desirable to use a test statistic that is invariant under location and scale parameters. This condition is satisfied by the sample skewness coefficient (see (2.7)).

---

[11] Hosking, J.R.M. (1984). Testing whether the shape is zero in the generalized extreme–value distribution. Biometrika 71, 367–374.

[12] See, e.g, Barndorff–Nielsen, O.E. and Cox, D.R. (1994). Inference and Asymptotics. Chapmann & Hall, London.

[13] Marohn, F. (2000). Testing extreme value models. Extremes 3, 362–384.

Tests based on the sample skewness coefficient are well known in the statistical literature. It is evident that the null–hypothesis of a Gumbel distribution is accepted for those distributions with skewness parameter close to 1.14, cf. page 21.

For tests based on spacings (differences of order statistics) we refer to Otten and van Montfort[14].

Goodness–of–fit tests for the Gumbel or EV model are obtained by χ^2–test statistics as mentioned in Section 3.3. A class of goodness–of–fit tests for the Gumbel model was investigated by Stephens[15]. Test statistics of the form $\sum_{i \leq n} \left(F(x_i) - \widehat{F}_n(x_i) \right)^2$ are utilized, where $\widehat{F}_n$ is the sample df and F is the Gumbel df with unknown parameters replaced by MLEs.

4.3   Extended Extreme Value Models and Related Models

We present several distributions, such as Wakeby, two–component extreme value and gamma distributions (also see [4], a book about the gamma and related distributions), which are also used for the modeling of maxima.

A Wakeby Construction

Distributions with a certain tail behavior may also be designed by using qfs. Note that every real–valued, nondecreasing and left continuous function on the interval $(0,1)$ defines a qf. Therefore,

$$Q(q) = a\big(1 - (1-q)^b\big) + c\big(-1 + (1-q)^{-d}\big) + e, \qquad 0 < q < 1$$

with $a, b, c, d > 0$, is a qf.

We see that the left and right tails of the distribution are related to those of a beta and, respectively, a Pareto distribution[16].

Two–Component Extreme Value Distributions

In cases where one observes a seasonal variation in the data, it is advisable to apply a seasonal separation to obtain identically distributed data. This was done

---

[14]Otten, A. and van Montfort, M.A.J. (1978). The power of two tests on the type of the distributions of extremes. J. Hydrology 37, 195–199.

[15]Stephens, M.A. (1977). Goodness of fit for the extreme value distribution. Biometrika 64, 583–588.

[16]Houghton, J.C. (1978). Birth of a parent: The Wakeby distribution for modeling flows. Water Resour. Res. 14, 1105–1110, and

Hosking, J.R.M., Wallis, J.R. and Wood, E.F. (1985). An appraisal of the regional flood frequency procedure in the UK Flood Studies Report. Hydrol. Sci. J. 30, 85–109.

by Todorovic and Rousselle[17] for flood data (in addition see [42] and Davison and Smith[18]). Then, the annual maximum can be regarded as the maximum of, e.g., monthly or certain seasonal maxima. Subsequently, we merely deal with two–component distributions. Then, the modeling by means of the product $G_1(x)G_2(x)$ of two EV dfs is adequate (called two–component EV distribution).

Such a modeling can also be used in conjunction with mixture distributions. Before going into details we present an example of tropical and non–tropical wind speeds (a data set of a similar type, without a classification of extraordinarily large data, consists of the annual floods of the Blackstone River in Woonsocket, Rhode Island, from 1929 to 1965 (stored in the file ht–black.dat)).

EXAMPLE 4.3.1. (Annual Maximum Wind Speeds for Jacksonville, Florida.) The data in Table 4.6 (stored in the file em–jwind.dat) were recorded by M.J. Changery[19] and further discussed in [34]. By reviewing weather maps for the days on which the annual maxima occurred, Changery identified storms of two types, namely tropical and non–tropical.

TABLE 4.6. Annual maximum wind speeds (in mph) from 1950 to 1979 with tropical storms indicated by T.

year	mph	type	year	mph	type	year	mph	type	year	mph	type	year	mph	type
1950	65	T	1956	44		1962	49		1968	47	T	1974	48	
1951	38	T	1957	42		1963	56		1969	53		1975	68	
1952	51		1958	38		1964	74	T	1970	40		1976	46	
1953	47		1959	34		1965	52	T	1971	51		1977	36	T
1954	42		1960	42	T	1966	44	T	1972	48		1978	43	
1955	42		1961	44		1967	69		1973	53		1979	37	

A preliminary statistical analysis is carried out separately for the two subsamples. A Gumbel modeling can be acceptable for both subsamples. The MLE(EV0) is $(\mu, \sigma) = (43.6, 6.7)$ for the tropical and $(\mu, \sigma) = (44.1, 9.0)$ for the non–tropical data. For a continuation see Example 6.1.2.

Tropical and non–tropical winds are selected at random by nature so that a modeling of a daily maximum wind speed by means of a mixture distribution is

---

[17]Todorovic, P. and Rousselle, J. (1971). Some problems of flood analysis. Water Resour. Res. 7, 1144–1150

[18]Davison, A.C. and Smith, R.L. (1990). Models for exceedances over high thresholds. J. R. Statist. Soc. B 52, 393–442.

[19]Changery, M.J. (1982). Historical extreme winds for the United States—Atlantic and Gulf of Mexico coastlines. U.S. Nuclear Regulatory Commission, NUREG/CR–2639.

adequate. Next, we follow the line of arguments by Rossi et al.[20] which suggests a modeling of the annual maximum by means of a two–component EV distribution.

Assume that $X_1, \ldots, X_{N(1)}$ and $Y_1, \ldots, Y_{N(2)}$ are maxima over a certain threshold. Assume that the X_i and Y_i are distributed according to F_1 and F_2. Under independence conditions and the condition that $N(1)$ and $N(2)$ are Poisson random variables one can verify (e.g., using the superposition of two Poisson processes) that both sequences can be jointly modeled by a sequence $Z_1, \ldots, Z_N$, where N is a Poisson random variable with parameter $\lambda = \lambda_1 + \lambda_2$ and the Z_i have the common mixture distribution $F(x) = p_1 F_1(x) + p_2 F_2(x)$ with $p_1 = \lambda_1/\lambda$ and $p_2 = \lambda_2/\lambda$. According to (1.54), the maximum of the Z_i has the df

$$\exp\big(-\lambda(1 - F(x))\big) = \exp\big(-\lambda_1(1 - F_1(x))\big)\exp\big(-\lambda_2(1 - F_2(x))\big)$$

for $x \geq 0$. This df is close to a two–component EV distribution if λ_1, λ_2 are small and F_1, F_2 are close to GP dfs.

Presently, we do not have direct access to statistical inference in the two–component EV model (beyond a visual one). Some further consideration will be made in Section 6.1, where the identification of the single components will be reconsidered within the framework of censored data.

Gamma (Pearson–Type III) and χ^2 Distributions

The gamma (Pearson–type III) distribution with shape parameter $r > 0$ was already introduced in (3.42) as a prior for a Bayesian estimation procedure. Gamma distributions may also be employed as mixing distributions in various cases and for the direct modeling of distributions of actual maxima.

The density of the standard gamma distribution with shape parameter $r > 0$ is

$$\text{Gamma:} \qquad h_r(x) = \tfrac{1}{\Gamma(r)} x^{r-1} e^{-x}, \qquad x > 0. \tag{4.6}$$

The exponential density is a special case for $r = 1$. From the relation $\Gamma(r + 1) = r\Gamma(r)$, it follows that the mean and variance of the gamma distribution are both equal to r.

Gamma dfs H_r are sum–reproductive in so far as the convolution is of the same type: we have $H_r^{m*} = H_{mr}$. Generally, the sum of m independent gamma random variables with parameters r_i is a gamma random variable with parameter $r_1 + \cdots + r_m$. It is apparent that gamma dfs approach the normal df as $r \to \infty$. For positive integers $r = n + 1$, one obtains the distribution of the sum of $n + 1$ iid exponential random variables X_i. In this case, the gamma df can be written

$$H_{n+1}(x) = 1 - \exp(-x)\sum_{i=0}^{n}\frac{x^i}{i!}, \qquad x > 0. \tag{4.7}$$

---

[20]Rossi, F., Fiorenti, M. and Versace, P. (1984). Two–component extreme value distribution for flood frequency analysis. Water Resour. Res. 20, 847–856.

The gamma df is also called the Erlang df in this special case.

It is noteworthy that χ^2–distributions with n degrees of freedom are gamma distributions with shape and scale parameters $r = n/2$ and $\sigma = 2$. If $X_1, \ldots, X_n$ are iid standard normal random variables, then $\sum_{i \leq n} X_i^2$ is a χ^2 random variable with n degrees of freedom.

Logistic Distributions as Mixtures of Gumbel Distributions

If $\{F_\vartheta\}$ is a family of dfs and h is a probability density, then

$$F_h(x) = \int F_\vartheta(x) h(\vartheta) \, d\vartheta \qquad (4.8)$$

is another df (the mixture of the dfs F_ϑ with respect to h). If f_ϑ is the density of F_ϑ, then

$$f_h(x) = \int f_\vartheta(x) h(\vartheta) \, d\vartheta$$

is the density of F_h. We refer to Section 8.1 for a detailed discussion of the heuristics behind the concept of mixing.

We show that a certain mixture of Gumbel distributions is a generalized logistic distribution, if the mixing is done with respect to a gamma distribution (a result due to Dubey[21]). Therefore, the generalized logistic model can be an appropriate model for distributions of maxima in heterogeneous populations. Likewise, one may deal with minima.

A Gumbel df with location and scale parameters μ and $\sigma > 0$ can be reparameterized by

$$
\begin{aligned}
G_{0,\mu,\sigma}(x) &= \exp(-\exp(-(x-\mu)/\sigma)) \\
&= \exp(-\vartheta \exp(-x/\sigma)),
\end{aligned}
$$

where $\vartheta = \exp(\mu/\sigma)$.

Mixing over the parameter ϑ with respect to the gamma density $h_{r,\beta}$, with shape and scale parameters r, $\beta > 0$, one obtains the mixture df

$$
\begin{aligned}
F_{\beta,r,\sigma}(x) &= \int_0^\infty \exp(-\vartheta \exp(-x/\sigma)) h_{r,\beta}(\vartheta) \, d\vartheta \\
&= (1 + \beta \exp(-x/\sigma))^{-r} \qquad (4.9)
\end{aligned}
$$

which is a generalized logistic df with two shape parameters β, $r > 0$ and the scale parameter $\sigma > 0$. If $r = 1$ and $\beta = 1$ and, thus, the mixing is done with respect to the standard exponential df, then the mixture is a logistic df

$$F_{1,1,\sigma}(x) = (1 + \exp(-x/\sigma))^{-1} \qquad (4.10)$$

---

[21]Dubey, S.D. (1969). A new derivation of the logistic distribution. Nav. Res. Logist. Quart. 16, 37–40.

with scale parameter $\sigma > 0$.

If the Gumbel df is replaced by the Gompertz df $\widetilde{G}_{0,\mu,\sigma}(x) = 1 - G_{0,\mu,\sigma}(-x)$ and $\vartheta = \exp(-\mu/\sigma)$, then the mixture is the converse generalized logistic df

$$\widetilde{F}_{\beta,r,\sigma}(x) = 1 - (1 + \beta \exp(x/\sigma))^{-r}. \qquad (4.11)$$

If $\beta = 1/r$, then the Gumbel df $G_{0,0,\sigma}$ and, respectively, the Gompertz df $\widetilde{G}_{0,0,\sigma}$ are the limits of the (converse) generalized logistic dfs in (4.9) and (4.11) as $r \to \infty$.

Log–Gamma (Log–Pearson–Type III) and GP–Gamma Distributions

If $\log(X)$ has the gamma density $h_r(x/\sigma)/\sigma$ for some scale parameter $\sigma > 0$, then X has the density

$$\text{Log–Gamma:} \qquad f_{1,r,\alpha}(x) = \frac{\alpha^r}{\Gamma(r)}(\log x)^{r-1}x^{-(1+\alpha)}, \qquad x > 1, \qquad (4.12)$$

for $\alpha = 1/\sigma$. We see that the log–gamma distribution has two shape parameters $r, \alpha > 0$. For $r = 1$, one obtains the standard Pareto density with shape parameter α.

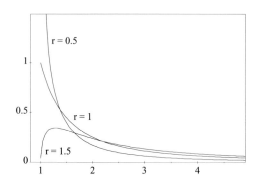

FIG. 4.3. Log–gamma densities for $\alpha = 1$ and $r = 0.5, 1, 1.5$.

Figure 4.3 indicates that log–gamma distributions have upper tails similar to those of Pareto (GP1) distributions, yet the shape can be completely different near the left endpoint.

In addition, if X is a log–gamma random variable, then $-1/X$ has the density

$$\text{GP–Gamma 2:} \qquad f_{2,r,\alpha}(x) = \frac{|\alpha|^r}{\Gamma(r)}(-\log|x|)^{r-1}(-x)^{-(1+\alpha)}, \qquad -1 < x < 0,$$

for parameters $\alpha < 0$, where we again changed the sign of the shape parameter. The standard beta (GP2) densities are special cases for $r = 1$.

Within this system, the gamma density may be denoted by $f_{0,r}$. The following diagram shows the relationships between the different models.

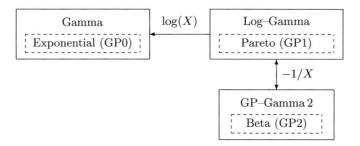

A unified model is achieved by using the von Mises approach. For $r > 0$ and $\gamma \neq 0$, we have

GP–Gamma: $f_{r,\gamma}(x) = \frac{1}{\Gamma(r)} \left(\frac{\log(1+\gamma x)}{\gamma} \right)^{r-1} (1 + \gamma x)^{-(1+1/\gamma)}$

for $0 < x$, if $\gamma > 0$, and $0 < x < 1/|\gamma|$, if $\gamma < 0$. The gamma densities $f_{r,0}$ are obtained in the limit as $\gamma \to 0$.

Fitting Gamma and Log–Gamma Distributions to Maxima

If $r < 1$ and $\gamma < -1$, then GP–gamma densities have a pole at zero and a monotonic decrease. Thus, due to the second property, there is a certain similarity to GP densities.

If $r > 1$, there is a greater similarity to EV densities:

- gamma densities have an exponential type upper tail and are tied down near zero by means of a factor x^{r-1},

- log–gamma densities with location parameter $\mu = -1$ have a Pareto type upper tail and are tied down near zero by

$$(\log(1 + x))^{r-1} \approx x^{r-1}.$$

Therefore, gamma and log–gamma densities have shapes that correspond visually to Fréchet densities if $r > 1$.

A further extension of the preceding models is achieved if we start with generalized gamma instead of gamma distributions; e.g., logarithmic half–normal distributions are included in such models. Yet, in the subsequent lines we deal with a different kind of extensions starting with converse gamma and converse generalized gamma distributions.

Generalized Gamma Distributions

If X is a gamma random variable with parameter $r > 0$, then $X^{1/\beta}$ is a generalized gamma random variable with density

$$\tilde{h}_{r,\beta}(x) = \frac{\beta}{\Gamma(r)} x^{\beta r - 1} \exp(-x^\beta), \qquad x \geq 0,$$

for $\beta > 0$. This transformation corresponds to that on page 38, where Weibull (EV2) dfs were deduced from converse exponential dfs. We see that the generalized gamma model includes the converse Weibull model (for $r = 1$). For $\beta = 2$ and $r = 1/2$ we have the half–normal distribution.

The normal distribution is included if we also take double generalized gamma distributions with densities of the form $\tilde{h}_{r,\beta}(|x|)/2$.

EV–Gamma Distributions

We indicate certain extensions of EV models by a diagram with transformations as in the EV family.

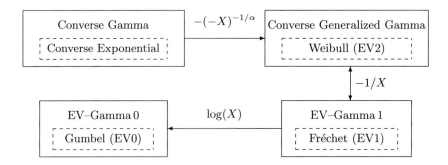

Within the EV–Gamma 1 model one obtains the Lévy distribution (cf. (6.18)) which is a sum–stable density with index $1/2$ and a certain skewness parameter. A von Mises representation is again possible. The applicability of such models must be still explored.

The Two–Parameter Beta Family

We mention also the beta distributions in the usual parameterization that are given by

$$f_{a,b}(x) = \frac{x^{a-1}(1-x)^{b-1}}{B(a,b)}, \qquad 0 \le x \le 1, \tag{4.13}$$

for shape parameters $a, \ b > 0$, where

$$B(a,b) = \int_0^1 x^{a-1}(1-x)^{b-1}\, dx$$

is the beta function. Notice that the beta densities in the GP2 model constitute a special case; we have $w_{1,\alpha,1,1} = f_{1,-\alpha}$. Recall that $B(a,b) = \Gamma(a)\Gamma(b)/\Gamma(a+b)$, where Γ is the gamma function.

Chapter 5

Generalized Pareto Models

This chapter deals once again with the central topic of this book, namely with exceedances (in other words, peak–over–threshold values) over high thresholds and upper order statistics. One may argue that this chapter is richer and more exciting than the preceding one concerning maxima. The role of extreme value (EV) dfs is played by generalized Pareto (GP) dfs.

5.1   Estimation in Generalized Pareto Models

Let $x_1, \ldots, x_n$ be the original data which are governed by a df F. We deal with upper extremes which are either

- the exceedances $y_1, \ldots, y_k$ over a fixed threshold u, or

- the k upper ordered values $\{y_1, \ldots, y_k\} = \{x_{n-k+1:n}, \ldots, x_{n:n}\}$, where k is fixed.

Recollect from (2.4) that a nonparametric estimate of the exceedance df $F^{[u]}$ is given by the sample df $\widehat{F}_k = \widehat{F}_k(\boldsymbol{y}; \cdot)$ based on the exceedances $y_1, \ldots, y_k$ over u. Likewise, the exceedance density or qf can be estimated by means of a sample exceedance density f_k or the sample exceedance qf $\widehat{F}_k^{-1}$.

Now we want to pursue a parametric approach. We assume that the actual exceedance df $F^{[u]}$ can be replaced by some GP df W with left endpoint u. For example, we estimate the exceedance df $F^{[u]}$ by means of a GP df $W_{\gamma_k, u, \sigma_k}$ (in the γ–representation), where γ_k and σ_k are estimates of the shape and scale parameters. It is advisable to add the threshold u to the sample when a parametric estimation based on exceedances is executed.

We also deal with likelihood–based estimator, namely MLEs and Bayesian estimators, based on exceedances. For that purpose, exceedances are dealt with like iid data. A justification for this approach can be found in Section 8.1, page 234.

Recall that the threshold value u is replaced by $x_{n-k:n}$ in the case of upper ordered values. Another random threshold is obtained for a data–driven choice of k, see page 137.

The Exponential Model GP0(u)

The statistical model for the exceedances over the threshold u consists of all exponential dfs with left endpoint u.

GP0(u): $\{W_{0,u,\sigma} : \sigma > 0\}.$

Estimators for the GP0(u) model:

- MLE(GP0): the MLE of the scale parameter is $\sigma_k = \frac{1}{k}\sum_{i \leq k}(y_i - u)$. It is apparent that the MLE is also a moment estimator.

- M–Estimator (GP0): an M–estimate of the scale parameter σ, which may be regarded as a robustified MLE (see (3.7)), is obtained as the solution to

$$\sum_{i \leq k}\left(\exp\left(-\frac{y_i - u}{\sigma b}\right) - \frac{b}{1+b}\right) = 0.$$

For $b \to \infty$, the likelihood equation is attained.

We refer to Section 3.5 for Bayesian estimators in the exponential model.

EXAMPLE 5.1.1. (Continuation of Example 4.1.1 about the Annual Floods of the Feather River.) According to the insight gained from Example 4.1.1, we may already expect that an exponential distribution can be fitted to the upper part of the annual maximum flood (which is indeed confirmed). From the MLE(GP0) with $k = 25$, we get the exponential distribution with location and scale parameters $\mu_k = 36,577$ and $\sigma_k = 45,212$. The 50 and 100–year flood levels are $u(50) = 213,450$ and $u(100) = 244,789$. We see that the 100–year flood level is exceeded by none of the 59 annual maximum discharges.

If one is not sure whether the exponential model is acceptable, one should alternatively carry out the estimation within the Pareto model or the unified generalized Pareto model which are handled below.

The Restricted Pareto Model GP1($u, \mu = 0$)

This is just the exponential model GP0(u) transformed by $T(x) = \log x$ (as outlined on page 37). Therefore, the estimators correspond to those in the exponential model. We consider a one–dimensional Pareto model parameterized by the shape parameter α, namely,

GP1($u, \mu = 0$): $\{W_{1,\alpha,0,u} : \alpha > 0\}.$

This is an appropriate model for exceedances over the threshold u because the left endpoints of these Pareto dfs are equal to u.

The shape parameter α can be represented as a functional parameter because

$$\int \log(x/u)\, dW_{1,\alpha,0,u}(x) = 1/\alpha \ . \tag{5.1}$$

Estimators of the shape parameter α within the model GP1$(u, \mu = 0)$:

- Hill[1]: this is the in the restricted Pareto model GP1$(u, \mu = 0)$. We have

$$\alpha_k = k \left/ \sum_{i \leq k} \log(y_i/u) \right. . \tag{5.2}$$

Recall that the threshold u is replaced by $x_{n-k:n}$ in the case of upper ordered values. Then, the estimate can be written

$$\alpha_k = k \left/ \sum_{i \leq k} \log(x_{n-i+1:n}/x_{n-k:n}) \right. \tag{5.3}$$

which is the Hill estimator in the original form. The Hill estimator may be introduced as a sample functional

$$\alpha_k = 1 \left/ \int \log(x/u)\, d\widehat{F}_k(x) \right. ,$$

where $\widehat{F}_k$ is the sample exceedance df (replacing $W_{1,\alpha,0,u}$ in equation (5.1)). For a discussion about the performance of the Hill estimator we refer to the subsequent page and the lines around the Figures 5.1 and 5.2.

- M–Estimator in the restricted Pareto model GP1$(u, \mu = 0)$: the estimate[2] of the shape parameter α that corresponds to the M–estimate of σ in the GP0 model is the solution to

$$\sum_{i \leq k} \left(\left(\frac{y_i}{u} \right)^{-\alpha/b} - \frac{b}{1+b} \right) = 0.$$

For $b \to \infty$, the Hill estimate is received.

- Bayes estimators for the model GP1$(u, \mu = 0)$[3]: take the gamma density

$$p(\alpha) = h_{s,d}(\alpha), \tag{5.4}$$

[1]Hill, B.M. (1975). A simple general approach to inference about the tail of a distribution. Ann. Statist. 3, 1163–1174.

[2]Reiss, R.–D., Haßmann, S. and Thomas, M. (1994). XTREMES: Extreme value analysis and robustness. In [15], Vol. 1, 175–187.

[3]Early references are Hill[1] and Rytgaard, M. (1990). Estimation in the Pareto distribution. ASTIN Bulletin 20, 201–216.

cf. (3.42), as a prior for the shape parameter α. This Bayes estimator corresponds to the one in the exponential model with unknown reciprocal scale parameter, see page 102. The likelihood function satisfies

$$L(\boldsymbol{y}|\alpha) \propto \alpha^k \exp\left(-\alpha \sum_{i \leq k} \log(y_i/u)\right), \qquad y_i > u, \qquad (5.5)$$

in α. Consequently, the posterior density satisfies $p(\alpha|\boldsymbol{y}) \propto L(\boldsymbol{y}|\alpha)h_{s,d}(\alpha)$, and one obtains

$$p(\alpha|\boldsymbol{y}) = h_{s',d'}(\alpha) \qquad (5.6)$$

for the posterior density which is the gamma density with parameters $s' = s + k$ and $d' = d + \sum_{i \leq k} \log(y_i/u)$. The Bayes estimate, as the mean of the posterior density, is

$$
\begin{aligned}
\alpha_k^*(\boldsymbol{y}) &= \int \alpha p(\alpha|\boldsymbol{y})\, d\alpha \\
&= \frac{s+k}{d + \sum_{i \leq k} \log(y_i/u)} \, .
\end{aligned}
\qquad (5.7)
$$

We see that the Bayes estimate is close to the Hill estimate if s and d are small or/and k is large.

The explicit representation of the Hill and the Bayes estimators facilitated the development of an asymptotic theory. In addition, these estimators possess an excellent performance if this one–dimensional model is adequate. Supposedly, to check the performance of these estimators, simulations were usually run under standard Pareto distributions with location parameter $\mu = 0$.

Also, the Hill estimator has been applied to data in a lot of research papers. Yet, this does not necessarily imply that the Hill estimator and related estimators should be used in applications. We refer to the comments around Figures 5.1 and 5.2 for a continuation of this discussion.

What should be done if the diagnostic tools—like the sample mean or median excess function—in Chapter 2 indicate that the GP1$(u, \mu = 0)$ model is significantly incorrect, as, e.g., in the case of financial data in Chapter 16. It is likely that one of these estimators was applied because a heavy–tailed or, specifically, a Pareto–like tail of the underlying df was conjectured.

It suggests itself to carry out the statistical inference within the full model GP1(u) of Pareto dfs with left endpoint u which are introduced in the next subsection.

For dfs in the full Pareto GP1(u) model we have

$$W_{1,\alpha,\mu,\sigma} = W_{1,\alpha,\mu,u-\mu} \approx W_{1,\alpha,0,u}, \qquad (5.8)$$

if u is sufficiently large. Therefore, estimators especially tailored for the submodel GP1$(u, \mu = 0)$ are of a certain relevance for the full Pareto model GP1(u) in asymptotic considerations. Yet, for finite sample sizes the approximation error in (5.8) may cause a larger bias and, thus, an unfavorable performance of the estimator.

Also see Section 6.5, where theoretical results are indicated that the approximation error is of a smaller order for small shape parameters α.

The Full Pareto GP1(u) Model: (α, μ, σ)–Parameterization

The model of all Pareto dfs (in the α–parameterization) with shape parameter $\alpha > 0$ and left endpoint u is given by

$$\text{GP1}(u): \qquad \{W_{1,\alpha,\mu,\sigma} : \alpha, \sigma > 0, \ \mu \text{ real}, \ \mu + \sigma = u\}.$$

It is apparent that the restricted Pareto model GP1$(u, \mu = 0)$ is a submodel of the model GP1(u) with $\mu = 0$.

The MLE in the GP1(u) model corresponds to the MLE in the generalized Pareto model and does not exist in the present Pareto model with a certain positive probability.

The Full Pareto GP1(u) Model: (α, η)–Parameterization

The GP1(u) model can be parameterized by two parameters $\alpha, \eta > 0$. For every df $W_{1,\alpha,\mu,\sigma}$ with $\mu + \sigma = u$,

$$
\begin{aligned}
W_{1,\alpha,\mu,\sigma}(x) &= W_{1,\alpha,\mu,u-\mu} \\
&= W_{1,\alpha,u(1-\eta),u\eta}(x) \\
&= 1 - \left(1 + \frac{(x-u)/u}{\eta}\right)^{-\alpha}, \qquad x \geq u, \qquad (5.9)
\end{aligned}
$$

where $\eta = 1 - \mu/u$ has the nice property of being a scale parameter for the normalized excesses $(x - u)/u$.

One obtains a reparameterized full Pareto model (for exceedances over u)

$$(\alpha, \eta)\text{–parameterization:} \qquad \{W_{1,\alpha,u(1-\eta),u\eta} : \alpha, \eta > 0\}. \qquad (5.10)$$

In this representation, the restricted Pareto model GP1$(u, \mu = 0)$ is a submodel with $\eta = 1$.

Within the reparameterized model of Pareto dfs one gets some closer insight why the Hill estimator, and related estimators such as the Bayes estimator in (5.7), are inaccurate even in such cases where a Paretian modeling is acceptable.

Notice that a larger scale parameter η puts more weight on the upper tail of the df. If such a Pareto df is estimated, yet $\eta = 1$ is kept fixed in the modeling, then the heavier weight in the tail, due to a larger parameter η, must be compensated by a smaller estimated shape parameter α.

In the following illustration, we show simulated densities of the Hill estimator based on $k = 45$ largest Pareto data out of $n = 400$. The number of simulations is $N = 4000$. Whereas the shape and scale parameters are kept fixed, we vary the location parameter of the underlying Pareto distribution.

As expected, the density of the Hill estimator of α is nicely centered around the true shape parameter $\alpha = 10$ if $\mu = 0$ because the underlying df belongs to the restricted Pareto model $GP1(u, \mu = 0)$.

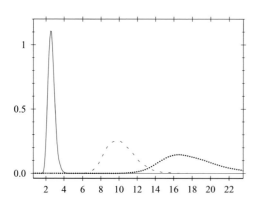

FIG. 5.1. Simulated densities of Hill estimator under the Pareto distributions with shape parameter $\alpha = 10$, $\sigma = 1$ and $\mu = -1$, 0, 1 from left to right.

The Hill estimator has a bad performance for the other cases (for a continuation see Fig. 5.2); e.g., for $\mu = -1$ it centers around a value 3.

Bayes Estimation in the Full Pareto Model, (α, η)–Parameterization

Bayes estimators for the model $GP1(u)$, in the (α, η)–parameterization, were dealt with by Reiss and Thomas[4]. For a more recent article see Diebolt et al.[5]

As a prior density take

$$p(\alpha, \eta) = h_{s,d}(\alpha)f(\eta), \qquad (5.11)$$

where $h_{s,d}$ is a gamma density taken as a prior for α, and f, as a prior for η, is another probability density which will be specified later. The likelihood function for the present model is determined by

$$L(\boldsymbol{y}|\alpha, \eta) \propto (\alpha/\eta)^k \exp\left(-(1+\alpha)\sum_{i \le k}\log\left(1 + \frac{y_i - u}{\eta u}\right)\right), \qquad y_i > u. \quad (5.12)$$

[4]Reiss, R.-D. and Thomas, M. (1999). A new class of Bayesian estimators in Paretian excess–of–loss reinsurance. ASTIN Bulletin 29, 339–349.

[5]Diebolt, J., El–Aroui, M.–A., Garrido, M. and Girard, S. (2005). Quasi–conjugate Bayes estimates for GPD parameters and application to heavy tails modelling. Extremes 8, 57–78.

Deduce the posterior density

$$p(\alpha, \eta | \boldsymbol{y}) = h_{s', d'(\eta)}(\alpha) \tilde{f}(\eta), \qquad (5.13)$$

where $h_{s', d'(\eta)}$ is again a gamma density with parameters $s' = s + k$ and $d'(\eta) = d + \Sigma_{i \leq k} \log(1 + (y_i - u)/(\eta u))$. The probability density $\tilde{f}$ is characterized by

$$\tilde{f}(\eta) \propto \eta^{-k}(d'(\eta))^{-s'} \exp\left(-\sum_{i \leq k} \log\left(1 + \frac{y_i - u}{\eta u}\right)\right) f(\eta). \qquad (5.14)$$

Simple calculations yield that the Bayes estimates of α and η are

$$\alpha_k^*(\boldsymbol{y}) = \int \alpha p(\alpha, \eta | \boldsymbol{y}) \, d\alpha d\eta = \int \frac{s'}{d'(\eta)} \tilde{f}(\eta) \, d\eta, \qquad (5.15)$$

and

$$\eta_k^*(\boldsymbol{y}) = \int \eta p(\alpha, \eta | \boldsymbol{y}) \, d\alpha d\eta = \int \eta \tilde{f}(\eta) \, d\eta. \qquad (5.16)$$

We see that the Bayesian estimator of α is just the estimator in (5.7), if the prior distribution—and, thus, also the posterior—is a point measure with mass equal to one at $\sigma = 1$.

We remark that gamma priors were chosen for the parameter α in the restricted model because they possess the nice property of being conjugate priors. This property still holds in the full model in so far that the conditional posterior for α is again a gamma distribution. Such a natural choice seems not to exist for the scale parameter σ. As priors for η one may also take, e.g., reciprocal gamma distributions as introduced in (3.43). For a continuation see Section 8.3.

If the Pareto modeling is not adequate for the actual fat or heavy–tailed df, then one may choose one of the models introduced in Section 5.5.

The Beta Model (GP2)

We shortly mention the model of beta (GP2) distributions with upper endpoint equal to zero and scale parameter u. The left endpoint of the distribution is $-u$. Recall from Section 1.4 that these distributions form a submodel of the family of beta distributions as dealt with in the statistical literature. In addition, a negative shape parameter is taken.

$$\text{GP2(u):} \qquad \{W_{2, \alpha, 0, u} : \alpha < 0\}.$$

The GP1 model can be transformed to the GP2 model by means of the transformation $T(x) = -1/x$ (cf. page 37 again). Thus, an MLE corresponding to the Hill estimator in the GP1 model, therefore called Hill(GP2) estimator, becomes applicable. Shift the data below zero if an upper endpoint other than zero is plausible.

The Generalized Pareto Model

The given model is

$$\text{GP(u):} \qquad\qquad \{W_{\gamma,u,\sigma} : \gamma \text{ real}, \sigma > 0\},$$

where u and σ are the location and scale parameters. Notice that the Pareto df $W_{1,\alpha,\mu,u-\mu}$ in (5.8) is equal to $W_{\gamma,u,\sigma}$ with $\sigma = \gamma(u - \mu)$.
Estimators γ_k of γ include:

- MLE(GP): the MLE in the GP model must be evaluated by an iteration procedure[6]. The remarks about the MLE(EV) also apply to the present estimator.

- Moment(GP): the moment estimator[7] takes the form $\gamma_{1,k} + \gamma_{2,k}$, where $\gamma_{1,k} = 1/\alpha_k$ is the reciprocal of the Hill estimator. The second term $\gamma_{2,k}$ is constructed by means of $l_{j,k} = \sum_{i \le k}(\log(y_i/u))^j/k$ for $j = 1, 2$. We have

$$\gamma_{2,k} = 1 - 1/\big(2\big(1 - l_{1,k}^2/l_{2,k}\big)\big).$$

Roughly speaking, $\gamma_{1,k}$ (respectively, $\gamma_{2,k}$) estimates the shape parameter γ if $\gamma \ge 0$ (if $\gamma \le 0$) and $\gamma_{1,k}$ (respectively, $\gamma_{2,k}$) is close to zero if $\gamma \le 0$ (if $\gamma \ge 0$).

> **Warning!** The moment estimator has an excellent performance in general, yet it should not be applied to a full data set of exceedances. In that case, one obtains irregular estimates.

- Drees–Pickands(GP): an LRSE of the shape parameter γ in the GP model can be defined in similarity to that in the EV model, cf. page 111. Let

$$\gamma_k = \log(\hat{r})/\log(1/a)$$

with $q_0 = 1 - q$, $q_1 = 1 - aq$ and $q_3 = 1 - a^2 q$, where $0 < q, a < 1$. By taking $q = 4i/k$ and $a = 1/2$, one obtains the Pickands[8] estimate, which corresponds to a Dubey estimate (cf. page 111) in the EV model[9]. We have

$$\gamma_{k,i} = \log\left(\frac{y_{k-i:k} - y_{k-2i:k}}{y_{k-2i:k} - y_{k-4i:k}}\right)/\log 2$$

---

[6]Prescott, P. and Walden, A.T. (1980). Maximum likelihood estimation of the parameters of the generalized extreme–value distribution. Biometrika 67, 723–724.

[7]Dekkers, A.L.M., Einmahl, J.H.J. and de Haan, L. (1989). A moment estimator for the index of an extreme–value distribution. Ann. Statist. 17, 1833–1855.

[8]Pickands, J. (1975). Statistical inference using extreme order statistics. Ann. Statist. 3, 119–131.

[9]A closely related estimator may be found in the article by Weiss (1971). Asymptotic inference about a density function at an end of its range. Nav. Res. Logist. Quart. 18, 111–114.

for $i \leq [k/4]$. A linear combination $\sum_{i \leq [k/4]} c_{k,i} \gamma_{k,i}$ with estimated optimal scores $c_{k,i}$ was studied by H. Drees[10]. Estimates of γ based on two and, respectively, five Pickands estimates were dealt with by M. Falk[11] and T.T. Pereira[12].

- Slope(GP): the slope $\beta_{1,k}$ of the least squares line (see (2.40)), fitted to the mean excess function right of the k largest observations, is close to $\gamma/(1-\gamma)$ and, therefore, $\gamma_k = \beta_{1,k}/(1+\beta_{1,k})$ is a plausible estimate of γ.

- *L*–Moment Approach: This method will be introduced in Section 14.4 in conjunction with flood frequency analysis.

In addition, σ may be estimated by a least squares estimator σ_k (a solution to (2.40) for $\mu = u$ with Φ replaced by W_{γ_k}), where γ_k is the Moment, Drees–Pickands or slope estimator.

If $\gamma > -1/2$, then one should apply the MLE(GP) or Moment(GP). The Drees–Pickands estimator possesses remarkable asymptotic properties, yet the finite sample size performance can be bad for $\gamma > 0$. Yet, we believe that the concept of LRSEs has also the potential to provide accurate estimators for smaller sample sizes in such cases. In addition, estimators of that type can be especially tailored such that certain additional requirements are fulfilled. The Slope(GP) estimator is only applicable if $\gamma < 1$.

Estimation Based on the Original Data

Let α_k, γ_k and σ_k be the estimates of α, γ and σ based on

(a) the exceedances $y_1, \ldots, y_k$ over u, or

(b) the upper ordered values $y_i = x_{n-i+1:n}$,

as defined in the preceding lines. Moreover, we assume that the original data $x_1, \ldots, x_n$ are available. Put $\alpha_{k,n} = \alpha_k$ and $\gamma_{k,n} = \gamma_k$. Derive $\mu_{k,n}$ and $\sigma_{k,n}$ from u and σ_k as described on page 58. Then, one may use $W_{\gamma_{n,k},\mu_{n,k},\sigma_{n,k}}$ as a parametric estimate of the upper tail of the underlying df F. Recall that the sample df $F_n(x; \cdot)$ is a global estimator of F.

Likewise, use the pertaining parametric densities and qfs for estimating the underlying density f and qf F^{-1} near the upper endpoints. Thereby, we obtain parametric competitors of the kernel density and the sample qf.

---

[10]Drees, H. (1995). Refined Pickands estimators of the extreme value index. Ann. Statist. 23, 2059–2080.

[11]Falk, M. (1994). Efficiency of convex combinations of Pickands estimator of the extreme value index. Nonparametric Statist. 4, 133–147.

[12]Pereira, T.T. (2000). A spacing estimator for the extreme value index. Preprint.

Diagram of Estimates

The choice of the threshold u or, likewise, the number k of upper extremes corresponds to the bandwidth selection problem for the kernel density in Section 2.1. Such a choice can be supported visually by a diagram. Thereby, estimates $\gamma_{k,n}$, $\mu_{k,n}$ and $\sigma_{k,n}$ are plotted against the number k of upper ordered values.

The following properties become visible:

- if k is small, then there is a strong fluctuation of the values $\gamma_{k,n}$ for varying k;

- for an intermediate number k of extremes, the values of the estimates $\gamma_{k,n}$ stabilize around the true value γ,

- finally, if k is large, then the model assumption may be strongly violated and one observes a deviation of $\gamma_{k,n}$ from γ.

In other words, if k is small, then the variance of the estimator is large and the bias is small (and vice versa); in between, there is a balance between the variance and the bias and a plateau becomes visible.

We start with a negative result: the Hill diagram for 400 Pareto data generated under the shape, location and scale parameters $\alpha = 10$, $\mu = -1$, $\sigma = 1$ provides a smooth curve, yet it is difficult to detect the aforementioned plateau in Fig. 5.2 (left). This is due to the fact that the Hill estimator is inaccurate for larger α and $\mu \neq 0$, also see (6.38).

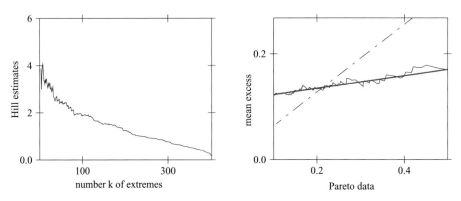

FIG. 5.2. (left.) Hill diagram for 400 Pareto data generated under the shape, location and scale parameters $\alpha = 10$, $\mu = -1$, $\sigma = 1$. (right.) Pareto mean excess function pertaining to Hill (dashed) and MLE(GP) (dotted) estimates based on $k = 45$ upper extremes; sample mean excess function (solid).

For real data, one may compare, e.g., the mean excess function, which belongs to the estimated parameters, with the sample mean excess function in order to check the adequacy of a parametric estimate. In Fig. 5.2 (right), one recognizes again that the Hill estimate is not appropriate. The MLE(GP) with values $\alpha = -20.23$ (respectively, $\gamma = -0.05$) is acceptable.

In Fig. 5.2, the estimation was carried out within the ideal model for the MLE(GP) and, hence, the performance of the MLE(GP) can be improved when the number k is equal to the sample size $n = 400$. The situation changes for Fréchet data. Recall from (1.47) that a Fréchet df is close to a Pareto df in the upper tail.

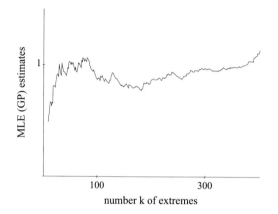

FIG. 5.3. MLE(GP) estimates $\alpha_{k,n}$ plotted against $k = 1, \dots, 400$ for standard Fréchet data under the shape parameter $\alpha = 1$.

In Fig. 5.3, a plateau with right endpoint $k = 80$ becomes visible. Therefore, we suggest to base the estimation on this number of upper extremes.

Automatic Choice of the Number of Extremes

The selection of the right endpoint of a plateau may also be done in an automatic manner. The following ad hoc procedure works reasonably well. Let $\gamma_{k,n}$ be estimates of the shape parameter γ based on the k upper extremes (likewise, one may deal with estimates $\alpha_{k,n}$ of α). Denote the median of $\gamma_{1,n}, \dots, \gamma_{k,n}$ by $\text{med}(\gamma_{1,n}, \dots, \gamma_{k,n})$. Choose k^* as the value that minimizes

$$\frac{1}{k} \sum_{i \leq k} i^{\beta} |\gamma_{i,n} - \text{med}(\gamma_{1,n}, \dots, \gamma_{k,n})| \tag{5.17}$$

with $0 \leq \beta < 1/2$. A slight smoothing of the series of estimates improves the performance of the procedure for small and moderate sample sizes. Modifications are obtained, for example, by taking squared deviations and $\gamma_{k,n}$ in place of the median.

Other selection procedures may be found in the papers by Beirlant et al.[13]

---

[13]Beirlant, J., Vynckier, P. and Teugels, J.L. (1996). Tail index estimation, Pareto quantile plots, and regression diagnostics. JASA 91, 1659–1667.

and by Drees and Kaufmann[14] and in the literature cited therein. We also refer to Csörgő et al.[15] for the smoothing of tail index estimators.

Peaks–Over–Threshold and Annual Maxima Methods

Besides selecting exceedances or upper ordered values from the original data $x_1, \ldots, x_n$, there is another method of extracting extremes, namely by taking maxima out of blocks (see (1.5) and Chapter 4). Therefore, there are two different approaches to the estimation of the tail index α or γ of a df:

- based on exceedances or upper extremes as dealt with in this section (pot method),

- based on maxima within certain subperiods (annual maxima method).

For the second approach, one must assume that the original data $x_1, \ldots, x_n$ are observable. Let

$$\{x_{(j-1)m+1}, \ldots, x_{jm}\}, \qquad j = 1, \ldots, k, \tag{5.18}$$

be a partition of $x_1, \ldots, x_n$, where $n = mk$, and take the maximum y_j out of the jth block.

Assume that the x_i are governed by a df F. If the block size m is large, then an EV df can be accurately fitted to the actual df F^m of the maximum. Yet, one must cope with the disadvantage that the number k of maxima is small (and vice versa).

According to foregoing remarks, the EV df G has the same shape parameter as the GP df W corresponding to the exceedances. For discussions concerning the efficiency of both approaches, we refer to [42], Section 9.6 and to Rasmussen et al.[16].

Random Thresholds

If the k largest values are selected, then $x_{n-k:n}$ may be regarded as a random threshold. The index is also random either if the sample sizes are random or as in the case of automatically selected thresholds (in (5.17)). Other choices are

- the smallest annual maximum,

---

[14]Drees, H. and Kaufmann, E. (1998). Selection of the optimal sample fraction in univariate extreme value estimation. Stoch. Proc. Appl. 75, 149–172

[15]Csörgő, S., Deheuvels, P. and Mason, D.M. (1985). Kernel estimates of the tail index of a distribution. Ann. Statist. 13, 1050–1078.

[16]Rasmussen, P.F., Ashkar, F., Rosbjerg, D. and Bobée, B. (1994). The pot method for flood estimation: a review, 15–26 (in [13]).

- the threshold such that, on the average, the number of selected extremes in each year is equal to 3 or 4.

For the latter two proposals, we refer to a paper by Langbein[17].

Extrapolation

In this section, a GP df was fitted to the upper tail of the sample df, thus also to the underlying df F in the region where the exceedances are observed. Using this parametric approach, it is possible to construct estimates of certain functional parameters of F with variances smaller than those of nonparametric estimates. This advantage is possibly achieved at the cost of a larger bias if the parametric modeling is inaccurate.

There will be a new maximum or observations larger than the present ones in the near future. These observations are outside of the region where a GP df was fitted to the data. Statistical questions concerning future extremes should be approached with the following pragmatic attitude. Of course we do not know whether the insight gained from the data can be extrapolated to a region where no observation has been sampled. Keep your fingers crossed and evaluate the risk of future extreme observations under the estimated df.

In the long run, more information will be available and the modeling should be adjusted again.

Estimation of Parameters When a Trend Must be Dealt With

In Section 4.1, pages 113–116, we studied the question of estimating location parameters $\mu(t_i; \beta_0, \ldots, \beta_p)$ of EV dfs which depend on the time t_i. Thus we assumed that there is certain trend in the data y_i that are measurements at time t_i. For that purpose we

- estimated the trend in the data by the least squares method;

- employed parametric estimators in the EV model to the residuals,

- deduced estimators of the parameters $\mu(t_i; \beta_0, \ldots, \beta_p)$.

Now, we assume that exceedances y_i over a threshold u are observed within a certain time period. The exceedances y_i at the exceedance times t_i follow GP dfs $W_{\gamma, u, \sigma(t_i; \beta_0, \ldots, \beta_p)}$. Thus, we allow a trend in the scale parameter σ.

The aim is to estimate γ and a trend function $\sigma(t; \beta_0, \ldots, \beta_p)$ in the scale parameter. Let n_j be the number of exceedances within a jth sub–period. We implicitly assume stationarity for the exceedances in each sub–period. We proceed in the following manner:

---

[17]Langbein, W.B. (1949). Annual floods and the partial duration flood. Transactions Geophysical Union 30, 879–881.

- determine estimates $\hat{\gamma}_j$ and $\hat{\sigma}_j$ for each sub–period;

- take $\hat{\gamma} = \sum_j n_j \hat{\gamma}_j / \sum_j n_j$ as an estimate of the stationary parameter γ,

- evaluate a trend function for the scale parameter by fitting a regression line to the σ_j.

EXAMPLE 5.1.2. (Ozone Cluster Maxima[18].) The data set used in this study has been collected by the Mexico City automatic network for atmospheric monitoring (RAMA) (from the south monitoring station called Pedregal). The data are the daily maxima of ozone levels measured in parts per million from July 1987 to April 1999, stored in mexozone.dat.

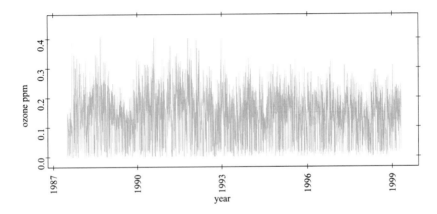

FIG. 5.4. Time series of daily maximum ozone levels from Pedregal Monitoring Station from July 1987 to April 1999.

We take eleven twelve month periods from July to June and one last period from July to April. The data y_i are cluster maxima taken according to the run length definition with a minimum gap of two days. In Table 5.1 we listed the estimates of the scale parameters σ_j for the threshold $u = 0.22$.

TABLE 5.1. Scale parameters σ of estimated GP distributions for the threshold $u = .22$ and estimated shape parameter $\gamma = -.35$.

						year						
	87	88	89	90	91	92	93	94	95	96	97	98
σ	.066	.061	.066	.082	.092	.068	.050	.051	.046	.041	.043	.028

---

[18]Villaseñor, J.A., Vaquera, H. and Reiss, R.–D. (2001). Long–trend study for ground level ozone data from Mexico City. In: 2nd ed. of this book, Part V, pages 353–358.

The regression curve fitted to the σ_j is $\exp(-0.003t^2 - 0.34t - 0.01)$. Similar calculations can be carried out within the exponential model where the shape parameter γ is equal to zero.

This data set is also studied in Example 6.5.2 in conjunction with the question of penultimate approximations.

Estimation Based on the Annual k Largest Order Statistics.

Up to now we investigated two different methods of extracting upper extremes from a set of data, namely, to take maxima out of blocks or to take exceedances over a higher threshold, where the latter includes the selection of the largest k values out of the data set. One may as well select the k largest values out of each block. If $k = 1$, this reduces to the usual blocks method, whereas for larger k this method is closer to the pot–method.

A parametric statistical model for the k largest order statistics is suggested by limiting distributions as in the case of maxima and exceedances (see, e.g., [42], Section 5.4). One may take submodels, which are the Gumbel, Fréchet and Weibull models if $k = 1$, or a unified model in the von Mises parameterization.

EXAMPLE 5.1.3. (Exceptional Athletic Records.) The data (communicated by Richard Smith), illustrated in Fig. 5.5 and stored in records.dat, concern women's 1500m and 3000m best performances from 1972 to 1992. M.E. Robinson and J.A. Tawn[19] and R.L. Smith[20] analyzed these data in view of an improvement of the 3000 m record from 502.62 to 486.11 seconds in the year 1993.

Confidence intervals and, respectively, prediction intervals showed evidence that the outstanding record in 1993 is inconsistent with the previous performances. Some insight into the nature of the record in 1993 would be also gained by computing T–year records based on the historical data.

From the scatterplots one recognizes trends and an "Olympic year" effect in the data. In view of the similar nature of both data sets and the small sample sizes, one could take into account a pooling of the data, cf. Section 14.3.

Asymptotic Distributions of Estimators and Confidence Intervals in the Generalized Pareto Model

We first deal with the case of iid random variables with common GP df.

---

[19]Robinson, M.E. and Tawn, J.A. (1995). Statistics for exceptional athletics records. Appl. Statist. 44, 499–511.

[20]Smith, R.L. (1997). Statistics for exceptional athletics records: letter to the editor. Appl. Statist. 46, 123–127.

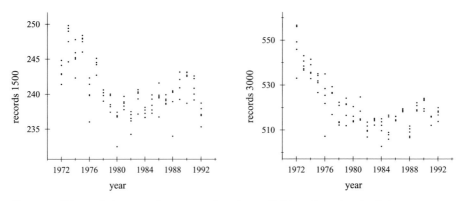

FIG. 5.5. Women's best performances for 1500m (left) and 3000m (right).

- (The MLE in the GP1 Model.) We formulate the asymptotic normality of the MLE $\hat{\alpha}_k$ in the GP1 model in terms of the γ–parameterization to make the result easily comparable to the performance of the MLE in the GP model. For iid random variables with common df $W_{1,\alpha,0,\sigma}$, the reciprocal MLE $1/\hat{\alpha}_k$ is asymptotically normal with mean $\gamma = 1/\alpha$ and variance γ^2/k.

- (The MLE in the GP Model.) The MLE $(\hat{\gamma}_k, \hat{\sigma}_k)$ of (γ, σ) in the GP model is asymptotically normal with mean vector (γ, σ) and covariance matrix Σ/k, where

$$\Sigma = (1+\gamma)\begin{pmatrix} (1+\gamma) & \sigma \\ \sigma & 2\sigma^2 \end{pmatrix}$$

if $\gamma > 1/2$. In particular, the asymptotic variance of $\hat{\gamma}_k$ is $(1+\gamma)^2/k$. This result is taken from the paper by Davison and Smith mentioned on page 121.

We see that the asymptotic variance of $1/\hat{\alpha}_k$ is much smaller than that of $\hat{\gamma}_k$ for γ close to zero. Yet, as noted on page 136, the MLE in the GP1 model can have a large bias if the special Pareto assumption is violated.

The estimation based on random exceedances or upper order statistics can be reduced to the preceding problem concerning iid random variables by using the conditioning devices in Section 8.1, page 234.

In the first case, one must deal with a random number K of exceedances over a threshold u: apply the preceding results conditioned on $K = k$. Then, the asymptotic normality holds for the estimators $1/\hat{\alpha}_K$ and $(\hat{\gamma}_K, \hat{\sigma}_K)$ and variances and covariance matrices with k replaced by $E(K)$. Secondly, u is replaced by the order statistic $X_{n-k:n}$ of n iid random variables X_i. The MLEs are computed conditioned on $X_{n-k:n} = u$. The resulting estimators have the same performance as $1/\hat{\alpha}_k$ and $(\hat{\gamma}_k, \hat{\sigma}_k)$. Such calculations can be carried out for a larger class of estimators.

We remark that $1/\hat{\alpha}_K$ and $(\hat{\gamma}_K, \hat{\sigma}_K)$ are MLEs in a pertaining point process model (see [43]). Moreover, these results remain valid under distributions which are sufficiently close (in the upper tail) to a GP df. We refer to (6.35) for an appropriate technical condition.

Based on the asymptotic normality result for estimators it is straightforward to construct confidence intervals for the parameters (as it was done in Section 3.2). It would also be of interest to deal with confidence bands and testing in conjunction with reliability functions such as the mean excess function. Yet, this goal is not further pursued in this book.

5.2 Testing Within Generalized Pareto Models

For testing an exponential upper tail of a distribution against other GP–type tails one may again use the sample skewness coefficient. Subsequently, we prefer the sample coefficient which leads to test statistic introduced by Hashofer and Wang[21] We also include the LR test for the exponential model. The tests are either based on exceedances over a threshold u or on the $k + 1$ largest order statistics.

Use the conditioning arguments given at the end of Section 8.1 to fix the critical values.

A Test Based on the Sample Coefficient of Variation

The coefficient of variation $\mathrm{var}_F^{1/2}/m_F$ is the standard deviation divided by the mean. In the following we use the reciprocal squared coefficient. This value is equal to $1 - 2\gamma$ for GP distributions $W_{\gamma,0,\sigma}$ with $\gamma < 1/2$. For exceedances $y_1, \ldots, y_k$ over a threshold u one may take the scale invariant test statistic $(\bar{y}_k - u)^2/s_k^2$. Such a test statistic was proposed by Gomes and van Montfort[22].

Using ordered values one obtains the corresponding (location and scale invariant) test statistic

$$\frac{(\bar{x}_k - x_{n-k:n})^2}{\frac{1}{k-1} \sum_{i \leq k} \left(x_{n-i+1:n} - \bar{x}_k\right)^2}, \tag{5.19}$$

where $\bar{x}_k = \frac{1}{k} \sum_{i \leq k} x_{n-i+1:n}$. This is the test statistic introduced by Hashofer and Wang for testing $\gamma = 0$ against $\gamma \neq 0$. The asymptotic considerations by Hashofer and Wang were based on results in a paper by Weissman[23] which is of interest in

---

[21]Hashofer, A.M. and Wang, Z. (1992). A test for extreme value domain of attraction. JASA 87, 171–177.

[22]Gomes, M.I. and van Montfort, M.A.J. (1986). Exponentiality versus generalized Pareto, quick tests. Proc. 3rd Internat. Conf. Statist. Climatology, 185–195.

[23]Weissman, I. (1978). Estimation of parameters and large quantiles based on the k largest observations. JASA 73, 812–815.

its own right; alternatively, one may use results in [48], pages 136–137, and the conditioning argument.

The local, asymptotic optimality of tests based on such test statistics was proven by Marohn (article cited on page 119).

Likelihood Ratio (LR) Test for the Exponential (GP0) Model

Within the GP model, one tests $H_0 : \gamma = 0$ against $H_1 : \gamma \neq 0$ with unknown scale parameters $\sigma > 0$. Thus, the exponential hypothesis is tested against other GP distributions. For a given vector $\boldsymbol{y} = (y_1, \ldots, y_k)$ of exceedances over the threshold u, the likelihood ratio (LR) statistic is

$$T_{\mathrm{LR}}(\boldsymbol{y}) = 2 \log \Big(\prod_{i \leq k} w_{\hat{\gamma}, u, \hat{\sigma}}(y_i) \Big/ \prod_{i \leq k} w_{0, u, \tilde{\sigma}}(y_i) \Big)$$

with $(\hat{\gamma}, \hat{\sigma})$ and $\tilde{\sigma}$ denoting the MLEs in the GP and GP0 models. The p–value is

$$p_{\mathrm{LR}}(\boldsymbol{y}) = 1 - \chi_1^2(T_{\mathrm{LR}}(\boldsymbol{y})).$$

The p–value, modified with a Bartlett correction, is

$$p(\boldsymbol{y}) = 1 - \chi_1^2\big(T_{\mathrm{LR}}(\boldsymbol{x})/(1 + b/k)\big)$$

for $b = 4.0$. Recall that u is replaced by $x_{n-k:n}$ if the largest ordered values are taken instead of exceedances.

5.3   Testing Extreme Value Conditions with Applications

co–authored by J. Hüsler and D. Li[24]

We introduce two methods of testing one dimensional extreme value conditions and apply them to two financial data sets and a simulated sample.

Introduction and the Test Statistics

Extreme value theory (EVT) can be applied in many fields of interest as hydrology, insurance, finance, if the underlying df F of the sample is in the max–domain of attraction of an extreme value distribution G_γ, denoted by $F \in \mathcal{D}(G_\gamma)$. We suppose commonly that the random variables $X_1, X_2, \ldots, X_n$ are iid with df F such that for some real γ there exist constants $a_n > 0$ and reals b_n with

$$\lim_{n \to \infty} F^n(a_n x + b_n) = G_\gamma(x) := \exp\Big(-(1 + \gamma x)^{-1/\gamma}\Big) \qquad (5.20)$$

---

[24]both at the Department of Math. Statistics, University of Bern.

where $1 + \gamma x > 0$. In case of $\gamma = 0$, $G_0(x)$ is interpreted as $\exp(-e^{-x})$, cf. also (1.25).

Under the extreme value condition (5.20), the common statistical approach consists of estimating the extreme value index γ and the normalizing constants b_n and a_n which are the location and scale parameters. Then based on these estimators, one can deduce quantiles, tail probabilities, confidence intervals, etc. The goodness–of–fit is usually analyzed in a qualitative and subjective manner by some graphical devices, as, e.g., by Q–Q plots and excess functions, which can easily be drawn with Xtremes, cf. also the remarks about "the art of statistical modeling" on page 61.

However the extreme value condition does not hold for all distributions. For example, the Poisson distribution and the geometric distribution are not in $\mathcal{D}(G_\gamma)$ for any real γ (see, e.g., C.W. Anderson[25], or Leadbetter et al. [39]). Thus, before applying extreme value procedures to maxima or exceedances one should check the basic null–hypothesis

$$H_0: \quad F \in \mathcal{D}(G_\gamma) \quad \text{for some real } \gamma.$$

We are going to introduce two tests in (5.21) and, respectively, (5.22) which we call test E and test T.

Dietrich et al.[26] present the following test E assuming some additional second order conditions. They propose to use the test statistic

$$E_n := k \int_0^1 \left(\frac{\log X_{n-[kt],n} - \log X_{n-k,n}}{\hat{\gamma}_+} - \frac{t^{-\hat{\gamma}_-} - 1}{\hat{\gamma}_-}(1 - \hat{\gamma}_-) \right)^2 t^\eta \, dt \qquad (5.21)$$

for some $\eta > 0$. It is shown that E_n converges in distribution to

$$E_\gamma := \int_0^1 \left((1 - \gamma_-)(t^{-\gamma_- -1}W(t) - W(1)) - (1 - \gamma_-)^2 \frac{t^{-\gamma_-} - 1}{\gamma_-} P \right.$$
$$\left. + \frac{t^{-\gamma_-} - 1}{\gamma_-} R + (1 - \gamma_-)R \int_t^1 s^{-\gamma_- -1} \log s \, ds \right)^2 t^\eta \, dt,$$

where $\gamma_+ = \max\{\gamma, 0\}$, $\gamma_- = \min\{\gamma, 0\}$, W is Brownian motion, and the random variables P and R are some integrals involving W (for details see Dietrich et al.). The estimates $\hat{\gamma}_+$ and $\hat{\gamma}_+$ for γ_+ and γ_-, respectively, are fixed to be the moment estimator, cf. page 134.

Thus the limiting random variable E_γ depends in general on γ and η, only. However, if $\gamma \geq 0$, the limiting distribution does not depend on γ, since the terms $(t^{-\gamma_-} - 1)/\gamma_-$ are to be interpreted as $-\log t$. Dietrich et al. state the result for

---

[25] Anderson, C.W. (1970). Extreme value theory for a class of discrete distributions with application to some stochastic processes. J. Appl. Probab. 7, 99–113.

[26] Dietrich, D., de Haan, L. and Hüsler, J. (2002). Testing xtreme value conditions. Extremes 5, 71–85.

$\eta = 2$, but it is easy to extend the result to any $\eta > 0$. This approach is based on sample qfs and of G_γ. They propose in addition another test statistic which is valid for $\gamma > 0$. But it seems that this test statistic is no so good as the test statistic E_n.

For testing H_0, one needs to choose an η and to derive the quantiles of the limiting random variable E_γ, which can be done by simulations for some γ, see Table 5.2 below.

TABLE 5.2. Quantiles $Q_{p,\gamma}$ of the limiting random variable E_γ for the test E with $\eta = 2$.

				p				
γ	0.10	0.30	0.50	0.70	0.90	0.95	0.975	0.99
≥ 0	0.028	0.042	0.057	0.078	0.122	0.150	0.181	0.222
-0.1	0.027	0.041	0.054	0.074	0.116	0.144	0.174	0.213
-0.2	0.027	0.040	0.053	0.072	0.114	0.141	0.169	0.208
-0.3	0.027	0.040	0.054	0.073	0.113	0.140	0.168	0.206
-0.4	0.027	0.040	0.054	0.073	0.114	0.141	0.169	0.207
-0.5	0.027	0.040	0.054	0.073	0.115	0.141	0.169	0.208
-0.6	0.027	0.040	0.054	0.074	0.116	0.144	0.173	0.212
-0.7	0.028	0.041	0.055	0.074	0.118	0.147	0.176	0.218

Then in applications one continues as follows:

- First, estimate $\hat\gamma_+$ and $\hat\gamma_-$ by the moment estimator and calculate the value of the test statistic E_n.

- Secondly, determine the corresponding quantile $Q_{1-\alpha,\hat\gamma}$ by linear interpolation using the values of Table 5.2. Here $\hat\gamma = \hat\gamma_+ + \hat\gamma_-$ and α is usually 0.05. Moreover we use $Q_{1-\alpha,\hat\gamma} = Q_{1-\alpha,-0.7}$ if $\hat\gamma < -0.7$.

- Finally, compare the value of E_n with the quantile $Q_{1-\alpha,\hat\gamma}$. If $E_n > Q_{1-\alpha,\hat\gamma}$, then reject H_0 with nominal type I error α.

Drees et al.[27] propose a test T restricted to $\gamma > -1/2$, assuming also some additional second order conditions. For each $\eta > 0$, their test statistic

$$T_n := k \int_0^1 \left(\frac{n}{k} \bar F_n \left(\hat a_{n/k} \frac{x^{-\hat\gamma} - 1}{\hat\gamma} + \hat b_{n/k} \right) - x \right)^2 x^{\eta-2} \, dx \tag{5.22}$$

converges in distribution to $T_\gamma := \int_0^1 \left(W(x) + L_\gamma(x) \right)^2 x^{\eta-2} \, dx$, where W is Brownian motion, and the process L_γ depends on the asymptotic distribution of $(\hat\gamma, \hat a, \hat b)$,

---

[27]Drees, H., de Haan, L. and Li, D. (2006). Approximations to the tail empirical distribution function with application to testing extreme value conditions. J. Statist. Plann. Inf. 136, 3498–3538.

which is some $\sqrt{k}$–consistent estimator of (γ, a, b). Now proceed as in case of test E based on the quantiles in Table 5.3.

TABLE 5.3. Quantiles $Q_{p,\gamma}$ of the limiting random variable T_γ for the test T with $\eta = 1.0$.

					p				
γ	0.10	0.30	0.50	0.70	0.90	0.95	0.975	0.99
4	0.086	0.123	0.161	0.212	0.322	0.393	0.462	0.558
3	0.085	0.120	0.156	0.205	0.307	0.372	0.440	0.532
2	0.083	0.116	0.150	0.195	0.286	0.344	0.402	0.489
1.5	0.082	0.115	0.148	0.192	0.282	0.340	0.400	0.480
1	0.082	0.114	0.146	0.189	0.276	0.330	0.388	0.466
0.5	0.083	0.116	0.149	0.194	0.285	0.343	0.404	0.481
0.25	0.085	0.119	0.153	0.120	0.295	0.355	0.415	0.499
0	0.089	0.126	0.163	0.213	0.319	0.388	0.455	0.542
−0.1	0.091	0.129	0.168	0.221	0.330	0.400	0.471	0.569
−0.2	0.093	0.133	0.174	0.231	0.350	0.425	0.500	0.604
−0.3	0.096	0.139	0.183	0.242	0.369	0.449	0.531	0.653
−0.4	0.100	0.145	0.192	0.256	0.393	0.484	0.576	0.690
−0.45	0.103	0.150	0.199	0.320	0.416	0.511	0.605	0.735
−0.499	0.107	0.157	0.210	0.338	0.439	0.546	0.652	0.799

The exact formulas of L_γ and T_γ are known for the MLE (for MLEs see Smith, cf. page 111, and Drees et al.[28]) which depend on γ and η, only. Similar to the test E, the test T can be applied for testing $F \in \mathcal{D}(G_\gamma)$, but only if $\gamma > -1/2$. Note that this second approach is based on the sample df.

Hüsler and Li[29] discuss the tests E and T by extensive simulations. If $F \notin \mathcal{D}(G_\gamma)$, then the power depends on how 'close' F is to some $F^* \in \mathcal{D}(G_\gamma)$. For instance, if $F = \text{Poisson}(\lambda)$ with λ small, the tests detect with high power the alternative. However, if λ is large, so the Poisson df could be approximated by a normal one, then the power is getting small. In summary, they suggest to

1. choose $\eta = 2$ for the test E and $\eta = 1$ (or maybe also $\eta = 2$) for the test T;

2. if the extreme value index γ seems to be positive, then both tests can be applied equally well to test H_0; otherwise the test E is preferable.

---

[28] Drees, H., Ferreira, A. and de Haan, L. (2004). On maximum likelihood estimation of the extreme value index. Ann. Appl. Probab. 14, 1179–1201.

[29] Hüsler, J. and Li, D. (2006). On testing extreme value conditions. Extremes 9, 69–86.

Data and Analysis

We apply the two tests to financial data. First, two examples consider log–returns, cf. Section 16.1, of daily equity over the period 1991–2003 for ABN ARMO and for the ING bank with sample size $n = 3283$. The other data are log–returns of daily exchange rates Dutch guilder–U.S. dollar and British pound–U.S. dollar during the period from January 1, 1976 to April 28, 2000 with sample size $n = 6346$.

For each data mentioned, we first calculate the maximum likelihood estimator (MLE) and moment estimator (Mon.) of the extreme value index, which are presented in the upper right plots in the figures for varying k. Secondly we derive the test statistics T_n and E_n and their corresponding 0.95 quantiles, which are shown in the two lower plots in each Fig. 5.6 to 5.9, also with varying k.

Fig. 5.6 and 5.7 show the data, the ML and the moment estimators for γ depending on the chosen number k and the two tests E and T applied to the largest k observations for the ING and the ABN AMRO example.

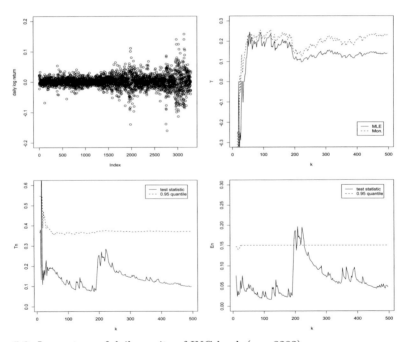

FIG. 5.6. Log–return of daily equity of ING bank ($n = 3283$).

It is evident that the test E is more sensitive to the selection of k; with large k usually the null–hypothesis is rejected by the test E. Each test provides a hint on the largest possible selection of k, preventing us to select a too large k. Note that we should select k such that $k/n \to 0$ as $n \to \infty$. Thus, a k up to 200 or up to 300 may be chosen in the ING and ABN ARMO examples, respectively. This is either based on the visual inspection of the γ plots or much better by the behavior of the tests shown in the test plots. In this perspective, the test E seems be more sensitive. Since both T_n and E_n are much smaller than their corresponding 0.95 quantiles for a large range of k (k/n is from 3% to 10%), both tests tend not to reject the null–hypothesis in both examples for the mentioned k's. So we may assume for each example that the underlying $F \in \mathcal{D}(G_\gamma)$.

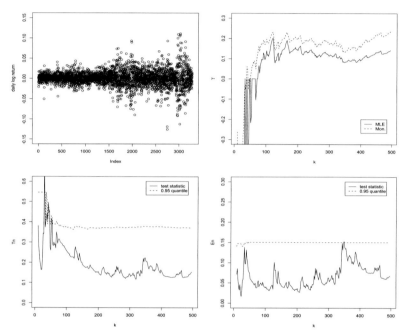

FIG. 5.7. Log–return of daily equity of ABN ARMO bank ($n = 3283$).

The situation is similar for the daily exchange rate examples, see Fig. 5.8 and 5.9. Both tests reject the null–hypothesis for large k, with $k/n > 10\%$ in the British pound–U.S. dollar exchange rate example. But again the test T is less sensitive with respect to k. This test T does not reject the null–hypothesis even for very large k in the Dutch guilder–U.S. dollar example.

We also apply the two tests to a simulated sample from Poisson(λ) distribution with sample size $n = 3000$ and $\lambda = 10$ and 100 (see Figures 5.10 and 5.11). As mentioned before $F =$ Poisson(10) or Poisson(100) are not in the max–domain and are not close to a normal distribution. Again the data and the MLE and the moment estimators are given. For k not small, both tests reject the null–hypothesis, so indicating that both Poisson(10) and Poisson(100) are not in the max–domain.

Note that the test E has a smoother behavior than the test T which may be influenced by the strange behavior of the MLE for these data. Also we see that the test E rejects the null–hypothesis already for smaller k than the test T.

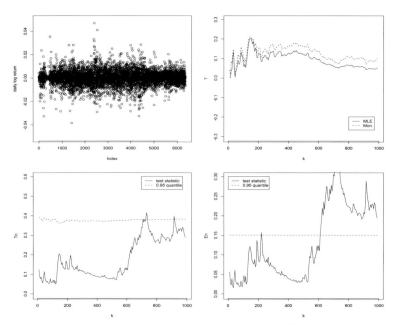

FIG. 5.8. Log–return of daily exchange rate British pound–U.S. dollar.

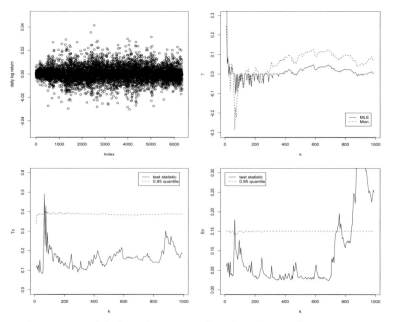

FIG. 5.9. Log–return of daily exchange rate Dutch guilder–U.S. dollar.

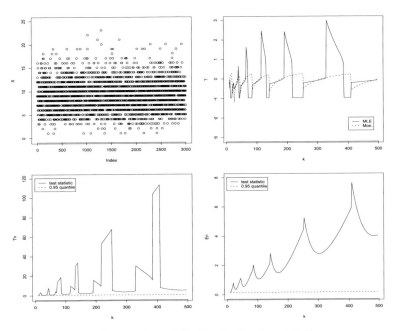

FIG. 5.10. One sample from Poisson(10) distribution ($n = 3000$).

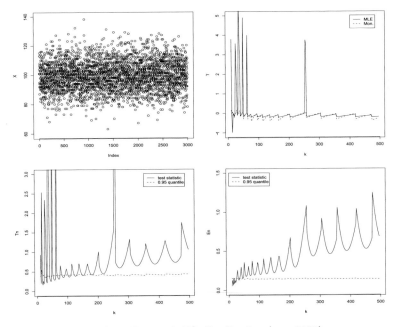

FIG. 5.11. One sample from Poisson(100) distribution ($n = 3000$).

5.4 Statistics in Poisson–GP Models

Assume that the original data $x_1, \ldots, x_n$ are governed by the df F. In contrast to the preceding sections, the exceedances over a given threshold u are grouped within certain cells.

Let u be a high threshold and consider a partition

$$u = u_1 < u_2 < \cdots < u_m < u_{m+1} = \infty \qquad (5.23)$$

of the interval $[u, \infty)$. The following considerations will be only based on the frequencies $n(i)$ of data in the m upper cells $[u_i, u_{i+1})$. Notice that $k = \sum_{i \leq m} n_i$ is the number of exceedances over u.

Note that the frequencies $n(i)$ are part of a multinomial scheme with cell probabilities $p_i := F(u_{i+1}) - F(u_i)$. If $1 - F(u)$ is sufficiently small, so that a Poisson approximation corresponding to (1.4) is applicable, then one may assume that the $n(1), \ldots, n(m)$ are mutually independent and each $n(i)$ is governed by a Poisson distribution P_{λ_i} with parameter $\lambda_i = np_i$.

We study the estimation within certain submodels which are deduced from Pareto (GP1) or generalized Pareto (GP) dfs.

The Poisson–GP1 Model

Assume that F is sufficiently close to some Pareto df $W_{1,\alpha,0,\sigma}$ above the threshold u and, hence, $F^{[u]}$ is close to $W_{1,\alpha,0,u}$. Then, the parameters $\lambda_i = np_i$ can be replaced by

$$\lambda_i(\alpha, \sigma) = n\sigma^\alpha (u_i^{-\alpha} - u_{i+1}^{-\alpha}), \qquad i = 1, \ldots, m,$$

where $u_{m+1}^{-\alpha} = 0$.

The MLEs α_m and σ_m of the parameters α and σ in the Poisson–GP1 model must be evaluated numerically, see [16], page 140.

EXAMPLE 5.4.1. (Increasing the Upper Limit in Liability Insurance.) Subsequently, we consider the question that there is an upper limit of one million, which makes the whole affair relatively harmless, and it is intended to increase the covering to the amount of four millions. Therefore, we are interested in the tail probabilities $p_1 = F(4) - F(1)$ and $p_2 = 1 - F(4)$, where F is the claim size distribution of those claim sizes over the priority (threshold) of 0.03 millions. Our calculations will be based on a data set of grouped claim sizes from the year 1983 (see Table 5.4; stored in im–claim.dat).

TABLE 5.4. Frequencies of claims within priorities in millions

priorities u_i	0.03	0.05	0.1	0.2	0.3	0.4	0.5	1
frequencies $n(i)$	147	89	35	11	7	5	4	3

According to the routine visualization of the data, it is plausible to accept the Pareto assumption. By applying the MLE in the Poisson–GP1 model, one obtains the estimate $\hat{\alpha} = 1.28$. The pertaining estimates of the tail probabilities are $\hat{p}_1 = 0.0093$ and $\hat{p}_2 = 0.0019$. These considerations indicate that an increase of the premiums by a smaller amount is sufficient. This calculation would lead to a disaster if the covering is removed, because very large claim sizes may occur with a probability which cannot be neglected.

The MLE in the Poisson–GP1 model has the same deficiencies as the Hill estimator. Therefore, we also compute the MLE in an enlarged model.

The Poisson–GP Model

This is a model deduced from generalized Pareto (GP) dfs $W_{\gamma,\mu,\sigma}$ with shape, location and scale parameters γ, μ and σ. If F is sufficiently close to $W_{\gamma,\mu,\sigma}$, then the truncated df $F^{[u]}$ is close to $W_{\gamma,u,\eta}$ with $\eta = \sigma + \gamma(u - \mu)$ (cf. (1.45)). This indicates that the problem of computing an MLE can be reduced to maximizing a likelihood function in the parameters γ and η.

First, the parameters $\lambda_i = np_i$ are replaced by

$$\lambda_i(\gamma, \mu, \sigma) = n\sigma^{1/\gamma}\varphi_i(\gamma, \eta), \qquad i = 1, \ldots, m,$$

where

$$\varphi_i(\gamma, \eta) = \big(\eta + \gamma(u_i - u)\big)^{-1/\gamma} - \big(\eta + \gamma(u_{i+1} - u)\big)^{-1/\gamma}, \qquad i = 1, \ldots, m-1,$$

and $\varphi_m(\gamma, \eta) = \big(\eta + \gamma(u_m - u)\big)^{-1/\gamma}$. Because $\sum_{i\leq m}\varphi_i(\gamma, \eta) = \eta^{-1/\gamma}$, the log–likelihood function as a function in γ, η and σ is

$$l(\gamma, \eta, \sigma) = \frac{k\log\sigma}{\gamma} + \sum_{i\leq m} n_i \log\varphi_i(\gamma, \eta) - n\left(\frac{\sigma}{\eta}\right)^{1/\gamma} + C, \qquad (5.24)$$

where C is a constant. The solution to the likelihood equation $(\partial/\partial\sigma)l = 0$ is

$$\sigma = \eta(k/n)^\gamma. \qquad (5.25)$$

Furthermore, the parameter μ can be regained from γ and η by

$$\mu = u - \eta\big(1 - (k/n)^\gamma\big)/\gamma. \qquad (5.26)$$

By inserting the right–hand side of (5.25) in (5.24), we see that γ and η must be determined as the values maximizing

$$\tilde{l}(\gamma, \eta) = k(\log\eta)/\gamma + \sum_{i\leq m} n_i \log\varphi_i(\gamma, \eta). \qquad (5.27)$$

This must be done numerically.

5.5   The Log–Pareto Model and Other Pareto–Extensions

First, we deal with mixtures of exponential, Pareto and converse Weibull distributions with respect to the Gamma distribution. We obtain the log–Pareto and the Burr model as an extension of the Pareto model. Secondly, the Benktander II and the truncated converse Weibull models are introduced which include the exponential and Pareto models.

Mixtures of Generalized Pareto Distributions

Recall from (4.8) that

$$F_h(x) = \int F_\vartheta(x) h(\vartheta)\, d\vartheta$$

is the mixture of the dfs F_ϑ with respect to density h. We give two examples in the present context:

- (Pareto Distributions as Mixtures of Exponential Distributions.) First, we show that mixtures of exponential dfs

$$F_\vartheta(x) = 1 - e^{-\vartheta x}, \qquad x > 0,$$

  with mean $1/\vartheta > 0$ are Pareto dfs, if the parameter ϑ is determined by sampling with respect to a gamma density $h_\alpha(x) = x^{\alpha-1} e^{-x}/\Gamma(\alpha)$, $x > 0$, with shape parameter $\alpha > 0$, cf. (3.42). The mixture

$$\int_0^\infty F_\vartheta(x) h_\alpha(\vartheta)\, d\vartheta = 1 - (1+x)^{-\alpha}, \qquad x > 0, \qquad (5.28)$$

  is a Pareto df with shape, location and scale parameters α, -1 and 1. For a continuation of this topic we refer to (16.30).

- (Log–Pareto Distributions as Mixtures of Pareto Distributions.) The log–Pareto df $L_{\alpha,\beta}$ with shape parameters $\alpha, \beta > 0$ possesses a super–heavy upper tail. The term "super–heavy" is characterized by the property that the log–transformation leads to a df with a heavy tail. We have

$$
\begin{aligned}
L_{\alpha,\beta}(x) &= W_{1,\alpha,-1,1/\beta}(\log x) \\
&= 1 - (1 + \beta \log x)^{-\alpha}, \qquad x \ge 1,
\end{aligned}
$$

  where $W_{1,\alpha,-1,1/\beta}$ is the Pareto df with shape, location and scale parameters α, -1 and $1/\beta$.

  The log–Pareto df $L_{\alpha,\beta}$ can be represented as a mixture of Pareto dfs with respect to the gamma density $h_{\alpha,\beta}$ with shape and scale parameters $\alpha, \beta > 0$.

Straightforward computations yield

$$\int_0^\infty W_{1,z}(x) h_{\alpha,\beta}(z) dz = L_{\alpha,\beta}(x), \qquad x \geq 1.$$

Note that mixing with respect to a Dirac measure leads to a Pareto distribution again.

Log–Pareto distributions within a generalized exponential power model are studied by Desgagné and Angers [30]. In a technical report[31] associated to an article by Diebolt et al., cited on page 132, these authors mention another mixture distribution, different from the log–Pareto one, with super–heavy tails. Log–Pareto random variables as innovations in an autoregressive process are studied by Zeevi and Glynn[32].

The Full Log–Pareto Model as an Extension of the Pareto Model

We present another parametrization of log–Pareto distributions with left endpoint of the support being equal to zero, and add a scale parameter, thus getting the full log–Pareto model.

Let X be a random variable with df

$$\widetilde{W}_{\alpha,\beta}(x) = 1 - (1 + x/\beta)^{-\alpha}, \quad x > 0, \tag{5.29}$$

which is a Pareto df with shape and scale parameters $\alpha > 0$ and $\beta > 0$.

Then, the transformed random variable $(\sigma/\beta)(\exp(X) - 1)$ has the log–Pareto df

$$\tilde{L}_{\alpha,\beta,\sigma}(x) = 1 - \left(1 + \frac{1}{\beta}\log\left(1 + \frac{\beta x}{\sigma}\right)\right)^{-\alpha}, \quad x > 0, \tag{5.30}$$

with shape parameters α, $\beta > 0$ and scale parameter $\sigma > 0$. We have

$$\tilde{L}_{\alpha,\beta,\sigma}(x) \rightarrow_{\beta \to 0} \widetilde{W}_{\alpha,\sigma}(x), \quad x > 0, \tag{5.31}$$

and, therefore, the log–Pareto model can be regarded as an extension of the Pareto model.

By repeating this procedure one gets models of loglog–Pareto dfs and, generally, iterated log–Pareto dfs with "mega–heavy" tails as extensions of the Pareto model. Notice that log–moments and iterated log–moments of such distributions are infinite.

---

[30]Desgagné, A. and Angers, J.–F. (2005). Importance sampling with the generalized exponential power density. Statist. Comp. 15, 189–195.

[31]see http://www.inria.fr/rrt/rr-4803.html

[32]Zeevi, A. and Glynn, P.W. (2004). Recurrence properties of autoregressive processes with super–heavy–tailed innovations. J. Appl. Probab. 41, 639–653.

The Burr Model as an Extension of the Pareto Model

Notice that $X^{1/\beta}$ is a standard Pareto random variable with shape parameter $\alpha\beta$, if $\beta > 0$ and X is a standard Pareto random variable with shape parameter α. Such a conclusion is no longer valid if a location parameter is added. The random variable $(X - 1)^{1/\beta}$ has a Burr df

$$F_{\alpha,\beta}(x) = 1 - (1 + x^\beta)^{-\alpha}, \qquad x \geq 0, \tag{5.32}$$

with shape parameters $\alpha, \beta > 0$. The density is

$$f_{\alpha,\beta}(x) = \alpha\beta x^{\beta-1}(1 + x^\beta)^{-(1+\alpha)}, \qquad x \geq 0.$$

Burr distributions can be represented as mixtures of converse Weibull dfs. For $\beta, \vartheta > 0$, let

$$\widetilde{G}_{\beta,\vartheta}(x) = 1 - \exp(-\vartheta x^\beta), \qquad x > 0,$$

be a Weibull df on the positive half–line. By mixing again with respect to the gamma density h_α over the parameter ϑ, one easily obtains

$$\int_0^\infty \widetilde{G}_{\beta,\vartheta}(x)h_\alpha(\vartheta)\,d\vartheta = 1 - (1 + x^\beta)^{-\alpha} = F_{\alpha,\beta}(x), \qquad x > 0. \tag{5.33}$$

For $\beta = 1$, one finds the Pareto df as a mixture of exponential distributions.

Burr distributions with parameter $\beta > 1$ have shapes that visually correspond to Fréchet densities.

Benktander II Distributions

We include two further families of distributions for modeling the tails of a distribution, namely the Benktander II and truncated converse Weibull distributions. Both types of distributions are well known within insurance mathematics and may be of interest in general.

Formula (2.28) for mean excess functions can be employed to design a family of dfs—called Benktander II dfs[33]—so that the mean excess functions are equal to x^b/a for $a > 0$ and $-a \leq b < 1$. Recall that such functions are attained by the mean excess functions of converse Weibull dfs with shape parameter $\alpha = b - 1$ as $u \to \infty$.

The standard Benktander II dfs with left endpoint equal to 1 are given by

$$F_{a,b}(x) = 1 - x^{-b}\exp\left(-\frac{a}{1-b}(x^{1-b} - 1)\right), \qquad x \geq 1, \tag{5.34}$$

where $a > 0$ and $-a \leq b < 1$ are two shape parameters.

---

[33]Benktander, G. (1970). Schadenverteilung nach Grösse in der Nicht–Leben–Versicherung. Mitteil. Schweiz. Verein Versicherungsmath., 263–283

Benktander II dfs have the interesting property that truncations $F_{a,b}^{[u]}$ are Benktander II dfs again in the form $F_{au^{1-b},b,0,u}$ with location and scale parameters $\mu = 0$ and $\sigma = u$. This is the pot–reproductivity of Benktander II distributions. Also,

- for $b = 0$, the Benktander II df is the exponential df with location and scale parameters $\mu = 1$ and $\sigma = 1/a$, and

- for $b \to 1$, one reaches the standard Pareto df with shape parameter $\alpha = 1+a$ in the limit.

Truncated Converse Weibull Distributions

It is analytically simpler to work with truncations of converse Weibull dfs themselves. These dfs have the additional advantage that we need not restrict our attention to Pareto dfs with shape parameter $\alpha > 1$.

Consider dfs

$$H_{1,\alpha,\beta}(x) = 1 - \exp\left(-\frac{\alpha}{1-\beta}\left(x^{1-\beta} - 1\right)\right), \qquad x \geq 1, \qquad (5.35)$$

where $\alpha > 0$ and β are two shape parameters.

By truncating converse Weibull dfs with shape and scale parameters $\tilde{\alpha} < 0$ and $\sigma > 0$ left of $u = 1$, one obtains dfs $H_{1,\alpha,\beta}$ with $\beta = 1 + \tilde{\alpha}$ and $\alpha = |\tilde{\alpha}|\sigma^{\tilde{\alpha}}$. Remember that converse Weibull densities have a pole at zero and a monotonic decrease if $-1 < \tilde{\alpha} < 0$ (which is equivalent to $0 < \beta < 1$).

The density and qf are

$$h_{1,\alpha,\beta}(x) = \alpha x^{-\beta} \exp\left(-\frac{\alpha}{1-\beta}\left(x^{1-\beta} - 1\right)\right), \qquad x \geq 1 \qquad (5.36)$$

and

$$H_{1,\alpha,\beta}^{-1}(q) = \left(1 - \frac{1-\beta}{\alpha}\log(1-q)\right)^{1/(1-\beta)}. \qquad (5.37)$$

In agreement with Benktander II dfs, the df $H_{1,\alpha,\beta}$ is

- an exponential df with location and scale parameters $\mu = 1$ and $\sigma = 1/\alpha$, if $\beta = 0$, and

- a standard Pareto df with shape parameter α in the limit as $\beta \to 1$.

In particular, it is understood that $H_{1,\alpha,1}$ is defined as such a limit in (5.35). Truncated converse Weibull dfs are pot–reproductive in so far as a truncation is of the same type. We have $H_{1,\alpha,\beta}^{[u]} = H_{1,\alpha u^{1-\beta},\beta,0,u}$.

We present a model for exceedances y_i over a threshold u:

$$\{H_{1,\alpha,\beta,0,u} : \alpha > 0,\, 0 \leq \beta \leq 1\}. \qquad (5.38)$$

For computing the MLE in that model, deduce from the first likelihood equation for the parameters α and $0 < b < 1$ that

$$\frac{1}{\alpha} = \frac{1}{k} \sum_{i \le k} \frac{(y_i/u)^{1-\beta} - 1}{1 - \beta}.$$

By inserting the value of α in the second likelihood equation, one finds

$$\left(\sum_{i \le k} \log(y_i/u) \right) \left(\frac{1}{k} \sum_{i \le k} \frac{(y_i/u)^{1-\beta} - 1}{1 - \beta} \right) + \sum_{i \le k} \frac{\partial}{\partial \beta} \frac{(y_i/u)^{1-\beta} - 1}{1 - \beta} = 0.$$

This equation must be solved numerically.

In the first equation, the results are that we gain

- the MLE for σ in the exponential model if $\beta = 0$, and

- the Hill estimate for α in the limit as $\beta \to 1$.

If the estimated β is sufficiently close to 1, then a Pareto hypothesis is justified. The threshold u may be replaced by an upper ordered value $x_{n-k:n}$ again. As in the GP case, a truncated converse Weibull distribution can be fitted to the upper tail of a distribution.

Including a Scale Parameter
in the Truncated Converse Weibull Model

A scale parameter $\sigma > 0$ should be included in the truncated converse Weibull model. This will be done in conjunction with a representation as in the GP case. Recall that the Pareto df W_γ in the γ–parameterization is the Pareto df $W_{1,\alpha,-\alpha,\alpha}$, where $\gamma = 1/\alpha > 0$. Likewise, let $H_{\gamma,\beta}$ be a truncated converse Weibull df $H_{1,\alpha,\beta,-\alpha,\alpha}$ with location and scale parameters $-\alpha$ and α, where $\gamma = 1/\alpha > 0$. We have

$$H_{\gamma,\beta}(x) = 1 - \exp\left(-\frac{1}{\gamma(1-\beta)} \left((1 + \gamma x)^{1-\beta} - 1 \right) \right), \qquad x \ge 0,$$

for shape parameters $\gamma > 0$ and $0 \le \beta \le 1$. It is apparent that $H_{\gamma,\beta}$ is

- the exponential df W_0, if $\beta = 0$,

- the Pareto df W_γ in the limit as $\beta \to 1$, and

- W_0 in the limit as $\gamma \to 0$.

Consider

$$\{H_{\gamma,\beta,u,\sigma} : \gamma, \sigma > 0, 0 < \beta \le 1\} \qquad\qquad (5.39)$$

as a statistical model for exceedances over a threshold u. A pot–reproductivity also holds in the extended framework.

Chapter 6

Advanced Statistical Analysis

Section 6.1 provides a discussion about non–random and random censoring. Especially, the sample df is replaced by the Kaplan–Meier estimate in the case of randomly censored data. In Section 6.2 we continue the discussion of Section 2.7 about the clustering of exceedances by introducing time series models and the extremal index. The insight gained from time series such as moving averages (MA), autoregressive (AR) and ARMA series will also be helpful for the understanding of time series like ARCH and GARCH which provide special models for financial time series, see Sections 16.7 and 16.8.

Student distributions provide further examples of heavy–tailed distributions besides Fréchet and Pareto distributions, see Section 6.3. Gaussian distributions are limits of Student distributions when the shape parameter α goes to infinity. Further distributions of that kind are sum–stable distributions with index $\alpha < 2$ which will be discussed in Section 6.4. Gaussian distributions are sum–stable with index $\alpha = 2$.

Rates of convergence towards the limiting distribution of exceedances and the concept of penultimate distributions are dealt with in Section 6.5. We also indicate a relationship between the concepts of pot–stability and regularly varying functions.

Higher order asymptotics for extremes is adopted in Section 6.6 to establish a bias reduction for estimators. This method is, e.g., useful to repair the Hill estimator.

6.1   Non–Random and Random Censoring

This section deals with the nonparametric estimation of the df by means of the Kaplan–Meier estimator (also called the product–limit estimator) and with esti-

mators in EV and GP models based on randomly censored data. Smoothing the
Kaplan–Meier estimator by a kernel leads to nonparametric density estimators.
We start with remarks about the fixed censoring.

Fixed Censoring

In the fixed censorship model, one distinguishes two different cases:

- (Type–I Censoring.) The data below and/or above a certain threshold u are
  omitted from the sample.

- (Type–II Censoring.) Lower and/or upper order statistics are omitted from
  the sample.

One also speaks of left/right censoring, if merely the lower/upper part of the
sample is omitted. Taking exceedances over a threshold or taking the k largest
ordered values is a left censoring (of Type–I or Type–II) of the sample. It would
be of interest to deal with such censorship models in conjunction with EV models[1].

Fixed Right Censoring in Generalized Pareto Models

Let y_i be the exceedances over a threshold u, yet those values above $c > u$ are
censored. Thus, we just observe $z_i = \min(y_i, c)$. A typical situation was described
in Example 5.4.1, where c is the upper limit of an insurance policy.
    This question can be dealt with in the following manner:

- if the number of $y_i \geq c$ is small and the statistical procedure for continuous
  data—as described in the preceding sections—is robust against rounding, we
  may just neglect the censoring,

- the number of censored data is the number of y_i in the cell $[c, \infty)$. If we take
  a partition of $[u, \infty)$ as in (5.23) with $u_m = c$, we may apply the procedures
  already presented in Section 5.4.

Randomly Censored Data

First, let us emphasize that our main interest concerns the df F_s under which the
survival times $x_1, \ldots, x_n$ are generated. Yet, data $(z_1, \delta_1), \ldots, (z_n, \delta_n)$ are merely
observed, where

- $z_i = \min(x_i, y_i)$ is x_i censored by some censoring time y_i,

- $\delta_i = I(x_i \leq y_i)$ provides the information as to whether censoring took place,
  where $\delta_i = 1$ if x_i is not censored, and $\delta_i = 0$ if x_i is censored by y_i.

---

[1]For relevant results see Balakrishnan, N. and Cohen, A.C. (1991). Order Statistics
and Inference. Academic Press, Boston.

Note that this procedure may be regarded as a right censoring with respect to a random threshold. A left censoring, where $z_i = \max(x_i, y_i)$, can be converted to a right censoring by changing the signs. Then, we have $-z_i = \min(-x_i, -y_i)$.

EXAMPLE 6.1.1. (Stanford Heart Transplantation Data.) We provide an example of censored survival times referring to the Stanford heart transplantation data. These data consist of the survival times (in days) of $n = 184$ patients who underwent a heart transplantation during the period from $t_0 \equiv$ Oct. 1, 1967 to $t_1 \equiv$ Feb. 27, 1980 at Stanford. We have

- the survival times $x_1, \ldots, x_n$, which are of primary interest, and

- the censoring data $y_i = t_1 - u_i$, where $u_i \in [t_0, t_1]$ is the date of operation. The survival time x_i is censored by y_i if the patient is alive at time t_1.

TABLE 6.1. Survival times x_i and censoring variables δ_i.

survival (in days)	0	1	1	1	2	3	·	·	·	2805	2878	2984	3021	3410	3695
censoring value	1	1	1	0	0	1	·	·	·	0	1	0	0	0	0

The full data set, taken from the collection by Andrews and Herzberg (cf. page 85), is stored in the file mc–heart.dat.

Subsequently, let the x_i and y_i be generated independently from each other under the survival df F_s and the censoring df F_c. In the preceding example, one may assume that F_c is the uniform df on the interval $[0, t_1 - t_0]$. Additionally, F_c is a degenerate df in the fixed censorship model. Our aim is to estimate the df F_s based on the z_i and δ_i. As in (1.6), check that the censored values z_i are governed by the df

$$
\begin{aligned}
H &= 1 - (1 - F_s)(1 - F_c) \\
&= F_s + (1 - F_s)F_c.
\end{aligned}
\tag{6.1}
$$

Note further that $\omega(H) = \min(\omega(F_s), \omega(F_c))$. One may distinguish the following situations.

- The fraction of censored data is small and the censoring may be regarded as some kind of a contamination that is negligible (such a case occurs when $\omega(F_c) \geq \omega(F_s)$ and $F_c(x)$ is small for $x < \omega(F_s)$). Yet, in this case, it can still be desirable to adopt procedures according to the present state–of–the–art censorship models.

- If the fraction of censored data is large, then the usual sample df $\widehat{F}_n$ based on the censored data z_i will not accurately estimate the target df F_s. It is necessary to apply procedures specially tailored for censored data.

- When using the Kaplan–Meier estimate one can recognize from the data whether $\omega(F_c) < \omega(F_s)$. Then we have $F(x) < 1 = H(x)$ for $\omega(F_c) \leq x < \omega(F_s)$. Notice that the observed z_i and δ_i contain no information about the form of the survival time df F_s beyond $\omega(F_c)$. One may try to estimate $F_s(x)$ for $x \leq \omega(F_c)$. Then, extrapolate this result to the region beyond $\omega(F_c)$ by means of a parametric modeling.

The Kaplan–Meier Estimate

Assume again that the survival times $x_1, \ldots, x_n$ are generated under the df F_s. Let $z_{1:n} \leq \ldots \leq z_{n:n}$ denote the ordered $z_1, \ldots, z_n$. If there are ties in the z_i's, the censored values are ranked ahead of the uncensored ones. Define the concomitant $\delta_{[i:n]}$ of $z_{i:n}$ by $\delta_{[i:n]} := \delta_j$ if $z_{i:n} = z_j$. The role of the usual sample df is played by the Kaplan–Meier estimate F_n^*. We have

$$
\begin{aligned}
F_n^*(x) &= 1 - \prod_{z_{i:n} \leq x} \left(1 - \frac{\delta_{[i:n]}}{n - i + 1} \right) \\
&= \sum_{i \leq n} w_{i,n} I(z_{i:n} \leq x),
\end{aligned}
\tag{6.2}
$$

where

$$
w_{i,n} = \frac{\delta_{[i:n]}}{n - i + 1} \prod_{j \leq i-1} \left(\frac{n - j}{n - j + 1} \right)^{\delta_{[j:n]}}.
$$

The $w_{i,n}$ are constructed in the following manner. At the beginning, each $z_{i:n}$ has the weight $1/n$. Let $z_{i_1:n} \leq \ldots \leq z_{i_m:n}$ be the censored ordered values. Then, let $w_{i,n} = 1/n$ for $i = 1, \ldots, i_1 - 1$ and $w_{i_1,n} = 0$. The weight $1/n$, initially belonging to $z_{i_1:n}$, is uniformly distributed over $z_{i_1+1:n} \leq z_{i_1+2:n} \leq \ldots \leq z_{n:n}$. Each of the remaining ordered values will then have the total weight $1/n + 1/(n(n - i_1))$. This procedure is continued for $i_2, \ldots, i_m$. We see that $\sum_{i \leq n} w_{i,n} = 1$ if $\delta_{[n:n]} = 1$. The weight

$$
\prod_{j \leq n-1} ((n - j)/(n - j + 1))^{\delta_{[j:n]}}
$$

eventually put on $z_{n:n}$ is omitted if $i_m = n$ [which is equivalent to $\delta_{[n:n]} = 0$]. Then, $\sum_{i \leq n} w_{i,n} < 1$ yielding $\lim_{x \to \infty} F_n^*(x) < 1$.

From this construction of the Kaplan–Meier estimate, it is apparent which part of the information contained in the data is still available. One can recapture the ordered uncensored data and the number of censored data between consecutive ordered uncensored data.

In analogy to (2.2), the Kaplan–Meier estimate $F_n^*(x)$ is approximately equal to the survival df $F_s(x)$, $x < \omega(H)$, for sufficiently large sample sizes n, written briefly[2]

$$\boxed{F_n^*(x) \approx F_s(x), \qquad x < \omega(H).}$$ (6.3)

This relation still remains valid for $x = \omega(H)$ if F_s is continuous at this point. Then, we also have $F_n^*(z_{n:n}) \approx F_s(\omega(H))$, which settles the question as to whether one can tell from the data that $F_s(\omega(H)) < 1$. This entails that the z_i and δ_i contain no information about an upper part of the survival time df F_s.

The Kernel Density Based on Censored Data

A kernel estimate of the density f of F_s is obtained by a procedure corresponding to that in Section 2.1 with weights $1/n$ replaced by $w_{i,n}$. Let

$$f_{n,b}(x) = \frac{1}{b} \sum_{i \le n} w_{i,n} \, k \left(\frac{x - z_{i:n}}{b} \right).$$

According to our preceding explanations, $f_{n,b}$ is an estimate of the survival time density f for $x < \omega(H)$. It is advisable to employ a right–bounded version.

EXAMPLE 6.1.2. (Continuation of Example 4.3.1 about Annual Maximum Wind Speeds for Jacksonville, Florida.) We want to identify more closely the distribution of the tropical maximum annual wind speeds dealt with in Example 4.3.1. These data can be regarded as randomly left censored by the non–tropical wind speeds. As mentioned before the left censoring can be converted to right censoring by changing signs (these data are stored in the file ec–jwind.dat). Xtremes is applied to these negative data, yet our following arguments concern the original wind speed data.

Plotting a kernel density only based on the tropical wind speeds and the kernel density for censored data we see that the distribution is now shifted to the left. This reflects the fact that we take into account tropical wind speeds which were censored by non–tropical ones. Another effect is that the distribution of tropical wind speeds can now be better described by a Fréchet density. Thus, the distribution is shifted to the left, yet the (visual) estimate indicates a heavier upper tail. This may have serious consequences for the forecast of catastrophic tropical storms.

Up to now there are no automatic procedures available that concern estimation within EV models based on censored data. First steps were done in the GP model.

---

[2] See, e.g., Shorack, G.R. and Wellner, J.A. (1986). Empirical Processes with Applications to Statistics. Wiley, New York.

Estimation in Generalized Pareto Models

Given $(z_1, \delta_1), \ldots, (z_n, \delta_n)$, consider the exceedances $\tilde{z}_1, \ldots, \tilde{z}_k$ over a threshold u. Denote the δ-values belonging to the $\tilde{z}_i$ by $\tilde{\delta}_1, \ldots, \tilde{\delta}_k$. It is clear that the $(\tilde{z}_i, \tilde{\delta}_i)$ only depend on the exceedances of the x_i and y_i over the threshold u and, hence, the exceedance dfs $F_s^{[u]}$ and $F_c^{[u]}$ are merely relevant. Assume that $F_s^{[u]}$ can be substituted by some GP df W if u is sufficiently large.

Maximize the likelihood function (made independent of the censoring df F_c)

$$(\gamma, \sigma) \to \log\left(\prod_{i \leq k} (w_{\gamma,\sigma}(\tilde{z}_i))^{\tilde{\delta}_i} (1 - W_{\gamma,\sigma}(\tilde{z}_i))^{1-\tilde{\delta}_i} \right),$$

where $w_{\gamma,\sigma}$ is the density of $W_{\gamma,\sigma}$. The likelihood function is computed under the condition that the x_i's and y_i's are independent replicates. Likewise, this can be performed in the α–parameterization.

GP1 MODEL: The MLE in the GP1 model of α for censored data is given by

$$\hat{\alpha}_k = \left(\sum_{i \leq k} \tilde{\delta}_i \right) \bigg/ \left(\sum_{i \leq k} \log(\tilde{z}_i/u) \right).$$

This estimator is related to the Hill estimator for non–censored data.

GP MODEL: The likelihood equations are solved numerically.

Extreme value analysis of randomly censored data is still in a research stadium. The preceding considerations should be regarded as a first step.

6.2   Models of Time Series, the Extremal Index

The purpose of discussing certain time series is twofold. Firstly, we want to provide examples for which the extremal index will be specified. Secondly, we make some preparations for Chapter 16, where series of speculative asset prices will be investigated within a time series framework.

Time series are based on white–noise series $\{\varepsilon_k\}$ which are sequences of uncorrelated random variables ε_k satisfying the conditions

$$E(\varepsilon_k) = 0 \quad \text{and} \quad E(\varepsilon_k^2) = \sigma^2.$$

Stochastic Linear Difference Equations

First, we deal with first–order difference equations for random variables Y_k starting at time zero. The value of Y_k is related to the value at the previous period by the equation

$$Y_k = \phi_0 + \phi_1 Y_{k-1} + \varepsilon_k$$

for certain constants ϕ_0, ϕ_1 with $|\phi_1| < 1$ and innovations ε_k.

Notice that the random variables Y_k satisfy these equations if, and only if, the random variables $X_k = Y_k - \phi_0/(1 - \phi_1)$ satisfy

$$X_k = \phi_1 X_{k-1} + \varepsilon_k, \qquad k = 1, 2, 3, \ldots . \tag{6.4}$$

By recursive substitution, one obtains

$$X_k = \phi_1^k X_0 + \sum_{i=0}^{k-1} \phi_1^i \varepsilon_{k-i}$$

as a solution to (6.4). If the innovations ε_k have zero–mean, then

$$EX_k = \phi_1^k EX_0 \to 0, \quad k \to \infty,$$

and

$$
\begin{aligned}
V(X_k) &= \phi_1^{2k} V(X_0) + \sigma^2 \big(\phi_1^{2k} - 1\big) \big/ \big(\phi_1^2 - 1\big) \\
&\to \sigma^2 / \big(1 - \phi_1^2\big), \quad k \to \infty,
\end{aligned}
\tag{6.5}
$$

if, in addition, $X_0, \varepsilon_1, \varepsilon_2, \varepsilon_3, \ldots$ are pairwise uncorrelated.

Weakly and Strictly Stationary Time Series

Besides weakly stationary processes—that are covariance–stationary processes (cf. page 72)—we also work with strictly stationary sequences $\{X_k\}$, where k runs over all integers or over the restricted time domain of nonnegative or positive integers.

For example, if in addition to the conditions specified around (6.5) the exact relations $EX_0 = 0$ and $V(X_0) = \sigma^2 / \big(1 - \phi_1^2\big)$ hold, then $\{X_k\}$ is a weakly stationary series with

$$\mathrm{Cov}(X_0, X_h) = \phi_1^h V(X_0). \tag{6.6}$$

We see that the solution of the stochastic linear equations converges to a stationary series.

A sequence $\{X_k\}$ is strictly stationary if the finite–dimensional marginal distributions are independent of the time lag h, that is, the joint distributions of $X_{k_1}, \ldots, X_{k_m}$ and, respectively, $X_{k_1+h}, \ldots, X_{k_m+h}$ are equal for all $k_1, \ldots, k_m$ and $h > 1$. One speaks of a Gaussian time series if all finite–dimensional marginals are Gaussian. For a Gaussian time series, weak and strict stationarity are equivalent.

EXAMPLE 6.2.1. (Continuation of Example 2.5.3 about Gaussian AR(1) Series.) The Gaussian time series $X_k = \phi_1 X_{k-1} + \varepsilon_k$ in (2.58) is a special case with $\phi_1 = d$ and $\varepsilon_k = (1 - d^2)^{1/2} Y_k$, where $Y_1, Y_2, Y_3, \ldots$ are iid standard Gaussian random variables. Because X_{k-1} and ε_k are independent we know, see (8.13), that the conditional df of X_k given $X_{k-1} = x_{k-1}$ is the df of $d x_{k-1} + (1 - d^2)^{1/2} Y_k$. This yields that the conditional expectation $E(X_k|x_{k-1})$ and conditional variance $V(X_k|x_{k-1})$ are equal to dx_{k-1} and, respectively, $1-d^2$.

Example 6.2.1 should be regarded as a first step towards the modeling of financial data by means of ARCH and GARCH series, see Sections 16.7 and 16.8.

AR(p) Time Series

Let the set of all integers k be the time domain. Given white noise $\{\varepsilon_k\}$ and constants $\phi_1,\ldots,\phi_p$, define an autoregressive (AR) time series $\{X_k\}$ of order p as a stationary solution to the AR equations

$$X_k = \phi_1 X_{k-1} + \cdots + \phi_p X_{k-p} + \varepsilon_k. \tag{6.7}$$

The time series in (6.6) and, specifically, in Example 6.2.1 with time domain restricted to the nonnegative or positive integers may also be addressed as AR(1) series.

MA(q) Time Series

An MA(q) time series $\{X_k\}$ for certain parameters $\theta_1,\ldots,\theta_q$ is defined by

$$X_k = \varepsilon_k + \theta_1 \varepsilon_{k-1} + \cdots + \theta_q \varepsilon_{k-q}. \tag{6.8}$$

This process is apparently covariance–stationary. The autocovariance function is

$$r(h) = \begin{cases} E\big(\varepsilon_1^2\big) \sum_{i=0}^{q-h} \theta_i \theta_{i+h} & h \le q, \\ & \text{for} \\ 0 & h > q, \end{cases}$$

where $\theta_0 = 1$.

If the ε_k are independent, then $\{X_k\}$ is strictly stationary. We mention two examples.

- (Gaussian MA(q) Series.) One obtains a Gaussian time series in the case of iid Gaussian random variables ε_k.

- (Cauchy MA(q) Series.) The construction in (6.8) can also be employed if the expectations or variances of the innovations ε_k do not exist. Then, one does not obtain an MA(q) process in the usual sense. Yet, the resulting process may still be strictly stationary. For example, one gets a strictly stationary sequence of Cauchy random variables with scale parameter $\sigma \sum_{i=0}^{q} \theta_i$, if the ε_k are iid Cauchy random variables with scale parameter σ and $\theta_i > 0$ for $i = 1,\ldots,q$, because the convolution of two Cauchy random variables with scale parameters c_1 and c_2 is a Cauchy random variable with scale parameter $c_1 + c_2$.

We remark that two values, namely 668.3 and 828.3 at 0.07 and 0.075, are not visible in the illustration.

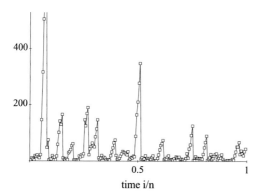

FIG. 6.1. Scatterplot of $n = 200$ MA(5) data of standard absolute Cauchy random variables with $c_i = i$.

ARMA(p,q) Time Series

By combining (6.8) and (6.7), one obtains the ARMA equations

$$X_k - \phi_1 X_{k-1} - \cdots - \phi_p X_{k-p} = \varepsilon_k + \theta_1 \varepsilon_{k-1} + \cdots + \theta_q \varepsilon_{k-q}. \qquad (6.9)$$

This equation may be written $\phi(B)X_k = \theta(B)\varepsilon_k$, where $B^j X_k = X_{k-j}$ is the backshift operator, and $\theta(B)$ and $\phi(B)$ are defined by means of the MA and AR polynomials

$$\theta(z) = 1 + \theta_1 z + \cdots + \theta_q z^q$$

and

$$\phi(z) = 1 - \phi_1 z - \cdots - \phi_p z^p.$$

An ARMA(p,q) time series $\{X_k\}$ is a stationary solution to these equations. Notice that MA(q) and AR(p) processes are special ARMA processes.

If the AR polynomial ϕ satisfies $\phi(z) \neq 0$ for all complex numbers z with $|z| \leq 1$, then it is well known that a stationary solution to the ARMA equations is found in the MA(∞) time series

$$X_k = \sum_{j=0}^{\infty} \psi_j \varepsilon_{k-j}, \qquad (6.10)$$

where the coefficients ψ_j are determined by the equation

$$\sum_{j=0}^{\infty} \psi_j z^j = \frac{\theta(z)}{\phi(z)}, \qquad |z| \leq 1.$$

The ARMA(p, q) process is said to be causal. We have $EX_k = 0$, and the autocovariance function is

$$r(h) = E\big(\varepsilon_1^2\big) \sum_{j=0}^{\infty} \psi_j \psi_{j+|h|}.$$

A series $\{Y_k\}$ is an ARMA(p, q) process with mean μ, if $X_k = Y_k - \mu$ is a causal ARMA(p, q) process.

One obtains a Gaussian time series $X_1, X_2, X_3, \ldots$ if the ε_k are iid Gaussian random variables. For that purpose, check that $(X_{k,n})_{k \leq m} = (\sum_{j=0}^{n} \psi_j \varepsilon_{k-j})_{k \leq m}$ is a Gaussian vector that converges pointwise to $(X_k)_{k \leq m}$, and the distribution of $(X_{k,n})_{k \leq m}$ converges weakly to a Gaussian distribution as n goes to ∞. For the simulation of such an ARMA process, one may use the innovation algorithm (cf. [5], page 264).

Estimation in ARMA-Models

We shortly describe estimators in AR and ARMA models. We refer to time–series books such as [26] and [5] for a more detailed description.

- (AR(p): Yule–Walker.) This is an estimator designed for the AR(p)–model. The Yule–Walker estimator computes estimates of the coefficients $\phi_1, \ldots, \phi_p$ of the AR(p) process and the variance σ^2 of the white noise series. The estimation is based on the Yule–Walker equations. Parameter estimates are obtained by replacing the theoretical autocovariances in the Yule–Walker equations by their sample counterparts.

- (ARMA(p, q): Hannan–Rissanen.) The Hannan–Rissanen algorithm uses linear regression to establish estimates for the parameters and the white noise variance of an ARMA(p, q) process. For this purpose, estimates of the unobserved white noise values $\epsilon_k, \ldots, \epsilon_{k-q}$ are computed.

- (ARMA(p, q): Innovations Algorithm.) One obtains estimates of the parameters and the white noise variance of a causal ARMA(p, q) process.

- (ARMA(p, q): MLE.) To compute the MLE of the parameters of a causal ARMA(p, q) process, one must use one of the preceding estimators to get an initial estimate.

The Extremal Index and the EV Modeling Revisited

The EV modeling for maxima can still be employed when the iid condition is considerably relaxed. For a stationary sequence of random variables X_i with common df F it is well known that (1.27) still holds under Leadbetter's mixing conditions D and D' (see, e.g., [39]) which concern a long range and a local mixing of the random variables.

If condition D' is weakened, one may still have

$$P\{\max\{X_1, \ldots, X_n\} \leq x\} \approx F^{\theta n}(x) \tag{6.11}$$

with $0 \leq \theta \leq 1$ for larger values x. The constant θ is the extremal index. This index is equal to 1 for iid random variables. A condition D'' which guarantees that

(6.11) holds was introduced by Leadbetter and Nandagopalan[3] and weakened to conditions $D^{(k)}$ by Chernick et al.[4] If (6.11) holds, then an EV modeling for dfs of maxima is still possible because G^θ is an EV df if G is an EV df. In (1.27), the location and scale parameters are merely changing. As already noted in Section 2.7, the extremal index is also the limiting reciprocal mean cluster size.

In the statistical literature, the extremal index was calculated for various stationary sequences. We mention only two examples concerning MA(q) and AR(1) sequences.

- (MA(q) Sequence.) If the innovations ε_i in (6.8) have a Pareto type upper tail with tail index $\alpha > 0$, then the extremal index is equal to[5]

$$\theta = \left(\max_{i \leq q} \{\theta_i\} \right)^\alpha \bigg/ \sum_{i \leq q} \theta_i^\alpha.$$

- (AR(1) Sequence.) Examples of AR(1) sequences with Cauchy marginals and $\theta < 1$ are dealt with in the aforementioned paper by Chernick et al.

For the Gaussian AR(1) sequences dealt with in Section 2.5, condition D' holds, and, therefore, the extremal index is equal to 1. This yields that maxima and exceedances over high thresholds asymptotically behave as those of iid random variables. This asymptotic result somewhat contradicts the clustering of exceedances in Fig. 2.21 for $d = 0.8$. Yet, this scatterplot only exhibits that the rate of convergence in the asymptotics is exceedingly slow. The illustration would not be drastically different when the sample size is increased to $n = 30,000$. An asymptotic formulation for that question that represents the small sample behavior of maxima (also that of the clustering) was provided by Hsing et al.[6]

Aggregation and Self–Similarity

Let $\{Z_i\}$ be a strictly stationary sequence of random variables. Consider the moving average

$$Z_k^{(m)} = \frac{1}{m} \sum_{i=(k-1)m+1}^{km} Z_i, \qquad k = 1, 2, 3, \dots \tag{6.12}$$

[3]Leadbetter, M.R. and Nandagopalan, S. (1989). On exceedance point processes for stationary sequences under mild oscillation restrictions. In [14], pp. 69–80.

[4]Chernick, M.R., Hsing, T. and McCormick, W.P. (1991). Calculating the extremal index for a class of stationary sequences. Adv. Appl. Prob. 23, 835–850.

[5]Davis, R.A. and Resnick, S.I. (1985). Limit theory for moving averages of random variables with regular varying tail probabilities. Ann. Prob. 13, 179–195.

[6]Hsing, T., Hüsler, J. and Reiss, R.–D. (1996). The extremes of a triangular array of normal random variables. Ann. Appl. Prob. 6, 671–686.

which is called aggregated sequence with level of aggregation m. It is evident that the aggregated sequence is again strictly stationary. In that context, one also speaks of self–similarity when the aggregated sequence is distributional of the same type as the original one. For example, if the X_i are iid standard normal, then

$$m^{1/2} X_k^{(m)} \stackrel{d}{=} X_1, \tag{6.13}$$

 that is we have equality of both random variables in distribution. This can be extended to all sum–stable random variables such as Cauchy and Levy random variables, see Section 6.4. Another example is provided in the subsequent Section 6.3 about Student distributions.

6.3   Statistics for Student Distributions

In (1.62) we introduced the standard Student distribution with shape parameter $\alpha > 0$ and density

$$f_\alpha(x) = c(\alpha) \left(1 + \frac{x^2}{\alpha} \right)^{-(1+\alpha)/2} \tag{6.14}$$

where $c(\alpha) = \Gamma((1 + \alpha)/2)/(\Gamma(\alpha/2)\Gamma(1/2)\alpha^{1/2})$. Apparently, such a Student distribution has lower and upper tails with both tail indices equal to α. We see that the Cauchy distribution is a special case for $\alpha = 1$, cf. page 27.

For positive integers $\alpha = n$, we obtain in (6.14) the well–known Student distribution (t–distribution) with n degrees of freedom. This distribution can be represented by

$$X/(Y/n)^{1/2}, \tag{6.15}$$

where X and Y are independent random variables distributed according to the standard normal df and the χ^2–df with n degrees of freedom, cf. page 123. Recall that a χ^2 random variable with n degrees of freedom is a gamma random variable with shape parameter $r = n/2$ and scale parameter $\sigma = 2$.

A Stochastic Representation of Student Distributions

Using gamma random variables one may find a representation as in (6.15) for all Student distributions.

Consider the ratio $X/(Y/r)^{1/2}$, where again X and Y are independent, X is standard normal and Y is a standard gamma random variable with parameter r, cf. (4.6). The density is given by

$$
\begin{aligned}
g_r(z) &= \int_0^\infty \left(\vartheta \varphi(\vartheta z) \right) \left(2r\vartheta h_r(r\vartheta^2) \right) d\vartheta \\
&= \frac{\Gamma((2r + 1)/2)}{\Gamma(r)(2\pi r)^{1/2}} \left(1 + \frac{z^2}{2r} \right)^{-(2r+1)/2}.
\end{aligned}
\tag{6.16}
$$

This is the Student density in (6.14), if $r = \alpha/2$.

Notice that the Student density f_α is the reciprocal scale mixture of normal densities taken with respect to the distribution of $(2Y/\alpha)^{1/2}$, where Y is a gamma random variable with parameter $r = \alpha/2$. Alternatively, one may speak of a scale mixture of normal densities with respect to the distribution of $(2Y/\alpha)^{-1/2}$.

Properties of Student Distributions

- (Asymptotic Normality.) Using the relation

$$(1 + x^2/\alpha)^{\alpha/2} \to \exp(-x^2/2)$$

  as $\alpha \to \infty$ one may prove that f_α is the standard Gaussian density in the limit as $\alpha \to \infty$.

- (Construction of a Self–Similar Sequence.) Let the $\{X_i\}$ be iid standard normal random variables, and let Y be a gamma random variable with parameter $r = \alpha/2$ which is independent of the sequence $\{X_i\}$. Then, the

$$Z_i = X_i/(2Y/\alpha)^{1/2}$$

  are Student random variables with parameter α. Consider again moving averages $Z_k^{(m)}$ in (6.12). As a direct consequence of (6.13) one obtains

$$m^{1/2} Z_k^{(m)} \stackrel{d}{=} Z_1$$

  and, therefore, the Z_i are self–similar.

Maximum Likelihood Estimation in the Student Model

This is one of the earliest examples of the use of the ML method (R.A. Fisher (1922) in the article cited on page 47).

The likelihood equations have no explicit solution and must be numerically computed. In such cases, we will apply a Newton–Raphson iteration procedure by which the initial value of the iteration is an estimate determined by another estimation method. A local maximum of the likelihood function is occasionally computed instead of a global maximum.

Including a Skewness Parameter

Including a location parameter δ in the standard normal random variable X one obtains a noncentral Student distribution with noncentrality parameter δ.

An extension to the multivariate framework may be found in Section 11.3.

6.4   Statistics for Sum–Stable Distributions
co–authored by J.P. Nolan[7]

We add some statistical results for non–degenerate, sum–stable distributions which are different from the Gaussian ones. Sum–stable distributions are heavy–tailed, like Student distributions, if the shape parameter α (index of stability) is smaller than 2. All sum–stable distributions have unimodal densities.

Generally, a df F is sum–stable, if

$$F^{m*}(a_m x + b_m) = F(x) \tag{6.17}$$

for a certain choice of constants $a_m > 0$ and b_m. It is well known that $a_m = m^{1/\alpha}$ for some α with $0 < \alpha \leq 2$, also see page 31.

Thus, we have $\alpha = 2$ for the normal df and $\alpha = 1$ for the Cauchy df. The Lévy distribution is another example of a sum–stable distribution with $\alpha = 1/2$, for which a simple representation is feasible. The Lévy density is

$$f(x) = (2\pi)^{-1/2} x^{-3/2} \exp(-(2x)^{-1}), \qquad x \geq 0, \tag{6.18}$$

which is a special reciprocal gamma density, see (3.43).

Check that

$$F(x) = 2\big(1 - \Phi\big(\sqrt{1/x}\,\big)\big)$$

is the pertaining df. In contrast to the normal and Cauchy distributions, the Lévy distribution is not symmetric. The heavy tails and the possible asymmetry make stable laws an attractive source of models.

The stability property (6.17) is closely related to the Generalized Central Limit Theorem (GCLT). The classical Central Limit Theorem applies to terms having finite variance: if $X_1, X_2, \ldots$ have finite variance, then

$$m^{-1/2}(X_1 + X_2 + \cdots + X_m) - m^{1/2} E X_1$$

converges in distribution to a normal distribution. If the terms $X_1, X_2, \ldots$ have infinite variance, then the normalized sums

$$c_m(X_1 + X_2 + \cdots + X_m) - d_m$$

converge in distribution to a stable law, where the term c_m must be of the form $m^{-1/\alpha}$. Hence, stable laws are the only possible limiting distributions of normalized sums of iid terms.

Until recently, stable distributions were inaccessible for practical problems because of the lack of explicit formulas for the densities and distribution functions. However, new computer programs make it feasible to use stable distributions in applications.

---

[7] American University, Washington DC

All computations with sum–stable distributions were done using a DLL–version of STABLE[8].

Some Basic Facts About Characteristic Functions

Since there are no explicit formulas for stable densities, they are usually described through their characteristic functions or Fourier transforms. Let $x + iy$ be the usual representation of complex numbers and recall that

$$\exp(ix) = \cos(x) + i\sin(x).$$

The expectation of a complex variable $X + iY$ is defined as

$$E(X + iY) = E(X) + iE(Y).$$

The characteristic function χ_X of a real–valued random variable X is defined as

$$\chi_X(t) = E\big(\exp(itX)\big). \tag{6.19}$$

It is well known that there is a one–to–one relationship between the distribution of X and the characteristic function χ_X. The importance of characteristic functions becomes apparent from the fact that

$$\chi_{X+Y} = \chi_X \chi_Y$$

for independent random variables X and Y.

Because $\exp(u + w) = \exp(u)\exp(w)$ for complex numbers u and w we have

$$\begin{aligned}
\chi_{\mu+\eta X}(t) &= E\big(\exp(i\eta tX)\big)\exp(i\mu t) \\
&= \chi_X(\eta t)\exp(i\mu t) \tag{6.20}
\end{aligned}$$

for real numbers μ and η. Therefore, it suffices to specify the characteristic functions of standard variables for the construction of statistical models.

Symmetric Sum–Stable Distributions

We first introduce the standard sum–stable dfs $S(\alpha)$ with index of stability $0 < \alpha \leq 2$ which are symmetric about zero. These dfs can be represented by the real–valued characteristic functions

$$\chi_\alpha(t) = \exp(-|t|^\alpha). \tag{6.21}$$

The support of the df $S(\alpha)$ is the real line.

We use the following notation for

---

[8]The basic algorithms are described in Nolan, J.P. (1997). Numerical computation of stable densities and distribution functions. Commun. Statist.–Stochastic Models 13, 759–774. Note that the STABLE package is no longer included in Xtremes.

Symmetric Sum–Stable Distributions: $S(\alpha, 0, \sigma, \mu; 0),$

if scale and location parameters $\sigma > 0$ and μ are included. At the second position there is a skewness parameter which is equal to zero in the case of symmetric sum–stable distributions. The zero after the semicolon indicates a certain type of parameterization which will become important for skewed stable distributions.

Observe that the characteristic functions in (6.21) are continuous in the parameter α. The Fourier inverse formula for densities implies that this also holds for the pertaining densities and dfs. This property remains valid when location and scale parameters are added.

It is well known that $S(2, 0, 1/\sqrt{2}, 0; 0)$ is the standard normal df Φ. Thus, (6.20) yields

$$S(2) = S(2, 0, 1, 0; 0) = \Phi_{0, \sqrt{2}}$$

which is the normal df with location parameter zero and scale parameter $\sigma = \sqrt{2}$. In addition, $S(1, 0, 1, 0; 0)$ is the standard Cauchy df.

In Fig. 6.2 we plot some symmetric sum–stable densities varying between the normal and the Cauchy density.

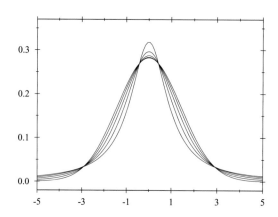

FIG. 6.2. Plots of symmetric, sum–stable densities with index of stability α equal to 2 (normal), 1.75, 1.5, 1.25 and 1 (Cauchy).

We see that the most significant difference between the normal density and other sum–stable symmetric densities is the weight in the tails. One could get a greater similarity in the center of the distributions by varying the scale parameter.

The formulas become a bit more complicated when a skewness parameter is included. In that context we deal with two different parameterizations.

Sum–stable distributions have Pareto–like tails. Yet, it is well known that estimators of the tail index $\alpha < 2$ are inaccurate if α is close to 2. This phenomenon will be clarified by computing the remainder term in the Pareto approximations.

From Christoph and Wolf [7] we know that the expansion

$$f_\alpha(x) = \frac{1}{\pi} \sum_{j \leq m} (-1)^{j+1} \frac{\Gamma(1 + j\alpha)}{j!} \sin(j\alpha\pi/2) |x|^{-(1+j\alpha)} + O\big(A_m(x)\big), \qquad (6.22)$$

holds for the density f_α of $S(\alpha)$, where

$$A_m(x) = O\big(|x|^{-(1+(m+1)\alpha)}\big), \qquad |x| \to \infty.$$

For $m = 3$, we have

$$f_\alpha(x) = \frac{1}{\sigma(\alpha)} w_{1,\alpha}\Big(\frac{x}{\sigma(\alpha)}\Big)\Big(1 + C(\alpha)\Big(\frac{x}{\sigma(\alpha)}\Big)^{-\alpha} + O\Big(\frac{x}{\sigma(\alpha)}\Big)^{-2\alpha}\Big) \qquad (6.23)$$

for certain scale parameters $\sigma(\alpha)$ and constants $C(\alpha)$, where $w_{1,\alpha}$ is the standard Pareto density with shape parameter α. The constant $C(\alpha)$ in the second order term satisfies

$$C(\alpha) = 24/(2 - \alpha) + O(1), \qquad \alpha \to 2, \qquad (6.24)$$

which shows that these constants are very large for α close to 2.

Adding a Skewness Parameter, a Continuous Parameterization

To represent the family of all non–degenerate, sum–stable distributions one must include a skewness parameter $-1 \le \beta \le 1$ in addition to the index of stability $0 < \alpha \le 2$. For $\alpha = 2$ one gets the normal df $S(2)$ for each β.

We choose a parameterization so that the densities and dfs vary continuously in the parameters. Such a property is indispensable for statistical inference and visualization techniques.

The continuous location–scale parameterization introduced by Nolan[9] is a variant of the (M) parameterization of Zolotarev[10]. Let

$$\chi_X(t) = \begin{cases} \exp\Big(-|t|^\alpha\big(1 + i\beta \tan\big(\frac{\alpha\pi}{2}\big)\mathrm{sign}(t)\big(|t|^{1-\alpha} - 1\big)\big)\Big) & \alpha \ne 1, \\ & \text{if} \\ \exp\Big(-|t|\big(1 + i\beta\frac{2}{\pi}\mathrm{sign}(t)\log(|t|)\big)\Big) & \alpha = 1. \end{cases}$$
$$(6.25)$$

Using this parameterization one gets a family of characteristic functions which is continuous in the parameters α and β. To prove the continuity at $\alpha = 1$, notice that $\tan(\pi/2 + x) = -1/x + o(x)$ and, therefore, according to (1.66)

$$\tan\Big(\frac{\alpha\pi}{2}\Big)\big(|t|^{1-\alpha} - 1\big) = \frac{2}{\pi}\frac{|t|^{1-\alpha} - 1}{1 - \alpha} + o(1 - \alpha)$$

$$\to (2/\pi)\log|t|, \qquad \alpha \to 1.$$

If $\beta = 0$, then one gets the family of characteristic functions in (6.21) for symmetric sum–stable distributions. The dfs pertaining to the characteristic functions in (6.25) are denoted by $S(\alpha, \beta; 0)$. We write

---

[9]Nolan, J.P. (1998). Parameterizations and modes of stable distributions. Statist. Probab. Letters 38, 187-195.

[10]Zolotarev, V.M. (1986). One–Dimensional Stable Distributions. Translations of Mathematical Monographs, Vol. 65, American Mathematical Society.

Continuous Parameterization: $S(\alpha, \beta, \sigma, \mu; 0)$,

if scale and location parameters $\sigma > 0$ and μ are included. As in the case of symmetric sum–stable distributions one obtains families of dfs and densities which are continuous in all parameters.

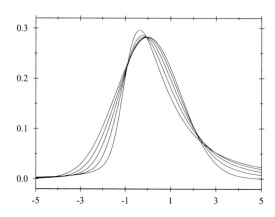

FIG. 6.3. Plots of sum–stable densities with index of stability α equal to 2 (normal), 1.75, 1.5, 1.25 and 1 (Cauchy) and skewness parameter $\beta =$ 0.75.

Deduce from (6.20) that a random variable $-X$ has the df $S(\alpha, -\beta; 0)$, if X has the df $S(\alpha, \beta; 0)$. The support of $S(\alpha, \beta; 0)$ is the real line, if $-1 < \beta < 1$.

The dfs $S(\alpha, 1; 0)$ and $S(\alpha, -1; 0)$ have the supports $(-\tan(\pi\alpha/2), \infty)$ and $(-\infty, \tan(\pi\alpha/2))$, respectively. Moreover, $S(1/2, 1, 1, 1; 0)$ is the Lévy df, as defined by (6.18).

The tails of most stable laws are Pareto–like: when $\beta \in (-1, 1]$, the tail probability and density satisfy as $x \to \infty$,

$$
\begin{aligned}
P(X > x) &\sim c_\alpha (1 + \beta) x^{-\alpha}, \\
f(x) &\sim \alpha c_\alpha (1 + \beta) x^{-(\alpha+1)},
\end{aligned}
\qquad (6.26)
$$

where $c_\alpha = \Gamma(\alpha) \sin(\pi\alpha/2)/\pi$. (The $\beta = -1$ case decays faster than any power.) This asymptotic power decay has been used to estimate stable parameters, but in most cases one must go extremely far out on the tails to see this tail behavior, e.g., Fofack and Nolan[11] and Section 6.5, so this approach is of limited usefulness in practice.

Adding a Skewness Parameter, the Conventional Parameterization

We also mention a parameterization which is more commonly used and which is particularly convenient for mathematical proofs (see, e.g., [7] and [46]). Given a

---

[11] Fofack, H. and Nolan, J.P. (1999). Tail behavior, modes and other characteristics of stable distributions. Extremes 2, 39–58.

random variable X with df $S(\alpha, \beta; 0)$ let

$$Y = \begin{cases} \gamma\big(X + \beta \tan(\pi\alpha/2)\big) + \delta, & \alpha \neq 1, \\ & \text{if} \\ \gamma\big(X + (2/\pi)\beta \log(\gamma)\big) + \delta, & \alpha = 1, \end{cases} \tag{6.27}$$

for parameters $\gamma > 0$ and δ. The pertaining dfs are denoted by

Conventional Parameterization: $\qquad S(\alpha, \beta, \gamma, \delta; 1)$.

Notice that γ and δ are scale and location parameters if $\alpha \neq 1$. The parameter δ is a location parameter for the dfs $S(\alpha, \beta, \gamma, 0; 1)$. If $\beta = 0$, then the continuous and convential parameterizations coincide. According to (6.20) and (6.25) the pertaining characteristic functions are

$$\chi_Y(t) = \begin{cases} \exp\left(-\gamma^\alpha |t|^\alpha \big(1 - i\beta \tan\big(\tfrac{\alpha\pi}{2}\big)\mathrm{sign}(t)\big) + i\delta t\right) & \alpha \neq 1, \\ & \text{if} \\ \exp\left(-\gamma |t| \big(1 + i\beta\tfrac{2}{\pi}\mathrm{sign}(t)\log(|t|)\big) + i\delta t\right) & \alpha = 1. \end{cases} \tag{6.28}$$

This parameterization is not continuous at $\alpha = 1$ and, therefore, less useful for statistical inference. The reader is cautioned that several other parameterizations are used for historical and technical reasons.

Scale Mixtures of Normal Distributions

As in the case of Student distributions one may represent symmetric, sum-stable distributions as scale mixtures of normal distributions. Now, the mixing df is the sum–stable df with skewness parameter $\beta = 1$ and index $\alpha < 1$.

Let X and Y be independent random variables, where X is standard normal and Y has the df $S(\alpha/2, 1, 1, 0; 1)$ for some $\alpha < 2$. Notice that Y has the support $(0, \infty)$. Then,

$$Y^{1/2}X \tag{6.29}$$

has the df $S(\alpha, 0, \gamma, 0; 1)$, where $\gamma = 2^{-1/2}\sec(\pi\alpha/4)^{1/\alpha}$ with $\sec(x) = 1/\cos(x)$ denoting the secant function.

For example, if Y is the Lévy random variable, having the index $\alpha = 1/2$, then the scale mixture is the Cauchy df.

This construction will be of importance in the multivariate setup for the characterization of multivariate sum–stable, spherically or elliptically contoured dfs, see Section 11.4.

Estimating Stable Parameters

For estimating the four parameters in the $S(\alpha, \beta, \sigma, \mu; 0)$–parameterization several methods are applicable. We mention

- Quantile Method: McCulloch[12] uses five sample quantiles (with $q = 0.05$, 0.25, 0.5, 0.75, 0.95) and matches the observed quantile spread with the exact quantile spread in stable distributions.

- Empirical Characteristic Function Method: Kogon and Williams[13] use the known form of the characteristic function to estimate stable parameters. The empirical/sample characteristic function is computed from the sample $X_1, \ldots, X_n$ by

$$\hat{\chi}(t) = \Big(\sum_{j \leq n} \exp(itX_j) \Big) \Big/ n.$$

The sample parameters are estimated by regression techniques from the exact form for $\chi(t)$ given in (6.25).

- MLE: This method uses initial estimate of $(\alpha, \beta, \sigma, \mu)$ from the quantile method, and then maximizes the likelihood by a numerical search in the 4–dimensional parameter space. Large sample confidence interval estimates have been computed, see Nolan[14].

The estimated parameters may be converted to the conventional parameterization.

Diagnostics with Kernel Densities

In principle, it is not surprising that one can fit a data set better with the 4 parameter stable model than with the 2 parameter normal model. The relevant question is whether or not the stable fit actually describes the data well.

The diagnostics we are about to discuss are an attempt to detect non–stability. As with any other family of distributions, it is not possible to prove that a given data set is or is not stable. The best we can do is to determine whether or not the data are consistent with the hypothesis of stability. These tests will fail if the departure from stability is small or occurs in an unobserved part of the range.

The first step is to do a smoothed density plot of the data. If there are clear multiple modes or gaps in the support, then the data cannot come from a stable distribution. For density plots, we smoothed the data with the Epanechnikov or a Gaussian kernel with standard deviation given by a width parameter. In addition to the automatic choice we used trial and error to find a width parameter that was

---

[12]McCulloch, J.H. (1986). Simple consistent estimators of stable distribution parameters. Commun. Statist. Simulations 15, 1109–1136.

[13]Kogon, S.M. and Williams, D.B. (1998). Characteristic function based estimation of stable distribution parameters. In: A Practical Guide to Heavy Tails, R.J. Adler et all (eds.), Birkhäuser, Basel, 311-335.

[14]Nolan, J.P. (2001). Maximum likelihood estimation of stable parameters. In: Lévy Processes, ed. by Barndorff–Nielsen, Mikosch, and Resnick, Birkhäuser, Basel, 379–400.

as small as possible without showing too much oscillations from individual data points.

EXAMPLE 6.4.1. (Distributions of Black Market Exchange Rate Returns.) We examine two data sets of consecutive monthly log-returns (cf. Chapter 16) from Colombia and Argentina[15] of a relatively small small sample size of 119 studied by Akgiray et al.[16] and further investigated by Koedijk et al.[17]. Akgiray et al. estimated parameters of sum–stable and Student distributions for several Latin American exchange rate series.

First one must note that these series exhibit the typical properties of financial data, namely a certain dependency which becomes visible by periods of tranquility and volatility. This entails that one cannot expect a good performance of estimation procedures.

For the Colombia data we estimated a sum–stable distribution $S(\alpha, \beta, \sigma, \mu; 0) = S(1.21, -0.14, 0.0059, 0.0096; 0)$ and a Student distribution with shape, location and scale parameters $\alpha = 1.83$, $\mu = 0.0094$ and $\sigma = 0.0072$. It is remarkable that in both cases one gets a shape parameter α less that 2.

In Fig. 6.4 (left) we plot the estimated sum–stable density and a kernel density based on the Epanechnikov kernel and the width parameter $b = 0.007$. The estimated Student density can hardly be distinguished from the sum–stable density and is, therefore, omitted.

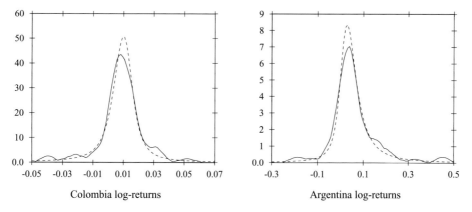

FIG. 6.4. Kernel densities and estimated sum–stable densities for Colombia (left) and for Argentina (right) exchange rate data.

For the Argentina exchange rate data the ML estimation procedures gives the sum–stable distribution $S(\alpha, \beta, \sigma, \mu; 0) = S(1.093, -0.44, 0.036, 0.0036; 0)$. This density is strongly skewed to the right. For a continuation of this example see Example 6.4.2.

---

[15]Communicated by S. Caserta and C.G. de Vries and stored in blackmarket.dat.

[16]Akgiray, V., Booth, G.G. and Seifert, B. (1988). Distribution properties of Latin American black market exchange rates. J. Int. Money and Finance 7, 37–48.

[17]Koedijk, K.G., Stork, P.A. and Vries, de C.G. (1992). Differences between foreign exchange rate regimes: the view from the tail. J. Int. Money and Finance 11, 462–473.

The density plots give a good sense of whether the fit matches the data near the mode of the distribution, but generally they are uninformative on the tails where both the fitted density and the smoothed data density are small.

Diagnostics with Q–Q Plots

If the smoothed density is plausibly stable, proceed with a stable fit and compare the fitted distribution with the data using Q–Q and P–P plots and more refined statistical tools coming from extreme value analysis.

We observed practical problems with Q–Q plots for heavy tailed data. While using Q–Q plots to compare simulated stable data sets with the exact corresponding cumulative df, we routinely had two problems with extreme values: (1) most of the data is visually compressed to a small region and (2) on the tails there seems to be an unacceptably large amount of fluctuation around the theoretical straight line.

For heavy tailed stable distributions we should expect such fluctuations: if $X_{i:n}$ is the ith order statistic from an iid stable sample of size n with common df F, $q = (i - 1/2)/n$ and $F^{-1}(q)$ is the q–quantile, then for n large, the distribution of $X_{i:n}$ is approximately normal with expectation $EX_{i:n} \approx F^{-1}(q)$ and variance

$$V(X_{i:n}) \approx q(1 - q)/nf(F^{-1}(q))^2.$$

The point is that Q–Q plots may appear non–linear on the tails, even when the data set is stable.

Diagnostics with Trimmed Mean Excess Functions

One should be aware that the mean excess function does not exist, if the tail index α of the underlying df is smaller than 1. In addition, the sample mean excess function provides an inaccurate estimator, if $\alpha \leq 1.5$. Therefore, we employ trimmed mean excess functions, cf. Section 2.2, in the subsequent analysis.

In conjunction with sum–stable dfs there is another difficulty, namely that the constant in the remainder term of the Pareto approximation is large for α close to 2, see (6.24). One cannot expect that linearity in the upper tail of the underlying trimmed mean excess function becomes clearly visible in the sample version.

In Fig. 6.5 we plot the underlying trimmed mean excess function with trimming factor $p = 0.8$—simulated under the gigantic sample size of 100,000 data—for standard, symmetric sum–stable dfs with parameters $\alpha = 1.5$ and $\alpha = 1.8$, and in both cases three different sample trimmed mean excess functions for sample sizes $n = 1000$.

For the parameter $\alpha = 1.5$ (respectively, $\alpha = 1.8$) the linearity of the "theoretical" trimmed mean excess function becomes visible beyond the threshold $u = 1.5$ ($u = 2.5$) which corresponds to the 85% quantile (95% quantile). This indicates

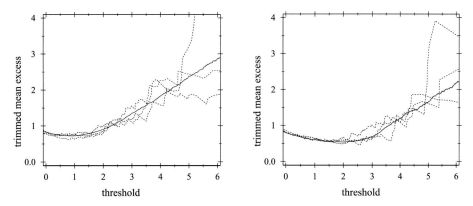

FIG. 6.5. "Theoretical" trimmed mean excess functions (solid) and sample versions (dotted) based on $n = 1000$ simulated data for the parameter $\alpha = 1.5$ (left) and $\alpha = 1.8$ (right).

why larger sample sizes are required to detect the linearity in the upper tail of the trimmed mean excess function, if α is close to 2.

EXAMPLE 6.4.2. (Continuation of Example 6.4.1.) The following considerations concern the Argentina exchange rate data which were studied in Example 6.4.1. Recall that the estimated sum-stable distribution was $S(\alpha, \beta, \sigma, \mu; 0) = S(1.093, -0.44, 0.036, 0.0036; 0)$.

In Fig. 6.6 we plot the sample trimmed mean excess function with trimming factor 0.8 and the "theoretical" trimmed mean excess function simulated under the estimated parameters (under the Monte Carlo sample size of 50,000).

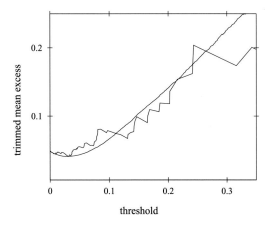

FIG. 6.6. Sample trimmed mean excess function based on Argentina exchange rate data and "theoretical" trimmed mean excess function.

The number of data above the threshold $u = 1.3$ is $k = 24$. The MLE and moment estimates in the full Pareto model based on the 24 largest data are around $\alpha = 2$. This also confirms that there is a distribution with a very heavy tail.

The overall impression from the analysis of data and of simulations is that upper tails of sum–stable dfs can be successfully analyzed by sample trimmed mean excess functions, if α is not too close to 2 and the sample size is sufficiently large.

6.5   Ultimate and Penultimate GP Approximation

co–authored by E. Kaufmann[18]

From this section some theoretical insight can be gained why the modeling of dfs of maxima or exceedances by means of EV or GP dfs leads to inaccurate results in certain cases although the underlying df F belongs to the max or pot–domain of attraction of an EV or GP df.

This question is handled by computing remainder terms in the limit theorems (1.27) and (1.46) for dfs of maxima and exceedance dfs. This book is devoted to exceedances to some larger extent and, therefore, we focus our attention on the approximation to exceedance dfs in order not to overload this section. Our arguments are closely related to those in [16], pages 11–45.

At the beginning we start with some explanations about the relationship between the pot–stability of GP dfs and the concept of regularly varying functions. In the second part of this section one may find some results about von Mises bounds for such approximations.

The POT–Stability and Regularly Varying Functions

In the following lines, we merely study the case of Pareto dfs $W_{1,\alpha,\mu,\sigma}$ in the α–parametrization. Similar results hold for GP dfs in general. From (1.44) we know that for exceedance dfs of Pareto dfs the relation

$$W_{1,\alpha,\mu,\sigma}^{[u]}(x) = W_{1,\alpha,\mu,u-\mu}(x), \quad x \geq u,$$

holds for thresholds $u > \mu + \sigma$, which is the pot–stability of Pareto dfs.

Recall from (1.12) that for any df F,

$$\overline{F^{[u]}}(x) = \bar{F}(x)/\bar{F}(u), \quad x \geq u,$$

where $\bar{F}$ is the survivor function of F. Rewriting the pot–stability of Pareto dfs $F = W_{1,\alpha,\mu,\sigma}$ in terms of survivor functions and replacing x by ux, one gets

$$\bar{F}(ux)/\bar{F}(u) = \overline{F^{[u]}}(ux) = \left(\frac{x - \mu/u}{1 - \mu/u} \right)^{-\alpha}$$

$$\rightarrow_{u \to \infty} x^{-\alpha}, \qquad x \geq 1. \tag{6.30}$$

---

[18] University of Siegen.

In the special case of $F = W_{1,\alpha,0,\sigma}$—this is the case of a Pareto df with location parameter $\mu = 0$—we have equality in (6.30), that is,

$$\bar{F}(ux)/\bar{F}(u) = x^{-\alpha}, \qquad x > 1, \tag{6.31}$$

holds. This indicates that pot–stability is closely related to the property that the survivor function $\bar{F}$ is regularly varying at ∞.

Generally, a measurable function $U : (0,\infty) \to (0,\infty)$ is called regularly varying at infinity with index (exponent of variation) ρ, if

$$U(tx)/U(t) \to_{t\to\infty} x^{\rho}, \qquad x > 0, \tag{6.32}$$

see, e.g., [44], Section 0.4. The canonical ρ–varying function is $U(x) = x^{\rho}$.

If $\rho = 0$, then U is called slowly varying. One speaks of regularly varying at zero if (6.32) holds with $t \to \infty$ replaced by $t \downarrow 0$.

Recall from Section 1.3, page 19, that a df F belongs to the max–domain of attraction of the EV df G_{γ} with $\gamma > 0$ if, and only if, $\omega(F) = \infty$ and the survivor function $\bar{F}$ is regularly varying at ∞ with index $\rho = -1/\gamma$.

If $F = W_{1,\alpha,\mu,\sigma}$ is a Pareto df with location parameter $\mu \neq 0$, then condition (6.32) is satisfied for $\bar{F}$ with $\rho = -\alpha$, yet the rate is exceedingly slow if α is a larger index. In the following lines we make this remark more precise by computing remainder terms.

Computing the Remainder Term in the GP Approximation

We start with a condition imposed on the survivor function $\bar{F} = 1 - F$ of a df F under which the remainder term of the approximation in (1.46) of an exceedance df $F^{[u]}$ by means of a GP df W can be calculated easily.

Our interest is primarily focused on GP dfs, yet the subsequent formulas (6.33) to (6.37) are applicable to any df W as an approximate df. Assume that

$$\bar{F}(x) = \overline{W}(x) \left(1 + O\left(\overline{W}^{1/\delta}(x)\right)\right) \tag{6.33}$$

holds[19] for some $\delta > 0$. Thus, F is close to the df W in the upper tail. If this condition holds, we say that a df F belongs to the δ–neighborhood of the df W. Notice that the remainder term is small if δ is small. In addition, we have $\omega(F) \leq \omega(W)$ for the right–hand endpoints of the supports of F and W.

Condition (6.33), with W being a GP df, is satisfied for $\delta = 1$ by many of the distributions mentioned in this book such as Fréchet, Weibull, Gumbel, generalized Cauchy, Student, Burr and sum–stable distributions with $\alpha < 2$ and various mixtures.

Straightforward calculations show that the relation

$$\left|F^{[u]}(x) - W^{[u]}(x)\right| = O\left(\overline{W}^{1/\delta}(u)\right), \qquad u \leq x < \omega(F), \tag{6.34}$$

---

[19]This condition can be attributed to L. Weiss (1971), see the article cited on page 134.

holds uniformly over u for the exceedance dfs of F and W. A special case was mentioned in (1.48). Recall that the exceedance df $W^{[u]}$ of a GP df W is again a GP df (which is the pot–stability of GP dfs). Also, a df belongs to the pot–domain of attraction of a GP df, if it belongs to the δ–neighborhood of a GP df.

Condition (6.33) reformulated in terms of densities is

$$f(x) = w(x) \left(1 + O\left(\overline{W}^{1/\delta}(x) \right) \right). \tag{6.35}$$

By integration one obtains that (6.35) implies (6.33). In addition, simple calculations yield that

$$f^{[u]}(x) = w^{[u]}(x) \left(1 + O\left(\overline{W}^{1/\delta}(u) \right) \right), \qquad u \le x < \omega(F), \tag{6.36}$$

holds for the exceedance densities $f^{[u]}$ and $w^{[u]}$ uniformly in the thresholds u.

In Section 9.4 we apply a reformulation of (6.36) in terms of the Hellinger distance (introduced in (3.4)). (6.36) implies

$$H\left(f^{[u]}, w^{[u]} \right) = O\left(\overline{W}^{1/\delta}(u) \right), \tag{6.37}$$

where H is the Hellinger distance between $f^{[u]}$ and $w^{[u]}$.

Approximations to actual distributions of exceedances or maxima can still be satisfactorily accurate for small sample sizes, if the constant in the remainder term is small, with hardly any improvement when the sample size increases. For a longer discussion see [42], pages 172–175.

Comparing the Tail Behavior of Certain Pareto Distributions

We also want to discuss the poor performance of the Hill estimator, as illustrated in Figure 5.2 (left), by computing a refinement of (6.33). The Hill estimator is an MLE in the $GP1(u, \mu = 0)$ model, where the location parameter μ is equal to zero. If $\mu \ne 0$, then it is well known that the Hill estimator is still consistent (as outlined in Section 5.1).

We compare the tail behavior of Pareto dfs $W_{1,\alpha,\mu,\sigma}$ with location parameter $\mu \ne 0$ to that for $\mu = 0$. For that purpose, an expansion of length 2 is computed. By applying a Taylor expansion one gets

$$\overline{W}_{1,\alpha,\mu,\sigma}(x) = \overline{W}_{1,\alpha,0,\sigma}\left(1 + \frac{\alpha\mu}{\sigma} \overline{W}_{1,\alpha,0,\sigma}^{1/\alpha}(x) + O\left(\overline{W}_{1,\alpha,0,\sigma}^{2/\alpha}(x) \right) \right). \tag{6.38}$$

We see that Pareto dfs with $\mu \ne 0$ belong to the δ–neighborhood of a Pareto df with $\mu = 0$, yet $\delta = \alpha$ can be large which yields a large remainder term in the approximation.

This indicates that estimators of the shape parameter α, which are especially tailored for the $GP1(u, \mu = 0)$ model such as the Hill estimator, are inaccurate, if the parameters α is large. Of course, the constant μ/σ is important as well.

A stronger deviation of the actual distribution from the GP1$(u, \mu = 0)$ model yields a larger bias of the Hill estimator a fact which was already observed by Goldie and Smith[20].

The Hall Condition

A refinement of the Weiss condition (6.33), with the factor

$$1 + O\big(\overline{W}^{1/\delta}(x)\big)$$

replaced by an expansion

$$1 + C(1 + o(1))\overline{W}^{1/\delta}(x), \qquad (6.39)$$

is the Hall condition[21]. See (6.38) for an example. Another example is provided by (6.23) which is an expansion for sum–stable distributions: the Hall condition holds for $\gamma = 1/\alpha$ and $\delta = 1$, yet the constant—hidden in the remainder term—is very large if α is close to 2.

In terms of densities we get a

refined δ–neighborhood: $\qquad f(x) = w(x)\left(1 + \widetilde{C}(1 + o(1))\overline{W}^{1/\delta}(x)\right), \quad (6.40)$

where $\widetilde{C} = C(1 + 1/\delta)$.

The expansion for sum–stable densities in the book by Christoph and Wolf also includes those with a skewness parameter $\beta \neq 0$. If $\alpha \neq 1, 2$, then the sum–stable densities generally satisfy condition (6.35) with $\delta = 1$; if $\alpha = 1$ and $\beta \neq 0$, then this relation holds for $\delta = 1 + \varepsilon$ for every $\varepsilon > 0$.

Under such a condition the optimal number k of extremes was studied by Hall and Welsh[22]. Early results on the bias–correction of estimators were established under this condition. The latter topic will separately be studied in Section 6.6.

Such Hall type conditions are also of interest in conjunction with the automatic choice of k (see page 137).

Von Mises Bounds for the Remainder Term in the GP Approximation to Exceedance DFs

Conditions (6.33) and (6.35) are not satisfied by the normal, log–normal, gamma, Gompertz and converse Weibull distributions, for example. Yet, by using the von

---

[20]See, e.g., Goldie, C.M. and Smith, R.L. (1987). Slow variation with remainder term: A survey of the theory and its applications. Quart. J. Math. Oxford Ser. 38, 45–71.

[21]Hall, P. (1982). On estimating the endpoint of a distribution. Ann. Statist. 10, 556–568.

[22]Hall, P. and Welsh, A.H. (1985). Adaptive estimates of parameters of regular variation. Ann. Statist. 13, 331–341.

Mises condition (2.32), one may prove that these distributions belong to the pot–domain of attraction of the exponential df (and, likewise, to the max–domain of attraction of the Gumbel df). We refer to [20], pages 67–68, for the required computations in the case of the log–normal distribution.

Subsequently, we make use of the von Mises condition (2.32), which is equivalent to the following

von Mises condition: $\qquad M(t) := \left(\dfrac{1-F}{f}\right)'(F^{-1}(1-t)) - \gamma \to 0, \quad t \downarrow 0,$ (6.41)

for some real parameter γ.

An upper bound related to that in (6.37) will be determined by the remainder function M. Note that M is independent of location and scale parameters of the df F. Also,

- if $M = 0$, then F is a GP df,

- if

$$M(t) \sim ct^{1/\delta} \qquad \text{as } t \downarrow 0,$$

then f satisfies condition (6.35) for some GP density w (that is, f belongs to the δ–neighborhood of a GP density w with shape parameter γ).

In addition, assume that

$$\frac{M(tx)}{M(t)} \to x^{\rho}, \qquad t \downarrow 0, \quad x > 0, \tag{6.42}$$

holds for some $\rho \geq 0$. This condition says that the remainder function M is regularly varying at zero, if $\rho > 0$, and slowly varying at zero, if $\rho = 0$, see page 183. Note that, (6.42) implies (6.41), if $\rho > 0$.

The case $\rho = 0$ includes distributions for which poor rates of convergence are achieved in (6.37). For the standard Gompertz df $\tilde{G}_0$ we have $\gamma = 0$ and

$$M(t) = -(\log t)^{-1}. \tag{6.43}$$

Such an order is also achieved by M for normal, log–normal, gamma and converse Weibull dfs.

We compute an upper bound for the remainder term in the GP approximation which is related to that in (6.37). Under the conditions (6.41) and (6.42) we have[23]

$$H(f^{[u]}, w_{\gamma,u,\sigma(u)}) \sim C|M(\bar{F}(u))|, \tag{6.44}$$

where

$$\sigma(u) = \begin{cases} (1-F(u))/f(u) & \quad \gamma \geq 0, \\ & \text{if} \\ (\omega(F) - u)|\gamma| & \quad \gamma < 0. \end{cases} \tag{6.45}$$

---

[23]Kaufmann, E. and Reiss, R.–D. (2002). An upper bound on the binomial process approximation to the exceedance process. Extremes 5, 253–269.

For a continuation of this topic, within the framework of exceedance processes, we refer to Section 9.4.

Penultimate Modeling

By computing remainder terms one is also able to discuss the concept of penultimate distributions. Within EV or GP models one may find dfs which are more accurate approximations of the actual df F^m or $F^{[u]}$ than the limiting ones. Such EV or GP dfs are called penultimate dfs.

Early results of that type are contained in the paper by Fisher and Tippett cited on page 18. Further references concerning the penultimate EV approximation are given at the end of this section.

The starting points are the conditions (6.41) and (6.42) which determine an upper bound for the ultimate rate. Under these conditions a penultimate approximation exists if, and only if, M is slowly varying[24].

We assume that the von Mises condition (6.41) and condition[24],

$$\frac{M(tx) - M(t)}{A(t)} \to c \log x, \qquad t \downarrow 0, \quad x > 0, \tag{6.46}$$

hold, where $A : (0,1) \to (0,\infty)$ is another auxiliary function and $c = -1, 1$.

This condition implies that M and A are slowly varying at 0. Hence, (6.42) holds for M and A with $\rho = 0$. The rate of convergence in (6.42) is of order $A(t)/M(t)$ which shows that A is of smaller order than M.

We compute an upper bound for the remainder term in the penultimate GP approximation which correspond to that in (6.44). Under the conditions (6.41) and (6.46) we have[24]

$$H(f^{[u]}, w_{\gamma(u),u,\sigma(u)}) \sim CA(\bar{F}(u)) \tag{6.47}$$

with $\sigma(u)$ as in (6.45), and γ replaced by

$$\gamma(u) := \gamma + M(\bar{F}(u)). \tag{6.48}$$

Recall that $M(t) = -(\log t)^{-1}$ for the Gompertz df. In addition, we have

$$A(t) = (\log t)^{-2}$$

for the second auxiliary function A in this case.

The existence of penultimate distributions can have serious consequences for the estimation of GP distributions (and, likewise, for EV distributions) and the interpretation of such results.

EXAMPLE 6.5.1. (Extreme Life Spans.) Our analysis concerns extreme life spans of women born before and around the year 1900 and later living in West Germany.

---

[24]Kaufmann, E. (2000). Penultimate approximations in extreme value theory. Extremes 3, 39–55.

The given data are the ages of female persons at death in West–Germany in the year 1993[25]. None of the persons died at an age older than 111 years. The analysis is based on the ages at death of 90 or older.

TABLE 6.2. Frequencies of life spans (in years).

life span	90	91	92	93	94	95	96	97	98	99	100
frequency	12079	10273	8349	6449	5221	3871	2880	1987	1383	940	579
life span	101	102	103	104	105	106	107	108	109	110	111
frequency	340	207	95	63	36	16	9	4	1	1	1

Gompertz dfs (see page 54) are one of the classical life span dfs. A Gompertz density with location and scale parameters $\mu = 83.0$ and $\sigma = 9.8$ fits well to the histogram of the life span data above 95 year. According to (6.43)–(6.45), an exponential df provides an ultimate approximation. The penultimate approach, see (6.47) and (6.48), leads to a beta dfs with $\gamma(u) \uparrow 0$ as $u \to \infty$. For more details we refer to Section 19.1.

The possibility of a penultimate approximation must also be taken in account when the shape parameter $\gamma(u)$ varies with the threshold u.

EXAMPLE 6.5.2. (Ozone Cluster Maxima.) We discuss one further aspect of the case study mentioned in Example 5.1.2. In this case we do not have a classical distribution for all of the data in mind. However, for the increasing thresholds $u = -0.20, 0.22, 0.24$ one obtains as estimates of the shape parameter γ the values $\hat{\gamma}(.20) = -0.41$, $\hat{\gamma}(.22) = -.35$ and $\hat{\gamma}(.24) = -.33$. This indicates the possibility that we estimated penultimate approximations to the actual distribution with the exponential distribution providing an ultimate approximation.

This short discussion underlines the importance of penultimate approximation in extreme value theory. The situation is characterized by the facts that there is a bad approximation rate by means of the ultimate distribution and a slight improvement of the accuracy of the approximation by means of penultimate distributions.

The possibility of occurance of penultimate distributions should not be ignored in statistically oriented applications, e.g., in conjunction with speculations about the right endpoint of life–span distribution.

---

[25]Stored in the file um–lspge.dat (communicated by G.R. Heer, Federal Statistical Office).

The Remainder Term in the EV Approximation

To obtain sharp bounds on the remainder term of the EV approximation to the joint distribution of several maxima, one must compute the Hellinger distance between such distributions, see [42] and [16]. Such bounds can be easily computed if F belongs to a δ–neighborhood of a GP distribution.

In the literature about probability theory one finds bounds on the maximum deviation of dfs or, in exceptional cases, for the variational distance between distributions in the case of a single maximum.

From the Falk–Marohn–Kaufmann theorem[26], it is known that an algebraic rate of convergence—that is a rate of order $O(n^{-c})$ for some $c > 0$—can be achieved if, and only if, condition (6.33) is satisfied. This clarifies why one gets inaccurate results in certain cases. For example, the large MSEs in Table 4.2 for normal data can be explained by the fact that the normal distribution does not satisfy condition (6.33).

An upper bound, which involves the von Mises condition, was established by Radtke[27]. A closely related upper bound may be found in an article by de Haan and Resnick[28]. The condition formulated by de Haan and Resnick is better adjusted to approximations of sample maxima.

First results concerning the penultimate approximation to distributions of maxima are due to Fisher and Tippett (in the article cited on page 18). These results were considerably extended by Gomes[29]. Exact penultimate rates were recently established by Gomes and de Haan[30] and in the afore mentioned article by Kaufmann[24].

---

[26] Falk, M. and Marohn, F. (1993). Von Mises conditions revisited. Ann. Probab. 21, 1310–1328, and Kaufmann, E. (1995). Von Mises conditions, δ–neighborhoods and rates of convergence for maxima. Statist. Probab. Letters 25, 63–70.

[27] Radtke, M. (1988). Konvergenzraten und Entwicklungen unter von Mises Bedingungen der Extremwerttheorie. Ph.D. Thesis, University of Siegen, (also see, e.g., [42], page 199).

[28] de Haan, L. and Resnick, S.I. (1996). Second order regular variation and rates of convergence in extreme value theory. Ann. Probab. 24, 97-124.

[29] Gomes, M.I. (1984). Penultimate limiting forms in extreme value theory. Ann. Inst. Statist. Math. 36, Part A, 71–85, and Gomes, M.I. (1994). Penultimate behaviour of the extremes. In [15], Vol. 1, 403–418.

[30] Gomes, M.I. and Haan, L. de (1999). Approximation by penultimate extreme value distributions. Extremes 2, 71–85.

6.6   An Overview of Reduced–Bias Estimation
co–authored by M.I. Gomes[31]

The semi–parametric estimators of first order parameters of extreme or even rare events—like the tail index, the extremal index, a high q–quantile, a return period of a high level, and so on—may be based on the k top order statistics $X_{n-i+1:n}$, $i = 1, \ldots, k$, of iid random variables with common df F. Before dealing with second order parameters, which are decisive for the bias reduction, we recall some basic facts about the asymptotic theory for heavy tails in the first order framework. Next, we formulate a basic condition for the second order framework, and

- introduce the concept of bias reduction;

- provide details about the jackknife methodology;

- study an approximate maximum likelihood approach, together with the introduction of simple bias–corrected Hill estimators,

- provide an application to data in the field of finance.

This section is concluded with remarks about some recent literature on bias reduction.

The First Order Framework Revisited

The estimators for an extreme events' parameter, say η, are consistent if the df F is in the domain of attraction of an EV df G_γ, and k is intermediate, i.e.,

$$k = k_n \to \infty \quad \text{and} \quad k = o(n) \text{ as } n \to \infty. \tag{6.49}$$

We shall assume here that we are working with heavy tails and, thus, with shape parameters $\gamma > 0$ in the unified model. Then, according to (1.29), a df F is in the max–domain of attraction of G_γ if, and only if, for every $x > 0$,

$$\bar{F}(tx)/\bar{F}(t) \to_{t\to\infty} x^{-1/\gamma}, \text{ or equivalently, } U(tx)/U(t) \to_{t\to\infty} x^\gamma, \tag{6.50}$$

where $\bar{F} = 1 - F$ is the survivor function, and $U(t) = F^{-1}(1 - 1/t)$ for $t \geq 1$.

Thus, according to (6.32), the survivor function is regularly varying at infinity with a negative index $-1/\gamma$, or equivalently, U is of regular variation with index γ. We shall here concentrate on these Pareto–type distributions, for which (6.50) holds.

---

[31]University of Lisbon, this research was partially supported by FCT/POCTI and POCI/FEDER.

A Second Order Condition and
a Distributional Representation of the Estimators

Under the first order condition (6.50), the asymptotic normality of estimators for an extreme events' parameter η is attained whenever we assume a second order condition, i.e., when we assume to know the rate of convergence towards zero of, for instance, $\log U(tx) - \log U(t) - \gamma \log x$, as $t \to \infty$.

Such a second order condition may be written as

$$\frac{\log U(tx) - \log U(t) - \gamma \log x}{A(t)} \to_{t \to \infty} \frac{x^\rho - 1}{\rho}, \qquad (6.51)$$

where $\rho \leq 0$ and $A(t) \to 0$ as $t \to \infty$. More precisely,

$$|A(tx)/A(t)| \to_{t \to \infty} x^\rho \quad \text{for all } x > 0;$$

that is, $|A|$ is regularly varying at infinity with index ρ.

For any classical semi–parametric estimator $\eta_{n,k}$, which is consistent for the estimation of the parameter η, and under the second order condition (6.51), there exists a function $\varphi(k)$, converging towards zero as $k \to \infty$, such that the following asymptotic distributional representation

$$\eta_{n,k} \stackrel{d}{=} \eta + \sigma\varphi(k)P_k + b_\rho A(n/k)\left(1 + o_p(1)\right), \qquad (6.52)$$

holds (thus, we have equality in distribution with one term converging to zero in probability). Here the P_k are asymptotically standard normal random variables, $\sigma > 0$, b_ρ is real and $\neq 0$, and $A(\cdot)$ the function in (6.51).

We may thus provide approximations to the variance and the bias of $\eta_{n,k}$ given by $(\sigma\varphi(k))^2$ and $b_\rho A(n/k)$, respectively. Consequently, these estimators exhibit the same type of peculiarities:

- high variance for high thresholds $X_{n-k:n}$, i.e., for small values of k;

- high bias for low thresholds, i.e., for large values of k;

- a small region of stability of the sample path (plot of the estimates versus k as it is done in Fig. 5.3), as a function of k, making problematic the adaptive choice of the threshold, on the basis of any sample paths' stability criterion;

- a "very peaked" mean squared error, making difficult the choice of the value k_0 where the mean squared error function $\mathrm{MSE}(\eta_{n,k})$ attains its minimum.

The Concept of Reduced–Bias Estimators

The preceding peculiarities have led researchers to consider the possibility of dealing with the bias term in an appropriate manner, building new estimators $\eta_{n,k}^R$, the so–called reduced–bias estimators.

Under the second order condition (6.51) and for k intermediate, i.e., whenever (6.49) holds, the statistic $\eta^R_{n,k}$, a consistent estimator of a functional of extreme events $\eta = \eta(F)$, based on the k top order statistics in a sample from a heavy tailed df F, is said to be a reduced–bias semi–parametric estimator of η, whenever

$$\eta^R_{n,k} \stackrel{d}{=} \eta + \sigma_R \varphi(k) P^R_k + o_p(A(n/k)), \qquad (6.53)$$

where the P^R_k are asymptotically standard normal and $\sigma_R > 0$, with $A(\cdot)$ and $\varphi(\cdot)$ being the functions in (6.51) and (6.52).

Notice that for the reduced–bias estimators in (6.53), we no longer have a dominant component of bias of the order of $A(n/k)$, as in (6.52). Therefore,

$$\sqrt{k}\left(\eta^R_{n,k} - \eta\right)$$

is asymptotically normal with null mean value not only when $\sqrt{k}\, A(n/k) \to 0$ (as for the classical estimators), but also when $\sqrt{k}\, A(n/k) \to \lambda$, finite and non–null. Such a bias reduction provides usually a stable sample path for a wider region of k–values and a reduction of the mean squared error at the optimal level.

Such an approach has been carried out in the most diversified manners, and from now on we shall restrict ourselves to the tail index estimation, i.e., we shall replace the generic parameter η by the tail index γ, in (6.50). As a consequence, $\varphi(k) = 1/\sqrt{k}$. The key ideas are either to find ways of getting rid of the dominant component $b_\rho A(n/k)$ of bias in (6.52), or to go further into the second order behavior of the basic statistics used for the estimation of γ, like the log–excesses or the scaled log–spacings.

Historical Remarks

We mention some pre–2000 results about bias–corrected estimators in extreme value theory. Such estimators may be dated back to Reiss[32], Gomes[33], Drees[34] and Peng[35], among others. Gomes uses the Generalized Jackknife methodology in Gray and Schucany [23], and Peng deals with linear combinations of adequate tail index estimators, in a spirit quite close to the one associated to the Generalized Jackknife technique.

---

[32]Reiss, R.-D. (1989). Extreme value models and adaptive estimation of the tail index. In [14], 156–165.

[33]Gomes, M.I. (1994). Metodologias Jackknife e Bootstrap em Estatística de Extremos. In Mendes-Lopes et al. (eds.), Actas II Congresso S.P.E., 31–46.

[34]Drees, H. (1996). Refined Pickands estimators with bias correction. Comm. Statist. Theory and Meth. 25, 837–851.

[35]Peng, L. (1998). Asymptotically unbiased estimator for the extreme–value index. Statist. Probab. Letters 38, 107–115.

The latter technique was also used in Martins et al.[36], where convex mixtures of two Hill estimators, computed at two different levels, are considered. Within the second order framework, Beirlant et al.[37] and Feuerverger and Hall[38] investigate the accommodation of bias in the scaled log–spacings and derive approximate "maximum likelihood" and "least squares" reduced–bias tail index estimators.

The Jackknife and Related Methodologies

First we provide some details about the Jackknife methodology, due to J. Tukey. This is a nonparametric resampling technique, essentially in the field of exploratory data analysis, whose main objective is the reduction of bias of an estimator, by means of the construction of an auxiliary estimator based on B. Quenouille's re-sampling technique, and the consideration of a suitable combination of the two estimators.

The Generalized Jackknife statistics of Gray and Schucany [23] are more generally based on two different estimators of the same functional, with similar bias properties. More precisely, and as a particular case of the Jackknife theory, if we have two different biased consistent estimators $\eta_n^{(1)}$ and $\eta_n^{(2)}$ of the functional parameter $\eta(F)$, such that $E\left(\eta_n^{(1)}\right) = \eta + \varphi(\eta)\, d_1(n)$ and $E\left(\eta_n^{(2)}\right) = \eta + \varphi(\eta)\, d_2(n)$, then, denoting by

$$q_n := \frac{\mathrm{BIAS}\left(\eta_n^{(1)}\right)}{\mathrm{BIAS}\left(\eta_n^{(2)}\right)} = \frac{d_1(n)}{d_2(n)},$$

the Generalized Jackknife statistic associated to $\left(\eta_n^{(1)}, \eta_n^{(2)}\right)$ is

$$\eta_n^G\left(\eta_n^{(1)}, \eta_n^{(2)}\right) = \frac{\eta_n^{(1)} - q_n \eta_n^{(2)}}{1 - q_n},$$

which is an unbiased consistent estimator of $\eta(F)$, provided that $q_n \neq 1$ for all n.

Generalized Jackknife Estimators of the Tail Index

Whenever we are dealing with semi–parametric estimators of the tail index, or even other parameters of extreme events, we have usually information about the asymptotic bias of these estimators. We may thus choose estimators with similar asymptotic properties, and build the associated Generalized Jackknife statistic.

---

[36] Martins, M.J., Gomes, M.I. and Neves, M. (1999). Some results on the behavior of Hill estimator. J. Statist. Comput. and Simulation 63, 283–297.

[37] Beirlant, J., Dierckx, G., Gogebeur, Y. and Matthys, G. (1999). Tail index estimation and an exponential regression model. Extremes 2, 177–200.

[38] Feuerverger, A. and Hall, P. (1999). Estimating a tail exponent by modelling departure from a Pareto distribution. Ann. Statist. 27, 760–781.

This methodology has first been used by Martins et al., as noted above, and still later on, in Gomes et al.[39]. These authors suggest several Generalized Jackknife estimators of the tail index γ. We shall here refer only to the one based on the classical Hill estimator for γ, namely

$$\gamma_{n,k}^{(1)} := \frac{1}{k} \sum_{i \leq k} \left(\log X_{n-i+1:n} - \log X_{n-k:n} \right), \tag{6.54}$$

cf. (5.3), and on the alternative estimator

$$\gamma_{n,k}^{(2)} := \frac{M_{n,k}^{(2)}}{2M_{n,k}^{(1)}},$$

where, related to the $l_{j,k}$ with a fixed threshold on page 134,

$$M_{n,k}^{(j)} := \frac{1}{k} \sum_{i \leq k} (\log X_{n-i+1:n} - \log X_{n-k:n})^j , \quad j \geq 1. \tag{6.55}$$

Under the second order condition (6.51), and with $P_k^{(1)}$ and $P_k^{(2)}$ asymptotically standard normal random variables, we have individually and jointly the validity of the distributional representations

$$\gamma_n^{(1)}(k) \stackrel{d}{=} \gamma + \frac{\gamma P_k^{(1)}}{\sqrt{k}} + \frac{A(n/k)}{1-\rho} + o_p(A(n/k)), \tag{6.56}$$

$$\gamma_n^{(2)}(k) \stackrel{d}{=} \gamma + \frac{\sqrt{2}\,\gamma P_k^{(2)}}{\sqrt{k}} + \frac{A(n/k)}{(1-\rho)^2} + o_p(A(n/k)), \tag{6.57}$$

where one may choose

$$P_k^{(1)} = \sqrt{k} \left(\sum_{i \leq k} \eta_i/k - 1 \right) \quad \text{and} \quad P_k^{(2)} = \frac{\sqrt{2}}{2} \left(\frac{\sqrt{k}}{2} \left(\sum_{i \leq k} \eta_i^2/k - 2 \right) - P_k^{(1)} \right),$$

with η_i, $i \geq 1$ being iid standard exponential random variables. Consequently, $\mathrm{Cov}\left(P_k^{(1)}, P_k^{(2)}\right) = \sqrt{2}/2$.

The ratio between the dominant components of bias of $\gamma_{n,k}^{(1)}$ and $\gamma_{n,k}^{(2)}$ is equal to $1 - \rho$, and we thus get the Generalized Jackknife estimator

$$\gamma_{\rho,k}^{GJ} := \left(\gamma_{n,k}^{(1)} - (1-\rho)\,\gamma_{n,k}^{(2)} \right)/\rho, \tag{6.58}$$

---

[39] Gomes, M.I., Martins, M.J. and Neves, M. (2000). Alternatives to a semi–parametric estimator of parameters of rare events—the Jackknife methodology. Extremes 3, 207–229, and Gomes, M.I., Martins, M.J. and Neves, M. (2002). Generalized Jackknife semi–parametric estimators of the tail index. Portugaliae Mathematica 59, 393–408.

where ρ must still be replaced by an estimator $\hat{\rho}$. Note that this estimator is exactly the estimator studied by Peng, cf. page 192, who claimed that no good estimator for the second order parameter ρ was then available, and considered a new ρ–estimator, alternative to the ones in Feuerverger and Hall, cf. page 193, Beirlant et al.[40] and Drees and Kaufmann, cf. page 138.

We formulate two asymptotic results for the estimator in (6.58).

- Under the second order condition (6.51) and k intermediate,

$$\gamma_{\rho,k}^{GJ} \stackrel{d}{=} \gamma + \frac{\gamma P_k^{GJ}\sqrt{2\rho^2 - 2\rho + 1}}{|\rho|\sqrt{k}} + o_p(A(n/k)), \qquad (6.59)$$

where P_k^{GJ} is an asymptotically standard normal random variable. This result comes directly from the expression of $\gamma_{\rho,k}^{GJ}$, together with the distributional representations in (6.56) and (6.57).

- The result in (6.59) remains true for the Generalized Jackknife estimator $\gamma_{\hat{\rho},k}^{GJ}$ in (6.58) with ρ replaced by $\hat{\rho}$, provided that $\hat{\rho} - \rho = o_p(1)$ for all k on which we base the tail index estimation, i.e., whenever $\sqrt{k}\, A(n/k) \to \lambda$, finite. For these values of k, we have that $\sqrt{k}\left(\gamma_{\hat{\rho},k}^{GJ} - \gamma\right)$ is asymptotically normal with mean zero and variance

$$\sigma_{GJ}^2 = \gamma^2\Big(1 + \Big(\frac{1-\rho}{\rho}\Big)^2\Big). \qquad (6.60)$$

This result comes from the fact that

$$\frac{d\gamma_{\rho,k}^{GJ}}{d\rho} = \frac{\gamma_{n,k}^{(2)} - \gamma_{n,k}^{(1)}}{\rho^2} = O_p\big(1/\sqrt{k}\big) + O_p\big(A(n/k)\big),$$

and

$$\gamma_{\hat{\rho},k}^{GJ} \stackrel{d}{=} \gamma_{\rho,k}^{GJ}(k) + \big(\hat{\rho} - \rho\big)\Big(O_p\big(1/\sqrt{k}\big) + O_p\big(A(n/k)\big)\Big)(1 + o_p(1)). \quad (6.61)$$

A closer look at (6.61) reveals that it does not seem convenient to compute $\hat{\rho}$ at the same level k we use for the tail index estimation. Indeed, if we do that, and since we can have $\hat{\rho} - \rho = O_p\big(1/(\sqrt{k}\, A(n/k))\big)$, we are going to have a change in the asymptotic variance of the tail index estimator, because $(\hat{\rho} - \rho)\,A(n/k)$ is then a term of the order of $1/\sqrt{k}$.

Gomes et al., see the Extremes–article mentioned on page 194, have indeed suggested the misspecification of ρ at $\rho = -1$, and the consideration of the estimator $\gamma_{n,k}^{GJ} := 2\,\gamma_{n,k}^{(2)} - \gamma_{n,k}^{(1)}$ which is a reduced–bias estimator, in the sense herewith defined, i.e., in the sense of (6.53), if and only if $\rho = -1$. This was essentially due to the high bias and variance of the existing estimators of ρ at that time, together with the idea of considering $\hat{\rho} = \hat{\rho}_k$.

---

[40]Beirlant, J., Vynckier, P. and Teugels, J.L. (1996). Excess function and estimation of the extreme–value index. Bernoulli 2, 293–318.

The Estimation of ρ

We shall consider here special members of the class of estimators of the second order parameter ρ proposed by Fraga Alves et al.[41] Under adequate general conditions, they are semi–parametric asymptotically normal estimators of ρ, whenever $\rho < 0$, which show highly stable sample paths as functions of k, the number of top order statistics used, for a wide range of large k–values. Such a class of estimators is parameterized by a tuning real parameter τ, and may be defined as,

$$\hat{\rho}_{\tau,k} := -\left|3(T_{n,k}^{(\tau)} - 1)/(T_{n,k}^{(\tau)} - 3)\right|, \tag{6.62}$$

where

$$T_{n,k}^{(\tau)} := \begin{cases} \dfrac{\left(M_{n,k}^{(1)}\right)^{\tau} - \left(M_{n,k}^{(2)}/2\right)^{\tau/2}}{\left(M_{n,k}^{(2)}/2\right)^{\tau/2} - \left(M_{n,k}^{(3)}/6\right)^{\tau/3}} & \text{if } \tau \neq 0, \\[2em] \dfrac{\log\left(M_{n,k}^{(1)}\right) - \frac{1}{2}\log\left(M_{n,k}^{(2)}/2\right)}{\frac{1}{2}\log\left(M_{n,k}^{(2)}/2\right) - \frac{1}{3}\log\left(M_{n,k}^{(3)}/6\right)} & \text{if } \tau = 0, \end{cases}$$

with $M_{n,k}^{(j)}$ given in (6.55).

We shall here summarize a few results proved in Fraga Alves et al., now related to the asymptotic behavior of the estimators of ρ in (6.62):

- For $\rho < 0$, if (6.49) and (6.51) hold, and if $\sqrt{k}\, A(n/k) \to \infty$, as $n \to \infty$, the statistic $\hat{\rho}_{\tau,k}$ in (6.62) converges in probability towards ρ, as $k \to \infty$, for any real τ.

- Under additional restrictions on k related to a third order framework which is not discussed here, $\sqrt{k}\, A(n/k)(\hat{\rho}_{\tau,k} - \rho)$ is asymptotically normal with mean zero and variance

$$\sigma_{\rho}^2 = \left(\frac{\gamma(1-\rho)^3}{\rho}\right)^2 (2\rho^2 - 2\rho + 1).$$

- For large levels k_1, of the type $k_1 = [n^{1-\epsilon}]$ with $\epsilon > 0$ small, and for a large class of heavy tailed models, we can guarantee that,

$$\left(\hat{\rho}_{\tau,k_1} - \rho\right)\log n = o_p(1) \quad \text{as } n \to \infty.$$

---

[41] Fraga Alves, M.I., Gomes, M.I. and de Haan, L. (2003). A new class of semi–parametric estimators of the second order parameter. Portugaliae Mathematica 60, 193–213.

The Estimation of ρ in Action

The theoretical and simulated results in the above mentioned article by Fraga Alves et al., as well as further results in Caeiro et al.[42], lead to the proposal of the following algorithm for the ρ–estimation:

Algorithm for ρ–estimation.

1. Consider any level
$$k_1 = \min\left(n - 1, [n^{1-\epsilon}] + 1\right), \quad \text{with } \epsilon \text{ small, say } \epsilon = 0.05. \tag{6.63}$$

2. Given a sample $(X_1, X_2, \ldots, X_n)$, plot, for $\tau = 0$ and $\tau = 1$, the estimates $\hat{\rho}_{\tau,k}$ in (6.62), $1 \leq k < n$.

3. Consider $\{\hat{\rho}_{\tau,k}\}_{k \in \mathcal{K}}$, for large k, say $k \in \mathcal{K} = \left([n^{0.995}], [n^{0.999}]\right)$, and compute their median, denoted χ_τ. Next choose the tuning parameter
$$\tau := \begin{cases} 0 & \text{if } \sum_{k \in \mathcal{K}} (\hat{\rho}_{0,k} - \chi_0)^2 \leq \sum_{k \in \mathcal{K}} (\hat{\rho}_{1,k} - \chi_1)^2, \\ 1 & \text{otherwise.} \end{cases}$$

A few comments:

- Step 3 of the algorithm leads in almost all situations to the tuning parameter $\tau = 0$ whenever $|\rho| \leq 1$ and $\tau = 1$, otherwise. Such a expert's guess usually provides better results than a possibly "noisy" estimation of τ, and is highly recommended in practice. For details on this and similar algorithms for the ρ–estimation, see Gomes and Pestana[43]. The choice of the level k_1 in (6.63), and the ρ–estimator
$$\hat{\rho}_{\tau 1} := -\left| \frac{3\left(T_{n,k_1}^{(\tau)} - 1\right)}{T_{n,k_1}^{(\tau)} - 3} \right|, \quad k_1 = [n^{0.995}], \quad \tau = \begin{cases} 0 & \text{if } \rho \geq -1, \\ 1 & \text{if } \rho < -1, \end{cases} \tag{6.64}$$
is a sensible one.

- It is however possible to consider in Steps 2 and 3 of the algorithm a set $\mathcal{T}$ of τ–values larger than the set $\mathcal{T} = \{0, 1\}$, to draw sample paths of $\hat{\rho}_{\tau,k}$ in (6.62) for $\tau \in \mathcal{T}$, as functions of k, selecting the value of τ which provides higher stability for large k, by means of any stability criterion. A possible choice on the lines of the algorithm is thus $\tau^* := \arg\min_\tau \sum_{k \in \mathcal{K}} (\hat{\rho}_{\tau,k} - \chi_\tau)^2$.

[42]Caeiro, F., Gomes, M.I. and Pestana, D. (2005). Direct reduction of bias of the classical Hill estimator. Revstat 3, 111–136.

[43]Gomes, M.I. and Pestana, D. (2004). A simple second order reduced–bias' tail index estimator. J. Statist. Comp. and Simulation, in press.

Representations of Scaled Log–Spacings

Let us consider the representations of scaled log–spacings

$$U_i := i \left(\log X_{n-i+1:n} - \log X_{n-i:n} \right), \quad 1 \le i \le k.$$

which are useful for the bias reduction and of interest in its own right.

Under the second order condition (6.51), and for $\rho < 0$, Beirlant et al., cf. page 193, motivated the following approximation for the scaled log–spacings:

$$U_i \sim \left(\gamma + A(n/k) \, (i/k)^{-\rho} \right) \eta_i, \quad 1 \le i \le k, \tag{6.65}$$

where η_i, $i \ge 1$, denotes again a sequence of iid standard exponential random variables. In the same context, Feuerverger and Hall, cf. page 193, considered the approximation,

$$U_i \sim \gamma \exp \left(A(n/k) \, (i/k)^{-\rho} / \gamma \right) \eta_i = \gamma \exp \left(A(n/i)/\gamma \right) \eta_i, \quad 1 \le i \le k. \tag{6.66}$$

The representation (6.65), or equivalently (6.66), has been made more precise, in the asymptotic sense, in Beirlant et al.[44], in a way quite close in spirit to the approximations established by Kaufmann and Reiss[45] and Drees et al.[46].

The Hall Condition, Again

We shall here further assume just as in Feuerverger and Hall, cf. page 193, that we are in Hall's class of Pareto–type models, with a survivor function

$$\bar{F}(x) = C x^{-1/\gamma} \left(1 + D x^{\rho/\gamma} + o\left(x^{\rho/\gamma} \right) \right) \quad \text{as} \quad x \to \infty,$$

$C > 0$, D real, $\rho < 0$. Notice that this condition is (6.39) with $\overline{W}(x) = C x^{-1/\gamma}$ and $\rho = -1/\delta$. Then, (6.51) holds and we may choose

$$A(t) = \alpha t^\rho =: \gamma \beta t^\rho, \quad \beta \text{ real}, \quad \rho < 0. \tag{6.67}$$

The Maximum Likelihood Estimation
Based on the Scaled Log–Spacings

The use of the approximation in (6.66) and the joint maximization in γ, β and ρ of the approximate log–likelihood of the scaled log–spacings, i.e., of

$$\log L(\gamma, \beta, \rho; U_i, 1 \le i \le k) = -k \log \gamma - \beta \sum_{i \le k} (i/n)^{-\rho} - \frac{1}{\gamma} \sum_{i \le k} e^{-\beta(i/n)^{-\rho}} U_i,$$

[44] Beirlant, J., Dierckx, G., Guillou, A. and Stărică, C. (2002). On exponential representations of log–spacings of extreme order statistics. Extremes 5, 157–180.

[45] Kaufmann, E. and Reiss, R.–D. (1998). Approximation of the Hill estimator process. Statist. Probab. Letters 39, 347–354.

[46] Drees, H., de Haan, L. and Resnick, S.I. (2000). How to make a Hill plot. Ann. Statist. 28, 254–274.

led A. Feuerverger and P. Hall, cited on page 193, to an explicit expression for $\hat{\gamma}$, as a function of $\hat{\beta}$ and $\hat{\rho}$, given by

$$\hat{\gamma} = \hat{\gamma}^{FH}_{\hat{\beta},\hat{\rho},k} := \frac{1}{k} \sum_{i \leq k} e^{-\hat{\beta}(i/n)^{-\hat{\rho}}} U_i. \tag{6.68}$$

Then, $\hat{\beta} = \hat{\beta}^{FH}_k$ and $\hat{\rho} = \hat{\rho}^{FH}_k$ are numerically obtained, through

$$(\hat{\beta},\hat{\rho}) := \arg\min_{(\beta,\rho)} \left\{ \log\left(\tfrac{1}{k}\sum_{i \leq k} e^{-\beta(i/n)^{-\rho}} U_i\right) + \beta\left(\tfrac{1}{k}\sum_{i \leq k}(i/n)^{-\rho}\right) \right\}. \tag{6.69}$$

If (6.49) and the second order condition (6.51) hold, it is possible to state the following results.

- If we assume ρ to be known,

$$\gamma^{FH}_{\hat{\beta},\rho,k} \stackrel{d}{=} \gamma + \gamma\left(\frac{1-\rho}{\rho}\right)\frac{\Gamma_k}{\sqrt{k}} + o_p(A(n/k)),$$

where Γ_k is asymptotically standard normal.

- If ρ is unknown as well as β, as usually happens, and they are both estimated through (6.69), then

$$\gamma^{FH}_{\hat{\beta},\hat{\rho},k} \stackrel{d}{=} \gamma + \gamma\left(\frac{1-\rho}{\rho}\right)^2 \frac{\Gamma^*_k}{\sqrt{k}} + o_p(A(n/k))$$

holds, where Γ^*_k is an asymptotically standard normal random variable.

- Consequently, even when $\sqrt{k}\,A(n/k) \to \lambda$, non–null, we have an asymptotic normal behavior for the reduced–bias tail index estimator, with a null asymptotic bias, but at the expenses of a large asymptotic variance, ruled by $\sigma^2_{FH} = \gamma^2\left((1-\rho)/\rho\right)^4 > \sigma^2_{GJ}$ for $|\rho| < 3.676$, with σ^2_{GJ} provided in (6.60). Indeed, $\sqrt{k}\left(\gamma^{FH}_{\hat{\beta},\hat{\rho},k} - \gamma\right)$ is asymptotically normal wit mean zero and variance

$$\sigma^2_{FH} = \gamma^2\left(\frac{1-\rho}{\rho}\right)^4. \tag{6.70}$$

A Simplified Maximum Likelihood Tail Index Estimator and the External Estimation of ρ

The ML estimators of β and ρ in (6.69) are the solution of the ML system of equations

$$\hat{b}_{10}\hat{B}_{00} - \hat{B}_{10} = 0 \quad \text{and} \quad \hat{B}_{11} - \hat{b}_{11}\hat{B}_{00} = 0,$$

where for non–negative integers j and l,

$$\hat{b}_{jl} \equiv \hat{b}_{jl}(\hat{\rho}) := \frac{1}{k} \sum_{i \leq k} \left(\frac{i}{n}\right)^{-j\hat{\rho}} \left(\log \frac{i}{n}\right)^l =: \left(\frac{k}{n}\right)^{\hat{\rho}} \hat{c}_{jl},$$

and

$$\hat{B}_{jl} \equiv \hat{B}_{jl}(\hat{\rho}, \hat{\beta}) := \frac{1}{k} \sum_{i \leq k} \left(\frac{i}{n}\right)^{-j\hat{\rho}} \left(\log \frac{i}{n}\right)^l e^{-\hat{\beta}(i/n)^{-\hat{\rho}}} U_i =: \left(\frac{k}{n}\right)^{\hat{\rho}} \hat{C}_{jl}.$$

The first ML equation may then be written as,

$$\sum_{i \leq k} i^{-\hat{\rho}} \exp\left(-\hat{\beta}(i/n)^{-\hat{\rho}}\right) U_i = \hat{\gamma}\left(\sum_{i \leq k} i^{-\hat{\rho}}\right),$$

with $\hat{\gamma}$ given in (6.68). The use of $e^x \sim 1 + x$ as $x \to 0$, led Gomes and Martins[47] to an explicit estimator for β, given by

$$\hat{\beta}_{\hat{\rho},k}^{GM} := \left(\frac{k}{n}\right)^{\hat{\rho}} \frac{\hat{c}_{10}\hat{C}_{00} - \hat{C}_{10}}{\hat{c}_{10}\hat{C}_{10} - \hat{C}_{20}}, \quad \hat{C}_{j0} = \hat{C}_j = \frac{1}{k} \sum_{i \leq k} \left(\frac{i}{k}\right)^{-j\hat{\rho}} U_i, \qquad (6.71)$$

and the following approximate maximum likelihood estimator for the tail index γ,

$$\hat{\gamma}_{\hat{\rho},k}^{GM} := \frac{1}{k} \sum_{i \leq k} U_i - \hat{\beta}_{\hat{\rho},k}^{GM} \left(\frac{n}{k}\right)^{\hat{\rho}} \hat{C}_1, \qquad (6.72)$$

based on an adequate, consistent estimator for ρ. The estimator in (6.72) is thus a bias–corrected Hill estimator, i.e., the dominant component of the bias of the Hill estimator, provided in (6.56) and equal to $A(n/k)/(1 - \rho) = \gamma\beta(n/k)^{\rho}/(1 - \rho)$ is estimated though $\hat{\beta}_{\hat{\rho},k}^{GM} (n/k)^{\hat{\rho}} \hat{C}_1$, and directly removed from the Hill estimator in (6.54), which can also be written as $\gamma_{n,k}^H = \sum_{i \leq k} U_i/k$.

If the second order condition (6.51) holds, if $k = k_n$ is a sequence of intermediate positive integers, and $\sqrt{k} A(n/k) \to \lambda$, finite, as $n \to \infty$, with $\hat{\rho}$ any ρ–estimator such that $\hat{\rho} - \rho = o_p(1)$ for any k such that $\sqrt{k}A(n/k) \to \lambda$, finite, one obtains that $\sqrt{k}(\gamma_{\hat{\rho},k}^{GM} - \gamma)$ is asymptotically normal with mean zero and variance

$$\sigma_{GM}^2 = \frac{\gamma^2(1 - \rho)^2}{\rho^2}. \qquad (6.73)$$

---

[47]Gomes, M.I. and Martins, M.J. (2002). "Asymptotically unbiased" estimators of the tail index based on external estimation of the second order parameter. Extremes 5, 5–31.

External Estimation of β and ρ

Gomes et al.[48] suggest the computation of the β estimator $\hat{\beta}_{\hat{\rho},k}^{GM}$ at the level k_1 in (6.63), the level used for the estimation of ρ. With the notation $\hat{\beta} := \hat{\beta}_{\hat{\rho},k_1}^{GM}$, they suggest thus the replacement of the estimator in (6.72) by

$$\hat{\gamma}_{\hat{\beta},\hat{\rho},k}^{\overline{M}} \equiv \overline{M}_{\hat{\beta},\hat{\rho},k} := \gamma_{n,k}^H - \hat{\beta} \left(\frac{n}{k}\right)^{\hat{\rho}} \hat{C}_1, \qquad (6.74)$$

where $\gamma_{n,k}^H$ denotes the Hill estimator in (6.54), and $(\hat{\beta}, \hat{\rho})$ are adequate consistent estimators of the second order parameters (β, ρ). With the same objectives, but with a slightly simpler analytic expression, we shall here also consider the estimator

$$\hat{\gamma}_{\hat{\beta},\hat{\rho},k}^{\overline{H}} \equiv \overline{H}_{\hat{\beta},\hat{\rho},k} := \gamma_{n,k}^H \left(1 - \hat{\beta} (n/k)^{\hat{\rho}} /(1 - \hat{\rho})\right), \qquad (6.75)$$

studied in Caeiro et al., cf. page 197. Notice that the dominant component of the bias of the Hill estimator is estimated in (6.75) through $\gamma_{n,k}^H \hat{\beta}(n/k)^{\hat{\rho}}/(1 - \hat{\rho})$, and directly removed from Hill's classical tail index estimator.

The estimation of β and ρ at the level k_1 in (6.63), of a higher order than the level k used for the tail index estimation, enables the reduction of bias without increasing the asymptotic variance, which is kept at the value γ^2, the asymptotic variance of Hill's estimator.

Denoting by $\hat{\gamma}_{\hat{\beta},\hat{\rho},k}^{\bullet}$ any of the estimators in (6.74) and (6.75), we now formulate two asymptotic results for these estimators.

- Under the second order condition (6.51) and k intermediate, further assuming that $A(\cdot)$ can be chosen as in (6.67), we have

$$\hat{\gamma}_{\beta,\rho,k}^{\bullet} \stackrel{d}{=} \gamma + \frac{\gamma P_k^{\bullet}}{\sqrt{k}} + o_p(A(n/k)), \qquad (6.76)$$

  where $P_k^{\bullet}$ is an asymptotically standard normal random variable. This is an immediate consequence of the representation of $\hat{\gamma}_{\beta,\rho,k}^{\bullet}$.

- The result in (6.76) remains true for the reduced–bias tail index estimators estimators in (6.74) and (6.75), provided that $\hat{\beta} - \beta = o_p(1)$ and $(\hat{\rho} - \rho) \log(n/k) = o_p(1)$ for all k on which we base the tail index estimation, i.e., whenever $\sqrt{k} \, A(n/k) \to \lambda$, finite.

  For these values of k, one gets that $\sqrt{k}(\hat{\gamma}_{\hat{\beta},\hat{\rho},k}^{\bullet} - \gamma)$ is asymptotically normal with mean zero and variance

$$\sigma_{\bullet}^2 = \gamma^2. \qquad (6.77)$$

---

[48]Gomes, M.I., Martins, M.J. and Neves, M. (2005). Revisiting the second order reduced bias "maximum likelihood" extreme value index estimators. Notas e Comunicações CEAUL 10/2005. Submitted.

This is immediate from

$$\hat{\gamma}^{\bullet}_{\hat{\beta},\hat{\rho},k} - \hat{\gamma}^{\bullet}_{\beta,\rho,k} \sim -\frac{A(n/k)}{1-\rho}\left(\frac{\hat{\beta}-\beta}{\beta} + (\hat{\rho}-\rho)\log(n/k)\right).$$

In Fig. 6.7, we illustrate the differences between the sample paths of the estimator in (6.75), denoted $\overline{H}_{\hat{\beta},\hat{\rho}}$ for the sake of simplicity. We have considered a sample of size $n = 10,000$ from a Fréchet model, with $\gamma = 1$, when we compute $\hat{\beta}$ and $\hat{\rho}$ at the same level k used for the estimation of the tail index γ (left), when we compute only $\hat{\beta}$ at that same level k, being $\hat{\rho}$ computed at a larger k-value, let us say an intermediate level k_1 such that $\sqrt{k_1}\,A(n/k_1) \to \infty$, as $n \to \infty$ (center) and when both $\hat{\rho}$ and $\hat{\beta}$ are computed at that high level k_1 (right).

We have estimated β through $\hat{\beta}_{0,k} = \hat{\beta}^{GM}_{\hat{\rho}_{01},k}$ in (6.71), computed at the level k used for the estimation of the tail index, as well as computed at the level $k_1 = [n^{0.995}]$ in (6.63), the one used for the estimator $\hat{\rho}_{01}$ in (6.64). We use the notation $\hat{\beta}_{01} = \hat{\beta}_{0,k_1}$. The estimates of β and ρ have been incorporated in the $\overline{H}$–estimator, leading to $\overline{H}_{\hat{\beta}_{0,k},\hat{\rho}_{0,k}}$ (left), $\overline{H}_{\hat{\beta}_{0,k},\hat{\rho}_{01}}$ (center) and $\overline{H}_{\hat{\beta}_{01},\hat{\rho}_{01}}$ (right).

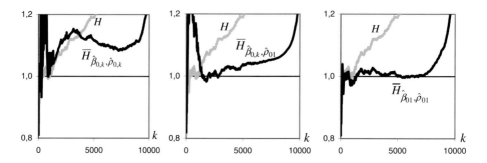

FIG. 6.7. External estimation of (β,ρ) at a larger fixed level k_1 (right) versus estimation at the same level both for β and ρ (left) and only for β (center).

From the pictures in Fig. 6.7, as well as from the asymptotic results in (6.60), (6.70), (6.73) and (6.77), we thus advise, in practice:

- The direct estimation of the dominant component of the bias of Hill's estimator of a positive tail index γ. The second order parameters in the bias should be computed at a fixed level k_1 of a larger order than that of the level k at which we compute the Hill estimator.

- Such an estimated bias should then be directly removed from the classical Hill estimator.

- Doing this, we are able to keep the asymptotic variance of the new reduced-bias tail index estimator equal to γ^2, the asymptotic variance of the Hill estimator.

A Simulation Experiment

We have here implemented a simulation experiment, with 1000 runs, for an underlying Burr parent,

$$F(x) = 1 - \left(1 + x^{-\rho/\gamma}\right)^{1/\rho}, \ x \geq 0,$$

with $\rho = -0.5$ and $\gamma = 1$. For these Burr models, $\beta = \gamma$ for any ρ. We have again estimated β through $\hat{\beta}_{0,k}$, computed at the level k used for the estimation of the tail index, as well as computed at the level k_1 in (6.63), the one used for the estimator $\hat{\rho}_{01}$ in (6.64). We use the same notation as before. The simulations show that the tail index estimator $\overline{H}_{\hat{\beta}_{01},\hat{\rho}_{01},k}$ seems to work reasonably well, as illustrated in Fig. 6.8.

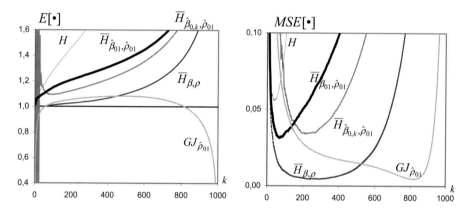

FIG. 6.8. Mean values and mean squared errors of estimators the Hill and reduced–bias estimators, for samples of size $n = 1000$, from a Burr parent with $\gamma = 1$ and $\rho = -0.5$ ($\beta = 1$).

The discrepancy between the behavior of the estimator $\overline{H}_{\hat{\beta}_{01},\hat{\rho}_{01}}$ and the random variable $\overline{H}_{\beta,\rho}$ suggests that some improvement in the estimation of second order parameters may be still welcome, but the behavior of the mean squared error of the $\overline{H}$–estimator is rather interesting: the new estimator $\overline{H}_{\hat{\beta}_{01},\hat{\rho}_{01},k}$ is better than the Hill estimator not only when both are considered at their optimal levels, but also for every sub–optimal level k, and this contrarily to what happens with the classical reduced–bias estimators $GJ_{\hat{\rho}_{01}}$ and $\overline{H}_{\hat{\beta}_{0,k},\hat{\rho}_{01}}$, as we may also see in Fig. 6.8.

Some overall conclusions

- If we estimate the first order parameter at a level k, and use that same level k for the estimation of the second order parameter ρ and β in $A(t) = \gamma \beta t^\rho$, we get a much higher asymptotic variance than when we compute ρ at a larger level k_1, computing β at the same level k. And we may still decrease the asymptotic variance of the tail index reduced–bias estimators, if we estimate both second order parameters, β and ρ, at a larger level k_1 than the one used for the estimation of the first order parameter.

- The main advantage of the new reduced–bias estimators in (6.74) and (6.75) lies on the fact that we can estimate β and ρ adequately through $\hat{\beta}$ and $\hat{\rho}$ so that the MSE of the new estimator is smaller than the MSE of Hill's estimator for all k, even when $|\rho| > 1$, a region where has been difficult to find alternatives for the Hill estimator. And this happens together with a higher stability of the sample paths around the target value γ. The pioneering paper on a new type of reduced–bias estimators of a positive tail index, i.e., the ones with an asymptotic variance equal to γ^2, is by Gomes et al.[49], who introduce a weighted Hill estimator, linear combination of the log–excesses.

- To obtain information on the asymptotic bias of these reduced–bias estimators we should have gone further into a third order framework, specifying the rate of convergence in the second order condition in (6.51). This is however beyond the scope of this chapter. Interested readers may look into Gomes et al.[50]

The Estimation of β and γ in Action

We go on with the following:

Algorithm (β and γ estimation).

4. Chosen τ in step 3, work then with $(\hat{\rho}, \hat{\beta}) = (\hat{\rho}_{\tau 1}, \hat{\beta}_{\tau 1}) := (\hat{\rho}_{\tau, k_1}, \hat{\beta}_{\hat{\rho}_{\tau, k_1}})$, $\hat{\rho}_{\tau, k}$, k_1 and $\hat{\beta}_{\hat{\rho}, k}$ given in (6.62), (6.63) and (6.71), respectively.

5. Estimate the optimal level for the estimation through the Hill, given by $\hat{k}_0^H = \left((1 - \hat{\rho})n^{-\hat{\rho}}/(\hat{\beta}\sqrt{-2\hat{\rho}})\right)^{2/(1-2\hat{\rho})}$, compute $k_{min} = \hat{k}_0^H/4$, $k_{max} = 4\hat{k}_0^H$ and plot the classical Hill estimates H_k, $k_{min} \leq k \leq k_{max}$.

6. Plot also, again for $k_{min} \leq k \leq k_{max}$, the reduced-bias tail index estimates $\overline{H}_{\hat{\beta}, \hat{\rho}, k}$ and $\overline{M}_{\hat{\beta}, \hat{\rho}, k}$, associated to the estimates $(\hat{\rho}, \hat{\beta})$ in step 4.

---

[49]Gomes, M.I., de Haan, L. and Rodrigues, L. (2004). Tail index estimation through accommodation of bias in the weighted log–excesses. Notas e Comunicações CEAUL 14/2004. Submitted.

[50]Gomes, M.I., Caeiro, F. and Figueiredo, F. (2004). Bias reduction of a tail index estimator through an external estimation of the second order parameter. Statistics 38, 497–510.

7. Obtain χ_H, $\chi_{\overline{H}}$ and $\chi_{\overline{M}}$, the medians of H_k, $\overline{H}_{\hat{\beta},\hat{\rho},k}$ and $\overline{M}_{\hat{\beta},\hat{\rho},k}$, respectively, for $k_{min} \leq k \leq k_{max}$. Compute the indicators, $I_H := \sum_{k_1 \leq k \leq k_2} \left(H_k - \chi_H \right)^2$, $I_{\overline{H}} := \sum_{k_1 \leq k \leq k_2} \left(\overline{H}_{\hat{\beta},\hat{\rho},k} - \chi_{\overline{H}} \right)^2$ and $I_{\overline{M}} := \sum_{k_1 \leq k \leq k_2} \left(\overline{M}_{\hat{\beta},\hat{\rho},k} - \chi_{\overline{M}} \right)^2$.

8. Let T be the estimator (among H, $\overline{H}$ and $\overline{M}$) providing the smallest value among I_H, $I_{\overline{H}}$ and $I_{\overline{M}}$. Consider $\hat{\gamma}_T = \chi_T$ as estimate of γ.

Financial Data Analysis

We now provide a data analysis of log–returns associated to the Euro–British pound daily exchange rates, collected from January 4, 1999, until November 17, 2005.

The number of positive log–returns of these data is $n_0 = 385$. The sample paths of the ρ–estimates associated to the tuning parameter $\tau = 0$ and $\tau = 1$ lead us to choose, on the basis of any stability criterion for large k, like the one suggested in step 3. of the Algorithm, the estimate associated to $\tau = 0$. The estimates obtained are $\left(\hat{\rho}_0, \hat{\beta}_0 \right) = (-0.686, 1.047)$ obtained at $k_1 = 808$.

In Fig. 6.9, for the Euro–British pound data, we picture the sample paths of the estimators of the second order parameters ρ (left) and the sample paths of the classical Hill estimator H in (6.54), the second order reduced–bias tail index estimators $\overline{H}_0 = \overline{H}_{\hat{\beta}_{01},\hat{\rho}_{01}}$ and $\overline{M}_0 = \overline{M}_{\hat{\beta}_{01},\hat{\rho}_{01}}$, provided in (6.74) and (6.75), respectively (right). For this data set, the criterion in step 7. of the Algorithm led to the choice of $\overline{M}_0$ and to the estimate $\hat{\gamma}_{\overline{M}_0} = 0.289$ (associated to $k = 132$).

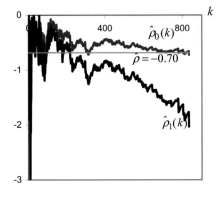 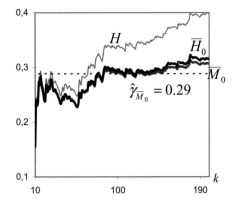

FIG. 6.9. Estimates of ρ, through $\hat{\rho}_{\tau,k}$ in (6.62), $\tau = 0$ and 1 (left), and of γ, through H, $\overline{M}$ and $\overline{H}$ in (6.54), (6.74) and (6.75) (right), for the positive log–returns on Euro–British pound data.

Additional Literature on Reduced–Bias Tail Index Estimation

Other approaches to reduced–bias estimation of the tail index can be found in Gomes and Martins[51], Fraga Alves[52] and Gomes et al.[53] [54], and the references therein.

---

[51]Gomes, M.I. and Martins M.J. (2001). Alternatives to Hill's estimator—asymptotic versus finite sample behaviour. J. Statist. Planning and Inference 93, 161–180.

[52]Fraga Alves, M.I. (2001). A location invariant Hill–type estimator. Extremes 4, 199–217.

[53]Gomes, M.I., Figueiredo, F. and Mendonça, S. (2005). Asymptotically best linear unbiased tail estimators under a second order regular variation condition. J. Statist. Plann. Inf. 134, 409–433

[54]Gomes, M.I., Miranda, C. and Viseu, C. (2006). Reduced bias tail index estimation and the Jackknife methodology. Statistica Neerlandica 60, 1–28.

Chapter 7

Statistics of Dependent Variables

coauthored by H. Drees[1]

Classical extreme value statistics is dominated by the theory for independent and identically distributed (iid) observations. In many applications, though, one encounters a non–negligible serial (or spatial) dependence. For instance, returns of an investment over successive periods are usually dependent, cf. Chapter 16, and stable low pressure systems can lead to extreme amounts of rainfall over several consecutive days. These examples demonstrate that a positive dependence between extreme events is often particularly troublesome as the consequences, which are already serious for each single event, may accumulate and finally result in a devastating catastrophe.

In Section 7.1 we discuss the impact of serial dependence on the statistical tail analysis. A more detailed analysis of the consequences for the estimation of the extreme value index and extreme quantiles is given in Section 7.2 and Section 7.3, respectively, under mild conditions on the dependence structure. Section 7.4 deals with (semi–)parametric time series models where the dependence structure is known up to a few parameters.

Throughout this chapter we assume that a strictly stationary univariate time series X_i, $1 \leq i \leq n$, with marginal df F is observed, cf. page 165.

---

[1] Universität Hamburg.

7.1 The Impact of Serial Dependence

The presence of serial dependence has two consequences for the modeling and the statistical analysis of the data. Firstly, in contrast to the iid case, the stochastic behavior is not fully determined by the df of a single observation, but the serial dependence structure must be taken into account, e.g., when the distribution of the sum of two successive observations is to be analyzed. Secondly, even if one is interested only in the tail behavior of the marginal distribution, then often the serial dependence strongly influences the accuracy of estimators or tests known from classical extreme value statistics for iid data.

If one does not assume a (semi–)parametric time series model, then relatively few statistical methods for the estimation of the dependence structure between extreme observations are available. One possible approach to the analysis of the dependence is to apply multivariate extreme value statistics to the vectors of successive observations. Although this approach has proved fruitful in probability theory, it is of limited value in the statistical analysis because, due to the 'curse of dimensionality', in practice only very few consecutive observations can be treated that way.

Alternatively, estimators for certain parameters related to the serial dependence have been proposed in the literature. Best known are estimators of the extremal index, that is, the reciprocal value of the mean cluster size, cf. Sections 2.7 and 6.2. Hsing[2] also proposed estimators for the cluster size distribution. Unfortunately, all these parameters bear limited information about the serial dependence structure. For example, Gomes and de Haan[3] proved that the probability of k successive exceedances of an ARCH(1) time series over a high threshold does not only depend on the extremal index but on the whole dependence structure of the ARCH process. of the marginal parameters that of the aforementioned dependence parameters. Since a more general approach to the statistical analysis of the dependence structure is still an open problem, here we will focus on the second aspect, that is, the influence of the serial dependence on estimators of marginal tail parameters like the extreme value index or extreme quantiles.

Recall from (6.11) that under weak conditions one has

$$P\Big\{ \max_{1 \leq i \leq n} X_i \leq x \Big\} \approx P\Big\{ \max_{1 \leq i \leq [n\theta]} \tilde{X}_i \leq x \Big\}$$

for large values of n where $\tilde{X}_i$ denote iid random variables with the same df F as X_1 and $\theta \in (0,1]$ is the extremal index. Hence, as far as maxima are concerned, the serial dependence reduces the effective sample size by a factor θ. Likewise, one may expect that a time series with a non–negligible serial dependence bears less

---

[2]Hsing, T. (1991). Estimating the parameters of rare events. Stoch. Processes Appl. 37, 117–139.

[3]Gomes, M.I. and de Haan, L. (2003). Joint exceedances of the ARCH process. Preprint, Erasmus University Rotterdam.

information on F than an iid sample of the same size, and that thus the estimation error for marginal parameters will be higher. Indeed, we will see that under mild conditions on the dependence structure the same estimators as in the classical iid setting can be used, but that their distribution is less concentrated about the true value. Generally, the application of standard estimation procedures relies on ergodic theory[4] [5]. However, if one does not account for this loss of accuracy, e.g., in flood frequency analysis, then it is likely that safety margins are chosen too low to prevent catastrophic events with the prescribed probability.

The construction of confidence intervals based on dependent data is substantially more complicated than in the iid setting, because the extent to which the estimation error is increased by the serial dependence does not only depend on a simple dependence parameter—like the extremal index—but on the whole dependence structure in a complex manner. Therefore, a completely new approach to the construction of confidence intervals is needed, see Sections 7.2 and 7.3. As a by-product, we will also obtain a new graphical tool for choosing the sample fraction on which the estimation is based, that may be useful also in the standard iid setting.

Instead of examining the accuracy of estimators of marginal parameters under dependence, one may also first try to decluster the observed time series and then to apply the classical statistical theory to the nearly independent data thus obtained, see also the end of Section 2.7. This approach, however, has two serious drawbacks. Firstly, the declustering is often a delicate task that must be done manually by subject-matter specialists. Secondly, although a cluster bears less information than iid data of the same size, taking into account only one observation from each cluster usually seems a gross waste of information which will often lead to even larger estimation errors for the parameter of interest. For these reasons, in what follows we will adopt are more refined approach.

7.2 Estimating the Extreme Value Index

Assume that the df F belongs to the max–domain of attraction of an extreme value distribution with index γ. Recall from (1.46) that then the conditional distribution of an exceedance over a high threshold u can be approximated by a GP distribution W_{γ,u,σ_u}. In Section 5.1 this approximation was used to derive estimators of the extreme value index γ which use only the k largest observations from analogous estimators in a parametric GP model. We will see that under mild conditions the same estimators can be used for time series data.

---

[4]Révész, P. (1968). The Laws of Large Numbers. Academic Press, New York.

[5]Pantula, S.G. (1988). Estimation of autoregressive models with ARCH errors. Sankhyā 50, 119–148.

Conditions on the Dependence Structure

The trivial time series $X_i = X_1$ for all i shows that, without further conditions, in general this approach is not justified any longer if the data exhibit serial dependence. To avoid problems of that type, we require that the dependence between the data observed in two time periods separated by l time points vanishes as l increases. More precisely, we assume that the so–called β–mixing coefficients tend to 0 sufficiently fast. For technical details of the conditions and results mentioned in this and the next section, we refer to a series of articles[6] [7] [8]. This mixing condition is usually satisfied for linear time series like ARMA models, and also for the GARCH models defined by (16.29) and (16.44)—introduced in conjunction with returns of random asset prices—and, more general, for Markovian time series under mild conditions. However, time series models with long range dependence are often ruled out.

Secondly, we need to assume that the serial dependence between exceedances over a high threshold u stabilizes as u increases. This very mild condition will be automatically satisfied if the vector (X_1, X_{1+h}) belongs to the max–domain of attraction of a bivariate extreme value distribution, see (12.2), for all $h > 0$. Finally, we assume that the sizes of clusters of extreme observations have a finite variance.

Normal Approximations and Confidence Intervals

Under these mild conditions, it can be shown that the k largest order statistics $X_{n-k+1:n}, \ldots, X_{n:n}$ can be approximated by the order statistics of GP random variables. More precisely, one can approximate the empirical tail quantile function pertaining to the observed order statistics by a GP quantile function plus a stochastic error term of order $k^{-1/2}$, provided k is sufficiently small relative to the sample size n, but not too small. The stochastic error term can be described in terms of a centered Gaussian process whose covariance function is determined by the tail dependence structure of the observed time series.

From this result, one can derive approximations to the distributions of the estimators introduced in Section 5.1. For example, the Hill estimator (in the case $\gamma > 0$) and the maximum likelihood estimator in the GP model based on the k largest order statistics (if $\gamma > -1/2$) are approximately normally distributed with mean γ and variance $\gamma^2 \sigma_0^2/k$ and $(1 + \gamma)^2 \sigma_0^2/k$, respectively, where σ_0^2 is

---

[6]Drees, H. (2000). Weighted approximations of tail processes for β–mixing random variables. Ann. Appl. Probab. 10, 1274–1301.

[7]Drees, H. (2002). Tail empirical processes under mixing conditions. In: H.G. Dehling, T. Mikosch und M. Sørensen (eds.), Empirical Process Techniques for Dependent Data, 325–342, Birkhäuser, Boston.

[8]Drees, H. (2003). Extreme quantile estimation for dependent data with applications to finance. Bernoulli 9, 617–657.

determined by the dependence structure. If the data are independent, then $\sigma_0^2 = 1$ while typically $\sigma_0^2 > 1$ if serial dependence is present in the data. Recall that this approximation is justified only if k is sufficiently small relative to the length of the time series, while for larger values of k the estimators exhibit a non–negligible bias if F is not exactly equal to a GP df or, for the Hill estimator, the location parameter in the GP1 model does not equal 0. Hence the Hill estimator need not outperform the ML estimator, despite its smaller variance, cf. the discussion around Fig. 5.1 and Fig. 5.2.

Unlike for iid samples, the estimators will not necessarily be asymptotically normal if k grows too slowly as the sample size n increases. This new effect can be explained as follows. If k is very small relative to n, then all order statistics used by the estimator stem from very few clusters of large observations. However, within one cluster the behavior of large observations is not determined by the tail df only, but it is also influenced by the specific serial dependence structure. In practice, the lower bound on the rate at which k tends to infinity does not cause major problems, because usually the β–mixing coefficients tend to 0 at an exponential rate. In this case, it is sufficient that k is of larger order than $\log^{2+\varepsilon} n$ for some $\varepsilon > 0$, which is anyway needed to obtain accurate estimates.

More generally, a large class of estimators $\hat{\gamma}_n^{(k)}$ that use only the k largest order statistics are approximately normally distributed with mean γ and variance σ^2/k with a $\sigma^2 > 0$ that is determined by the tail dependence structure of the time series. While this approximation allows a comparison of the performance of different estimators for a given time series model, it cannot be used directly for the construction of confidence intervals, because σ^2 is unknown.

To overcome this problem, we take advantage of the fact that the size of the random fluctuations of the estimates $\hat{\gamma}_n^{(k)}$ as k varies are proportional to σ. More concretely, one can show that, for a suitable chosen j, the estimator

$$\hat{\sigma}_n^2 := \sum_{i=j}^{k} (\hat{\gamma}_n^{(i)} - \hat{\gamma}_n^{(k)})^2 \Big/ \sum_{i=j}^{k} (i^{-1/2} - k^{-1/2})^2 \tag{7.1}$$

is consistent for σ^2. Simulation studies indicate that in practice one may choose j equal to the smallest number such that $\hat{\gamma}_n^{(j)}$ is well defined, e.g., $j = 2$ for the Hill estimator. Now approximative confidence intervals for γ can be easily constructed. For example,

$$\left[\hat{\gamma}_n^{(k)} - \Phi^{-1}(1 - \alpha/2)\hat{\sigma}_n k^{-1/2}, \hat{\gamma}_n^{(k)} + \Phi^{-1}(1 - \alpha/2)\hat{\sigma}_n k^{-1/2} \right] \tag{7.2}$$

defines a two–sided confidence interval with nominal coverage probability $1 - \alpha$.

Changes of Interest Rates—a Case Study, Part 1

Life insurance companies often invest a large proportion of the capital they manage in low risk bonds. If they guarantee a minimal interest rate to their customers, they are exposed to the risk that the interest rates of the low risk bonds drop below this level. On the other hand, a hike of the interest rates will let the value of the portfolio decrease rapidly, which may cause problems if many customers withdraw their capital from the company. In both cases a quick change of the interest rate is particularly troublesome, as it gives the company little time to adjust its investment strategy.

In this case study, we want to analyze the interest rates of the US treasury bonds with maturity in 10 years observed on a monthly basis from 1957 to 1999. The time series R_i, $1 \leq i \leq m$, of length $m = 515$ is assumed stationary. Given the long observational period, this assumption is somewhat questionable but no natural alternative model is at hand. We are interested in extreme quantiles of the change of the rate over one year. In order to not waste data, we analyze the interest rate changes over overlapping one year periods, namely

$$X_i := R_{i+12} - R_i, \quad 1 \leq i \leq n := m - 12 = 503,$$

which are clearly dependent, see Fig. 7.1.

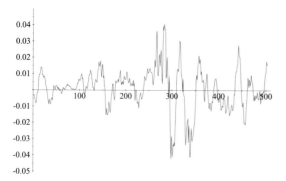

FIG. 7.1. Yearly changes of interest rates of 10 year treasury bonds.

As a first step in the analysis of the upper tail, we estimate the extreme value index. The left plot of Fig. 7.2 displays the Hill estimator, the ML estimator in the GP model and the moment estimator, cf. Section 5.1, as a function of the number k of largest order statistics. All three estimates are quite different: the Hill plot exhibits a clear (upward) trend almost from the beginning, while the curves of the ML estimates and the moment estimates are nearly parallel for k between 50 and 200. As a rule of thumb, such a behavior indicates that a Pareto model with non-vanishing location parameter fits the upper tail well. Indeed, after a shift of the data by 0.05, the ML estimator and the moment estimator yield almost identical values for $80 \leq k \leq 220$ which are relatively stable for $100 \leq k \leq 160$, and the Hill plot is very stable for $80 \leq k \leq 160$ with values close to the other estimates (Fig. 7.2, right plot). Again we see that the Hill estimator is particularly sensitive to shifts of the data, whereas the moment estimator is less strongly influenced and the ML estimator is invariant under such shifts, cf. Fig. 5.1. Here, after the shift, the Hill estimator with $k = 160$ seems a reasonable choice, that yields $\gamma \approx 0.111$. Other diagnostic tools, like a QQ–plot, confirm that the corresponding pure Pareto model fits the tail well after the shift of the data.

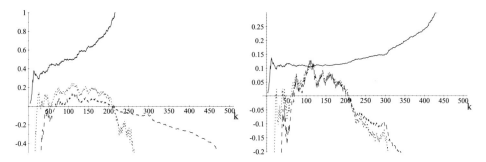

FIG. 7.2. Hill estimator (solid), ML estimator (dashed) and moment estimator (dotted) for the right tail of yearly interest rate changes (left) and the same data shifted by 0.05 (right).

Next we calculate the confidence intervals (7.2). The estimates of the asymptotic variance according to (7.1) are plotted in the left graph of Fig. 7.3. Again the curve is quite stable for k between 80 and 180. The right graph of Fig. 7.3 displays the Hill estimator together with the 95%–confidence intervals (7.2) as a function of k. For comparison, also the confidence intervals

$$\left[\hat{\gamma}_n^{(k)}(1 - \Phi^{-1}(1 - \alpha/2)k^{-1/2}), \hat{\gamma}_n^{(k)}(1 + \Phi^{-1}(1 - \alpha/2)k^{-1/2})\right] \qquad (7.3)$$

are plotted for $\alpha = 0.05$, that would be appropriate if the data were independent and hence the asymptotic variance was equal to γ^2. As one expects, the confidence intervals which take the serial dependence into account, are considerably wider: for $k = 160$ it equals $[0.086, 0.136]$ compared with the interval $[0.094, 0.128]$ derived from the theory for iid data. (However, note that these confidence intervals do not account for the shift of the data by 0.05, which is motivated by the achieved coincidence of the moment estimator and the ML estimator and is thus data driven. Taking this into account would lead to even wider confidence intervals.)

The analysis of the lower tail is more difficult. After shifting the data by 0.2 one obtains the ML estimates and the moment estimates shown in the left plot of Fig. 7.4. Here the true value may be negative, so that the Hill estimator cannot be used. For $k \leq 120$ both estimators are very unstable, but for k between 120 and 275 the curves are relatively stable and close together. A QQ–plot based on the moment estimator with $k = 270$ shows that the tail is fitted reasonably well, although the most extreme observations deviate from the ideal line. Simulations, however, show that for dependent data one must expect much bigger deviations in the QQ–plot than one typically observes for iid data. Unfortunately, because

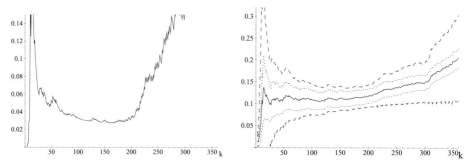

FIG. 7.3. Estimated asymptotic variance according to (7.1) (left) and the Hill estimator (solid) with 95%–confidence intervals (7.2) (dashed) and 95%–confidence intervals (7.3) assuming independence (dotted) (right).

of the large fluctuations of the estimators for small values of k, the estimates for the asymptotic variance are unreliable, and no reasonable confidence interval is available.

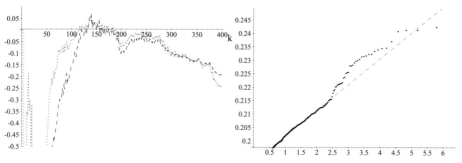

FIG. 7.4. ML estimates (dashed) and moment estimates (dotted) for the left tail (left) and QQ–plot (right).

Choice of the Sample Fraction Used for Estimation

The shape of the curve of variance estimates displayed in the left plot of Fig. 7.3 is quite common when the Hill estimator is used. Except for the very first values of k, the variance estimates are usually rather large when only few order statistics are used, then the curve decreases until it stabilizes for intermediate values of k, and finally at a certain point it sharply increases. Since the estimator $\hat{\sigma}_n^2$ just measures an average squared difference of Hill estimators based on different numbers of order statistics, it cannot distinguish between random fluctuations and systematic changes due to an increasing bias. Hence, for larger values of k, the growing bias leads to increasing variance estimates. Although in this range the variance estimates are clearly quite inaccurate, in practice this effect is to be welcomed for two reasons.

Firstly, much more clearly than the Hill plot, the sharp kink in the curve of variance estimates indicates the point from which on the bias significantly contributes to the error of the Hill estimator. Therefore, the plot of the variance estimates is a very useful tool for choosing the number of largest order statistics to be used by the Hill estimator even for iid data.

Secondly, the confidence intervals (7.2) start to widen again at the point where the bias kicks in, whereas the confidence intervals (7.3) motivated by the theory for iid data usually shrink with increasing k. Hence the latter give a totally false impression; indeed, in literature one can often find plots of "confidence intervals" which are disjoint for different values of k! In contrast, the confidence intervals derived from the theory for dependent data are qualitatively more reasonable for large values of k (though they are too conservative) in that the growing uncertainty due to the increasing bias is taken into account.

Unfortunately, as in the analysis of the lower tail in the case study, the variance estimates (7.1) are often unreliable when the estimator of the extreme value index are very unstable for small values of k. Sometimes it might help to start with a larger number of order statistics, that is, to choose a larger value for j, but this requires the number of observations to be large, which limits the applicability of this approach. In this case, one might think of robustified versions of the variance estimator. For example, the squared difference $(\hat{\gamma}_n^{(i)} - \hat{\gamma}_n^{(k)})^2$ could be replaced with the absolute difference and, accordingly, $(i^{-1/2} - k^{-1/2})^2$ with $|i^{-1/2} - k^{-1/2}|$, but it is not clear whether the resulting estimator is consistent for σ^2, too.

7.3 Extreme Quantile Estimation

In most applications, not the extreme value index (which is just a parameter in the limit model for maxima or exceedances) but, e.g., extreme quantiles $x_p :=$ $F^{-1}(1 - p)$ for small $p > 0$ are the main parameters of interest, see Chapter 16.

Estimation in the Restricted Pareto Model

The estimation of extreme quantiles is particularly simple if the largest order statistics approximately behave as in the restricted Pareto model with location parameter 0. Then x_p can be estimated by the so-called Weissman estimator

$$\hat{x}_{n,p} := \hat{x}_{n,p}^{(k)} := X_{n-k+1:n} \left(\frac{np}{k}\right)^{-\hat{\gamma}_n} \tag{7.4}$$

where $\hat{\gamma}_n = \hat{\gamma}_n^{(k)}$ denotes an estimator of γ that is based on the k largest observations. Of course, this approach only makes sense if k is much bigger than np, because else one could simply use the empirical quantile $X_{n-[np]+1:n}$.

In the setting of Section 7.2, one has under very mild extra conditions on p

$$\frac{1}{\log(k/(np))} \log \frac{\hat{x}_{n,p}}{x_p} = \frac{1}{\log(k/(np))}\left(\frac{\hat{x}_{n,p}}{x_p} - 1\right)(1 + o_P(1))$$
$$= (\hat{\gamma}_n - \gamma)(1 + o_P(1)).$$

Hence, if $\hat{\gamma}_n$ is approximately normal with mean γ and variance σ^2/k, then both $\log(\hat{x}_{n,p}/x_p)$ and $\hat{x}_{n,p}/x_p - 1$ are approximately normal with mean zero and variance $\log^2(k/(np))\sigma^2/k$. In particular if one uses the Hill estimator $\hat{\gamma}_n$, the normal approximation is much better for $\log(\hat{x}_{n,p}/x_p)$ than for the relative estimation error $\hat{x}_{n,p}/x_p - 1$, because $\log \hat{x}_{n,p}$ is a linear function of $\hat{\gamma}_n$, which in turn is a linear statistic and hence is often well approximated by a normal random variable (cf. Fig. 7.10). Therefore, confidence intervals shall be based on the normal approximation for $\log \hat{x}_{n,p}$ rather than the normal approximation for $\hat{x}_{n,p}$. If $\hat{\sigma}_n^2$ is a suitable estimator for σ^2, then

$$\Big[\hat{x}_{n,p} \exp\big(-\Phi^{-1}(1-\alpha/2)\hat{\sigma}_n k^{-1/2} \log(k/(np))\big),$$
$$\hat{x}_{n,p} \exp\big(\Phi^{-1}(1-\alpha/2)\hat{\sigma}_n k^{-1/2} \log(k/(np))\big)\Big] \tag{7.5}$$

is a confidence interval for x_p with approximate coverage probability $1 - \alpha$.

Here one may use the variance estimator (7.1) defined in terms of differences of γ–estimates, but it is more natural to define an analogous estimator in terms of logarithmic quantile estimators $\log \hat{x}_{n,p}^{(k)}$ for different numbers k of largest order statistics:

$$\hat{\sigma}_{n,p}^2 := \sum_{i=j}^{k}\left(\frac{\log(\hat{x}_p^{(i)}/\hat{x}_p^{(k)})}{\log(i/(np))}\right)^2 \bigg/ \sum_{i=j}^{k}\left(i^{-1/2} - \frac{\log(k/(np))}{\log(i/(np))}k^{-1/2}\right)^2, \tag{7.6}$$

with j greater than np. Unlike (7.1), this estimator measures directly the fluctuations of $\log \hat{x}_{n,p}^{(k)}$ as a function of k. Because the estimator (7.1), as an estimator for the asymptotic variance of the quantile estimator, relies on the aforementioned asymptotic relationship between the fluctuations of $\hat{\gamma}_n$ and of $\log \hat{x}_{n,p}^{(k)}$, one may expect that the estimator $\hat{\sigma}_{n,p}^2$ performs better for moderate sample sizes when this asymptotic approximation is inaccurate.

In an extensive simulation study[9] it has been shown that for the Hill estimator and several time series models, the confidence intervals (7.5) with the variance estimated by (7.6) are quite accurate, though a bit too conservative, because the asymptotic variance is over–estimated. To correct for this over–estimation, one may choose $k \geq k_0$ such that $\hat{\sigma}_{n,p}^2$ (or the width of the confidence intervals) is minimized, when k_0 is chosen such that the variance estimates seem reliable for $k \geq k_0$. This simple trick works surprisingly well for many time series models.

[9]Drees, H. (2003). Extreme quantile Estimation for Dependent Data with Applications to Finance. Bernoulli 9, 617–657.

Note that the asymptotic variance of the quantile estimators does not depend on p. Hence one may use an exceedance probability $\tilde{p}$ in the definition (7.6) of $\hat{\sigma}_{n,\tilde{p}}^2$ different from the exceedance probability p of the extreme quantile one is actually interested in. While this approach contradicts the general philosophy explained above (namely that one should use the same estimators for the estimation of the variance and for the estimation of the parameter of interest), it improves the performance of the confidence intervals significantly if p is so small such that $F^{-1}(1-p)$ lies far beyond the range of observations and thus the quantile estimators are quite inaccurate. In such a situation it is advisable to estimate the variance by $\hat{\sigma}_{n,\tilde{p}}^2$ such that the corresponding quantile $F^{-1}(1-\tilde{p})$ lies at the border of the sample, e.g., $\tilde{p} = 1/n$.

Changes of Interest Rates—a Case Study, Part 2

In the situation of the case study discussed above, we aim at estimating the maximal yearly change of interest rates of the 10 year US treasury bonds which is exceeded with probability $1/2000$, i.e., we want to estimate $x_{0.0005}$. Again we use the Hill estimator for the extreme value index applied to the observed yearly changes shifted by 0.05.

The left plot in Fig. 7.5 displays the variance estimates (7.6) with $j = 2$. (The estimates do not change much when one uses $\hat{\sigma}_{n,\tilde{p}}^2$ with $\tilde{p} = 1/500$ and $j = 3$.) As the variance estimates $\hat{\sigma}_n^2$ for the Hill estimator, see Fig. 7.3, the variance estimates $\hat{\sigma}_{n,p}^2$ are quite high if only few order statistics are used and they stabilize for k around 150, but here for $k = 160$ one obtains an estimated variance of 0.038, while for the Hill estimator one gets 0.027. This seems to contradict the asymptotic result, according to which these asymptotic variances are equal. However, a closer inspection of the proof of the asymptotic normality of the quantile estimator shows that the first term ignored in the asymptotic expansion is just of the order $1/\log(k/(np))$ smaller than the leading term. Thus for moderate sample sizes, the asymptotic result can be quite misleading.

The quantile estimates and the confidence intervals based on the variance estimates discussed above are shown in the right plot of Fig. 7.5 together with the confidence intervals when the asymptotic variance is estimated by $\hat{\gamma}_n^2$ as it is suggested by the asymptotic theory for iid data. (The estimates are corrected for the shift by 0.05 of the observed interest rate changes.) Here, for $k = 160$, the former interval $[0.044, 0.089]$ is more than 4 times longer than the latter interval $[0.060, 0.070]$! In such a situation, for two reasons it is questionable to use the standard asymptotic theory: firstly, one ignores the loss of accuracy due to the serial dependence, and secondly, the first order asymptotic approximation of the quantile estimator is rather crude for moderate sample sizes.

The point estimate for $x_{0.0005}$ is 0.065. Note that the confidence intervals are not symmetric about this point estimate, because we have used the normal approximation of $\log \hat{x}_{n,p}$ rather than that of $\hat{x}_{n,p}$ to construct the confidence intervals.

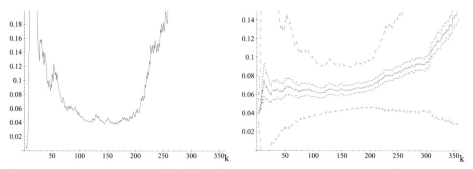

FIG. 7.5. Variance estimator (7.6) (left) and the estimated quantile $F^{-1}(1 - 1/2000)$ (solid) with confidence intervals (7.5) (dashed) and the confidence intervals derived from the asymptotic theory for iid data (dotted) (right).

Estimation in the Generalized Pareto Model

While in principle it is quite simple to calculate point estimates of extreme quantiles in the full GP model, the construction of reliable confidence intervals is a difficult task.

To estimate $x_p = F^{-1}(1 - p)$, one fits the GP model to the k largest order statistics, as it is described in Section 5.1. To this end, one may use the maximum likelihood estimator of the extreme value index and the scale parameter of the approximating GP model (provided $\gamma > -1/2$), or the moment estimators

$$\hat{\gamma}_n := l_{1,n} + 1 - \frac{1}{2}\left(1 - \frac{l_{1,n}^2}{l_{2,n}}\right)^{-1}, \quad \hat{s}_n := \frac{2l_{1,n}^3}{l_{2,n}}$$

with

$$l_{j,n} := \frac{1}{k-1}\sum_{i=1}^{k-1}\log^j \frac{X_{n-i+1:n}}{X_{n-k+1:n}}.$$

Then an estimator for x_p is given by

$$\tilde{x}_{n,p} := X_{n-k+1:n} + \hat{s}_n \frac{(np/k)^{-\hat{\gamma}_n} - 1}{\hat{\gamma}_n}. \tag{7.7}$$

To construct confidence intervals in an analogous way as in the restricted Pareto model, one needs suitable approximations to the distribution of these quantile estimators. Unfortunately, it turns out that the approximations have a quite different form for heavy tailed and light tailed distributions. If F is heavy tailed (i.e., $\gamma > 0$), then the extreme quantiles are unbounded and the relative estimation error $\hat{x}_{n,p}/x_p - 1$ (or $\log(\hat{x}_{n,p}/x_p)$) can be approximated by a normal random variable as in the restricted Pareto model. Since for light tailed distributions (i.e., $\gamma < 0$) both the estimate and the extreme quantile to be estimated are close to the right endpoint of the distribution, the relative estimation error is of little interest. In this case, a normal approximation can be established for the standardized absolute estimation error.

More precisely, assume that the extreme value index $\gamma < 0$ and the scale parameter of the GP approximation of the exceedances over $X_{n-k:n}$ are estimated by $\hat{\gamma}_n = \hat{\gamma}_n^{(k)}$ and $\hat{s}_n = \hat{s}_n^{(k)}$, respectively, using the maximum likelihood approach (if $\gamma > -1/2$) or the moment estimators mentioned above. Then

$$\frac{k_n^{1/2}}{\hat{s}_n}(\tilde{x}_{n,p} - x_p)$$

is approximately normally distributed with mean 0 and a variance σ^2 which again is determined by the serial dependence in a complex manner.[10] The limiting variance can be consistently estimated by

$$\hat{\sigma}_n^2 := \sum_{i=j}^{k} \left(\frac{\tilde{x}_{n,p}^{(i)} - \tilde{x}_{n,p}^{(k)}}{\hat{s}_n^{(i)}}\right)^2 \bigg/ \sum_{i=j}^{k} (i^{-1/2} - k^{-1/2})^2 \qquad (7.8)$$

with $\hat{s}_n^{(i)}$ denoting the estimator of the scale parameter of the GP approximation of the exceedances over $X_{n-i:n}$.

Note that the estimators (7.6) and (7.8) for the approximative variances of the quantile estimators look quite different in the cases $\gamma > 0$ and $\gamma < 0$, respectively, and likewise the resulting confidence intervals do. In principle, one could first test for the sign of the extreme value index, and then construct the confidence intervals accordingly, but apparently this procedure does not work well in practice if γ is close to 0. The problem of finding good estimators of the approximative variance is aggravated by the large variability of many estimators of the extreme value index based on a small number of order statistics (cf. part 1 of the case study discussed above). Thus, the construction of reliable confidence intervals remains an open problem if γ is small in absolute value.

7.4 A Time Series Approach

Often parametric dependence structures are used to model time series, because they facilitate a better interpretation of the dynamics of the time series than a non–parametric model. Moreover, usually they allow for more efficient estimators. The best known examples are ARMA(p, q) models, where it is assumed that $(X_t)_{1 \le t \le n}$ is a stationary time series satisfying

$$X_t - \sum_{i=1}^{p} \varphi_i X_{t-i} = Z_t + \sum_{j=1}^{q} \vartheta_j Z_{t-j}, \quad p+1 \le t \le n, \qquad (7.9)$$

---

[10]Drees, H. (2003). Extreme quantile stimation for dependent data with applications to finance. Bernoulli 9, p. 652.

with centered iid innovations Z_t and, in addition, unknown coefficients $\varphi_1, \ldots, \varphi_p$ and $\vartheta_1, \ldots, \vartheta_q$.

Suppose that the df F_Z of the innovations has balanced heavy tails, that is, $F_Z \in D(G_\gamma)$ for some $\gamma > 0$ and $(1 - F_Z(x))/F_Z(-x) \to p/(1-p)$ as $x \to \infty$ for some $p \in (0,1)$. Then the tail of the df of the observed time series is related to the tail of F_Z as follows:

$$1 - F_X(x) \sim c_{\varphi,\vartheta}(1 - F_Z(x)), \qquad x \to \infty;$$

see, e.g., Datta and McCormick (1998)[11]. Here $c_{\varphi,\vartheta}$ is a constant depending on $(\varphi_1, \ldots, \varphi_p, \vartheta_1, \ldots, \vartheta_q)$. In particular, $F_X \in D(G_\gamma)$ and

$$F_X^{-1}(1-p) \sim F_Z^{-1}\left(1 - \frac{p}{c_{\varphi,\vartheta}}\right), \qquad p \to 0. \tag{7.10}$$

Therefore, one may adopt two different approaches to the tail analysis of the time series:

- Estimate γ and $F_X^{-1}(1-p)$ directly from the observed time series as discussed in the Sections 7.2 and 7.3.

- First estimate $\varphi_1, \ldots, \varphi_p$ and $\vartheta_1, \ldots, \vartheta_q$ by some standard method for ARMA models, cf. [5].

  In the next step, calculate the residuals $\hat{Z}_t$ based on the fitted model; for example, in an AR(p) model (i.e., if (7.9) holds with $\vartheta_1 = \cdots = \vartheta_q = 0$) let

$$\hat{Z}_t := X_t - \sum_{i=1}^{p} \hat{\varphi}_i X_{t-i}, \quad p+1 \le t \le n.$$

  Finally, apply the classical theory for iid observations to the residuals to estimate γ and $F_Z^{-1}(1 - p/c_{\hat{\varphi},\hat{\vartheta}})$ (as an approximation to $F_X^{-1}(1-p)$).

For AR(p) time series, Resnick and Stărică[12] demonstrated that the Hill estimator performs better in the second approach. A similar result was recently proved by Ling and Peng (2004)[13]. To the best of our knowledge, analogous results about the corresponding quantile estimators are not known up to now. We expect that the model based tail analysis using the residuals also leads to more accurate quantile estimates, because the estimation error of the Hill estimator asymptotically determines the performance of the quantile estimators.

---

[11]Datta, S. and McCormick, W.P. (1998). Inference for the tail parameters of a linear process with heavy tail innovations. Ann. Inst. Statist. Math. 50, 337–359.

[12]Resnick, S.I. and Stărică, C. (1997). Asymptotic behavior of Hill's estimator for autoregressive data. Comm. Statist. Stochastic Models 13, 703–721.

[13]Ling, S. and Peng, L. (2004). Hill's estimator for the tail index of an ARMA model. J. Statist. Plann. Inference 123, 279–293.

However, the relationship (7.10) between the tails of F_X and F_Z heavily relies on the model assumptions. Hence, even a moderate deviation from the assumed model can lead to gross estimation errors. This drawback of the indirect model based approach is aggravated by the fact that, due to the lack of data, model deviations in the tails are particularly difficult to detect statistically.

Quantile Estimation in AR(1) Models: a Simulation Study

Consider the AR(1) model

$$X_t = \varphi X_{t-1} + Z_t, \quad 2 \leq t \leq n = 2000, \tag{7.11}$$

with $1 - F_Z(x) = F_Z(-x) = 0.5(1+x)^{-1/\gamma}$, $x \geq 0$, $\varphi = 0.8$, and $\gamma = 1/2$, i.e., the positive and the negative part of the innovations have a Pareto distribution with location parameter -1. Thus $p = 1/2$ and in this case $c_{\varphi, \vartheta} = 1/(2(1 - |\varphi_1|))$.

We aim at estimating the quantile $F_X^{-1}(1 - p)$ with $p = 0.001$, which usually lies near the boundary of the range of observations. To calculate the true quantile approximately, we simulate a time series of length $410\,000$ and determine the 400th largest of the last $400\,000$ observations (i.e., the first $10\,000$ simulated values are used as a burn-in period to approach the stationary distribution). We do this $m = 10\,000$ times, and finally approximate the unknown true quantile by the average of the m simulated order statistics, which yields $F_X^{-1}(1 - p) = 37.015$ (with estimated standard deviation 0.02).

To estimate the quantile using the model based approach, we first estimate φ by the sample autocorrelation at lag 1, which is the Yule-Walker-type estimator for AR(1) time series, cf. Section 6.2:

$$\hat{\varphi} := \frac{\sum_{t=2}^n X_{t-1} X_t}{\sum_{t=1}^n X_t^2}. \tag{7.12}$$

Note that $\hat{\varphi}$ is a sensible estimator for φ even if the variance of X_t is infinite and hence the autocorrelation does not exist[14].

Then we calculate the residuals

$$\hat{Z}_t := X_t - \hat{\varphi} X_{t-1}, \quad 2 \leq t \leq n. \tag{7.13}$$

Finally, estimate $F_X^{-1}(1 - p)$ by

$$\hat{x}_{n,p}^{(2)} := \hat{Z}_{n-k:n-1} \left(\frac{(n-1)2(1 - |\hat{\varphi}|)}{k} \right)^{-\tilde{\gamma}_{n-1}}$$

with

$$\tilde{\gamma}_{n-1} := \frac{1}{k-1} \sum_{i=1}^{k-1} \log \frac{\hat{Z}_{n-i:n-1}}{\hat{Z}_{n-k:n-1}}$$

---

[14]Davis, R.A. and Resnick, S.I. (1986). Limit theory for the sample covariance and correlation functions of moving averages. Ann. Statist. 14, 533–558.

denoting the Hill estimator based on the k largest order statistics of these residuals. In addition, we estimate $F_X^{-1}(1-p)$ directly using $\hat{x}_{n,p}$ defined in (7.4).

The empirical root mean squared error (rmse) of these estimators (obtained in $10\,000$ simulations) are displayed in Fig. 7.6 as a function of k. Since these estimates of the true rmse are somewhat unstable (i.e. they change from simulation to simulation), we also show the corresponding simulated L_1–errors $E|\hat{x}_{n,p}^{(2)} - F_X^{-1}(1-p)|$ and $E|\hat{x}_{n,p} - F_X^{-1}(1-p)|$, which are more reliable approximations to the true errors.

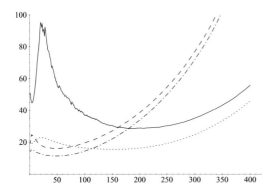

FIG. 7.6. Empirical root mean squared error of $\hat{x}_{n,p}$ (solid line) and $\hat{x}_{n,p}^{(2)}$ (dashed line) and the corresponding empirical L_1–errors (dotted line, resp. dash-dotted line) in the AR(1) model (7.11).

As expected, the estimator $\hat{x}_{n,p}^{(2)}$ based on the residuals is more accurate if k is chosen appropriately. Both the rmse and the L_1–error of the model based estimator $\hat{x}_{n,p}^{(2)}$ is minimal for $k = 50$ with minimal values 15.8 and 11.2, respectively. In contrast, the optimal k is much greater for the direct estimator $\hat{x}_{n,p}$ and the errors are about 80% and 40%, respectively, larger. (The minimal rmse 28.3 is attained for $k = 184$ and the minimal L_1–error 15.3 for $k = 159$.)

The situation changes dramatically if we perturb the linear AR(1) model by a logarithmic term:

$$X_t = \varphi X_{t-1} + \delta \cdot \operatorname{sgn}(X_{t-1}) \log\big(\max(|X_{t-1}|, 1)\big) + Z_t \qquad (7.14)$$

with F_Z as above and $\delta = 0.6$. Then the true quantile $F_X^{-1}(1-p)$ to be estimated is approximately equal to 41.37 (with estimated standard deviation 0.02).

Note that the logarithmic perturbation of the linear dependence structure, that affects only the observations greater than 1 in absolute value, is difficult to distinguish from an increase of the autoregressive parameter φ. Indeed, in the average the estimate $\hat{\varphi}$ defined in (7.12) equals 0.92 in the perturbed model, and the average relative difference between the perturbed autoregressive term $\varphi X_{t-1} + \delta \cdot \operatorname{sgn}(X_{t-1}) \log\big(\max(|X_{t-1}|, 1)\big)$ and the fitted AR(1) term $\hat{\varphi} X_{t-1}$ is just 5.2%. Therefore, in the scatterplot of X_t versus X_{t-1} displayed in Figure 7.7 for one simulated sample of size $n = 2\,000$ the deviation from the linear dependence can hardly be seen. Moreover, the model deviation is also difficult to detect by statistical tests. For example, the turning point test and the difference–sign test, see [5], p. 37f., to the nominal size 5% reject the null hypothesis that the residuals $\hat{Z}_t$ defined in (7.13) are iid with probability less than 6%. (Note that the more popular tests for the linear AR(1) dependence structure which are based on the sample autocorrelation function are not applicable in the present setting because the variance may be infinite.)

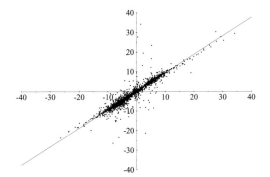

FIG. 7.7. Scatterplot of X_t versus X_{t-1} and line through origin with slope $\hat{\varphi}$.

Fig. 7.8 shows the rmse and L_1–errors of the direct quantile estimator $\hat{x}_{n,p}$ and the estimator $\hat{x}_{n,p}^{(2)}$ which is based on the (wrong) linear AR(1) model (7.11). Now the model based estimator has minimal rmse 31.7 and minimal L_1-error 22.8 (both attained at $k = 2$) while the simulated errors for the direct estimator equal 14.1 (for $k = 282$) and 10.0 (for $k = 305$), respectively. Thus, for the perturbed time series, the estimation error of the model based estimator is more than double as big as the error of the direct quantile estimator, which does not rely on a specific time series model. Another disadvantage of the model based estimator is the sensitivity of its performance to the choice of k. The smallest but one L_1–error is about 30% larger than the minimal value; it is attained for $k = 18$. Moreover, for $k > 50$ the estimation error increases rapidly.

Fig. 7.9 displays the estimated density of the distribution of the quantile estimators in the AR(1) model (7.11) (left plot) and the perturbed model (7.14) (right plot). Here we use kernel estimates based on the Epanechnikov kernel with bandwidth 2 applied to the simulated quantile estimates with minimal L_1–error. In the AR(1) model, the mode of the distribution of the model based estimator $\hat{x}_{n,p}^{(2)}$ is very close to the true value, which is indicated by the vertical dotted line. The distribution of $\hat{x}_{n,p}$ has a lower mode and is also a bit more spread out. In contrast, in the perturbed model (7.14) the mode of the density of $\hat{x}_{n,p}^{(2)}$ is much too low and the distribution is much more spread out than the distribution of $\hat{x}_{n,p}$. In addition, the estimated density of the model based estimator is also shown for $k = 18$ (i.e. the value leading to the smallest but one L_1–error). Here the mode is approximately equal to the true value, but the distribution is even more spread

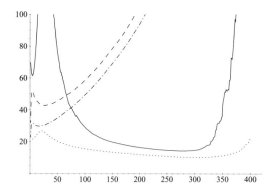

FIG. 7.8. Empirical root mean squared error of $\hat{x}_{n,p}$ (solid line) and $\hat{x}_{n,p}^{(2)}$ (dashed line) and the corresponding empirical L_1–errors (dotted line, resp. dash-dotted line) in the perturbed AR(1) model (7.14).

out.

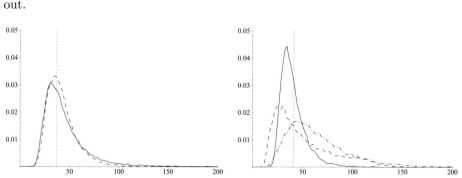

FIG. 7.9. Kernel density estimates for $\hat{x}_{n,p}$ (solid line) and $\hat{x}_{n,p}^{(2)}$ (dashed line) with minimal the L_1–error in the AR(1) model (7.11) (left plot) and the perturbed model (7.14) (right plot), where in addition, the estimated density is shown for $\hat{x}_{n,p}^{(2)}$ with the smallest but one L_1–error (dash-dotted line); the true values are indicated by the vertical lines.

From Fig. 7.9 it is also obvious that the distribution of the quantile estimator $\hat{x}_{n,p}$ is strongly skewed to the right. Hence it cannot be well fitted by a normal distribution. Fig. 7.10 demonstrates that the normal fit to the distribution of $\log \hat{x}_{n,p}$ is much better. Therefore, as explained above, one should use the normal approximation to $\log \hat{x}_{n,p}$ rather than a normal approximation to $\hat{x}_{n,p}$ to construct confidence intervals.

Non–linear Time Series

Non–linear time series models are particularly popular in financial applications. For example, various types of GARCH time series have been proposed to model returns of risky assets, cf. Section 16.7. Here we consider GARCH(1,1) time series, i.e., stationary solutions $\{R_t\}$, with t ranging over all natural numbers, of the

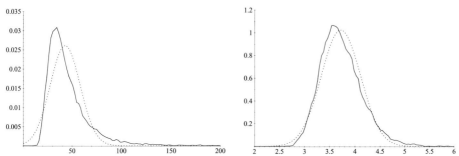

FIG. 7.10. Kernel density estimates for $\hat{x}_{n,p}$ (solid line, left plot) and $\log \hat{x}_{n,p}$ (solid line, right plot) with minimal the L_1–error in the AR(1) model (7.11) together with the densities of the normal distribution fitted by minimizing the Hellinger distance (cf. Section 3.1) (dotted lines).

stochastic recurrence equations

$$R_t = \tilde{\sigma}_t \varepsilon_t,$$
$$\tilde{\sigma}_t^2 = \alpha_0 + \alpha_1 R_{t-1}^2 + \beta_1 \tilde{\sigma}_{t-1}^2,$$

with $\alpha_0 > 0, \alpha_1, \beta_1 \geq 0$ and iid innovations ε_t with mean 0 and variance 1. Under mild conditions on the distribution of ε_t, the returns R_t are heavy–tailed with extreme value index $\gamma > 0$ determined by the equation

$$E(\alpha_1 \varepsilon_1^2 + \beta_1)^{1/(2\gamma)} = 1,$$

cf. page 398.

   If one fixes the distribution of innovations, e.g., one assumes standard normal innovations, then one might estimate the extreme value index from this relationship after replacing the unknown parameters by suitable estimates, e.g., quasi maximum likelihood estimates, cf. Section 16.7. If one does not fix the distribution of the innovations, then first one must fit some model to the residuals, in order to estimate γ this way. In any case, unlike for linear time series models, here the extreme value index depends on the whole distribution of the innovations. Thus any misspecification of this distribution may lead to a serious estimation error for γ. In an empirical study[15], Stărică and Pictet compared the estimates of the extreme value index obtained from fitting a GARCH(1,1) model with normal innovations to the returns of several exchange rates aggregated over different time intervals with the Hill estimators applied directly to these data sets. It turned out that for almost all levels of aggregation the approach based on the GARCH model apparently underestimates the tail thickness drastically, an effect which is most

[15]Stărică, C. and Pictet, O. (1997). The tales the tails of GARCH processes tell. Preprint, Chalmers University Gothenburg.

likely caused by a misspecification of the time series model (or by the choice of the estimator).

However, even if the assumed GARCH model is correct, this does not help much when it comes to extreme quantile estimation. While it is known that, for symmetric innovations,

$$P\{R_t > x\} \sim cx^{-1/\gamma} \quad \text{as} \quad x \to \infty,$$

no analytic expression in terms of α_0, α_1 and β_1 and the distribution of ε_t is known for the constant $c > 0$. Hence, in the present setting, we cannot copy the model based approach for the estimation of extreme quantiles discussed above for ARMA models. Of course, in principle, one can first fit the GARCH model and then obtain an approximation to the extreme quantile by simulation, but the simulation results obtained by Mikosch[16] indicate that it can be quite computer intensive to obtain accurate approximations of extreme quantiles this way even if the model would be exactly known.

In view of these drawbacks of the time series approach, the general method discussed in the Sections 7.2 and 7.3 seems the only feasible way to obtain reliable estimates of extreme quantiles. In fact, this conclusion holds true for most non–linear time series models, for which typically no simple asymptotic relationship between the tails of the innovations and the tails of the time series is available.

---

[16]Mikosch, T. (2003). Modeling dependence and tails of financial time series. In: Extreme Value Theory and Applications, Finkenstadt, B. and Rootzén, H. (eds.), Chapman and Hall, 185–286.

Chapter 8

Conditional Extremal Analysis

In this chapter we provide an overview of the conditioning concept including some theoretical applications. After some heuristic arguments and technical preparations in Section 8.1. we study conditional extremes in a nonparametric and, respectively, in a parametric framework, see Sections 8.2 and 8.3.

The Bayesian view towards the Bayesian estimation principle is outlined in Section 8.4. Especially, we give the standard Bayesian interpretation of the posterior density as a conditional density.

8.1 Interpretations and Technical Preparations

This section lays out heuristic arguments about the conditioning concept and collects some relevant auxiliary technical results. We make an attempt to clarify the relationships between joint distributions, conditional distributions and mixtures (shortly address as "roundabout"). In this section we also study the linear regression problem for a random design as a first application.

Conditional Distribution Functions

In Section 1.1 we introduced elementary conditional dfs, namely exceedance dfs $P(Y \leq y | Y > u)$ and excess dfs $P(Y - u \leq y | Y > u)$. In addition, the mean excess function (in other words, the mean residual life function) was determined by the means of excess dfs, cf. (2.20).

In the sequel, we consider conditional probabilities of a random variable Y conditioned on the event that $X = x$. Generally, the conditional df $F(y|x)$ of Y given $X = x$ is defined by an integral equation, see (8.3). In the special case, where

$P\{X = x\} > 0$, one may compute the conditional df by

$$F(y|x) = P(Y \leq y|X = x) = \frac{P\{X = x, Y \leq y\}}{P\{X = x\}}. \tag{8.1}$$

The joint random experiment with outcome (x, y), described by the random vector (X, Y), can be regarded as a two–step experiment:

Step 1 (initial experiment): observe x as an outcome of the random experiment governed by X.

Step 2 (conditional experiment): generate y under the conditional df $F(\cdot|x)$ given x.

Keep in mind that y can be interpreted as the second outcome in the joint experiment as well as the outcome of the conditional experiment.

If X has a discrete distribution, then the joint df $F(x, y)$ of X and Y can be regained from the distribution of X and the conditional df by

$$F(x, y) = \sum_{z \leq x} F(y|z)P\{X = z\}. \tag{8.2}$$

If X does not have a discrete distribution, then the conditional df $F(\cdot|x)$ is a solution of the integral equation

$$F(x, y) = \int_{-\infty}^{x} F(y|z)\, dF_X(z), \tag{8.3}$$

where F_X, usually written $F(x)$, denotes the df of X. The preceding interpretations just remain the same. Notice that (8.2) can be written in the form of (8.3).

It is apparent that the df of Y is given by

$$F(y) = \int F(y|z)\, dF_X(z). \tag{8.4}$$

In addition, a density $f(\cdot|x)$ of $F(\cdot|x)$ is referred to as conditional density of Y given $X = x$. Usually, we write $f(y|x)$ in place of $f(\cdot|x)$, etc.

Conditional Distributions

The conditional distribution pertaining to the conditional df $F(y|x)$ is denoted by

$$P(Y \in \cdot|X = x). \tag{8.5}$$

Apparently, $P(Y \leq y|X = x) = F(y|x)$.

Generally, a conditional distribution $P(Y \in \cdot|X = x)$ has the property

$$P\{X \in B, Y \in C\} = \int_B P(Y \in C|X = x)\, d\mathcal{L}(X)(x), \tag{8.6}$$

and, hence, also

$$P\{Y \in C\} = \int P(Y \in C | X = x) \, d\mathcal{L}(X)(x) \tag{8.7}$$

which reduces to (8.3) and (8.4) in the special case of dfs.

Basic Technical Tools for Conditional Distributions

In the following lines we formulate some concepts and results for conditional dfs and related conditional densities The experienced reader may elaborate these ideas in greater generality, namely, in terms of conditional distributions and conditional densities with respect to some dominating measure.

- (Computing the conditional density.) The conditional density of Y given $X = x$ can be written as

$$f(y | x) = \frac{f(x, y)}{f(x)} I(f(x) > 0). \tag{8.8}$$

  Therefore, the conditional density can be deduced from the joint density $f(x, y)$ because

$$f(x) = \int f(x, y) \, dy \qquad \text{and} \qquad f(y) = \int f(x, y) \, dx. \tag{8.9}$$

- (Computing the joint density.) If $f(x)$ is a density of $F(x)$, then

$$f(x, y) = f(y | x) f(x) \tag{8.10}$$

  is a density of $F(x, y)$; by induction one also gets the representation

$$f(x_1, \ldots, x_n) = f(x_1) \prod_{i=2}^{n} f(x_i | x_{i-1}, \ldots, x_1), \tag{8.11}$$

  where $f(x_1, \ldots, x_n)$ is the density of the df $F(x_1, \ldots, x_n)$. This indicates, extending thereby Step 2 above, that the data may be sequentially generated by conditional dfs.

- (Inducing conditional distributions.) For measurable maps g,

$$P(Y \in g^{-1}(B) | X = x) = P(g(Y) \in B | X = x). \tag{8.12}$$

- (Conditioning in the case of independent random variables.) If X and Y are independent, one can prove in a direct manner that the conditional df of $g(X, Y)$ given $X = x$ is the df of $g(x, Y)$. Thus we have

$$P(g(X, Y) \le z | X = x) = P\{g(x, Y) \le z\}. \tag{8.13}$$

  As a special case one gets that the conditional df of Y is the unconditional df, if X and Y are independent.

- (Conditioning in the case of random vectors.) Assume that (X_i, Y_i) are iid random vectors und put $\boldsymbol{X} = (X_1, \ldots, X_n)$ and $\boldsymbol{Y} = (Y_1, \ldots, Y_n)$. Then the conditional df of $\boldsymbol{Y}$ given $\boldsymbol{X} = \boldsymbol{x}$ is

$$P(\boldsymbol{Y} \le \boldsymbol{y} | \boldsymbol{X} = \boldsymbol{x}) = \prod_{i=1}^{n} P(Y_i \le y_i | X_i = x_i). \qquad (8.14)$$

To prove this result, verify that the right–hand side satisfies the required property of the conditional distribution[1].

The Concept of Conditional Independence

The random variables $Y_1, \ldots, Y_n$ are called conditionally independent given $\boldsymbol{X}$, if

$$P(\boldsymbol{Y} \le \boldsymbol{y} | \boldsymbol{X} = \boldsymbol{x}) = \prod_{i=1}^{n} P(Y_i \le y_i | \boldsymbol{X} = \boldsymbol{x}). \qquad (8.15)$$

Assume that the random vectors $(X_1, Y_1), \ldots, (X_n, Y_n)$ are iid. According to (8.14),

$$P(\boldsymbol{Y} \le \boldsymbol{y} | \boldsymbol{X} = \boldsymbol{x}) = \prod_{i=1}^{n} P(Y_i \le y_i | X_i = x_i) \qquad (8.16)$$

which implies that

$$P(Y_i \le y_i | \boldsymbol{X} = \boldsymbol{x}) = P(Y_i \le y_i | X_i = x_i). \qquad (8.17)$$

Therefore, the Y_i are conditionally independent with conditional distributions

$$P(Y_i \le y_i | \boldsymbol{X} = \boldsymbol{x}) = P(Y_i \le y_i | X_i = x_i).$$

Conditional Expectation, Conditional Variance, Conditional q–Quantile

Let Y be a real–valued random variable with $E|Y| < \infty$ and X any further random variable. Denote again by $F(y|x)$ the conditional df of Y given $X = x$. We recall some well–known functional parameters of the conditional df.

- The mean

$$E(Y|X = x) = \int y \, dF(\cdot|x)(y) \qquad (8.18)$$

of the conditional df is called the conditional expectation of Y given $X = x$. We shortly write $E(Y|x)$ in place of $E(Y|X = x)$ when no confusion can arise.

Notice that the conditional expectation $E(Y|x)$ is the mean within the conditional experiment described by Step 2 on page 228.

---

[1]Presently, we do not know an appropriate reference to the stochastical literature.

- We mention another conditional functional parameter, namely the conditional variance $V(Y|x)$ of Y given $X = x$ which is the variance of the conditional df. Thus, $V(Y|x)$ is the variance in the conditional experiment. We have

$$V(Y|x) := \int (y - E(Y|x))^2 \, dF(\cdot|x)(y). \qquad (8.19)$$

The conditional variance can be represented by means of conditional expectations. With the help of (8.12) one gets

$$V(Y|x) = E(Y^2|x) - E(Y|x)^2. \qquad (8.20)$$

- Later on we will also be interested in the conditional q–quantile

$$q(Y|x) := F(\cdot|x)^{-1}(q) \qquad (8.21)$$

of Y given $X = x$ as the q–quantile of the conditional df $F(y|x)$.

In addition, $E(Y|X) := g(X)$ is the conditional expectation of Y given X, where $g(x) = E(Y|x)$. Notice that $E(Y|X)$ is again a random variable. Likewise, one may speak of the conditional variance $V(Y|X)$ and conditional q–quantile $q(Y|X)$ of Y given X.

The Linear Regression Function for a Random Design

We characterize the linear regression model for a fixed design as the conditional model in the corresponding random design problem. It is pointed out that the linear regression function is a conditional expectation.

Later on, corresponding questions will be studied in the Sections 8.3, 9.5, 16.8 and Chapter 15, where we specify parametric statistical models in the conditional setup.

First we recall some basic facts from Section 2.5 about linear regression for a fixed design. Assume that the random variables Y_t have the representation

$$Y_t = \alpha + \beta x_t + \varepsilon_t, \qquad t = 1, \ldots, n, \qquad (8.22)$$

with the residuals ε_i being independent random variables with expectations $E\varepsilon_t = 0$, and the x_t are predetermined values. Recall that $g(x) = \alpha + \beta x$ is the linear regression function with intercept α and slope β.

This may be equivalently formulated in the following manner: let the Y_t be independent random variables with expectations

$$EY_t = \alpha + \beta x_t, \qquad t = 1, \ldots, n. \qquad (8.23)$$

In the subsequent example, the x_t are also random.

EXAMPLE 8.1.1. (Exercises and Tests.) The following illustration is based on the percentages x_t of solved exercises in a tutorial and the scores y_t in a written test (data set

TABLE 8.1. percentages x_t of solved exercises and scores y_t in test

x_t	41.1	42	24.5	50	56.1	50.7	56.1	82.5	90.8	...
y_t	0	8.5	6	3	9	5	15	11.5	17	...

exam96.dat). The maximum attainable score is 20. Some of the data are displayed in Table 8.1.

In Fig. 8.1 we plot the scores y_t against the percentages x_t and include the estimated least squares line.

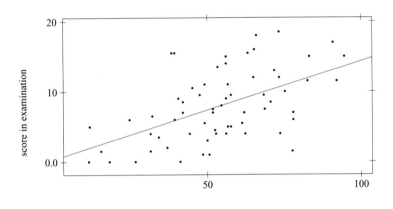

FIG. 8.1. Plotting scores in examination against percentages of solved exercises and fitted least squares line.

For the (academic) teacher it is very satisfactory that solving exercises has a positive effect for students on passing the test.

Subsequently, the x_t are regarded as outcomes of random variables X_t. This leads to the random design problem.

We assume that $(X_1, Y_1), \ldots, (X_n, Y_n)$ are iid copies of a random vector (X, Y) which has the representation

$$Y = \alpha + \beta X + \varepsilon, \tag{8.24}$$

where X, ε are independent and $E\varepsilon = 0$. Because X and ε are independent, we get from (8.13) that

$$P(Y \in \cdot \,|X = x) = \mathcal{L}(\alpha + \beta x + \varepsilon). \tag{8.25}$$

Thus, the Y_t conditioned on $X_t = x_t$ are distributed as in the fixed design problem.

For the conditional expectation we have

$$E(Y|x) = \alpha + \beta x \qquad (8.26)$$

which is again called linear regression function.

Put again $\boldsymbol{X} = (X_1, \ldots, X_n)$ and $\boldsymbol{Y} = (Y_1, \ldots, Y_n)$. From (8.16) and (8.26) we know that the Y_t, conditioned on $\boldsymbol{X} = \boldsymbol{x}$, are conditionally independent with conditional expectations

$$E(Y_t|x_t) = E(Y_t|X_t = x_t) = \alpha + \beta x_t. \qquad (8.27)$$

These properties correspond to those in the fixed design problem. This is the reason why results for the fixed design regression carry over to the random design regression.

Estimation and prediction in the linear regression model for a random design will be studied at the end of this section.

The Roundabout: Joint, Conditional, and Mixture Distributions

We want to explain the relationships between joint, conditional and mixture distributions in a "roundabout":

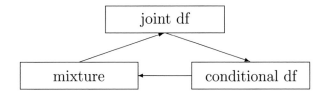

We may

(i) start with a joint df $F(x, y)$ of random variables X and Y;

(ii) decompose the joint df in the marginal df $F(x)$ of X and the conditional df $F(y|x)$ of Y given $X = x$, see (8.3);

(iii) regard the conditional df and, respectively, the marginal df of X as parameterized dfs $F(y|\vartheta)$ and mixing df $F(\vartheta)$, replacing the x by ϑ, and, finally, return to (i).

We provide some details concerning the last step: let $F(\cdot|\vartheta) = F_\vartheta$ be a family of dfs parameterized by ϑ, and let F be a mixing df. Then, the mixture of the F_ϑ with respect to F is

$$G(y) = \int F(y|\vartheta)\, dF(\vartheta) = \int F(y|\vartheta) f(\vartheta)\, d\vartheta, \qquad (8.28)$$

where the latter equality holds for a density f of F, see (4.8). If the integration in (8.28) is carried out from $-\infty$ to x, then one obtains a bivariate df $F(\vartheta, y)$ as in (8.3). Therefore, one may introduce random variables X and Y with common df $F(\vartheta, y)$. Replacing ϑ by x we are in the initial situation (i) with the mixing df as df of X and the mixture df as df of Y.

These arguments may be helpful to understand, e.g., the Bayesian view towards Bayesian statistics, see page 243.

Conditional Distributions of Exceedances and Order Statistics

Further insight in the statistical modeling of exceedances over a threshold u and upper order statistics may be gained by conditioning on the random number K of exceedances and, respectively, on the $(k+1)$st upper order statistic $X_{n-k:n}$:

- the random exceedances over u conditioned on $K = k$ are distributed as k iid random variables with common df $F^{[u]}$ (see [43] and [16]),

- the upper order statistics $X_{n-k+1:n} \leq \cdots \leq X_{n:n}$ conditioned on $X_{n-k:n} = u$ are distributed as the order statistics $Y_{1:k} \leq \cdots \leq Y_{k:k}$ of k iid random variables Y_i with common df $F^{[u]}$ (see [42]).

Consequently, the statistical analysis of exceedances and upper order statistics can be reduced to the standard model of iid random variables.

Analyzing Dependent Data, the Rosenblatt Transformation

All the exploratory tools, like QQ–plots or sample excess functions, concern iid data. In the case of independent, not necessarily identically distributed (innid) data one my use the probability transformation to produce iid, $(0,1)$–uniform data.

In the case of dependent random variables one has to apply the Rosenblatt[2] transformation to achieve such a result.

Let $(X_1, \ldots, X_d)$ be a random vector. Let

$$F_i\left(\cdot \mid x_{i-1}, \ldots, x_1\right)$$

be the conditional df of X_i given $X_{i-1} = x_{i-1}, \ldots, X_1 = x_1$. Assume that

$$P\{\omega \in \Omega : F_i\left(\cdot \mid X_{i-1}(\omega), \ldots, X_1(\omega)\right) \text{ continuous}, \ i = 1, \ldots, d\} = 1.$$

Then,

$$F_1(X_1), F_2(X_2 \mid X_1), \ldots, F_d(X_d \mid X_{d-1}, \ldots, X_1) \tag{8.29}$$

are iid, $(0,1)$–uniformly distributed random variables.

---

[2]Rosenblatt, M. (1952). Remarks on a multivariate transformation. Ann. Math. Statist. 23, 470–472.

The following short proof of the Rosenblatt transformation (8.29) in its general form is due to Edgar Kaufmann (personal communication), and it is worthwhile to indicate the basic idea: the joint integral is decomposed to an iterated integral (by a Fubini theorem) and the probability transformation is applied in the inner integral. First we mention these auxiliary results.

(a) Probability Transformation: according to the probability transformation, see page 38, we have

$$\int I\big(F(x) \in B)\big)\, dF(x) = P\{U \in B\}$$

for continuous dfs F where U is $(0,1)$–uniformly distributed

(b) Fubini Theorem[3]: for random variables X, Y and non–negative, measurable functions f,

$$\int f(x,y)\, d\mathcal{L}(X,Y)(x,y) = \int \left(\int f(x,y)\, P(Y \in dy | X = x) \right) d\mathcal{L}(X)(x)$$

which is the Fubini theorem for conditional distributions. If X and Y are independent, then the conditional distribution is the distribution of Y and one gets the usual Fubini theorem.

To prove the Rosenblatt transformation (8.29) it suffices to verify that $\mathbf{X} = (X_{d-1}, \ldots, X_1)$ and $F_d(X_d|\mathbf{X})$ are independent and $F_d(X_d|\mathbf{X})$ is $(0,1)$–uniformly distributed. For measurable sets A and B one gets

$$P\{\mathbf{X} \in A,\, F_d(X_d|\mathbf{X}) \in B\}$$
$$= \int I\big(\mathbf{x} \in A\big) I\big(F_d(x|\mathbf{x}) \in B\big)\, d\mathcal{L}(\mathbf{X}, X_d)(\mathbf{x}, x)$$
$$\overset{(b)}{=} \int I\big(\mathbf{x} \in A\big) \left(\int I\big(F_d(x|\mathbf{x}) \in B\big) F(dx|\mathbf{x}) \right) d\mathcal{L}(\mathbf{X})(\mathbf{x})$$
$$\overset{(a)}{=} P\{\mathbf{X} \in A\} \mathbf{P}\{U \in B\}$$

where U is $(0,1)$-uniformly distributed. The proof is concluded.

Predicting a Random Variable

In the following lines, we recall some basic facts about the concept of predicting (forecasting) a random variable. The aim is the prediction of a future value y of a random experiment based on an observable value $g(x)$, where x is a realization of another experiment, and the function g describes a fixed rule according to which the prediction is carried out.

---

[3]E.g., Gänssler, P. and Stute, W. (1977). Wahrscheinlichkeitstheorie. Springer, Berlin, Satz 1.8.13.

Clearly, one should be interested in the performance of the predictor. For that purpose, let Y and X be random variables which represent the two experiments with outcomes y and x. We predict Y by means of $g(X)$. It is desirable that

- the predictor $g(X)$ is unbiased; that is, $E(Y - g(X)) = 0$,

- the remainder term in the prediction is small, e.g., measured by the mean squared loss $E\big((Y - g(X))^2\big)$.

It is well known that the conditional expectation $E(Y|X)$ is the best predictor of Y based on X. Recall from page 230 that $E(Y|X) = g(X)$, where $g(x) = E(Y|x)$.

If g depends on an unknown parameter—this will usually be the case—one has to replace these parameters by estimators in the prediction procedure. We provide an example in conjunction with the linear regression model.

Estimation and Prediction in the Linear Regression Model for a Random Design

First we study once again the linear regression problem as specified in (8.23). Based on $\boldsymbol{x} = (x_1, \ldots, x_n)$ and $\boldsymbol{Y} = (Y_1, \ldots, Y_n)$ one gets the following least squares estimators

$$\hat{\beta}_{\boldsymbol{x},\boldsymbol{Y}} \;=\; \frac{\sum_{t=1}^{n}(Y_t - \bar{Y})(x_t - \bar{x})}{\sum_{i=1}^{n}(x_t - \bar{x})^2},$$

$$\hat{\alpha}_{\boldsymbol{x},\boldsymbol{Y}} \;=\; \bar{Y} - \hat{\beta}_{\boldsymbol{x},\boldsymbol{Y}}\bar{x}$$

of the intercept α and the slope β. Plugging in these estimators into the regression function, one also gets

$$\hat{g}_{\boldsymbol{x},\boldsymbol{Y}}(x) = \hat{\alpha}_{\boldsymbol{x},\boldsymbol{Y}} + \hat{\beta}_{\boldsymbol{x},\boldsymbol{Y}}x$$

as an estimator of the linear regression function $g(x) = \alpha + \beta x$.

In the random design problem, see (8.27), estimate the intercept α and the slope β by the unbiased least squares estimators

$$\hat{\beta}_{\boldsymbol{X},\boldsymbol{Y}} \;=\; \frac{\sum_{i=1}^{n}(Y_t - \bar{Y})(X_t - \bar{X})}{\sum_{i=1}^{n}(X_t - \bar{X})^2},$$

$$\hat{\alpha}_{\boldsymbol{X},\boldsymbol{Y}} \;=\; \bar{Y} - \hat{\beta}_{\boldsymbol{X},\boldsymbol{Y}}\bar{X} \tag{8.30}$$

to get the unbiased estimator

$$\hat{g}_{\boldsymbol{X},\boldsymbol{Y}}(x) = \hat{\alpha}_{\boldsymbol{X},\boldsymbol{Y}} + \hat{\beta}_{\boldsymbol{X},\boldsymbol{Y}}x \tag{8.31}$$

of the linear regression function $g(x) = \alpha + \beta x = E(Y|x)$.

Based on x we want to predict the future value y. As mentioned before the conditional expectation

$$E(Y|X) = \alpha + \beta X$$

provides the optimal predictor of Y, yet the parameters α and β are unknown. Plugging in the estimators one gets by means of $\hat{g}_{x,y}(x)$ in (8.31) a reasonable prediction of y based on the observed x, $\boldsymbol{x}$ and $\boldsymbol{y}$.

EXAMPLE 8.1.2. (Continuation of Example 8.1.1.) We want predict the future score y in the test based on the known percentage x of solved exercises in the tutorial. As a prediction use the pertaining value of the estimated regression line in Fig. 8.1.

Later on we will also write $E_{\alpha,\beta}(Y|x)$ and $E_{\alpha,\beta}(Y|X)$ to indicate the dependence of the conditional expectation on the parameters α and β.

The Notion of a Predictive Distribution

Occasionally, a conditional distribution is called predictive distribution, see, e.g., the book by Fan and Yao [18], page 454. One could extend this notion and generally address a random distribution as a predictive distribution. We shall use the term "predictive distribution" in a narrower sense.

The conditional distribution usually includes unknown parameters, which have to be estimated, so that one can deduce a prediction of a random functional parameter such as the conditional expectation. We shall refer to this random distribution with the unknown parameters replaced by estimates—the "estimated random distribution"—as predictive distribution.

The posterior distribution in the Bayesian framework, cf. Section 8.4, may also be addressed as predictive distribution, if the prior distribution does not depend on superparameters. A predictive distribution may also depend on the model choice or certain approximations, cf. Section 16.8.

EXAMPLE 8.1.3. (Continuation of Example 8.1.) In the linear regression set–up, where $Y = \alpha + \beta X + \varepsilon$ with X, ε being independent, we get as predictive distribution in the narrow sense (unknown parameters are replaced by the least squares estimators) by distributions of the form

$$\mathcal{L}(\hat{\alpha}_{\boldsymbol{x},\boldsymbol{y}} + \hat{\beta}_{\boldsymbol{x},\boldsymbol{y}}x + \varepsilon).$$

By the way, this predictive distribution is the conditional distribution of

$$\hat{\alpha}_{\boldsymbol{X},\boldsymbol{Y}} + \hat{\beta}_{\boldsymbol{X},\boldsymbol{Y}}X + \varepsilon \quad \text{given} \quad X = x, \boldsymbol{X} = \boldsymbol{x}, \boldsymbol{Y} = \boldsymbol{y},$$

if the random vector $(X, \boldsymbol{X}, \boldsymbol{Y})$ and the innovation ε are independent.

Notice that $(X, \boldsymbol{X}, \boldsymbol{Y})$ represents the past which can be observed by the statistician or hydrologist, etc. Therefore, the predictive distribution is observable, if, ε is specified, e.g., as a standard normal random variable.

On page 236 the performance of a predictor $g(X)$ of Y was measured by the mean squared loss. A related measurement is possible within the framework of predictive distributions. The role of Y and $g(X)$ is played by a random distribution $K(\cdot|X)$ and a predictive distribution $K^*(\cdot|X,Z)$, where Z represents the additional available information. Now, the performance is measured by the expected loss

$$E\big(d\big(K(\cdot|X), K^*(\cdot|X,Z)\big)\big),$$

where d is an appropriate distance.

Densities of predictive distributions are called predictive densities. Diebold et al.[4] address the actual conditional densities in (8.11) as "data generating process", and the pertaining predictive densities as a sequence of density forecasts (in accordance with our terminology). Likewise one can speak of a sequence of predictive distributions.

8.2 Conditional Extremes: a Nonparametric Approach

One of the primary aims of extreme value analysis is to evaluate a higher quantile (such as a T–year threshold) of a random variable Y. In the present section we discuss such questions under the condition that the value x of a covariate X is already known. Thus, we want to estimate, for example, a higher quantile of the conditional df $F(\cdot|x)$ of Y given $X = x$ (or some other parameter of the upper tail of the conditional df).

The message from this section is that one may deal with conditional extremes as if we had ordinary extremes from the conditional df. For that purpose we give an outline of some technical results from the book [16], entitled Laws of Small Numbers.

Let (x_i, y_i) be realizations of the random vector (X, Y). The basic idea is to base the estimation of the conditional df on those y_i which belong to the x_i close to the specified value x. The maximum, exceedances over a threshold u, or the k largest order statistics belonging to the selected y_i may be addressed as conditional extremes (conditional sample maximum, conditional exceedances or conditional order statistics).

There are two different possibilities of selecting the x_i.

- (Nearest Neighbor Approach.) Take those x_i which are the k closest neighbors to x.

- (Fixed Bandwidth Approach.) Select those x_i which are closer to x than a predetermined bandwidth b.

It is understood that distances are measured with respect to the Euclidian distance, yet the subsequent results are valid in greater generality. For applications in hydrology and insurance we refer to the Sections 14.2 and 17.3.

---

[4]Diebold, F.X., Gunther, T.A. and Tay, A.S. (1998). Evaluating density forecasts with applications to financial risk management. Int. Economic Review 39, 863–883.

The Nearest Neighbor Approach

Let $(X_1, Y_1), \ldots, (X_n, Y_n)$ be iid random vectors with common df $F(x, y)$ and let $(x_1, y_1), \ldots, (x_n, y_n)$ be realizations. Let $y'_1, \ldots, y'_k$ be the y_i for which the pertaining x_i are the k closest neighbors to x. The y'_i are taken in the original order of their outcome. Denote by $Y'_1, \ldots, Y'_k$ the pertaining random variables.

The Y'_i can be replaced by iid random variables $Y^*_1, \ldots, Y^*_k$ with common df $F(\cdot|x)$, if

$$k = o\big(n^{4/5}\big) \tag{8.32}$$

and, if a certain technical condition holds for the density of (X, Y) (for details we refer to [16], Theorem 3.5.2, where the replacement is formulated in terms of the Hellinger distance).

As a consequence we know that functionals like $\max(Y'_1, \ldots, Y'_k)$ can be replaced by $\max(Y^*_1, \ldots, Y^*_k)$. Now one may construct statistical models etc. as in the unconditional case.

Likewise, the exceedances of the $Y'_1, \ldots, Y'_k$ over a threshold u can be replaced by the exceedances of the $Y^*_1, \ldots, Y^*_k$ over u.

A result similar to that for the maxima was deduced by Gangopadhyay[5] under a condition $k = o\big(n^{2/3}\big)$ which is stronger than (8.32). On the other hand, Gangopadhyay merely requires a local condition on the density of (X, Y) compared to the overall condition imposed in [16], Theorem 3.5.2.

The Fixed Bandwidth Approach

Assume again that there are iid random vectors $(X_1, Y_1), \ldots, (X_n, Y_n)$ with common df $F(x, y)$. Let $y'_1, \ldots, y'_k$ be the y_i such that $|x_i - x| \leq b$, where b is predetermined. The y'_i are taken in the original order of their outcome. Denote by $Y'_1, \ldots, Y'_{K(n)}$ the pertaining random variables, where $K(n)$ is independent of the $Y'_1, Y'_2, Y'_3, \ldots$.

The Y'_i can be replaced by random variables $Y^*_1, \ldots, Y^*_{K^*(n)}$, where

- $Y^*_1, Y^*_2, Y^*_3, \ldots$ is a sequence of iid random variables with common df $F(\cdot|x)$,

- $K^*(n)$ is a binomial random variable which is independent of the Y^*_i,

if

$$b = o\big(n^{-1/5}\big) \tag{8.33}$$

and, if a certain technical condition holds for the density of (X, Y) (cf. [16], Corollary 3.1.6, where the replacement is formulated in terms of the Hellinger distance).

---

[5]Gangopadhyay, A.K. (1995). A note on the asymptotic behavior of conditional extremes. Statist. Probab. Letters 25, 163–170.

As a consequence we know that functionals like $\max(Y_1', \ldots, Y_{K(n)}')$ can be replaced by $\max(Y_1^*, \ldots, Y_{K^*(n)}^*)$. Now one may construct statistical models etc. as in the unconditional case.

One may also deal with exceedances over a higher threshold. In addition, the binomial random variable $K^*(n)$ can be replaced by a Poisson random variable with the same expectation as $K^*(n)$. Then, one arrives at a Poisson process as introduced in Section 9.1.

8.3   Maxima Under Covariate Information

Contrary to the preceding section, we assume that the conditional distributions belong to a parametric family. In addition, we do not necessarily assume a linear regression framework as in Section 8.1. For example, we may assume, corresponding to (8.24), that

$$Y = \mu(X) + \sigma(X)\varepsilon$$

where ε is a standard normal random innovation which is independent of a covariate (explanatory variable) X. This yields that the conditional df of Y given $X = x$ is normal with location and scale parameters $\mu(x)$ and $\sigma(x)$. Thus, we have

$$F(y|x) = P(Y \leq y|X = x) = \Phi_{\mu(x),\sigma(x)}(y).$$

Such a modeling could be appropriate in Example 8.1.1.

Subsequently, this idea is further developed within the framework of EV models in the present section and of GP models in Section 9.5. The use of such models was initiated in the celebrated article by Davison and Smith (cited on page 121).

A Conditional Extreme Value Model

Assume that the conditional df of Y given $X = x$ is an extreme value df, namely

$$F(y|x) = G_{\gamma(x),\mu(x),\sigma(x)}(y) \qquad (8.34)$$

where the shape, location and scale parameters $\gamma(x)$ $\mu(x)$ and $\sigma(x)$ depend on x.

To reduce the number of parameters we may assume that $\gamma(x)$ does not depend on x, and

$$\begin{aligned} \mu(x) &= \mu_0 + \mu_1 x, \\ \sigma(x) &= \exp(\sigma_0 + \sigma_1 x). \end{aligned} \qquad (8.35)$$

Therefore, we have to deal with the unknown parameters γ, μ_0, μ_1, σ_0 and σ_1.

Conditional Maximum Likelihood Estimation

Let again $\boldsymbol{X} = (X_1, \ldots, X_n)$ and $\boldsymbol{Y} = (Y_1, \ldots, Y_n)$. We assume that the Y_t are conditionally independent conditioned on $\boldsymbol{X}$ with $P(Y_t \leq y | \boldsymbol{X} = \boldsymbol{x}) = P(Y_t \leq y | X_t = x_t)$, see page 230.

The estimation of the unknown parameters γ, μ_0, μ_1, σ_0 and σ_1 may be carried out by means of a conditional maximum likelihood (ML) method. The conditional likelihood function is given by

$$L(\gamma, \mu_0, \mu_1, \sigma_0, \sigma_1) = \prod_{t=1}^{n} g_{\gamma, \mu(x_t), \sigma(x_t)}(y_t).$$

Based on the conditional MLE's $\hat{\gamma}_{\boldsymbol{x},\boldsymbol{y}}$, $\hat{\mu}_{0,\boldsymbol{x},\boldsymbol{y}}$, $\hat{\mu}_{1,\boldsymbol{x},\boldsymbol{y}}$, $\hat{\sigma}_{0,\boldsymbol{x},\boldsymbol{y}}$, $\hat{\sigma}_{1,\boldsymbol{x},\boldsymbol{y}}$ of the parameters γ, μ_0, μ_1, σ_0, σ_1 one also gets the estimates

$$\hat{\mu}_{\boldsymbol{x},\boldsymbol{y}}(x) = \hat{\mu}_{0,\boldsymbol{x},\boldsymbol{y}} + \hat{\mu}_{1,\boldsymbol{x},\boldsymbol{y}} x$$
$$\hat{\sigma}_{\boldsymbol{x},\boldsymbol{y}}(x) = \exp(\hat{\sigma}_{0,\boldsymbol{x},\boldsymbol{y}} + \hat{\sigma}_{1,\boldsymbol{x},\boldsymbol{y}} x)$$

of $\mu(x)$ and $\sigma(x)$. Now, replace the parameters in the EV df $G_{\gamma, \mu(x), \sigma(x)}$ by the estimated parameters to get an estimate of the conditional df $P(Y \leq y | X = x)$.

For related MLEs in the pot–framework we refer to Section 9.5.

Estimating Conditional Functional Parameters

In the subsequent lines we want to distinguish in a strict manner between estimation and prediction procedures. On the one hand, we estimate parameters, which may be real–valued or functions, and on the other hand, we predict random variables.

In that context, we merely consider the mean and the q–quantile as conditional functional parameters. Another important parameter is, of course, the variance.

Conditional Mean Function (Expectation):

$$E_{\gamma, \mu, \sigma}(Y | x) = \int z G_{\gamma, \mu(x), \sigma(x)}(dz). \tag{8.36}$$

Conditional q–Quantile Function:

$$q_{\gamma, \mu, \sigma}(Y | x) = G_{\gamma, \mu(x), \sigma(x)}^{-1}(q). \tag{8.37}$$

If the conditioning is based on a covariate we also speak of a covariate conditional quantile or a covariate conditional expectation. Likewise within a time series framework, one may speak of a serial conditional quantile or a serial conditional expectation. It is apparent that the conditioning may also be based on covariate as well as serial information (a case not treated in this book).

Estimate the conditional functions by replacing the unknown parameters by the conditional MLE's based on $\boldsymbol{x}$ and $\boldsymbol{y}$. For example,

$$q_{\hat{\gamma}\boldsymbol{x},\boldsymbol{y},\hat{\mu}\boldsymbol{x},\boldsymbol{y},\hat{\sigma}\boldsymbol{x},\boldsymbol{y}}(Y|x)$$

is an estimate of the conditional q–quantile $q_{\gamma,\mu,\sigma}(Y|x)$ for each x.

Predicting the Conditional q–Quantile

In place of the factorization of the conditional expectation $E_{\gamma,\mu,\sigma}(Y|x)$ of Y given $X = x$ consider the conditional expectation $E_{\gamma,\mu,\sigma}(Y|X)$ of Y given X.

Recall that the conditional expectation is the best prediction of Y based on X with respect to the quadradic loss function. In the present context, we have to replace the unknown parameters by estimates. By

$$E_{\hat{\gamma}\boldsymbol{X},\boldsymbol{Y},\hat{\mu}\boldsymbol{X},\boldsymbol{Y},\hat{\sigma}\boldsymbol{X},\boldsymbol{Y}}(Y|X)$$

one gets a predictor of the random variables $E_{\gamma,\mu,\sigma}(Y|X)$ and Y based on $\boldsymbol{X}, \boldsymbol{Y}$ and X.

Likewise,

$$q_{\hat{\gamma}\boldsymbol{X},\boldsymbol{Y},\hat{\mu}\boldsymbol{X},\boldsymbol{Y},\hat{\sigma}\boldsymbol{X},\boldsymbol{Y}}(Y|X) \tag{8.38}$$

is a predictor of the conditional q–quantile $q_{\gamma,\mu,\sigma}(Y|X)$ based on $\boldsymbol{X}, \boldsymbol{Y}$ and X.

In the same manner, predictions in the pot–framework, in conjunction with Poisson processes of exceedances, are addressed in Section 9.5.

8.4   The Bayesian Estimation Principle, Revisited

The Bayesian estimation principle—cf. Sections 3.5 and 5.1—will be formulated in a greater generality.

An Interpretation of the Posterior Density

In Section 3.5 we started with

- the prior density $p(\vartheta)$ as a mixing density, and

- densities $L(\cdot|\vartheta)$ of distributions parameterized by ϑ and expressed by the likelihood function.

Using these densities we may introduce the joint density $L(x|\vartheta)p(\vartheta)$ of a random vector (ξ, θ), see (8.8).

According to (8.8) and (8.10) one may interchange the role of ξ and θ. From the joint distribution of (ξ, θ) one may deduce the conditional density of θ given $\xi = x$ which leads to the posterior density in (3.39) as well as in (8.41) below.

The Bayesian Two–Step Experiment

The prior distribution for the parameter ϑ is regarded as the distribution of a random variable θ with outcome ϑ. This random parameter θ describes the initial step in a two–step experiment. The distributions, represented by ϑ, in the given statistical model are regarded as conditional distributions of another random variable ξ given $\theta = \vartheta$, cf. page 233 for the "roundabout" description.

Two special cases were dealt with in Section 3.5. In both cases a gamma distribution with parameters s and d served as a prior distribution.

- $P(\xi \in \cdot | \theta = \vartheta)$ is the common distribution of iid random variables $X_1, \ldots, X_k$ with df F_ϑ and density f_ϑ, and likelihood function $L(\boldsymbol{x}|\vartheta) = \prod_{i \leq k} f_\vartheta(x_i)$.

- The X_i are iid Poisson random variables with parameter λ and likelihood function $L(x|\lambda) = \prod_{i \leq k} P_\lambda\{x_i\}$.

- Both previous cases are combined in the present section within a Poisson process setting with a likelihood function given in a product form.

Because the prior distribution and the family of conditional distributions are specified by the statistician, the joint distribution $\mathcal{L}(\xi, \theta)$ of ξ and θ is also known, cf. (8.10). The joint density is $L(x|\vartheta)p(\vartheta)$, where $p(\vartheta)$ is a density of the prior distribution and $L(x|\vartheta)$ is the likelihood function for the given statistical model.

If a statistical model of prior distributions—instead of a fixed one—is specified by the statistician, then the present viewpoint also enables the estimation of the prior distribution from data.

In the preceding lines we reformulated and extended a technical problem discussed in Chapter 3 within a two–step stochastical procedure.

- In the initial stage there is a stochastic experiment governed by the prior distribution $\mathcal{L}(\theta)$ which generates an unobservable outcome, namely the parameter ϑ.

- Afterwards, an observation x is generated under ϑ. With respect to the total two–step experiment, the value x is governed by $\mathcal{L}(\xi)$.

Although θ is unobservable, one gets knowledge of this random parameter in an indirect manner, namely, by means of x. The information contained in x is added to the initial information which is represented by the prior distribution $\mathcal{L}(\theta)$. As a result, one gets the updated information expressed by the posterior distribution $P(\theta \in \cdot | \xi = x)$. This illuminates the importance of the posterior distribution in its own right.

Computing Bayesian Estimates

We compute the Bayes estimate within a general framework. Let again $\mathcal{L}(\theta)$ and $p(\vartheta)$ denote the prior distribution and prior density, respectively. We repeat some of the computations in Section 3.1 concerning the Bayes estimate.

The MSE of an estimator $\widehat{T}(X)$ of the functional parameter $T(\vartheta)$ can be written as an integral with respect to the conditional distribution of ξ given $\theta = \vartheta$. We have

$$E\big((\widehat{T}(X) - T(\vartheta))^2|\vartheta\big) = \int (\widehat{T}(x) - T(\vartheta))^2 P(\xi \in dx|\theta = \vartheta).$$

Therefore, the Bayes risk with respect to a prior distribution $\mathcal{L}(\theta)$ can be written as

$$
\begin{aligned}
R(p, \widehat{T}) &= \int \left(\int (\widehat{T}(x) - T(\vartheta))^2 P(\xi \in dx|\theta = \vartheta) \right) \mathcal{L}(\theta)(d\vartheta) \\
&= \int (\widehat{T}(x) - T(\vartheta))^2 \mathcal{L}(\theta, \xi)\,(d\vartheta dx) \\
&= \int \left(\int (\widehat{T}(x) - T(\vartheta))^2 P(\theta \in d\vartheta|\xi = x) \right) \mathcal{L}(\xi)(dx) \ . \quad (8.39)
\end{aligned}
$$

Now proceed as in (3.41) to get the Bayes estimate

$$T^*(x) = \int T(\vartheta) P(\theta \in d\vartheta|\xi = x). \qquad (8.40)$$

Thus, we get a representation of the Bayes estimate by means of the posterior distribution $P(\theta \in \cdot|\xi = x)$. If ϑ is one–dimensional and $T(\vartheta) = \vartheta$, then $T^* = E(\theta|\xi = x)$ is the conditional expectation of θ given $\xi = x$.

In fact, (8.40) is an extension of (3.38). To see this, one must compute the conditional density $p(\vartheta|x)$ of θ given $\xi = x$. Because $L(x|\vartheta)p(\vartheta)$ is the joint density of ξ and θ one obtains

$$p(\vartheta|x) = \frac{L(x|\vartheta)p(\vartheta)}{\int L(x|\vartheta)\mathcal{L}(\theta)\,(d\vartheta)} \ , \qquad (8.41)$$

which is the posterior density given x (see also Section 8.1, where the "roundabout" of joint, conditional and mixture dfs is discussed in detail).

Bayesian Estimation and Prediction

From (8.40) one recognizes that the Bayes estimate can be written as the conditional expectation $T^*(x) = E(T(\theta)|\xi = x)$. Moreover, the characteristic property of the Bayes estimate of minimizing the Bayes risk can be reformulated (cf. second line in (8.39)) as

$$E\big((T^*(\xi) - T(\theta))^2\big) = \inf_{\widehat{T}} E\big((\widehat{T}(\xi) - T(\theta))^2\big),$$

where the inf ranges over all estimators $\widehat{T}$. Of course, this is a well–known property of conditional expectations.

The property that T^* is the Bayes estimator for the functional T can be rephrased by saying that $T^*(\xi) = E(T(\theta)|\xi)$ is the best predictor of the random variable $T(\theta)$. This approach also allows the introduction of a linear Bayes estimator by taking the best linear predictor of $T(\theta)$.

The conditional (posterior) distribution of $T(\theta)$ given $\xi = x$ may be addressed as predictive distribution (cf. Section 8.1, page 237) in the Bayesian framework. This predictive distribution is known to the statistician as long as the prior distribution does not include superparameters, cf. Section 14.5.

Further Predictive Distributions in the Bayesian Framework

Let (ξ, θ) be a random vector where θ is distributed according to the prior density $p(\vartheta)$. Outcomes of the random variables ξ and θ are x and the parameter ϑ. The conditional density of ξ given $\theta = \vartheta$ is $p(x|\vartheta)$. From (8.41) we know that the posterior density (the conditional density of θ is given $\xi = x$) is given by

$$p(\vartheta|x) = p(x|\vartheta)p(\vartheta) \Big/ \int p(x|\vartheta)p(\vartheta)\,d\vartheta\,. \tag{8.42}$$

Predictive distributions may be dealt with in the following extended framework (as described in the book by Aitchison and Dunsmore[6]): Consider the random vector (ξ, η, θ) where θ has the (prior) density $p(\vartheta)$, and ξ and η are conditional independent (conditioned on $\theta = \vartheta$) with conditional densities $p(x|\vartheta)$ and $p(y|\vartheta)$. Thus, the conditional density of (ξ, η) given $\theta = \vartheta$ is $p(x, y|\vartheta) = p(x|\vartheta)p(y|\vartheta)$.

The predictive distribution is the conditional distribution of η given $\xi = x$ which has the density

$$p(y|x) = \int p(y|\vartheta)p(\vartheta|x)\,d\vartheta\,, \tag{8.43}$$

where $p(\vartheta|x)$ is the posterior density in (8.42). To establish (8.43) we make use of the representation $p(y|x) = p(x, y)/p(x)$ of the conditional density of η given $\xi = x$, where $p(x, y)$ and $p(x)$ are the densities of (ξ, η) and ξ. We have

$$
\begin{aligned}
p(y|x) &= p(x, y)/p(x) \\
&= \int p(x, y|\vartheta)p(\vartheta)\,d\vartheta \Big/ p(x) \\
&= \int p(x|\vartheta)p(y|\vartheta)p(\vartheta)\,d\vartheta \Big/ \int p(x|\vartheta)p(\vartheta)\,d\vartheta \\
&= \int p(y|\vartheta)p(\vartheta|x)\,d\vartheta\,.
\end{aligned}
$$

If $p(\vartheta)$ is a conjugate prior for both conditional densities $p(x|\vartheta)$ and $p(y|\vartheta)$, then the last integrand is proportional to a density which is of the same type as the prior $p(\vartheta)$ and the integrand can be analytically computed.

---

[6]Aitchison, J. and Dunsmore, I.R. (1975). Statistical Prediction Analysis. Cambridge University Press, Cambridge.

EXAMPLE 8.4.1. (Predictive distributions for the restricted Pareto model.) We assume that the statistical models related to the random variables ξ and η are both restricted Pareto models with thresholds u and v. Let

$$p(\boldsymbol{x}|\alpha) = \prod_{i \leq k} w_{1,\alpha,0,u}(x_i)$$

and

$$p(y|\alpha) = w_{1,\alpha,0,v}(y) \,.$$

As in Section 5.1 take the gamma density $p(\alpha) = h_{s,d}(\alpha)$ as a prior for the shape parameter α. Then the gamma density $p(\alpha|\boldsymbol{x}) = h_{s',d'}(\alpha)$ is the posterior with respect to the $p(\boldsymbol{x}|\alpha)$ where $s' = s + k$ and $d' = d + \sum_{i \leq k} \log(x_i/u)$.

For computing the predictive density $p(y|\boldsymbol{x})$ put $s^* = s' + 1$ and $d^* = d' + \log(y/v)$. Writing the integrand as a gamma density we get

$$
\begin{aligned}
p(y|\boldsymbol{x}) &= \int_0^\infty w_{1,\alpha,0,v}(y) h_{s',d'}(\alpha) \, d\alpha & (8.44)\\
&= d'^{s'} \Gamma(s')^{-1} y^{-1} \int_0^\infty \alpha^{s^*-1} \exp(-d^*\alpha) \, d\alpha \\
&= d'^{s'} d^{*-s^*} s' y^{-1} \\
&= (s+k)\Big(d + \sum_{i \leq k} \log \frac{x_i}{u}\Big)^{s+k} y^{-1} \Big(d + \sum_{i \leq k} \log \frac{x_i}{u} + \log \frac{y}{v}\Big)^{-(s+k+1)}
\end{aligned}
$$

for $y > v$.

We note an extension where η is also a vector of iid Paretian random variables. If $p(y|\alpha)$ is replaced by

$$p(\boldsymbol{y}|\alpha) = \prod_{i \leq m} w_{1,\alpha,0,v}(y_i) \,,$$

then the predictive density is

$$p(\boldsymbol{y}|\boldsymbol{x}) = d'^{s'} d^{*-s^*} \Big(\prod_{j \leq m} y_j\Big)^{-1} \prod_{j \leq m} (s' + j - 1)$$

with $s^* = s' + m$ and $d^* = d' + \sum_{j \leq m} \log(y_j/v)$ for $y_j > v$.

It would be desirable to give these ideas more scope within the extreme value setting.

Chapter 9

Statistical Models for Exceedance Processes

In this chapter, Poisson processes and related processes are studied. These processes are essential for hydrological, environmental, financial and actuarial studies in Chapters 14 to 17. In Section 9.1 the basic concepts are introduced. We particularly mention the modeling of exceedances and exceedance times by means of Poisson processes. Within the framework of Poisson processes, we reconsider the concept of a T–year level in Section 9.2. The maximum likelihood and Bayesian estimation within models of Poisson processes is addressed in Section 9.3. The explanations about the GP approximation of exceedance dfs, cf. Section 6.5, will be continued within the framework of binomial and Poisson processes in Section 9.4. An extension of the modeling by Poisson processes from the homogeneous case to the inhomogeneous one is investigated in Section 9.5.

9.1 Modeling Exceedances by Poisson Processes: the Homogeneous Case

In this section, we deal with observations occurring at random times. Especially, we have exceedance times modeled by Poisson processes. Let T_i denote the arrival time of the ith random observation X_i, where necessarily $0 \leq T_1 \leq T_2 \leq T_3 \leq \cdots$. The X_i will be addressed as marks.

Homogeneous Poisson and Poisson(λ, F) Processes

The most prominent examples of arrival processes are homogeneous Poisson processes with intensity λ, where the interarrival times $T_1, T_2 - T_1, T_3 - T_2, \ldots$ are iid random variables with common exponential df

$$F(x) = 1 - e^{-\lambda x}, \qquad x \geq 0,$$

with mean $1/\lambda$. Under this condition, the arrival time T_i is a gamma random variable. The numbers of observations occurring up to time t constitute the counting process

$$N(t) = \sum_{i=1}^{\infty} I(T_i \leq t), \qquad t \geq 0. \tag{9.1}$$

The arrival times $T_1 \leq T_2 \leq T_3 \leq \cdots$ as well as the counting process $N(t)$ are addressed as a homogeneous Poisson process with intensity λ (in short, as a Poisson(λ) process). Deduce from (4.7) that

$$
\begin{aligned}
P\{N(t) = k\} &= P\{T_k \leq t, T_{k+1} > t\} \\
&= P\{T_k \leq t\} - P\{T_{k+1} \leq t\} \\
&= \frac{(\lambda t)^k}{k!} e^{-\lambda t}.
\end{aligned}
$$

Hence, the number $N(t)$ of arrival times is a Poisson random variable with parameter λt. The mean value function—describing the mean number of observations up to time t—is

$$\Psi_\lambda(t) = E(N(t)) = \lambda t.$$

In addition, it is well known that the homogeneous Poisson process has independent and stationary increments, that is,

$$N(t_1), N(t_2) - N(t_1), \ldots, N(t_n) - N(t_{n-1}), \qquad t_1 < t_2 < \cdots < t_n,$$

are independent, and $N(t_j) - N(t_{j-1})$ has the same distribution as $N(t_j - t_{j-1})$. We see that the mean number of observations occurring within a time unit is

$$E\big(N(t+1) - N(t)\big) = E(N(1)) = \lambda$$

which gives a convincing interpretation of the intensity λ of a homogeneous Poisson process.

The sequence $\{(T_i, X_i)\}$ of arrival times and marks is a Poisson(λ, F) process if the following conditions are satisfied.

Poisson(λ, F) Conditions:

(a) the sequence of arrival times $0 \leq T_1 \leq T_2 \leq T_3 \leq \cdots$ is a Poisson(λ) process;

(b) the marks $X_1, X_2, X_3, \ldots$ are iid random variables with common df F,

(c) the sequences $T_1, T_2, T_3, \ldots$ and $X_1, X_2, X_3, \ldots$ are independent.

A justification of such a process in flood frequency studies with exponential exceedances was given by Todorovic and Zelenhasic[1]. To capture the seasonality of flood discharges, an extension of the present framework to inhomogeneous Poisson processes with time–dependent marks is desirable. Yet, in order not to overload this section, the introduction of such processes is postponed until Section 9.5.

Poisson Approximation of Exceedances and Exceedance Times

In Section 1.2, the exceedance times $\tau_1 \leq \tau_2 \leq \tau_3 \leq \cdots$ of iid random variables Y_i over a threshold u were described in detail. Moreover, if F is the common df of the Y_i, then we know that the exceedances are distributed according to the exceedance df $F^{[u]}$. Exceedance times and exceedances are now regarded as arrival times and marks. If $1 - F(u)$ is sufficiently small, then an adequate description of exceedance times and exceedances is possible by means of a Poisson$(1 - F(u), F^{[u]})$ process. For further details see Section 9.5 and the monographs [44] and [43].

Exceedances for Poisson(λ, F) Processes

Now we go one step further and deal with exceedances and exceedance times for a Poisson(λ, F) process of arrival times T_i and marks X_i. If only the exceedances of the X_i over a threshold v and the pertaining times are registered, then one can prove that exceedance times and exceedances constitute a Poisson$(\lambda(1 - F(v)), F^{[v]})$ process. We refer to Section 9.5 for a generalization of this result.

Mixed Poisson Processes, Pólya–Lundberg Processes

We mention another class of arrival processes, namely mixed Poisson processes. Our attention is focused on Pólya–Lundberg processes, where the mixing of homogeneous Poisson processes is done with respect to a gamma distribution. The marginal random variables $N(t)$ of a Pólya–Lundberg process are negative binomial random variables.

Let us write $N(t, \lambda)$, $t \geq 0$, for a homogeneous Poisson (counting) process with intensity $\lambda > 0$. By mixing such processes over the parameter λ with respect to some density f, one obtains a mixed Poisson process, say, $N(t)$, $t \geq 0$. Recollect that a mixed Poisson process represents the following two–step experiment: firstly, a parameter λ is drawn according to the distribution represented by the density f and, secondly, a path is drawn according to the homogeneous Poisson process with intensity λ. The marginals $N(t)$ are mixed Poisson random variables. We have

$$P\{N(t) = n\} = \int P_{\lambda t}\{n\} f(\lambda) \, d\lambda.$$

---

[1]Todorovic, P. and Zelenhasic, E. (1970). A stochastic model for flood analysis. Water Resour. Res. 6, 1641–1648.

In the special case where $f = f_{\alpha,\sigma}$ is a gamma density with shape and scale parameters $\alpha, \sigma > 0$, one obtains a Pólya–Lundberg process with parameters α and σ. Deduce from (3.31) that $N(t)$ is a negative binomial random variable with parameters α and $p = 1/(1 + \sigma t)$. In addition,

$$\Psi_{\alpha,\sigma}(t) = E(N(t)) = \alpha \sigma t, \qquad t \geq 0, \tag{9.2}$$

is the increasing mean value function of a Pólya–Lundberg process with parameters α and σ.

Arrival Times for Mixed Poisson Processes

Let $N(t)$, $t \geq 0$ be a counting process such as a mixed Poisson process. The ith arrival time can be written

$$T_i = \inf\{t > 0 : N(t) \geq i\}, \qquad i = 1, 2, 3, \ldots .$$

The df of the first arrival time T_1 has the representation

$$P\{T_1 \leq t\} = 1 - P\{N(t) = 0\}. \tag{9.3}$$

From (9.3) and (2.24) deduce

$$E(T_1) = \int_0^\infty P\{N(t) = 0\} \, dt.$$

In the special case of a Pólya–Lundberg process with parameters $\alpha, \sigma > 0$, one must deal with negative binomial random variables $N(t)$ for which

$$P\{N(t) = 0\} = (1 + \sigma t)^{-\alpha}. \tag{9.4}$$

Hence, the first arrival time T_1 is a Pareto random variable with shape, location and scale parameters α, $-1/\sigma$ and $1/\sigma$.

9.2   Mean and Median T–Year Return Levels

Recall that the T–year return level $u(T)$ of a sequence of random variables is that threshold u such that the first exceedance time $\tau_1 = \tau_{1,u}$ at u is equal to T in the mean, that is, $u(T)$ is the solution to the equation $E(\tau_{1,u}) = T$.

In Section 1.2, it was verified that the first exceedance time of iid random variables at a threshold u is a geometric random variable with parameter $p = 1 - F(u)$ which yields $u(T) = F^{-1}(1 - 1/T)$. In order to estimate $u(T)$, the unknown df F is usually replaced by an estimated EV or GP df.

The T–Year Return Level for Heterogeneous Variables

Next, we assume that $X_1, X_2, X_3, \ldots$ are independent, not necessarily identically distributed random variables. Let F_i denote the df of X_i. The probability that the first exceedance time τ_1 at the threshold u is equal to k is

$$P\{\tau_1 = k\} = (1 - F_k(u)) \prod_{i \leq k-1} F_i(u), \qquad k = 1, 2, 3, \ldots,$$

according to the first identity in (1.17). In the heterogeneous case, it may happen that, with a positive probability, the threshold u is never exceeded. In this instance, τ_1 is put equal to ∞.

To evaluate the T–year return level $u(T)$, one must solve $E(\tau_{1,u}) = T$ as an equation in u, which can be written

$$\sum_{k=1}^{\infty} k(1 - F_k(u)) \prod_{i \leq k-1} F_i(u) = T. \tag{9.5}$$

This equation must be solved numerically by a Newton iteration procedure. We remark that $u(T) \geq F^{-1}(1 - 1/T)$, if $F_i \leq F$. If $E(\tau_{1,u}) = \infty$, this approach is not applicable.

The Median T–Year Return Level

One may consider a different functional parameter of the first exceedance time $\tau_{1,u}$ to fix a T–year return level. To exemplify this idea, we compute a median of $\tau_{1,u}$ and determine the median T–year return level $u(1/2, T)$. The median of $\tau_{1,u}$ is

$$\mathrm{med}(\tau_{1,u}) = \min \left\{ m : \sum_{k \leq m} p(1 - p)^{k-1} \geq 1/2 \right\}$$

with $p = 1 - F(u)$. Because $\sum_{k \leq m} z^{k-1} = (1 - z^m)/(1 - z)$, one obtains

$$\mathrm{med}(\tau_{1,u}) = \langle \log(1/2) / \log(F(u)) \rangle, \tag{9.6}$$

where $\langle x \rangle$ is the smallest integer $\geq x$. The median T–year return level $u(1/2, T)$ is the solution to the equation $\mathrm{med}(\tau_{1,u}) = T$. Approximately, one gets

$$u(1/2, T) \approx F^{-1}\left(2^{-1/T}\right) \approx F^{-1}(1 - (\log 2)/T). \tag{9.7}$$

Because $\log(2) = 0.693\ldots$, the median T–year return level is slightly larger than the mean T–year return level in the iid case.

By generalizing this concept to q–quantiles of the first exceedance time $\tau_{1,u}$ one obtains a T–year return level $u(q, T)$ according to the q–quantile criterion (see also page 430).

Poisson Processes and T–Year Return Levels

When exceedances over u are modeled by a $\text{Poisson}\big(1 - F(u), F^{[u]}\big)$ process (cf. page 249), then the first exceedance time is an exponential random variable with expectation $1/(1 - F(u))$. Therefore, the expectation and the median of the first exceedance time are equal to a predetermined time span T for the thresholds

$$u = u(T) = F^{-1}(1 - 1/T)$$

and

$$u = u(1/2, T) = F^{-1}(1 - (\log 2)/T).$$

Thus, when employing a Poisson approximation of the exceedances pertaining to random variables X_i as in Section 1.2, one finds the same mean T–year return level and, approximately, the median T–year return level given in (9.7).

If we already start with a $\text{Poisson}(\lambda, F)$ process, then the mean and median T–year return level are

$$u(T) = F^{-1}(1 - 1/(\lambda T)) \tag{9.8}$$

and

$$u(1/2, T) = F^{-1}(1 - (\log 2)/(\lambda T)). \tag{9.9}$$

This can be verified in the following manner: As mentioned in Section 9.1, the marks exceeding v and the pertaining times constitute a $\text{Poisson}\big(\lambda(1 - F(v)), F^{[v]}\big)$ process and, hence, the first exceedance time is an exponential random variable with mean $1/(\lambda(1 - F(v)))$. This implies the desired result.

This suggests the following approach for estimating the T–year return level. Firstly, utilize a $\text{Poisson}(\lambda, W)$ modeling for a sufficiently large number of exceedances (over a moderately high threshold), where W is a GP df. Secondly, estimate the intensity by $\hat{\lambda} = N(t)/t$ and the GP df by $\widehat{W} = W_{\gamma_k, \mu_k, \sigma_k}$. Then,

$$\hat{u}(T) = \widehat{W}^{-1}(1 - 1/(\hat{\lambda} T)) \tag{9.10}$$

and

$$\hat{u}(1/2, T) = \widehat{W}^{-1}(1 - (\log 2)/(\hat{\lambda} T)) \tag{9.11}$$

are estimates of the mean and median T–year return level.

In the case of clustered data, proceed in a similar manner. The T–year return level of clustered data and of the pertaining cluster maxima (cf. page 78) are close to each other and, therefore, one may reduce the analysis to the cluster maxima.

9.3 ML and Bayesian Estimation in Models of Poisson Processes

In this section we study maximum likelihood (ML) and Bayes estimators for certain Poisson processes which provide a joint model for exceedance times and exceedances. The primary aim is to show that the estimators of Pareto and generalized Pareto (GP) parameters, as dealt with in Section 5.1, correspond to those in the Poisson process setting.

Models of Poisson Processes

We observe a random scenery up to time T. In applications this usually concerns the past T periods. For each parameter $\lambda > 0$ and each df F, let Poisson(λ, F, T) denote the Poisson(λ, F) process—introduced in Section 9.1—restricted to the time interval from 0 up to time T. Thus, there is a homogeneous Poisson process with intensity λ in the time scale with marks which have the common df F. Apparently, the number $N(T)$ of exceedances up to time T is a Poisson random variable with parameter λT. Assume that $F = F_\vartheta$ is a df with density f_ϑ. Thus, the Poisson(λ, F_ϑ, T) process is represented by the parameter vector (λ, ϑ).

In the present context the outcome of such a Poisson process is a sequence of pairs (t_i, y_i), for $i = 1, \ldots, k$, consisting of exceedance times t_i and the pertaining exceedances y_i over a threshold u. Notice that k is the outcome of the Poisson random variable $N(T)$ with parameter λT. If $k = 0$, then there are no observations, which happens with a positive probability.

The Likelihood Function for Poisson Process Models

By specifying a likelihood function, the ML and Bayesian estimation principles become applicable. It suffices to determine a likelihood function up to a constant to compute the ML and Bayes estimates.

One may prove (cf. [43], Theorem 3.1.1) that for every sample $\{(t_i, y_i)\}$ the likelihood function $L(\{(t_i, y_i)\}|\lambda, \vartheta)$ satisfies

$$L(\{(t_i, y_i)\}|\lambda, \vartheta) \propto L_1(k|\lambda)L_2(\boldsymbol{y}|\vartheta) \tag{9.12}$$

("$\propto$" again denotes that both sides are proportional), where

- $L_1(k|\lambda) = \big((\lambda T)^k/k!\big)\exp(-\lambda T)$ is the likelihood function for the Poisson model as given in (3.50),

- $L_2(\boldsymbol{y}|\vartheta) = \prod_{i \le k} f_\vartheta(y_i)$ is the likelihood function for the model of k iid random variables with common density f_ϑ as dealt with, e.g., in (3.36).

One recognizes that likelihood–based estimators of λ and ϑ only depend on the data by means of k and, respectively, $\boldsymbol{y} = (y_1, \ldots, y_k)$.

Maximum Likelihood Estimators for Poisson Processes with Generalized Pareto Marks

To get the MLE one must find the parameters λ and ϑ which maximize the likelihood function in (9.12). Due to the specific structure of the likelihood function in

(9.12), it is apparent that the ML procedure for Poisson processes splits up into those which were separately dealt with in the Sections 3.1 and 3.4.

Because the first factor $L_1(k|\lambda)$ is independent of ϑ it can be easily verified that the MLE of λ is equal to

$$\hat{\lambda}_k = k/T. \tag{9.13}$$

We distinguish two different models with respect to the parameter ϑ.

- (The Restricted Pareto Model GP1$(u, \mu = 0)$.) Let $\vartheta = \alpha$ be the unknown shape parameter in the restricted model GP1$(u, \mu = 0)$ of Pareto distributions (cf. page 116). Then, $(\hat{\lambda}_k, \hat{\alpha}_k(\boldsymbol{y}))$ is the MLE in the present Poisson process model, where $\hat{\alpha}_k(\boldsymbol{y})$ denotes the Hill estimate (cf. (5.1); also see [43], pages 142–143).

- (The Generalized Pareto Model GP(u).) Let $\vartheta = (\gamma, \sigma)$ be the parameter vector in the generalized Pareto model as described on page 134. The MLE in the pertaining Poisson process model is $(\hat{\lambda}_k, \hat{\gamma}_k(\boldsymbol{y}), \hat{\sigma}_k(\boldsymbol{y}))$, where $(\hat{\gamma}_k(\boldsymbol{y}), \hat{\sigma}_k(\boldsymbol{y}))$ is the MLE in the GP(u) model of iid random variables of a sample of size k.

Bayes Estimators for Poisson Processes with Pareto Marks

Next, we apply the concept of Bayes estimators, as outlined in Section 3.5 and further developed in the Sections 5.1 and 7.3, to certain statistical models of Poisson$(\lambda, F_\vartheta, T)$ processes, where F_ϑ is a Pareto df represented by ϑ. An extension of that concept is required to cover the present questions, yet the calculations remain just the same.

Recall from (9.12) that the likelihood function pertaining to the model of Poisson$(\lambda, F_\vartheta, T)$ processes satisfies $L(\{(t_i, y_i)\}|\lambda, \vartheta) \propto L_1(k|\lambda)L_2(\boldsymbol{y}|\vartheta)$. One must estimate the unknown parameters λ and ϑ.

The Bayes estimate of the intensity parameter λ of the homogeneous Poisson process of exceedance times can be computed independently of ϑ if the prior density can be written as the product

$$p(\lambda, \vartheta) = p_1(\lambda)p_2(\vartheta). \tag{9.14}$$

Under this condition, the posterior density is

$$p(\lambda, \vartheta|\{(t_i, y_i)\}) = p_1(\lambda|k)p_2(\vartheta|\boldsymbol{y}), \tag{9.15}$$

where $p_1(\lambda|k) \propto L_1(k|\lambda)p_1(\lambda)$ and $p_2(\vartheta|\boldsymbol{y}) \propto L_2(\boldsymbol{y}|\vartheta)p_2(\vartheta)$.

The Bayes estimator for a functional parameter

$$T(\lambda, \vartheta) = T_1(\lambda)T_2(\vartheta) \tag{9.16}$$

can be represented by

$$
\begin{aligned}
\widehat{T}(\{t_i, y_i\}) &= \int T_1(\lambda) T_2(\vartheta) p_1(\lambda|k) p_2(\vartheta|\boldsymbol{y}) \, d\lambda d\vartheta \\
&= \int T_1(\lambda) p_1(\lambda|k) \, d\lambda \int T_2(\vartheta) p_2(\vartheta|\boldsymbol{y}) \, d\vartheta \; .
\end{aligned}
\tag{9.17}
$$

We explicitly compute Bayes estimators of the intensity λ and the parameter ϑ.

- If λ is estimated, then $T_1(\lambda) = \lambda$, $T_2(\vartheta) = 1$ and $\int T_2(\vartheta) p_2(\vartheta|\boldsymbol{y}) \, d\vartheta = 1$. The Bayes estimate of λ is

$$
\lambda_k^* = \int \lambda p_1(\lambda|k) \, d\lambda.
\tag{9.18}
$$

  Specifically, if the prior $p_1(\lambda)$ is the gamma density with parameters r and c, then we know from (3.53) that

$$
\lambda_k^* = \frac{r+k}{c+T}
\tag{9.19}
$$

  is the Bayes estimate of the intensity λ.

- Bayes estimation of ϑ: Again, we deal with two different models, where $\vartheta = \alpha$ and, respectively, $\vartheta = (\alpha, \eta)$ are the parameters of Paretian models.

  - The restricted Pareto model GP1$(u, \mu = 0)$: Bayes estimators in this Poisson–Pareto model were dealt with by Hesselager[2]. One gets the Bayes estimate

$$
\alpha_k^*(\boldsymbol{y}) = \int \alpha p_2(\alpha|\boldsymbol{y}) \, d\alpha
\tag{9.20}
$$

    of the shape parameter α. If the prior $p_2(\alpha)$ is a gamma density with parameters s and d, then one obtains the Bayes estimate $\alpha_k^*(\boldsymbol{y})$ in (5.7).

  - The full Pareto model GP(u) in the (α, η)–parameterization: Let $\vartheta = (\alpha, \eta)$ be the parameter vector in the model of Pareto distributions in (5.10). The Bayes estimates of α and η can be written

$$
\alpha_k^*(\boldsymbol{y}) = \int \alpha p_2(\alpha, \eta|\boldsymbol{y}) \, d\alpha d\eta
\tag{9.21}
$$

    and

$$
\eta_k^*(\boldsymbol{y}) = \int \eta p_2(\alpha, \eta|\boldsymbol{y}) \, d\alpha d\eta
\tag{9.22}
$$

---

[2]Hesselager, O. (1993). A class of conjugate priors with applications to excess–of–loss reinsurance. ASTIN Bulletin 23, 77–90.

with $p_2(\alpha, \eta|\boldsymbol{y})$ as in (9.15). If

$$p_2(\alpha, \eta) = h_{s,d}(\alpha)f(\eta),$$

as in (5.11), then one obtains the Bayes estimates $\alpha_k^*(\boldsymbol{y})$ and $\eta_k^*(\boldsymbol{y})$ of α and η as in (5.15) and (5.16).

9.4 GP Process Approximations
co–authored by E. Kaufmann[3]

This section provides a link between the explanations in the Sections 1.2 and 5.1 about exceedances and exceedance dfs and Section 9.1, where exceedances and exceedance times are represented by means of Poisson processes.

Recall from Section 1.2 that the number of exceedances, of a sample of size n with common df F, over a threshold u is distributed according to a binomial distribution $B_{n,p}$ with parameter $p = \bar{F}(u) = 1 - F(u)$. In addition, the exceedances have the common df $F^{[u]} = (F(x) - F(u))/(1 - F(u))$ for $x \geq u$. In Section 5.1 the actual exceedance df $F^{[u]}$ was replaced by a GP df W, and in Section 6.5 the remainder term in this approximation was computed under certain conditions imposed on F. In the statistical context one must simultaneously deal with all exceedances over the given threshold u.

Binomial Processes Representation of Exceedances Processes

First, the exceedances will be represented by a binomial process.

Binomial$(n, p, F^{[u]})$ Conditions: Let

(a) $Y_1, Y_2, Y_3, \ldots$ be iid random variables with common df $F^{[u]}$,

(b) $K(n)$ be a binomial random variable with parameters n and $p = \bar{F}(u)$, which is independent of the sequence $Y_1, Y_2, Y_3, \ldots$.

The actual exceedances can be distributionally represented by the sequence

$$Y_1, Y_2, Y_3, \ldots, Y_{K(n)} \tag{9.23}$$

which can be addressed as binomial process.

Binomial Process Approximation

Next, the actual binomial process in (9.23) will be replaced by another binomial process

$$Z_1, Z_2, Z_3, \ldots, Z_{K(n)} \tag{9.24}$$

---

[3]University of Siegen; co–authored the 2nd edition.

where $F^{[u]}$ is replaced by $W^{[u]}$. Recall that $W^{[u]}$ is again a GP df if W is a GP df.

Let F and W have the densities f and w. In addition, assume that $\omega(F) = \omega(W)$. According to [16], Corollary 1.2.4 (iv), such an approximation holds with a remainder term bounded by

$$\Delta_F(n, u) := \left(n\bar{F}(u)\right)^{1/2} H\left(f^{[u]}, w^{[u]}\right), \tag{9.25}$$

where H is the Hellinger distance, cf. (3.4), between $f^{[u]}$ and $w^{[u]}$. More precisely, one gets in (9.25) a bound on the variational distance between the point processes pertaining to the sequences in (9.23) and (9.24).

Under condition (6.35) we have

$$H\left(f^{[u]}, w^{[u]}\right) = O\left(\overline{W}^{1/\delta}(u)\right), \tag{9.26}$$

and, therefore, because (6.33) also holds,

$$\Delta_F(n, u) = O\left(n^{1/2}\overline{W}^{(2+\delta)/2\delta}(u)\right). \tag{9.27}$$

For example, if u is the $(1 - k/n)$–quantile of W—with k denoting the expected number of exceedances over u—then the right–hand side in (9.27) is of order $k^{1/2}(k/n)^{1/\delta}$.

Von Mises Bounds

We mention the required modifications if condition (6.35) is replaced by the conditions (6.41) and (6.42).

The upper bound on the binomial process approximation, with $w^{[u]}$ replaced by $w_{\gamma,u,\sigma(u)}$, is

$$\begin{aligned} \Delta_F(n, u) &= \left(n\bar{F}(u)\right)^{1/2} H\left(f^{[u]}, w_{\gamma,u,\sigma(u)}\right) \\ &= O\left(\left(n\bar{F}(u)\right)^{1/2}\left|\eta\left(\bar{F}(u)\right)\right|\right) \end{aligned} \tag{9.28}$$

in analogy to (9.27) with $\sigma(u)$ as in (6.45).

Thus, the exceedances under the actual df $F^{[u]}$ can be replaced by GP random variables with common df $W_{\gamma,u,\sigma(u)}$ within the error bound in (9.28). For example, if F is the Gompertz df and $u = F^{-1}(1 - k/n)$, then

$$\Delta_F(n, u) = O\left(k^{-1/2}/\log n\right).$$

Penultimate Approximation

The corresponding upper bound for the penultimate approximation to the exceedances process, with $w_{\gamma,u,\sigma(u)}$ replaced by $w_{\gamma(u),u,\sigma(u)}$, is

$$\begin{aligned} \Delta_F(n, u) &= \left(n\bar{F}(u)\right)^{1/2} H\left(f^{[u]}, w_{\gamma(u),u,\sigma(u)}\right) \\ &= O\left(\left(n\bar{F}(u)\right)^{1/2}\tau\left(\bar{F}(u)\right)\right) \end{aligned} \tag{9.29}$$

under the conditions (6.41) and (6.46).

A Poisson Process Approximation

One gets a Poisson process instead of a binomial process if the binomial random variable with parameters n and p is replaced by a Poisson random variable with parameter $\lambda = np$. The remainder term of such an approximation, in terms of the variational distance, is bounded by p, see [43], Remark 1.4.1.

9.5   Inhomogeneous Poisson Processes, Exceedances Under Covariate Information

In this section we model

- frequencies of occurance times by means of an inhomogeneous Poisson process, and

- magnitudes by stochastically independent, time–dependent marks which are distributed according to generalized Pareto (GP) dfs.

A parametric modeling for the marks is indispensable to achieve the usual extrapolation to extraordinary large data.

Inhomogeneous Poisson and Poisson(Λ, F) Processes

We extend the concept of a Poisson(λ, F) process (cf. page 248) in two steps. Firstly, the intensity λ is replaced by a mean value function Λ or an intensity function (also denoted by λ) and, secondly, the df F of the marks is replaced by a conditional df $F = F(\cdot|\cdot)$.

Poisson(Λ, F) Conditions:

(a) (Poisson(Λ) Process.) Firstly, let $0 \le T_1 \le T_2 \le T_3 \le \cdots$ be a Poisson(1) process (a homogeneous Poisson process with intensity $\lambda = 1$, cf. page 248). Secondly, let Λ be a nondecreasing, right–continuous function defined on the positive half–line $[0, \infty)$ with $\Lambda(0) = 0$ and $\lim_{t \to \infty} \Lambda(t) = \infty$. Define the generalized inverse Λ^{-1} corresponding to the concept of a qf, see (1.64).

Then, the series $\tau_i = \Lambda^{-1}(T_i)$, $i \ge 1$ or, equivalently, the counting process

$$N(t) = \sum_{i=1}^{\infty} I(\tau_i \le t), \ t \ge 0,$$

can be addressed as an inhomogeneous Poisson process with mean value function Λ.

(b) (Conditional Marks.) Given occurrence times t_i consider random variables X_{t_i} with df $F(\cdot|t_i)$. Thus, we have marks which distributionally depend on the time at which they are observed.

One can verify that $N(t)$ is a Poisson random variable with expectation $\Lambda(t)$ which justifies the notion of Λ as a mean value function. Moreover notice that there is a homogeneous Poisson process with intensity λ if $\Lambda(t) = \lambda t$. If a representation $\int_a^b \lambda(x)\,dx = \Lambda(b) - \Lambda(a)$ holds, then $\lambda(x)$ is called the intensity function of Λ.

This two–step random experiment constitutes a Poisson(Λ, F) process. The reader is referred to [43], Corollary 7.2.2, where it is shown, in a more general setting, that such a process is a Poisson point process with intensity measure

$$\nu([0,t] \times [0,y]) = \int_0^t F(y|z)\,d\Lambda(z). \tag{9.30}$$

One obtains a Poisson(λ, F) process as a special case if $\Lambda(t) = \lambda t$ and $F(\cdot|t) = F$.

An Exceedance Process as an Inhomogeneous Poisson Process

Given a Poisson(Λ, F) process, the marks exceeding the threshold u and the pertaining exceedance times form a Poisson$(\Lambda_u, F^{[u]})$ process, where

$$\Lambda_u(t) = \int_0^t (1 - F(u|s))\,d\Lambda(s) \tag{9.31}$$

is the mean value function of the exceedance times, and

$$F^{[u]}(w|t) = \big(F(w|t) - F(u|t)\big)/\big(1 - F(u|t)\big), \qquad w \geq u, \tag{9.32}$$

are the conditional dfs of the exceedances.

In terms of Poisson point processes, cf. end of this section, this can be formulated in the following manner: if ν is the intensity measure pertaining to the Poisson(Λ, F) process, then the truncation of ν outside of $[0, \infty) \times [u, \infty)$ is the intensity measure pertaining to the Poisson$(\Lambda_u, F^{[u]})$ process (of the truncated process).

Densities of Poisson Processes

For the specification of likelihood functions one requires densities of Poisson processes with domain S in the Euclidean d–space.

Let N_0 and N_1 be Poisson processes on S with finite intensity measures ν_0 and ν_1; thus, $\nu_0(S) < \infty$ and $\nu_1(S) < \infty$ (cf. end of this section). If ν_1 has the ν_0–density h, then $\mathcal{L}(N_1)$ has the $\mathcal{L}(N_0)$–density g with

$$g(\{y_i\}) = \Big(\prod_{i=1}^k h(y_i)\Big) \exp\big(\nu_0(S) - \nu_1(S)\big), \tag{9.33}$$

where k is the number of points y_i, see, e.g., [43], Theorem 3.1.1, for a general formulation.

Statistical Modeling of the Exceedance Process

We start with independent random variables Y_t with dfs $F(y|t)$ and densities $f(y|t)$ for $t = 1, \ldots, n$. We implicitly assume that the dfs depend on some unknown parameter.

The discrete time points $t = 1, \ldots, n$ are replaced by a homogeneous Poisson process on the interval $[0, n]$ with intensity $\lambda = 1$ (this is motivated by weak convergence results for empirical point processes). The random variables Y_t are regarded as marks at t. Combining the random time points and the marks on gets a two–dimensional Poisson process with intensity measure $\nu([0, s] \times [0, y]) = \int_0^s F(y|t) \, dt$ as in (9.30).

The exceedances above the threshold u and the pertaining exceedance times form a Poisson$(\Lambda_u, F^{[u]})$ process as specified in (9.31) and (9.32) with $\Lambda(t) = t$. The intensity measure of the Poisson$(\Lambda_u, F^{[u]})$ process is given by

$$\nu_u([0, s] \times [u, y]) = \int_0^s F^{[u]}(y|t) \, d\Lambda_u(t)$$

and, therefore, $\nu_u(S) = \int_0^n (1 - F(u|t)) \, dt$ where $S = [0, n] \times [u, \infty)$. One may check that the Lebesgue density of ν_u is

$$h_u(t, y) = f(y|t), \quad (t, y) \in [0, n] \times [u, \infty).$$

by changing the order of integration.

Such a model will be applied in Section 14.2 to compute the T–year flood level in flood frequency analysis. Next we also specify a likelihood function for such processes which enables, e.g., the computing of MLEs in an exceedance model under covariate information.

A Likelihood Function for the Exceedance Process

The likelihood function of the Poisson$(\Lambda_u, F^{[u]})$ process, as a function of the parameter which is suppressed in our notation, is given by

$$L(\{(y_{t_i}, t_i)\}|\cdot) \propto \left(\prod_{i=1}^k f(y_{t_i}|t_i) \right) \exp\left(- \int_0^n (1 - F(u|t)) \, dt \right), \tag{9.34}$$

where k is the number of exceedances y_{t_i}, and the t_i are the pertaining exceedance times[4].

---

[4]To make (9.33) applicable, one has to use a density with respect to a finite measure which is equivalent to the Lebesgue measure. This entails that a factor is included in the representation of the density which does not depend on the given parameters and is, therefore, negligible in (9.34).

The dfs $F(y|t)$ are merely specified for $t = 1, \ldots, n$ and, therefore, the integral $\int_0^n (1 - F(u|t))\, dt$ in (9.34) is replaced by $\sum_{t=1}^n (1 - F(u|t))$. Notice that the term $1 - F(u|t)$ represents the exceedance probability at time t. Also notice that the likelihood function L is merely evaluated for exceedance times $t_i \in \{1, \ldots, n\}$ in conjunction with the pertaining exceedances y_i.

An Application to Exceedances under Covariate Information

Corresponding to (8.34), where EV dfs are studied, assume that the conditional df of Y given $X = x$ is equal (or close to) a generalized Pareto (GP) df for values y above a higher threshold u. More precisely, let

$$P(Y \le y | X = x) = W_{\gamma, \mu(x), \sigma(x)}(y), \qquad y > u, \qquad (9.35)$$

where the location and scale parameters $\mu(x)$ and $\sigma(x)$ are given as in (8.35). We take

$$
\begin{aligned}
\mu(x) &= \mu_0 + \mu_1 x, \\
\sigma(x) &= \exp(\sigma_0 + \sigma_1 x).
\end{aligned}
\qquad (9.36)
$$

Also assume that the Y_t are conditionally independent by conditioning on the covariate vector X with

$$
\begin{aligned}
P(Y_t \le y | X = x) &= P(Y_t \le y | X_t = x_t) \\
&= W_{\gamma, \mu(x_t), \sigma(x_t)}(y), \qquad y > u, \qquad (9.37)
\end{aligned}
$$

cf. (8.15) to (8.17), where such a condition is deduced for iid random pairs (X_t, Y_t).

Next the ML procedure in (9.34) is applied to dfs

$$F(y|t) = W_{\gamma, \mu(x_t), \sigma(x_t)}(y), \qquad y > u.$$

Put $y = (y_{t_1}, \ldots, y_{t_k})$ and $x = (x_1, \ldots, x_n)$, where the y_{t_i} are the exceedances above the threshold u. As in Section 8.3, one obtains conditional MLEs $\hat{\gamma}_{x,y}$, $\hat{\mu}_{x,y}(x)$, $\hat{\sigma}_{x,y}(x)$ of the parameters γ, $\mu(x)$ and $\sigma(x)$.

Other estimation procedures should be made applicable as well. We remind on the estimation of a trend

- in the location parameter of EV dfs, see pages 113–116,

- in the scale parameter of GP dfs, see pages 139–141.

In this context one should also study moment or L–moment estimators. Such estimators may serve as initial values in the ML procedure.

For larger q, one gets an estimate of the covariate conditional q–quantile

$$q_{\gamma, \mu, \sigma}(Y | x) = W_{\gamma, \mu(x), \sigma(x)}^{-1}(q) \qquad (9.38)$$

given $X = x$ by means of $q_{\hat{\gamma}_{x,y}, \hat{\mu}_{x,y}, \hat{\sigma}_{x,y}}(Y | x)$.

Predicting the Conditional q–Quantile

We may as well define the covariate conditional q–quantile $q_{\gamma,\mu,\sigma}(Y|X)$. Plugging in the estimators of the unknown parameters as in (8.38) one gets a predictor of $q_{\gamma,\mu,\sigma}(Y|X)$.

A Poisson Process Modeling of Financial Data

The modeling of higher excesses of returns by means of Poisson processes under covariate information (explanatory variables) was used by Tsay[5] to investigate the effect of changes in U.S. daily interest rates on daily returns of the S&P 500 index, and a covariate conditional Value–at–Risk (VaR). We refer to Section 16.8, where a serial conditional VaR is studied within the framework of GARCH time series.

An application of the present Poisson process modeling under covariate information to environmental sciences may be found in Chapter 15.

The General Notion of a Poisson Process

In Section 9.1, we started with the notion of a homogeneous Poisson process, denoted by Poisson(λ), with intensity λ. By adding marks one gets a Poisson(λ, F) process. Using a transformation in the time scale by means of a mean value function Λ one gets the inhomgeneous Poisson process Poisson(Λ, F). We also employed the notion of an intensity measure which presents the mean number of points in a measurable set.

Generally, one may define a Poisson process for every finite or σ–finite measure ν on a measurable space S, see [43]. This measure is again called intensity measure. The Poisson process is denoted by Poisson(ν). In textbooks it is usually assumed that the underlying space S is Polish or a locally compact Hausdorff space with a countable base, yet these conditions are superfluous for the definition of a Poisson process.

---

[5]Tsay, R.S. (1999). Extreme value analysis of financial data. Working paper, Graduate School of Business, University of Chicago; also see Section 7.7 in Tsay, R.S. (2002). Analysis of Financial Time Series. Wiley, New Jersey.

Part III

Elements of Multivariate
Statistical Analysis

Chapter 10

Basic Multivariate Concepts and Visualization

This chapter provides a short introduction to several probabilistic and statistical concepts in the multivariate setting such as, e.g., dfs, contour plots, covariance matrices and densities (cf. Section 10.1), and the pertaining sample versions (cf. Section 10.2) which may be helpful for analysing data.

Decomposition procedures of multivariate distributions by means of univariate margins and multivariate dependence functions, such as copulas, are studied in Section 10.3.

10.1 An Introduction to Basic Multivariate Concepts

This section provides the multivariate versions of distribution functions (dfs), survivor functions and densities besides the notation.

Notation

Subsequently, operations and relations for vectors are understood componentwise. Given row vectors $\boldsymbol{a} = (a_1, \ldots, a_d)$ and $\boldsymbol{b} = (b_1, \ldots, b_d)$, let

$$\max\{\boldsymbol{a}, \boldsymbol{b}\} = (\max\{a_1, b_1\}, \ldots, \max\{a_d b_d\}),$$

$$\min\{\boldsymbol{a}, \boldsymbol{b}\} = (\min\{a_1, b_1\}, \ldots, \min\{a_d b_d\}),$$

$$\boldsymbol{a}\boldsymbol{b} = (a_1 b_1, \ldots, a_d b_d),$$

$$\boldsymbol{a} + \boldsymbol{b} = (a_1 + b_1, \ldots, a_d + b_d),$$

$$\boldsymbol{a}/\boldsymbol{b} = (a_1/b_1, \ldots, a_d/b_d), \quad \text{if } b_i \neq 0.$$

If a real value is added to or subtracted from a vector, then this is done componentwise. Likewise, let $\boldsymbol{a}/b = (a_1/b, \ldots, a_d/b)$ for a real value $b \neq 0$.

Furthermore, relations "$\leq$", "$<$", "$\geq$", "$>$", etc. hold for vectors if they are valid componentwise; e.g., we have $\boldsymbol{a} \leq \boldsymbol{b}$, if $a_j \leq b_j$ for $j = 1, \ldots, d$ or $\boldsymbol{a} < \boldsymbol{b}$, if $a_j < b_j$ for $j = 1, \ldots, d$.

These summations and relations can be extended to matrices in a straightforward manner. Occasionally, we also use the multiplication of matrices that is different from the multiplication of row vectors as mentioned above: for a $d_1 \times d_2$–matrix $A = (a_{i,j})$ and a $d_2 \times d_3$–matrix $B = (b_{j,k})$, we have $AB = \left(\sum_{j \leq d_2} a_{i,j} b_{j,k} \right)$.

Especially, $\boldsymbol{a}\boldsymbol{b}' = \sum_{j \leq d} a_j b_j$ for vectors $\boldsymbol{a}$ and $\boldsymbol{b}$, where $\boldsymbol{b}'$ is the transposed vector of $\boldsymbol{a}$, that is, $\boldsymbol{b}$ is written as a column vector. More generally, A' denotes the transposed of a matrix A.

Multivariate Distribution and Survivor Functions

The reader should be familiar with the notions of a d–variate df, a d–variate density and the concept of iid random vectors $\boldsymbol{X}_1, \ldots, \boldsymbol{X}_m$ with common df F.

The df of a random vector $\boldsymbol{X} = (X_1, \ldots, X_d)$ is

$$F(\boldsymbol{x}) = P\{\boldsymbol{X} \leq \boldsymbol{x}\} = P\{X_1 \leq x_1, \ldots, X_d \leq x_d\}$$

for $\boldsymbol{x} = (x_1, \ldots, x_d)$. If $X_1, \ldots, X_d$ are independent, then

$$F(\boldsymbol{x}) = \prod_{j \leq d} F_{(j)}(x_j), \tag{10.1}$$

where $F_{(j)}$ is the df of X_j. If $X_1, \ldots, X_d$ are identically distributed and totally dependent—hence, the X_i are equal to X_1 with probability one—then

$$\begin{aligned} F(\boldsymbol{x}) &= P\{X_1 \leq x_1, \ldots, X_1 \leq x_d\} \\ &= F_{(1)}(\min\{x_1, \ldots, x_d\}). \end{aligned} \tag{10.2}$$

The d–variate survivor function corresponding to the df F is given by

$$\bar{F}(\boldsymbol{x}) = P\{\boldsymbol{X} > \boldsymbol{x}\} = P\{X_1 > x_1, \ldots, X_d > x_d\}.$$

Survivor functions will be of importance in multivariate extreme value analysis (as well as in the univariate setting, where a simpler representation of results for minima was obtained by survivor functions).

Mean Vector and Covariance Matrix

For characterizing the df of a random vector $\boldsymbol{X} = (X_1, \ldots, X_d)$, the mean vector

$$\boldsymbol{m} = (m_1, \ldots, m_d) = (EX_1, \ldots, EX_d)$$

and the covariance matrix $\Sigma = (\sigma_{j,k})$ are important, where the covariances are given by

$$\sigma_{j,k} = \mathrm{Cov}(X_j, X_k) = E\big((X_j - m_j)(X_k - m_k)\big),$$

whenever the covariance exists (cf. (2.52)). Notice that $\sigma_{j,j} = \sigma_j^2$ is the variance of X_j. One gets the correlation coefficients by

$$\rho_{j,k} = \sigma_{j,k}/(\sigma_j \sigma_k).$$

Note that the correlation coefficient $\rho_{j,k}$ ranges between -1 and 1. Random variables X_j and X_k are uncorrelated if $\rho_{j,k} = 0$. This condition holds if X_j and X_k are independent and the second moments are finite. If $X_j = X_k$ or $X_j = -X_k$ with probability 1, then $\rho_{j,k} = 1$ or $\rho_{j,k} = -1$.

Kendall's τ

For characterizing the dependence structure of a random vector $\boldsymbol{X} = (X_1, \ldots, X_d)$ one may also use Kendall's τ in place of the covariance. One of the advantages of Kendall's τ is that the second moments need not be finite.

Let $\boldsymbol{Y} = (Y_1, \ldots, Y_d)$ be another random vector which is distributional equal to $\boldsymbol{X}$, and let $\boldsymbol{X}$ and $\boldsymbol{Y}$ be independent. Then, define

$$\tau_{j,k} = P\{(Y_j - X_j)(Y_k - X_k) > 0\} - P\{(Y_j - X_j)(Y_k - X_k) < 0\}. \qquad (10.3)$$

Marginal Distribution and Survivor Functions

In (10.1) and (10.2), we already employed the jth margin df $F_{(j)}$ of a d–variate df F. Generally, let

$$F_K(\boldsymbol{x}) = P\{X_k \le x_k,\ k \in K\}$$

for any set K of indices between 1 and d. Thus, we have $F_{(j)}(x_j) = F_{\{j\}}(\boldsymbol{x})$. Apparently, F_K can be deduced from F by letting x_j tend to infinity for all indices j not belonging to K. Likewise, the margins $\bar{F}_K$ of a survivor function $\bar{F}$ are defined by

$$\bar{F}_K(\boldsymbol{x}) = P\{X_k > x_k,\ k \in K\}.$$

One reason for the importance of survivor functions is the following representation of a d–variate df F in terms of marginal survivor functions $\bar{F}_K$. We have

$$1 - F(\boldsymbol{x}) = \sum_{j \le d} (-1)^{j+1} \sum_{|K|=j} \bar{F}_K(\boldsymbol{x}). \qquad (10.4)$$

Likewise, we have

$$1 - \bar{F}(\boldsymbol{x}) = \sum_{j \le d} (-1)^{j+1} \sum_{|K|=j} F_K(\boldsymbol{x}). \qquad (10.5)$$

To verify (10.4) and (10.5), one must apply an additivity formula (see, e.g., (4.4) in [16]). Finally, we remark that F is continuous if the univariate margins $F_{(j)}$ are continuous. Deduce from (10.4) and (10.5) that

$$F(x_1, x_2) = F_1(x_1) + F_2(x_2) + \bar{F}(x_1, x_2) - 1 \tag{10.6}$$

and

$$\bar{F}(x_1, x_2) = \bar{F}_1(x_1) + \bar{F}_2(x_2) + F(x_1, x_2) - 1 \tag{10.7}$$

for bivariate dfs F with margins F_1 and F_2.

Multivariate Densities

If a df F has the representation

$$F(\boldsymbol{x}) = \int_{-\infty}^{\boldsymbol{x}} f(\boldsymbol{x})\, d\boldsymbol{x} = \int_{-\infty}^{x_d} \cdots \int_{-\infty}^{x_1} f(x_1, \ldots, x_d)\, dx_1 \cdots dx_d, \tag{10.8}$$

where f is nonnegative, then f is a (probability) density of F. A necessary and sufficient condition that a nonnegative f is the density of a df is

$$\int f(\boldsymbol{x})\, d\boldsymbol{x} := \int_{-\infty}^{\infty} f(\boldsymbol{x})\, d\boldsymbol{x} = 1. \tag{10.9}$$

If the random variables $X_1, \ldots, X_d$ are independent again and X_i possesses a density f_i, we see that

$$f(\boldsymbol{x}) = \prod_{i \leq d} f_i(x_i) \tag{10.10}$$

is a joint density of the $X_1, \ldots, X_d$.

Under certain conditions, a density can also be constructed from a given df F by taking partial derivatives[1]: if the d–fold partial derivative

$$f(\boldsymbol{x}) = \frac{\partial^d}{\partial x_1 \cdots \partial x_d} F(\boldsymbol{x}) \tag{10.11}$$

is continuous, then f is a density of F. It is sufficient that the given condition holds in an open rectangle $(\boldsymbol{a}, \boldsymbol{b}) = \prod_{j \leq d}(a_j, b_j)$—or, generally, an open set U— possessing the probability 1. Then, put $f = 0$ outside of $(\boldsymbol{a}, \boldsymbol{b})$ or U.

If the df F has a density f, then the survivor function $\bar{F}$ can be written

$$\bar{F}(\boldsymbol{x}) = \int_{\boldsymbol{x}}^{\infty} f(\boldsymbol{x})\, d\boldsymbol{x} = \int_{x_d}^{\infty} \cdots \int_{x_1}^{\infty} f(x_1, \ldots, x_d)\, dx_1 \cdots dx_d. \tag{10.12}$$

From this representation, as well as from (10.5), one realizes that the density f can also be deduced from the survivor function by taking the d–fold partial derivative.

---

[1] See Theorem A.2.2 in Bhattacharya, R.N. and Rao, R.R. (1976). Normal Approximation and Asymptotic Expansions. Wiley, New York.

Contour Plots of Surfaces, Bivariate Quantiles

Another indispensable tool for the visualization of a bivariate function f is the contour plot which consists of lines $\{(x, y) : f(x, y) = q\}$ for certain values q.

The illustration in Fig. 10.1 concerns a bivariate normal density which will be introduced in Section 11.1.

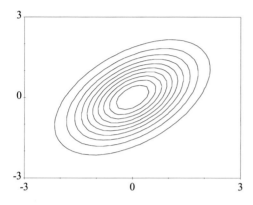

FIG. 10.1. Contour plot of the bivariate normal density with correlation coefficient $\rho = 0.5$.

For a bivariate df F, the line $\{(x, y) : F(x, y) = q\}$ may be addressed as the q–quantile of F. Therefore, a contour plot of a df F displays a certain collection of quantiles. Likewise, one may deal with upper p–quantiles $\{(x, y) : \bar{F}(x, y) = p\}$ pertaining to the survivor function.

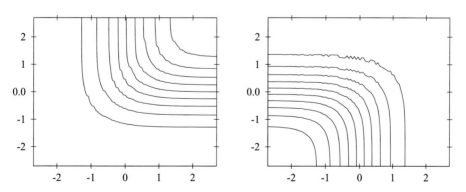

FIG. 10.2. Contour plots of the bivariate normal distribution with correlation coefficient $\rho = 0.5$. (left.) Quantiles of df for $q = i/10$ with $i = 1, \ldots, 9$. (right.) Quantiles of survivor function for $p = i/10$ with $i = 1, \ldots, 9$.

10.2   Visualizing Multivariate Data

In this section we introduce the sample versions of multivariate dfs, covariances, densities and also contour plots. Moreover, we display different versions of scatter-plots for trivariate data. Remarks about handling missing components in a vector of data are added.

Multivariate Sample Distribution and Survivor Functions

For d–variate data $\boldsymbol{x}_1, \ldots, \boldsymbol{x}_n$, the sample df is

$$F_n(\boldsymbol{x}) = \frac{1}{n} \sum_{i \leq n} I(\boldsymbol{x}_i \leq \boldsymbol{x}). \tag{10.13}$$

If $\boldsymbol{x}_1, \ldots, \boldsymbol{x}_n$ are governed by a d–variate df F, then the sample df $F_n(\boldsymbol{x})$ provides an estimate of $F(\boldsymbol{x})$ as in the univariate case. An illustration of a bivariate df may be found in Fig. 12.1.

A related remark holds for the sample survivor function which is defined by

$$\bar{F}_n(\boldsymbol{x}) = \frac{1}{n} \sum_{i \leq n} I(\boldsymbol{x}_i > \boldsymbol{x}). \tag{10.14}$$

The 2–dimensional sample df and survivor function must be plotted in a 3–D plot. A simplified representation is achieved by using sample contour plots as dealt with below.

Sample Mean Vector, Sample Covariances

The sample versions of the mean vector $\boldsymbol{m}$ and the covariance matrix $(\sigma_{j,k})$—based on d–variate data $\boldsymbol{x}_1, \ldots, \boldsymbol{x}_n$ with $\boldsymbol{x}_i = (x_{i,1}, \ldots, x_{i,d})$—are

- the vector $\boldsymbol{m}_n$ of the componentwise taken sample means

$$m_{j,n} = \frac{1}{n} \sum_{i \leq n} x_{i,j}, \tag{10.15}$$

- the matrix of sample covariances (cf. also (2.53))

$$s_{j,k,n} = \frac{1}{n-1} \sum_{i \leq n} (x_{i,j} - m_{j,n})(x_{i,k} - m_{k,n}). \tag{10.16}$$

For the sample variances we also write $s_{j,n}^2$ in place of $s_{j,j,n}$. In addition, the sample correlation correlations

$$\rho_{j,k,n} = s_{j,k,n} / (s_{j,j,n} s_{k,k,n})^{1/2}. \tag{10.17}$$

Bivariate Sample Contour Plots

The contour plots of sample df F_n and sample survivor function $\bar{F}_n$ are estimates of the contour plots of the underlying df F and survivor function $\bar{F}$. Thus, in the case of the sample survivor function one is computing lines of point t which admit an approximately equal number of points $x_i > t$.

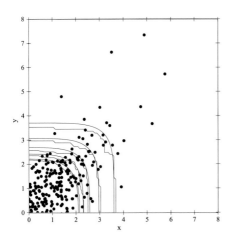

FIG. 10.3. 2–dimensional data and a contour plot of the sample survivor function.

As in the continuous case, one is evaluating the surface of the sample functions on a rectangular grid, and the approximate contours are computed by interpolation. For the details we refer to the book by Cleveland [8], pages 242–241.

Multivariate Kernel Density

Let k be a univariate kernel as presented in Section 2.1. Using such univariate kernels as factors one gets a d–variate kernel by $u(\boldsymbol{x}) = \prod_{j \leq d} k(x_j)$. In analogy to (2.14), define a d–variate kernel density $f_{n,b}$ at $\boldsymbol{x}$ by

$$f_{n,b}(\boldsymbol{x}) = \frac{1}{nb^d} \sum_{i \leq n} \prod_{j \leq d} k\left(\frac{x_j - y_{i,j}}{b}\right) \tag{10.18}$$

based on d–variate data $\boldsymbol{y}_i = (y_{i,1}, \ldots, y_{i,d})$, where $b > 0$ is a smoothing parameter. Note that (10.18) can be written

$$f_{n,b}(\boldsymbol{x}) = \frac{1}{nb^d} \sum_{i \leq n} u\left(\frac{\boldsymbol{x} - \boldsymbol{y}_i}{b}\right). \tag{10.19}$$

In this formula, the special kernel

$$u(\boldsymbol{x}) = \prod_{j \leq d} k(x_j)$$

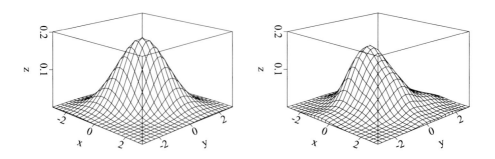

FIG. 10.4. (left.) 3–D plot of bivariate normal density with correlation coefficient $\rho = 0.5$. (right.) 3–D plot of kernel density based on 100 bivariate normal data with the bandwidth $b = .5$ (based on a bivariate normal kernel).

can be replaced by any d–variate probability density.

Modifications of this construction are necessary if the data exhibit a certain direction. Then, it is useful to consider a kernel u that mimics the shape of the data set. This question can also be dealt with in conjunction with data sphering and principle component analysis[2].

Bounded Kernel Density

Subsequently, we study bivariate data $\boldsymbol{y}_i = (y_{i,1}, y_{i,2})$ for $i = 1, \ldots, n$. As in the univariate case, we try to prevent a smoothing of the data beyond certain boundaries. One may assume that the first or second component is left or right–bounded.

First, let us assume that the first component is left bounded by a_1, that is, $y_{i,1} \geq a_1$ for $i = 1, \ldots, n$. For this case, a plausible kernel density may be constructed in the following manner:

1. reflect the $y_{i,1}$ in a_1 obtaining values $y'_{i,1} = 2a_1 - y_{i,1}$;

2. apply the preceding kernel density $f_{n,b}$ to the new enlarged data set consisting of $(y_{i,1}, y_{i,2})$ and $(y'_{i,1}, y_{i,2})$ for $i = 1, \ldots, n$,

3. restrict the resulting curve to $[a_1, \infty) \times (-\infty, \infty)$.

If the first component is also right–bounded by $a_2 > a_1$, reflect the y_{i1} also in a_2 and proceed as before. Of course, the order of these two steps can be interchanged. If this is necessary, continue this procedure in the second component, whereby one must start with the enlarged data set.

---

[2]Falk, M., Becker, R. and Marohn, F. (2002). Foundations of Statistical Analyses and Applications with SAS. Birkhäuser, Basel.

The 3–D Scatterplot

As in the 2–dimensional case the 3–dimensional scatterplots provide an indispensable tool to get a first impression of a data set.

We include two versions of a 3–dimensional scatterplot.

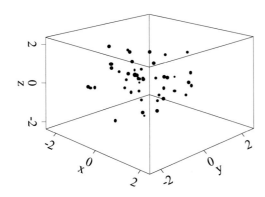

FIG. 10.5. A scatterplot of 50 trivariate normal data.

The facility to rotate such a scatterplot around the z–axis makes it particularly valuable.

Next, the 3–D scatterplot is once more applied to trivariate data where, however, the first and second components describe the site of a spatial measurement.

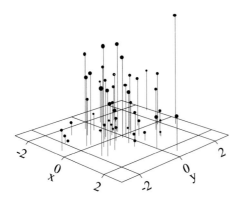

FIG. 10.6. A scatterplot of sites and measurements.

Contourplots, surface plots and scatterplots are available within the multivariate part of the menu system of Xtremes.

Missing Data

In conjunction with censored data, the question of handling missing data was solved by defining appropriate statistical models and by applying model–based procedures.

Subsequently, we deal with the situation where one or several components of an observation vector are missing (assuming that this happens in an irregular manner). Simply discarding the incompletely recorded observation vector will entail the loss of information, a consequence which should be avoided particularly for samples of smaller sizes. The proposed filling–in procedure for missing values is a nearest–neighbor method that is related to the stochastic regression imputation discussed in the book by Little and Rubin[3].

For simplicity, let us assume that $\boldsymbol{x}_l = (x_{l,1}, \ldots, x_{l,d-1}, \cdot)$ is a vector such that only the dth component is missing. Put $\boldsymbol{x}_l^{(d)} = (x_{l,1}, \ldots, x_{l,d-1})$. Fix a number k. From those vectors without a missing component, select $\boldsymbol{x}_{i_1}, \ldots, \boldsymbol{x}_{i_k}$ such that $\boldsymbol{x}_{i_1}^{(d)}, \ldots, \boldsymbol{x}_{i_k}^{(d)}$ are the k nearest neighbors of $\boldsymbol{x}_l^{(d)}$. Finally, a dth component $x_{l,d}$ is added to $\boldsymbol{x}_l$ which is randomly chosen from the dth components $x_{i_1,d}, \ldots, x_{i_k,d}$. Notice that this is a sampling procedure according to a conditional sample df (cf. [16], pages 96–98). Likewise, missing values in other components are filled in and the completed data are analyzed by standard methods.

EXAMPLE 10.2.1. (Annual Wind–Speed Maxima: Continuation of Examples 1.2.1 and 2.1.3.) We deal with the annual wind–speed maxima (in km/hr) based on hourly measurements at five Canadian stations (located at Vancouver, Regina, Toronto, Montreal and Shearwater) from 1947 to 1984. The wind–speeds at Shearwater from 1947 to 1949 are not recorded. Thus, if the incompletely recorded vectors are discarded, a sample of size 35 remains.

The following table only contains the data for the years 1947 to 1951. The missing values were generated by means of the nearest–neighborhood algorithm with $k = 5$.

TABLE 10.1. Annual wind–speed maxima from 1947 to 1951 with filled–in values.

	Maximum wind speed (km/h)				
Year	Vancouver	Regina	Toronto	Montreal	Shearwater
1947	79.5	88.8	79.5	120.2	(85.1)
1948	68.4	74.0	75.8	74.0	(79.5)
1949	74.0	79.5	88.8	64.8	(68.4)
1950	59.2	74.0	96.2	77.7	64.8
1951	74.0	79.5	72.1	61.0	79.5

---

[3]Little, R.J.A. and Rubin, D.B. (1987). Statistical Analysis with Missing Data. Wiley, New York.

10.3 Decompositions of Multivariate Distributions

Let $\boldsymbol{X} = (X_1, \ldots, X_d)$ be a random vector with df F. In this section we describe some procedures in which manner the df F can be decomposed into certain univariate dfs and a multivariate df where the latter df represents the dependence structure.

Thereby, one may separately analyze and estimate certain univariate components and the multivariate dependence structure. The piecing together of the different estimates yields an estimate of F.

Copulas

Assume that the univariate marginal dfs F_j are continuous (which implies that F is continuous). By applying the probability transformation as introduced on page 38, the random variables X_j can be transformed to $(0,1)$–uniform random variables. The df C of the vector $\boldsymbol{Y}$ of transformed random variables $Y_j = F_j(X_j)$ is called the copula pertaining to F. We have,

$$
\begin{aligned}
C(\boldsymbol{u}) &= P\{Y_1 \le u_1, \ldots, Y_d \le u_d\} \\
&= F\big(F_1^{-1}(u_1), \ldots, F_d^{-1}(u_d)\big),
\end{aligned}
\tag{10.20}
$$

where F_j^{-1} is the qf of F_j. In addition, $\boldsymbol{Y}$ may be addressed as copula random vector pertaining to $\boldsymbol{X}$.

Conversely, one may restore the original df F from the copula by applying the quantile transformation. We find

$$
F(\boldsymbol{x}) = C\big(F_1(x_1), \ldots, F_d(x_d)\big),
\tag{10.21}
$$

which is the desired decomposition of F into the copula C and the univariate margins F_j.

Likewise, the original univariate margins can be replaced by some other dfs. Thus, we may design a multivariate df with a given multivariate dependence structure (as, e.g., a copula) and predetermined univariate margins.

For a shorter introduction to various families of copula functions we refer to a paper by Joe[4]. There is a controversial discussion about the usefulness of copulas: we refer to Mikosch[5] and the attached discussion, as, e.g., the contribution by Genest and Rémillard[6].

---

[4]Joe, H. (1993). Parametric families of multivariate distributions with given marginals. J. Mult. Analysis 46, 262–282.

[5]Mikosch, T. (2006). Copulas: tales and facts. Extremes 9, 3–20.

[6]Genest, C. and Rémillard, B. (2006). Discussion of "Copulas: tales and facts", by Thomas Mikosch. Extremes 9, 27–36.

Data–Based Transformations, Empirical Copulas

In application one may carry out the transformations with estimated parametric univariate and multivariate dfs and and piece the univariate and multivariate components together, for details see page 389.

In the following example, the copula method is employed with estimated Gaussian copula and Student margins.

EXAMPLE 10.3.1. (Log-Returns Deutsche Bank against Commerzbank: Continuation of Example 9.4.1.) The scatterplot in Figure 10.7 again consists of the log-returns (with conversed signs) of the Deutsche Bank plotted against those of the Commerzbank on trading days from Jan. 1, 1992 to Jan. 11, 2002.

In addition, the contour lines of the estimated density of a Gaussian copula and univariate Student margins are plotted.

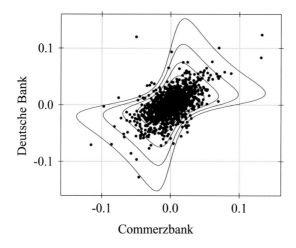

FIG. 10.7. Scatterplot of log-returns (with changed signs) Deutsche Bank against Commerzbank and contour plot of estimated density with Gaussian copula and univariate Student margins.

The exotic "butterfly" contour plot does not fit to the scatterplot which has an elliptical shape. We conclude that the modeling by means of a Gaussian copula and univariate Student margins is not correct.

A related illustration in conjunction with a scatterplot of daily differences of zero–bond–discount–factors may be found in Wiedemann et al., page 75[7]. These examples show that the copula method should be employed with utmost care. We are not aware whether there are general rules under which the copula method is always appropriate.

---

[7]Wiedemann, A., Drosdzol, A., Reiss, R.–D. and Thomas, M. (2005). Statistische Modellierung des Zinsänderungsrisikos. Teil 2: Multivariate Verteilungen. In: Modernes Risikomanagement (F. Romeike ed.). Wiley-VCH, Weinheim.

The transformations may also be based on kernel estimates of the dfs, see (2.17) and (10.19). In this case, the data are still transformed by means of continuous dfs.

In Section 13.3 we make use of univariate transformations of a different type, namely those by means of univariate sample dfs $\widehat{F}_n$ based on the sample vector $\boldsymbol{x} = (x_1, \ldots, x_n)$, see (2.1). Thus, the original data x_i in the margins are replaced by $y_i = \widehat{F}_n(x_i)$. Recall that the number of $x_j \leq x_i$ is the rank of x_i, denoted by $r_i(\boldsymbol{x})$, see, e.g., Hájek and Šidak, page 36[8]. The value $r_i(\boldsymbol{x})$ is the rank of x_i in the ordered sample $x_{1:n} < \ldots < x_{n:n}$. Therefore, the transformed data are relative ranks; we have

$$y_i = r_i(\boldsymbol{x})/n. \tag{10.22}$$

In that context one loosely speaks of empirical copulas. Concerning stochastical properties of order statistics and rank statistics we refer to the book by Hájek and Šidak.

Other Decompositions

The decomposition of the joint distribution of random variables X and Y in the marginal distribution of X and the conditional distribution of Y given $X = x$ was the topic of Section 8.1. Of course, this is the most important decomposition methodology for a random vector.

Notable is also the well–known decomposition of spherical random vectors in two independent random variables, namely the radial and angular components, see Section 11.2.

In conjunction with extreme values, we are interested in standard forms of multivariate distributions different from copulas, namely,

- the bivariate generalized Pareto (GP) dfs in their standard form, cf. Section 13.1, are closely related to copulas in so far that the margins are always equal to the uniform df on the interval $(-1, 0)$ which is the univariate GP df $W_{2,-1}$, thus, there is the slight modification of a shifted uniform df;

- the bivariate extreme value (EV) dfs in Section 12.1 are described by the Pickands dependence function and univariate margins equal to the reversed exponential df on the negative half–line which is the EV df $G_{2,-1}$ (by the way, other authors prefer the Frechét df $G_{1,1}$ as univariate margin);

- representations of a different type are employed in Section 12.4 where a bivariate dfs H on the left, lower quadrant $(-\infty, 0]^2$ is represented by a certain spectral decomposition. For that purpose one considers univariate dfs of the form

$$H_z(c) = H(cz, c(1 - z))$$

---

[8]Hájek, J. and Šidak, Z. (1967). Theory of Rank Tests. Academic Press, New York.

in the variable c for each z, where z and c may be regarded as the angular and, respectively, radial component.

If H is an EV df or, respectively, a GP df, then H_z is a reversed exponential df or, respectively, a uniform df depending on z,

- in analogy to the above mentioned decomposition for spherical random vectors, bivariate GP random vectors can be decomposed in independent radial and angular components, see Section 12.4.

Chapter 11

Elliptical and Related Distributions

In Section 11.1 we first recall some well–known, basic facts about multivariate Gaussian and log–normal models. The Gaussian distributions will serve as examples of spherical and elliptical distributions in Section 11.2. An introduction to multivariate Student and multivariate sum–stable distributions is separately treated in the Sections 11.3 and 11.4.

Thus, we deal with multivariate distributions, where the univariate margins have light, fat and heavy tails. Recall that Student distributions have Pareto–type tails with $\alpha > 0$, whereas the tail index of the non–Gaussian, sum–stable distributions is restricted to $\alpha < 2$.

11.1 Multivariate Gaussian Models

Corresponding to the univariate case, parametric models are built by starting with certain standard dfs and, subsequently, adding location and scale parameters in the single components. As examples we first discuss the multivariate normal and log–normal models.

In Section 12.1, the multivariate normal model will be converted into a multivariate EV model with univariate Gumbel margins.

Location and Scale Parameter Vectors

Let $\boldsymbol{X} = (X_1, \ldots, X_d)$ be a random vector with df F. Location and scale parameter vectors can be added to this df by considering random variables $Y_j = \mu_j + \sigma_j X_j$ for $j = 1, \ldots, d$. The joint df of the Y_j is given by

$$
\begin{aligned}
F_{\boldsymbol{\mu}, \boldsymbol{\sigma}}(\boldsymbol{x}) &= P\{X_1 \leq (x_1 - \mu_1)/\sigma_1, \ldots, X_d \leq (x_d - \mu_d)/\sigma_d\} \\
&= F\left(\frac{\boldsymbol{x} - \boldsymbol{\mu}}{\boldsymbol{\sigma}}\right),
\end{aligned}
\tag{11.1}
$$

where $\boldsymbol{\mu} = (\mu_1, \ldots, \mu_d)$ and $\boldsymbol{\sigma} = (\sigma_1, \ldots, \sigma_d)$ denote the location and scale vectors.

Multivariate Normal Distributions

The d–variate normal distribution serves as a first example of a multivariate distribution. Let $\boldsymbol{X} = (X_1, \ldots, X_d)$ be a vector of iid standard normal random variables. Linear changes of $\boldsymbol{X}$ are again normal vectors. If A is a $d \times d$–matrix $(a_{i,j})$ with determinant $\det A \neq 0$, then

$$\boldsymbol{Y}' = A\boldsymbol{X}' = \left(\sum_{j \leq d} a_{1,j} X_j, \ldots, \sum_{j \leq d} a_{d,j} X_j \right)' \tag{11.2}$$

is a d–variate normal vector with mean vector zero and covariance matrix $\boldsymbol{\Sigma} = (\sigma_{i,j}) = AA'$. The density of $\boldsymbol{Y}$ is given by

$$\varphi_{\boldsymbol{\Sigma}}(\boldsymbol{y}) = \frac{1}{(2\pi)^{d/2} \det(\boldsymbol{\Sigma})^{1/2}} \exp\left(-\frac{1}{2} \boldsymbol{y} \boldsymbol{\Sigma}^{-1} \boldsymbol{y}' \right), \tag{11.3}$$

where $\boldsymbol{\Sigma}^{-1}$ denotes the inverse of $\boldsymbol{\Sigma}$. Let

$$\Phi_{\boldsymbol{\Sigma}}(\boldsymbol{x}) = \int_{-\infty}^{\boldsymbol{x}} \varphi_{\boldsymbol{\Sigma}}(\boldsymbol{y}) \, d\boldsymbol{y} \tag{11.4}$$

be the df of $\boldsymbol{Y}$. We write $\varphi_{\boldsymbol{\mu}, \boldsymbol{\Sigma}}$ and $\Phi_{\boldsymbol{\mu}, \boldsymbol{\Sigma}}$ when a location vector $\boldsymbol{\mu}$ is added.

The density $\varphi_{\boldsymbol{\Sigma}}$ can be deduced from the density $\varphi_{\boldsymbol{I}}$ of the original normal vector $\boldsymbol{X}$ by applying the transformation theorem for densities (hereby, $\boldsymbol{I}$ denotes the unit matrix with the elements equal to one in the main diagonal and the elements are equal to zero otherwise). Moreover, marginal vectors $(Y_{j_1}, \ldots, Y_{j_k})$ with $1 \leq j_1 < j_2 < \cdots < j_k \leq d$ are normal again.

Notice that (11.2) is a special case of the following well–known result: if $\boldsymbol{Y}$ is a d–variate normal vector with mean vector $\boldsymbol{\mu}$ and covariance matrix $\boldsymbol{\Sigma}$ and A is a $k \times d$–matrix (with k rows and d columns and $k \leq d$) having rank k, then $\boldsymbol{Z}' = A\boldsymbol{Y}'$ is a k–variate normal vector with mean vector $A\boldsymbol{\mu}'$ and covariance matrix $A\boldsymbol{\Sigma}A'$.

Statistical Inference in the Multivariate Gaussian Model

The d–variate normal model is

$$\left\{ \Phi_{\boldsymbol{\mu}, \boldsymbol{\Sigma}} : \boldsymbol{\mu} = (\mu_1, \ldots, \mu_d), \ \boldsymbol{\Sigma} = (\sigma_{i,j}) \right\}, \tag{11.5}$$

where $\boldsymbol{\mu}$ is a location (mean) vector and $\boldsymbol{\Sigma}$ is the covariance matrix. Estimators are easily obtained by the sample mean vector $\boldsymbol{\mu}_n$ and the sample covariance matrix $\boldsymbol{\Sigma}_n$, cf. (10.15) and (10.16).

It is well known that the sample mean vector $\boldsymbol{\mu}_n$ and the sample covariance matrix $\boldsymbol{\Sigma}_n$ (with the factor $1/(n-1)$ replaced by $1/n$) are also the MLEs in the d–variate normal model.

Assume that $(Y_1, \ldots, Y_d)$ is a random normal vector with mean vector $\boldsymbol{\mu} = (\mu_1, \ldots, \mu_d)$ and covariance matrix $\boldsymbol{\Sigma} = (\sigma_{i,j})$. Then Y_j has the location and scale parameters μ_j and $\sigma_j = \sigma_{j,j}^{1/2}$. The correlation matrix is $(\rho_{i,j}) = (\sigma_{i,j}/(\sigma_i \sigma_j))$. Therefore, a normal distribution may be alternatively represented by the the correlation matrix $(\rho_{i,j})$ and the location and scale vectors $\boldsymbol{\mu} = (\mu_1, \ldots, \mu_d)$ and $\boldsymbol{\sigma} = (\sigma_1, \ldots, \sigma_d)$.

A Useful Result Concerning the Asymptotic Normality

The following result, which was applied in (4.4), is taken from [48], pages 118 and 122. Let

- $b_n^{-1}(\boldsymbol{X}_n - \boldsymbol{\mu})$ be distributed asymptotically according to $\Phi_{\boldsymbol{\Sigma}}$ (that is the pointwise convergence of the dfs) where $b_n \to 0$ as $n \to \infty$,

- g_i be real–valued functions with partial derivatives $\partial g_i / \partial x_j \neq 0$ at $\boldsymbol{\mu}$ for $i = 1, \ldots, m$ and $j = 1, \ldots, d$.

Let $g = (g_1, \ldots, g_m)$ and let $\boldsymbol{D}$ be the matrix of partial derivatives evaluated at $\boldsymbol{\mu}$. Then,

$$b_n^{-1}(g(\boldsymbol{X}_n) - g(\boldsymbol{\mu})) \tag{11.6}$$

is distributed asymptotically according to $\Phi_{\boldsymbol{D\Sigma D'}}$.

Multivariate Log–Normal Distributions

If the positive data $\boldsymbol{x}_1, \ldots, \boldsymbol{x}_n$ indicate a heavier upper tail than that of a multivariate normal distribution $\Phi_{\boldsymbol{\mu}, \boldsymbol{\Sigma}}$, then one may try to fit a multivariate log–normal df

$$F_{(\boldsymbol{\mu}, \boldsymbol{\Sigma})}(\boldsymbol{x}) = \Phi_{\boldsymbol{\mu}, \boldsymbol{\Sigma}}(\log(x_1), \ldots, \log(x_d)), \qquad \boldsymbol{x} > 0. \tag{11.7}$$

The values $\exp(\mu_i)$ are scale parameters and $\boldsymbol{\Sigma}$ is a matrix of shape parameters of the multivariate log–normal distribution. It is apparent that the univariate margins are univariate log–normal distributions as set forth in Section 1.6.

11.2 Spherical and Elliptical Distributions

First, we study multivariate normal distributions from the viewpoint of spherical and elliptical distributions.

Multivariate Normal Distributions, Revisited

We describe multivariate normal distributions as spherical and elliptical ones. Let again $\boldsymbol{X} = (X_1, \ldots, X_d)$ be a vector of iid standard normal random variables as in Section 11.3. The pertaining density is

$$\varphi_{\boldsymbol{I}}(\boldsymbol{x}) \;=\; \prod_{j \leq d} \varphi(x_j) \;=\; \frac{1}{(2\pi)^{d/2}} \exp\Big(-\frac{1}{2} \boldsymbol{x}\boldsymbol{x}' \Big), \tag{11.8}$$

where $\boldsymbol{I}$ is again the unit matrix.

Notice that $|\boldsymbol{x}| = \sqrt{\boldsymbol{x}\boldsymbol{x}'}$ is the Euclidean norm of the vector $\boldsymbol{x}$. It is apparent that the contour lines $\{\boldsymbol{x} : \varphi_{\boldsymbol{I}}(\boldsymbol{x}) = q\}$ are spheres. Therefore, one speaks of a spherical distribution. A random vector $\boldsymbol{X}$ with density $\varphi_{\boldsymbol{I}}$ has the following property: we have

$$\boldsymbol{A}\boldsymbol{X} \stackrel{d}{=} \boldsymbol{X}, \tag{11.9}$$

for any orthogonal matrix A; that is, for matrices $\boldsymbol{A}$ with $\boldsymbol{A}\boldsymbol{A}' = \boldsymbol{I}$. This can be analytically verified by means of equation (11.3) because $\boldsymbol{\Sigma}^{-1} = (\boldsymbol{A}\boldsymbol{A}')^{-1} = \boldsymbol{I}$ for orthogonal matrices $\boldsymbol{A}$. (11.9) is also evident from the fact that an orthogonal matrix causes the rotation of a distribution.

More generally, if $\boldsymbol{A}$ is a $d \times d$–matrix with $\det \boldsymbol{A} \neq 0$, then $\boldsymbol{A}\boldsymbol{X}$ is a normal vector with covariance matrix $\boldsymbol{\Sigma} = \boldsymbol{A}\boldsymbol{A}'$ and density $\varphi_{\boldsymbol{\Sigma}}$, cf. again (11.3). Now, the contour line $\{\boldsymbol{y} : \varphi_{\boldsymbol{\Sigma}}(\boldsymbol{y}) = q\}$ is an ellipse for every $q > 0$, and one speaks of an elliptical distribution.

Spherical and Elliptical Distributions

Generally, one speaks of a spherical random vector $\boldsymbol{X}$ if (11.9) is satisfied for any orthogonal matrix $\boldsymbol{A}$. Moreover, an elliptical random vector is the linear change of a spherical random vector under a matrix $\boldsymbol{A}$ with $\det \boldsymbol{A} \neq 0$.

In conjunction with multivariate Student and multivariate sum–stable distributions we will make use of the following construction: if $\boldsymbol{X}$ is a spherical random vector and Y is real–valued then $\boldsymbol{X}Y$ is again a spherical random vector because

$$\boldsymbol{A}(\boldsymbol{X}Y) = (\boldsymbol{A}\boldsymbol{X})Y \stackrel{d}{=} \boldsymbol{X}Y, \tag{11.10}$$

for any orthogonal matrix $\boldsymbol{A}$.

A Decomposition of Spherical Distributions

It is well known, see, e.g., [19], that a spherical random vector $\boldsymbol{X}$ can be written as

$$\boldsymbol{X} = R\boldsymbol{U},$$

where $R \geq 0$ is the radial component, $\boldsymbol{U}$ is uniformly distributed on the unit sphere $\mathbb{S} = \{\boldsymbol{u} : |\boldsymbol{u}| = 1\}$ and R, $\boldsymbol{U}$ are independent. Here $|\boldsymbol{x}| = \sqrt{\boldsymbol{x}\boldsymbol{x}'}$ is again the

Euclidean norm. We may call $\boldsymbol{U}$ the angular component (at least in the bivariate case).

If $P\{\boldsymbol{X} = \boldsymbol{0}\} = 0$, then we have $|\boldsymbol{X}| \overset{d}{=} R$ and $\boldsymbol{X}/|\boldsymbol{X}| \overset{d}{=} \boldsymbol{U}$.

11.3 Multivariate Student Distributions

We introduce another class of elliptical distributions, namely certain multivariate Student distributions, and deal with statistical questions. At the end of this section, we also introduce non–elliptical Student distributions.

Multivariate Student Distribution with Common Denominators

Recall from Section 6.3 that $Z = X/(2Y/\alpha)^{1/2}$ is a standard Student random variable with shape parameter $\alpha > 0$, where X is standard normal and Y is a gamma random variable with parameter $r = \alpha/2$.

Next, X will be replaced by a random vector $\boldsymbol{X} = (X_1, \ldots, X_d)$ of iid standard normal random variables X_i. Because $\boldsymbol{X}$ is spherical, we know from the preceding arguments that $\boldsymbol{X}/(2Y/\alpha)^{1/2}$ is a spherical random vector. In addition, if $\boldsymbol{A}$ is a matrix with $\det \boldsymbol{A} \neq 0$, then a linear change yields the elliptical random vector

$$\boldsymbol{Z} = \boldsymbol{A}\big(\boldsymbol{X}/(2Y/\alpha)^{1/2}\big) = (\boldsymbol{A}\boldsymbol{X})/(2Y/\alpha)^{1/2}, \tag{11.11}$$

where $\boldsymbol{A}\boldsymbol{X}$ is a normal random vector with covariance matrix $\boldsymbol{\Sigma} = \boldsymbol{A}\boldsymbol{A}'$.

This distribution of $\boldsymbol{Z}$ is the multivariate standard Student distribution with parameter $\alpha > 0$ and parameter matrix $\boldsymbol{\Sigma} = (\sigma_{i,j})$. We list some properties of these distributions.

- The density of $\boldsymbol{Z} = (Z_1, \ldots, Z_d)$ is

$$f_{\alpha,\boldsymbol{\Sigma}}(\boldsymbol{z}) = \frac{\Gamma((\alpha+d)/2)}{\Gamma(\alpha/2)(\alpha\pi)^{d/2}(\det \boldsymbol{\Sigma})^{1/2}} \big(1 + \alpha^{-1}\boldsymbol{z}\boldsymbol{\Sigma}^{-1}\boldsymbol{z}'\big)^{-(\alpha+d)/2}. \tag{11.12}$$

- According to (11.11), the marginal distributions are Student distributions again. In particular, Z_j is distributed according to the Student distribution with shape parameter $\alpha > 0$ and scale parameter $\sigma_j = \sigma_{j,j}^{1/2}$.

- If $\alpha > 2$, then $\frac{\alpha}{\alpha-2}\boldsymbol{\Sigma}$ is the covariance matrix. For $\alpha \leq 2$ the covariances do not exist.

> **Warning!** $\boldsymbol{\Sigma}$ is not the covariance matrix of the Student distribution with density $f_{\alpha,\boldsymbol{\Sigma}}$ for any α.

This was the reason why we spoke of a parameter matrix $\boldsymbol{\Sigma}$ instead of a covariance matrix in that context. In the sequel, we still address $\boldsymbol{\Sigma}$ as "covariance matrix"—in quotation marks—for the sake of simplicity.

Initial Estimation in the Student Model

The shape and scale parameters α and σ_j may be estimated within the univariate marginal models. Take, e.g., the arithmetic mean of the estimates of α in the marginal models as an estimate of α in the joint model. If α is known to be larger than 2, then one may utilize sample variances to estimate the "covariance matrix".

Otherwise, one may use an estimate based on Kendall's τ, see (10.3), We will estimate the "covariance matrix" $\mathbf{\Sigma} = (\sigma_{i,j})$ by estimating the "correlation matrix" $(\sigma_{i,j}/\sigma_i\sigma_j)$, where $\sigma_j = \sigma_{j,j}^{1/2}$ again. The operational meaning of this parameterization becomes evident by generating the Student random vector $\mathbf{Z}$ in (11.11) in two steps. Let D be the diagonal matrix with elements σ_j in the main diagonal: we have $\mathbf{Z} = D\tilde{\mathbf{Z}}$ where $\tilde{\mathbf{Z}} = (D^{-1}A\mathbf{X})/(2Y/\alpha)^{1/2}$ is the standardized Student random vector with "covariance matrix" $(\sigma_{i,j}/\sigma_i\sigma_j)$, and $\mathbf{Z}$ is the pertaining random vector with scale parameter vector $\boldsymbol{\sigma}$.

Maximum Likelihood Estimation in the Student Model

Based on initial estimates one may compute an MLE for the Student model using the Newton–Raphson procedure. We continue Example 10.3.1 where the application of the copula method produced a curious "butterfly" contour plot.

EXAMPLE 11.3.1. (Continuation of Example 10.3.1.) The following considerations are based on the daily stock indizes of the Commerzbank and the Deutsche Bank from Jan. 1, 1992 to Jan 11, 2002. The scatterplot in Figure 11.1 consists of the the pertaining log-returns on trading days with conversed signs.

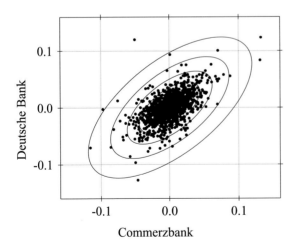

FIG. 11.1. Scatterplot of log-returns (with changed signs) Deutsche Bank against Commerzbank and contour plot of the estimated Student density.

The estimated bivarate Student distribution provides a better fit to the scatterplot due to its elliptical contour.

For a continuation we refer to page 390 where the Value–at–Risk is computed within the multivariate Student model.

Extending the Concept of Multivariate Student Distributions

In (6.16) we introduced Student distributions as those of $X/(Y/r)^{1/2}$, where X is standard normal and Y is a gamma random variable with parameter $r > 0$. This approach can be extended to the d–variate case dealing with random variables

$$X_i/(Y_i/r_i)^{1/2}, \qquad i = 1, \ldots, d,$$

where $\boldsymbol{X} = (X_1, \ldots, X_d)$ is multivariate normal with mean vector zero and correlation matrix $\boldsymbol{\Sigma}$, and $\boldsymbol{Y} = (Y_1, \ldots, Y_d)$ is a multivariate gamma vector. For example, the Y_i are totally dependent or independent (for details, see [32], Chapter 37). It is apparent that the univariate margins are Student random variables with parameter r_i.

A further extension to skewed distributions is achieved when the mean vector of $\boldsymbol{X}$ is unequal to zero. Then, the univariate margins are noncentral Student distributions.

11.4 Multivariate Sum–Stable Distributions
co–authored by J.P. Nolan[1]

In this section we give a short introduction to multivariate sum–stable distributions. We emphasis the statistical diagnostic which reduces the multivariate case to the univariate one which was outlined in Section 6.4.

Distributional Properties

A random vector $\boldsymbol{X} = (X_1, X_2, \ldots, X_d)$ is stable if for all $m = 2, 3, 4, \ldots$

$$\boldsymbol{X}_1 + \cdots + \boldsymbol{X}_m = a_m \boldsymbol{X} + \boldsymbol{b}_m,$$

in distribution, where $\boldsymbol{X}_1, \boldsymbol{X}_2, \ldots$ are iid copies of $\boldsymbol{X}$. The phrase "jointly stable" is sometimes used to stress the fact that the definition forces all the components X_j to be univariate sum–stable with the same α. Formally, if $\boldsymbol{X}$ is a stable random vector, then every one dimensional projection $\boldsymbol{u}\boldsymbol{X}' = \sum u_i X_i$ is a one dimensional stable random variable with the same index α for every $\boldsymbol{u}$, e.g.,

$$\boldsymbol{u}\boldsymbol{X}' \sim S(\alpha, \beta(\boldsymbol{u}), \gamma(\boldsymbol{u}), \delta(\boldsymbol{u}); 1). \tag{11.13}$$

The converse of this statement is true if the projections are all symmetric, or all strictly stable, or if $\alpha \geq 1$. Section 2.2 of [46] gives an example where $\alpha < 1$

---

[1]American University, Washington DC.

and all one dimensional projections are stable, but $\boldsymbol{X}$ is not jointly stable. In this case, a technical condition must be added to get a converse.

One advantage of (11.13) is that it gives a way of parameterizing multivariate stable distributions in terms of one dimensional projections. From (11.13) one gets the (univariate) characteristic function of $\boldsymbol{u}\boldsymbol{X}'$ for every $\boldsymbol{u}$, and hence the joint characteristic function of $\boldsymbol{X}$. Therefore α and the functions $\beta(\cdot)$, $\gamma(\cdot)$ and $\delta(\cdot)$ completely characterize the joint distribution. In fact, knowing these functions on the unit sphere $\mathbb{S} = \{\boldsymbol{u} : |\boldsymbol{u}| = 1\}$ in the Euclidean d–space characterizes the distribution.

The functions $\beta(\cdot)$, $\gamma(\cdot)$ and $\delta(\cdot)$ must satisfy certain regularity conditions. The standard way of describing multivariate stable distributions is in terms of a finite measure Λ on the sphere $\mathbb{S}$, called the spectral measure. It is typical to use the spectral measure to describe the joint characteristic function, we find it more convenient to relate it to the functions $\beta(\cdot)$, $\gamma(\cdot)$, and $\delta(\cdot)$.

Let $\boldsymbol{X} = (X_1, \ldots, X_d)$ be jointly stable with a representation as in (11.13). Then there exists a finite measure Λ on $\mathbb{S}$ and a d–variate vector $\boldsymbol{\delta}$ with

$$\gamma(\boldsymbol{u}) = \left(\int_{\mathbb{S}} |\boldsymbol{u}\boldsymbol{s}'|^\alpha \, \Lambda(d\boldsymbol{s}) \right)^{1/\alpha}$$

$$\beta(\boldsymbol{u}) = \frac{\int_{\mathbb{S}} |\boldsymbol{u}\boldsymbol{s}'|^\alpha \text{sign}\,(\boldsymbol{u}\boldsymbol{s}') \, \Lambda(d\boldsymbol{s})}{\int_{\mathbb{S}} |\boldsymbol{u}\boldsymbol{s}'|^\alpha \, \Lambda(d\boldsymbol{s})}$$

$$\delta(\boldsymbol{u}) = \begin{cases} \boldsymbol{\delta}\boldsymbol{u}' & \alpha \neq 1; \\[2mm] \boldsymbol{\delta}\boldsymbol{u}' - \frac{2}{\pi} \int_{\mathbb{S}} (\boldsymbol{u}\boldsymbol{s}') \ln |\boldsymbol{u}\boldsymbol{s}'| \, \Lambda(d\boldsymbol{s}) & \alpha = 1. \end{cases} \quad \text{if} \qquad (11.14)$$

It is possible for $\boldsymbol{X}$ to be non–degenerate, but singular. $\boldsymbol{X} = (X_1, 0)$ is formally a two dimensional stable distribution if X_1 is univariate stable, but it does not have a density. It can be shown that the following conditions are equivalent: (i) $\boldsymbol{X}$ is nonsingular, (ii) $\gamma(\boldsymbol{u}) > 0$ for all $\boldsymbol{u}$, (iii) the span of the support of Λ is the Euclidean d–space. If these conditions hold, then a smooth density exists. Relatively little is known about these densities—their support is a cone in general, and they are likely to be unimodal.

There are a few special cases of multivariate stable distributions where quantities of interest can be explicitly calculated. We concentrate on the bivariate case.

The Gaussian Case

The density function $\varphi_{\boldsymbol{\mu}, \boldsymbol{\Sigma}}$ of a multivariate Gaussian distribution was given in (11.3). The joint characteristic function of such a distribution is

$$\chi(\boldsymbol{u}) = \exp\left(-\boldsymbol{u}\boldsymbol{\Sigma}\boldsymbol{u}'/2 + \boldsymbol{u}\boldsymbol{\mu}' \right).$$

Hence the parameter functions are $\gamma(\boldsymbol{u}) = (\boldsymbol{u}\boldsymbol{\Sigma}\boldsymbol{u}'/2)^{1/2}$, $\beta(\boldsymbol{u}) = 0$ and $\delta(\boldsymbol{u}) = \boldsymbol{u}\boldsymbol{\mu}'$.

The Independent Case

Let $X_1 \sim S(\alpha, \beta_1, \gamma_1, \delta_1; 1)$ and $X_2 \sim S(\alpha, \beta_2, \gamma_2, \delta_2; 1)$ be independent, then $\boldsymbol{X} = (X_1, X_2)$ is bivariate stable. Since the components are independent, the joint density is

$$f(x_1, x_2) = g(x_1 | \alpha, \beta_1, \gamma_1, \delta_1; 1) g(x_2 | \alpha, \beta_2, \gamma_2, \delta_2; 1),$$

where $g(\cdot | \cdots)$ are the one dimensional stable densities. When there is an explicit form for a univariate density (Gaussian, Cauchy or Lévy cases), then there is an explicit form for the bivariate density. For example, the standardized Cauchy with independent components has density

$$f(x_1, x_2) = \frac{1}{\pi^2} \frac{1}{(1 + x_1^2)(1 + x_2^2)}. \tag{11.15}$$

For a general α the parameter functions are

$$\beta(u_1, u_2) = \frac{(\text{sign } u_1)\beta_1 |u_1 \gamma_1|^\alpha + (\text{sign } u_2)\beta_2 |u_2 \gamma_2|^\alpha}{|u_1 \gamma_1|^\alpha + |u_2 \gamma_2|^\alpha}$$

$$\gamma(u_1, u_2) = (|u_1 \gamma_1|^\alpha + |u_2 \gamma_2|^\alpha)^{1/\alpha}$$

$$\delta(u_1, u_2) = \begin{cases} u_1 \delta_1 + u_2 \delta_2, & \alpha \neq 1, \\ u_1 \delta_1 + u_2 \delta_2 - (2/\pi)(\beta_1 \gamma_1 u_1 \ln |u_1| + \beta_2 \gamma_2 u_2 \ln |u_2|), & \alpha = 1. \end{cases}$$

Note that $\beta(1, 0) = \beta_1$, $\gamma(1, 0) = \gamma_1$, and $\delta(1, 0) = \delta_1$, which corresponds to $(1, 0)\boldsymbol{X}' = X_1$ and $\beta(0, 1) = \beta_2$, $\gamma(0, 1) = \gamma_2$, and $\delta(0, 1) = \delta_2$ which corresponds to $(0, 1)\boldsymbol{X}' = X_2$. The corresponding spectral measure is discrete with four point masses $\gamma_1^\alpha (1 + \beta_1)/2$ at $(1, 0)$, $\gamma_1^\alpha (1 - \beta_1)/2$ at $(-1, 0)$, $\gamma_2^\alpha (1 + \beta_2)/2$ at $(0, 1)$, $\gamma_2^\alpha (1 - \beta_2)/2$ at $(0, -1)$.

Radially Symmetric Stable Distributions

Next we deal with radially symmetric (spherical) distributions. When $\alpha = 2$, they are Gaussian distributions with independent components. When $0 < \alpha < 2$, the components are univariate symmetric stable, but dependent. The Cauchy example has an explicit formula for the density, namely,

$$f(x_1, x_2) = \frac{1}{2\pi} \frac{1}{(1 + x_1^2 + x_2^2)^{3/2}}. \tag{11.16}$$

See Fig. 11.2 for an illustration.

That is not the same as the independent components case (11.15). When $\alpha \notin \{1, 2\}$, there is not a closed form expression for the density. The parameter functions are easy: $\beta(u_1, u_2) = 0$, $\gamma(u_1, u_2) = c$, and $\delta(u_1, u_2) = 0$.

The spectral measure for a radially symmetric measure is a uniform measure (a multiple of Lebesgue measure) on the unit circle. A linear change of variables gives elliptically contoured stable distributions, which have heavier tails than their Gaussian counterparts.

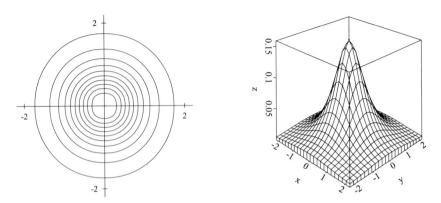

FIG. 11.2. Contour plot (left) and surface plot (right) for a radially symmetric Cauchy density.

Another Illustration

If the spectral measure has more mass in a certain arc, then the bivariate distribution bulges in that direction. This is illustrated in Fig. 11.3 for a bivariate sum–stable distribution with $\alpha = 0.8$, where the spectral measure has point masses of weight 0.125 at $(1,0)$, and weights 0.250 at $(\cos(\pi/3), \sin(\pi/3))$ and $(\cos(-\pi/3), \sin(-\pi/3))$.

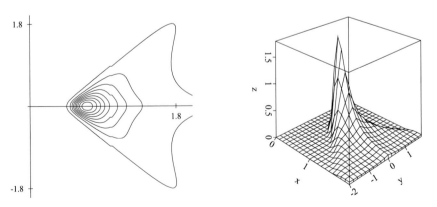

FIG. 11.3. Sum–stable density bulging in direction $\pi/3$ and $-\pi/3$ having also a smaller bulge along the x–axis corresponding to different masses.

Multivariate Estimation

For conciseness, we stated the above results in the conventional $S(\alpha, \beta, \gamma, \delta; 1)$ parameterization. As in the univariate case, one should use the $S(\alpha, \beta, \gamma, \delta; 0)$ parameterization in all numerical and statistical work to avoid the discontinuity at $\alpha = 1$.

There are several methods of estimating for multivariate stable distributions; in practice all involve some estimate of α and some discrete estimate of the spectral measure $\hat{\Lambda} = \sum_{k \leq m} \lambda_k 1_{\{s_k\}}$, $s_k \in \mathbb{S}$. Rachev and Xin[2] and Cheng and Rachev[3] use the fact that the directional tail behavior of multivariate stable distributions is Pareto, and base an estimate of Λ on this. Nolan, Panorska and McCulloch[4] define two other estimates of Λ, one based on the joint sample characteristic function and one based on one–dimensional projections of the data.

Another advantage of (11.14) is that it gives a way of assessing whether a multivariate data set is stable by looking at just one dimensional projections of the data. Fit projections in multiple directions using the univariate techniques described earlier, and see if they are well described by a univariate stable fit. If so, and if the α's are the same for every direction (if $\alpha < 1$, another technical condition holds), then a multivariate stable model is appropriate.

The parameters $(\alpha, \beta(\boldsymbol{u}), \gamma(\boldsymbol{u}), \delta(\boldsymbol{u}))$ may be more useful than Λ itself when two multivariate stable distributions are compared. This is because the distribution of $\boldsymbol{X}$ depends more on how Λ distributes mass around the sphere than exactly on the measure. Two spectral measures can be far away in the traditional total variation norm (e.g., one can be discrete and the other continuous), but their corresponding directional parameter functions and densities can be very close. Indeed, (11.13) shows that the only way Λ enters into the joint distribution is through the parameter functions.

Multivariate Diagnostics

The diagnostics suggested are:

- Project the data in a variety of directions $\boldsymbol{u}$ and use the univariate diagnostics described in Section 6.4 on each of those distributions. Bad fits in any direction indicate that the data is not stable.

- For each direction $\boldsymbol{u} \in \mathbb{S}$, estimate a value for the parameter functions $\alpha(\boldsymbol{u})$, $\beta(\boldsymbol{u})$, $\gamma(\boldsymbol{u})$, $\delta(\boldsymbol{u})$ by ML estimation. The plot of $\alpha(\boldsymbol{u})$ should be a constant,

[2]Rachev, S.T. and Xin, H. (1993). Test for association of random variables in the attraction of multivariate stable law. Probab. Math. Statist. 14, 125–141.

[3]Cheng, B.N. and Rachev, S.T. (1995). Multivariate stable future prices. Math. Finance 5, 133–153.

[4]Nolan, J.P., Panorska, A. and McCulloch, J.H. (2001). Estimation of stable spectral measures. Mathematical and Computer Modelling 34, 1113–1122.

significant departures from this indicate that the data has different decay rates in different directions and is not jointly stable. (Note that $\gamma(\cdot)$ will be a constant iff the distribution is isotropic.)

- Assess the goodness–of–fit by computing a discrete $\hat{\Lambda}$ by one of the methods above. Substitute the discrete $\hat{\Lambda}$ in (11.14) to compute parameter functions. If it differs from the one obtained above by projection, then either the data is not jointly stable, or not enough points were chosen in the discrete spectral measure approximation.

Chapter 12

Multivariate Maxima

Multivariate extreme value (EV) distributions are introduced as limiting distributions of componentwise taken maxima. In contrast to the univariate case, the resulting statistical model is a nonparametric one. Some basic properties and first examples of multivariate EV dfs are dealt with in Section 12.1.

In Section 12.2, we introduce the family of Gumbel–McFadden distributions which are perhaps the most important multivariate EV distributions. As an application, the question of economic choice behavior will be considered under disturbances which are multivariate EV distributed. In the special case of Gumbel–McFadden disturbances the choice probabilities are of a multinomial logit form.

Estimation in certain parametric EV submodels and will be investigated in Section 12.3.

In Section 12.4, we give an outline of a new spectral decomposition methodology in extreme value theory where bivariate dfs are decomposed in univariate dfs. For example, bivariate EV dfs are decomposed in univariate EV dfs.

12.1 Nonparametric and Parametric Extreme Value Models

Because the univariate margins of multivariate EV distributions are EV distributions, one may concentrate on the dependence structure of the multivariate distribution. Therefore, we particularly, study EV dfs with reversed exponential margins. In that context, the Pickands dependence function will be central for our considerations.

Multivariate Extreme Value Distributions, Max–Stability

The maximum of vectors $\boldsymbol{x}_i = (x_{i,1}, \ldots, x_{i,d})$ will be taken componentwise. We have $\max_{i \le m} \boldsymbol{x}_i = \left(\max_{i \le m} x_{i,1}, \ldots, \max_{i \le m} x_{i,d} \right)$. If $\boldsymbol{X}_1, \ldots, \boldsymbol{X}_m$ are iid random vectors with common d–variate df F, then

$$P\Big\{ \max_{i \leq m} \boldsymbol{X}_i \leq \boldsymbol{x} \Big\} = F^m(\boldsymbol{x}), \tag{12.1}$$

because $\max_{i \leq m} \boldsymbol{X}_i \leq \boldsymbol{x}$ holds if, and only if, $\boldsymbol{X}_1 \leq \boldsymbol{x}, \ldots, \boldsymbol{X}_m \leq \boldsymbol{x}$.

Corresponding to the univariate case, the actual df F^m will be replaced by a limiting df. If

$$F^m(\boldsymbol{b}_m + \boldsymbol{a}_m \boldsymbol{x}) \to G(\boldsymbol{x}), \qquad m \to \infty, \tag{12.2}$$

for vectors $\boldsymbol{b}_m$ and $\boldsymbol{a}_m > 0$, then G is called a d–variate EV df and F is said to belong to the max–domain of attraction of G.

One can prove that G is an EV df if, and only if, G is max–stable, that is,

$$G^m(\boldsymbol{b}_m + \boldsymbol{a}_m \boldsymbol{x}) = G(\boldsymbol{x}) \tag{12.3}$$

for certain vectors $\boldsymbol{b}_m$ and $\boldsymbol{a}_m > 0$.

Let $F_{(j)}$ denote the jth margin of F. From (12.2), it follows that

$$F_{(j)}^m(b_{m,j} + a_{m,j} x_j) \to G_{(j)}(x_j), \qquad m \to \infty, \tag{12.4}$$

where $G_{(j)}$ is the jth margin of G. Hence, the jth marginal df $G_{(j)}$ of a multivariate EV df is necessarily a univariate EV df. In addition, the vectors $\boldsymbol{b}_m = (b_{m,1}, \ldots, b_{m,d})$ and $\boldsymbol{a}_m = (a_{m,1}, \ldots, a_{m,d})$ are determined by the univariate convergence. Another consequence is the continuity of the multivariate EV dfs. Yet, in contrast to the univariate EV model, the d–variate one is a nonparametric model for $d \geq 2$.

Extreme Value Model for Independent Margins

If, in addition, the vectors $\boldsymbol{X}_i$ have independent margins, then the marginal maxima $\max_{i \leq m} X_{i,1}, \ldots, \max_{i \leq m} X_{i,d}$ are independent and, consequently,

$$F^m(\boldsymbol{b}_m + \boldsymbol{a}_m \boldsymbol{x}) = \prod_{j \leq d} F_{(j)}^m(b_{m,j} + a_{m,j} x_j) \to \prod_{j \leq d} G_{(j)}(x_j) \tag{12.5}$$

as $m \to \infty$ if (12.4) holds. For this special case, the limiting df G of $\max_{i \leq m} \boldsymbol{X}_i$ in (12.2), taken in the γ–parameterization, is

$$G(\boldsymbol{x}) = \prod_{j \leq d} G_{\gamma_j, \mu_j, \sigma_j}(x_j). \tag{12.6}$$

It is evident that the estimation of the parameters γ_j, μ_j and σ_j must be based on the data in the jth component.

The importance of the EV dfs as given in (12.6) comes from the fact that an EV df of this form may also occur if the random vectors $\boldsymbol{X}_i$ have dependent margins. We mention a prominent example.

(**Tiago de Oliveira, Geffroy, Sibuya.**) Let (X_i, Y_i) be iid normal vectors with correlation coefficient ρ strictly between -1 and 1. Under this condition, the maxima $\max_{i \leq m} X_i$ and $\max_{i \leq m} Y_i$ are asymptotically independent. We have

$$P\{\max_{i \leq m} X_i \leq b_m + a_m x, \ \max_{i \leq m} Y_i \leq d_m + c_m y\} \to G_0(x) G_0(y), \qquad m \to \infty, \quad (12.7)$$

with standardizing constants such that the univariate convergence holds.

To prove this result, one may verify a simple condition imposed on conditional probabilities which were already studied in Section 2.6. Let X and Y be random variables with common df F such that $F^m(b_m + a_m x) \to G(x)$ as $m \to \infty$ for all x. The condition

$$P(Y > u | X > u) \to 0, \qquad u \uparrow \omega(F), \quad (12.8)$$

is equivalent to the asymptotic independence of the pertaining maxima; that is,

$$P\{\max_{i \leq m} X_i \leq b_m + a_m x, \ \max_{i \leq m} Y_i \leq b_m + a_m y\} \to G(x) G(y), \qquad m \to \infty,$$

for iid copies (X_i, Y_i) of (X, Y), see Sibuya (1960) in the paper cited on page 74. For a proof of this result we refer to [20], page 258, or [42], page 235.

The asymptotic independence of the marginal maxima may be regarded as the intrinsic property of tail independence. It is equivalent to

$$\chi = \lim_{u \uparrow \omega(F)} P(Y > u | X > u) = 0$$

where χ is the tail dependence parameter introduced in Section 2.6.

Condition (12.8) also implies that the componentwise taken exceedances are independent if the thresholds are chosen in such a manner that the expected number of exceedances remains bounded when the sample size goes to infinity[1].

Thus, before using a complex model, one should check whether the EV model for independent margins or a related generalized Pareto model is applicable.

In addition, the pairwise asymptotic independence implies the joint asymptotic independence, see, e.g., [20] or [42].

Rates for the Asymptotic Independence of Maxima

Let F be a bivariate df with identical univariate margins $F_{(j)}$. Applying the representation (10.7) of a df by means of survivor functions one gets (see (7.2.11) in [42]) that

$$F^n(\boldsymbol{x}) = F_1^n(x_1) F_2^n(x_2) \exp\left(n \bar{F}(\boldsymbol{x})\right) + O(n^{-1}). \quad (12.9)$$

---

[1]Reiss, R.–D. (1990). Asymptotic independence of marginal point processes of exceedances. Statist. & Decisions 8, 153–165. See also [43], Section 6.2.

Therefore, the term $n\overline{F}(\boldsymbol{x})$ determines the rate at which the independence of the marginal maxima is attained. For a general result of this type we refer to [16], Lemma 4.1.3.

In the case of iid standard normal vectors with correlation coefficient ρ with $-1 < \rho < 1$ it was proven in [42] that

$$n\overline{F}(\boldsymbol{b}_n + \boldsymbol{b}_n^{-1}\boldsymbol{x}) = O\big(n^{-(1-\rho)/(1+\rho)}(\log n)^{-\rho/(1+\rho)}\big), \qquad (12.10)$$

where $\boldsymbol{b}_n = (b_n, b_n)$ and b_n satisfies the condition $b_n = \varphi(b_n)$. Thus, one obtains again the above mentioned result of the asymptotic independence of maxima of normal random variables. The rate of the asymptotic independence is slow when ρ is close to 1. For a continuation of this topic we refer to Section 13.3.

The Marshall–Olkin Model

The standard Marshall–Olkin distributions M_λ are bivariate EV distributions with exponential margins $G_{2,-1} = \exp(x)$, $x < 0$, where the dependence parameter λ ranges from 0 to 1. For $\lambda = 0$ we have independence, and for $\lambda = 1$ total dependence. We have

$$M_\lambda(x, y) = \exp\Big((1-\lambda)(x+y) + \lambda\min\{x, y\}\Big), \qquad x, y < 0. \qquad (12.11)$$

The df F_λ is necessarily continuous, yet a density does not exist for $\lambda > 0$, because standard Marshall–Olkin distributions have positive mass at the main diagonal which becomes visible in Fig. 12.1.

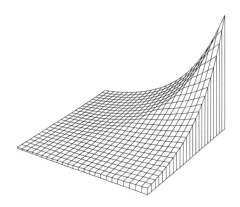

FIG. 12.1. 3–D plot of Marshall–Olkin df M_λ with $\lambda = 0.5$.

For $0 \leq \lambda \leq 1$, check that

$$M_\lambda(x, y) = P\bigg\{\max\bigg\{\frac{Z_2}{1-\lambda}, \frac{Z_1}{\lambda}\bigg\} \leq x, \ \max\bigg\{\frac{Z_3}{1-\lambda}, \frac{Z_1}{\lambda}\bigg\} \leq y\bigg\}, \qquad (12.12)$$

where Z_1, Z_2 and Z_3 are iid random variables with common exponential df $G_{2,-1}$.

This model can be extended to a d–variate one. Let $\Lambda = (\lambda_{i,j})$ be a $m \times d$–matrix such that $\lambda_{i,j} \geq 0$ and

$$\sum_{i \leq m} \lambda_{i,j} = 1$$

for each $j \leq d$. Let $Z_1, \ldots, Z_m$ be iid random variables with common exponential df $G_{2,-1}(x) = \exp(x)$, $x < 0$. Then,

$$
\begin{aligned}
M_\Lambda(\boldsymbol{x}) &= P\big\{\max_{i \leq m}\{Z_i/\lambda_{i,j}\} \leq x_j, \ j = 1, \ldots, d\big\} \\
&= \exp\Big(\sum_{i \leq m} \min_{j \leq d}(\lambda_{i,j} x_j)\Big), \qquad \boldsymbol{x} < 0, \qquad (12.13)
\end{aligned}
$$

is a d–variate max–stable df with univariate margins $G_{2,-1}$. In (12.12) there is a special case for $d = 2$ with $\lambda_{1,3} = \lambda_{2,2} = 0$.

Such dfs are also of theoretical interest (see [16], page 111) for a general representation of max–stable distributions.

Hüsler–Reiss Triangular Arrays

This model is constructed by taking limiting dfs of maxima of a triangular array of standard normal random vectors. The univariate margins are Gumbel dfs. We start with the bivariate case. Let Φ again denote the univariate standard normal df. For positive parameters λ we have

$$
\begin{aligned}
H_\lambda(x, y) &= \exp\Big(- e^{-x} - \int_y^\infty \Phi\Big(\frac{1}{\lambda} + \frac{\lambda(x-z)}{2}\Big) e^{-z}\, dz \Big) \qquad (12.14) \\
&= \exp\Big(- \Phi\Big(\frac{1}{\lambda} + \frac{\lambda(x-y)}{2}\Big) e^{-y} - \Phi\Big(\frac{1}{\lambda} + \frac{\lambda(y-x)}{2}\Big) e^{-x} \Big),
\end{aligned}
$$

where the second equality may be established by partial integration.

In addition, let

$$H_0(x, y) = G_0(x) G_0(y)$$

and

$$H_\infty(x, y) = G_0(\min\{x, y\})$$

which are the dfs describing the cases of independence and total dependence. We have

$$H_\lambda(x, y) \to H_0(x, y) \qquad \text{and} \qquad H_\lambda(x, y) \to H_\infty(x, y), \qquad \lambda \to \infty,$$

and, therefore, this model ranges continuously between independence and total dependence with λ indicating the degree of dependence.

The bivariate EV dfs H_λ in (12.14) generally occur as limiting dfs of maxima of iid random vectors $(X_{n,i}, Y_{n,i})$, $i = 1, \ldots, n$, where

$$(X_{n,i}, Y_{n,i}) \overset{d}{=} \left(S_1, \rho_n S_1 + \sqrt{1 - \rho_n^2} S_2\right) \tag{12.15}$$

with $\rho_n \uparrow 1$ at a certain rate as $n \to \infty$ and (S_1, S_2) being a spherical random vector, under the condition that the radius $R = \sqrt{S_1^2 + S_2^2}$ is in the max–domain of attraction of the Gumbel df G_0[2]. Notice that the $(X_{n,i}, Y_{n,i})$ are elliptically distributed, see Section 11.2. Further related results may be found in another article by Hashorva[3].

3–D plots of the density of H_λ with parameter $\lambda = 1$ are displayed in Fig. 12.2 with views from two different angles. On the left–hand side, one recognizes that h_λ is symmetric in x and y, cf. also (12.14). After a rotation of $90°$, the typical contours of the marginal Gumbel densities become visible.

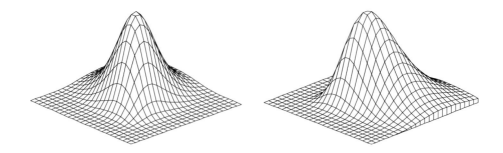

FIG. 12.2. (left.) 3–D plot with a view of the density h_λ along the main diagonal in the (x, y) plane for $\lambda = 1$. (right.) The density shown again after a rotation of $90°$ about the z–axis.

We compute the density $h_\lambda(x, y)$ and the conditional df $H_\lambda(y|x)$ for parameters $0 < \lambda < \infty$. The density can be computed by taking partial derivatives of H_λ, cf. (10.11). By employing the identity

$$\varphi\left(\lambda^{-1} + \lambda(x - y)/2\right) e^{-y} = \varphi\left(\lambda^{-1} + \lambda(y - x)/2\right) e^{-x},$$

one gets

$$\begin{aligned}
h_\lambda(x, y) &= H_\lambda(x, y) \Big(\Phi\left(\lambda^{-1} + \lambda(x - y)/2\right) \Phi\left(\lambda^{-1} + \lambda(x - y)/2\right) e^{-(x+y)} \\
&\quad + \lambda \varphi\left(\lambda^{-1} + \lambda(y - x)/2\right) e^{-x}/2 \Big).
\end{aligned} \tag{12.16}$$

[2]Hashorva, E. (2005). Elliptical triangular arrays in the max–domain of attraction of Hüsler–Reiss distribution. Statist. Probab. Lett. 72, 125–135.

[3]Hashorva, E. (2006). On the max–domain of attraction of bivariate elliptical arrays. Extremes 8, 225–233.

The conditional density is

$$h_\lambda(y|x) = h_\lambda(x, y)/g_0(x),$$

cf. (8.10). By integration, one obtains the conditional df

$$H_\lambda(y|x) = H_\lambda(x, y)\Phi\left(1/\lambda + \lambda(y - x)/2\right)/G_0(x). \qquad (12.17)$$

The conditional density and df are useful for generating data in two steps: firstly, generate x under the Gumbel df G_0 and, secondly, generate y under the conditional df $H_\lambda(\cdot\,|x)$.

There is a one–to–one relation between the dependence parameter λ and the correlation coefficient ρ of H_λ which is indicated in Fig. 12.3. We see that the correlation coefficient is practically zero when the shape parameter λ is smaller than .3.

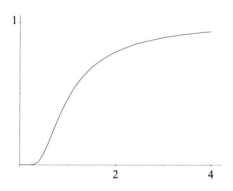

FIG. 12.3. Correlation coefficient ρ of H_λ plotted against the shape parameter λ.

Generally, let[4]

$$H_\Lambda(\boldsymbol{x}) = \exp\left(-\sum_{k\leq d} \int_{x_k}^\infty \Phi_{\Sigma(k)}\left(\left(\lambda_{i,k}^{-1} + \lambda_{i,k}(x_i - z)/2\right)_{i=1}^{k-1}\right)e^{-z}\,dz\right),$$

for a symmetric $d \times d$–matrix $\Lambda = (\lambda_{i,j})$ with $\lambda_{i,j} > 0$ if $i \neq j$ and $\lambda_{i,i} = 0$, and $\Phi_{\Sigma(k)}$ is a $(k-1)$–variate normal df (with the convention that $\Phi_{\Sigma(1)} = 1$). The mean vector of $\Phi_{\Sigma(k)}$ is zero and $\Sigma(k) = (\sigma_{i,j}(k))$ is a correlation matrix given by

$$\sigma_{i,j}(k) = \begin{cases} \lambda_{i,k}\lambda_{j,k}\left(\lambda_{i,k}^{-2} + \lambda_{j,k}^{-2} - \lambda_{i,j}^{-2}\right)/2 & 1 \leq i < j \leq k-1; \\ & \text{if} \\ 1 & i \neq j. \end{cases}$$

---

[4]Such a representation is given in Joe, H. (1994). Multivariate extreme–value distributions with applications to environmental data. Canad. J. Statist. 22, 47–64. Replacing $\Phi_{\Sigma(k)}$ by survivor functions (cf. (10.4)), one obtains the original representation.

Each pair of marginal random variables determines one parameter of the limiting df so that the dimension of the parameter vector is $d(d-1)/2$. For $d = 2$, the matrix Λ is determined by $\lambda = \lambda_{1,2}$ and one gets the bivariate case.

Extending and Modifying Standard Models

Let $\boldsymbol{Z} = (Z_1, \ldots, Z_d)$ be a random vector distributed according to one of the preceding multivariate EV dfs G. Denote by G_0 the common univariate df of the Z_j. Thus, G_0 is either an exponential or Gumbel df.

- (Location and Scale Vectors.) A first natural extension of the models is obtained by including location and scale parameters μ_j and σ_j. Thus, one deals with random variables $\mu_j + \sigma_j Z_j$ with joint df $\widetilde{G}(\boldsymbol{x}) = G\left((\boldsymbol{x} - \boldsymbol{\mu})/\boldsymbol{\sigma}\right)$ (cf. page 163).

- (Exchanging Marginal Dfs.) Next the marginal df G_0 is replaced by some other univariate EV df $G_{(j)}$ in the jth component. First remember that the $Y_j = G_0(Z_j)$ are $(0,1)$–uniformly distributed with common df

$$C(\boldsymbol{y}) = G\left(G_0^{-1}(y_1), \ldots, G_0^{-1}(y_d)\right)$$

which is the copula of G, cf. page 275.

Applying the quantile transformation one arrives at random variables $X_j = G_{(j)}^{-1}(Y_j)$ with marginal dfs $G_{(j)}$ and joint multivariate EV df

$$\begin{aligned}
\widetilde{G}(\mathbf{x}) &= C\left(G_{(1)}(x_1), \ldots, G_{(d)}(x_d)\right) \\
&= G\left(G_0^{-1}(G_{(1)}(x_1)), \ldots, G_0^{-1}(G_{(d)}(x_d))\right).
\end{aligned} \tag{12.18}$$

Using the latter approach, one may build multivariate EV models with specified univariate dfs such as, e.g., a Hüsler–Reiss model with Gumbel (EV 0) or EV margins.

As an example, the bivariate, standard Hüsler–Reiss df H_λ is transformed to a df $\widetilde{H}_\lambda$ with exponential margins $G_{(j)}(x) = G_{2,-1}(x) = \exp(x)$ for $x < 0$. Because $G_0(x) = \exp\left(-e^{-x}\right)$ is the standard Gumbel df, we have $G_0^{-1}(G_{(j)}(x)) = -\log(-x)$ and, therefore,

$$\widetilde{H}_\lambda(x,y) = \exp\left(\Phi\left(\frac{1}{\lambda} + \frac{\lambda}{2}\log\left(\frac{y}{x}\right)\right)y + \Phi\left(\frac{1}{\lambda} + \frac{\lambda}{2}\log\left(\frac{x}{y}\right)\right)x\right) \tag{12.19}$$

for $x, y < 0$, with $\widetilde{H}_0(x,y) = \exp(x+y)$ and $\widetilde{H}_\infty(x,y) = \exp(\min\{x,y\})$.

The Pickands Dependence Function for Distributions

A bivariate EV df with exponential marginals $G_{2,-1}(x) = \exp(x)$, $x < 0$, has the representation (cf. [20], 2nd edition, or [16]))

$$G(x,y) = \exp\left((x+y)D\left(x/(x+y)\right)\right), \qquad x, y \le 0, \tag{12.20}$$

where $D : [0,1] \to [0,1]$ is the Pickands dependence function which is convex and satisfies $\max(1-z, z) \le D(z) \le 1$ for $0 \le z \le 1$ with the properties of independence and complete dependence being characterized by $D = 1$ and, respectively, $D(z) = \max(1 - z, z)$.

We specify the explicit form of the Pickands dependence function for our standard examples.

- Marshall–Olkin df M_λ:

$$D_\lambda(z) = 1 - \lambda \min(z, 1 - z). \qquad (12.21)$$

- Hüsler–Reiss df $\widetilde{H}_\lambda$ in the version (12.19):

$$D_\lambda(z) = \Phi\left(\frac{1}{\lambda} + \frac{\lambda}{2} \log\left(\frac{z}{1-z}\right)\right) z + \Phi\left(\frac{1}{\lambda} + \frac{\lambda}{2} \log\left(\frac{1-z}{z}\right)\right)(1 - z). \qquad (12.22)$$

Recall that the Pickands dependence functions are convex and satisfies

$$\max(1 - z, z) \le D(z) \le 1 \qquad \text{for } 0 \le z \le 1.$$

The value $D(1/2)$ of the Pickands dependence function will be used to define a certain canonical representation.

Generally, a max–stable, d–variate df with univarite margins $G_{-2,1}$ has the representation

$$G(\boldsymbol{x}) = \exp\left(\int_S \min_{j \le d}(y_j x_j) \, d\mu(\boldsymbol{y})\right), \qquad \boldsymbol{x} < 0, \qquad (12.23)$$

where μ is a finite measure on the d–variate unit simplex $S = \{\boldsymbol{y} : \sum_{j \le d} y_j = 1, \ y_j \ge 1\}$ such that $\int_S y_j \, d\mu(\boldsymbol{y}) = 1$ for $j \le d$.

The Canonical Parameterization

For any bivariate EV df G with standard exponential margins $G_{2,-1}$, the Pickands dependence function D_G satisfies $1/2 \le D_G(1/2) \le 1$. Particularly, $D_G(1/2) = 1$ and $D_G(1/2) = 1/2$, if independence and, respectively, complete dependence holds. Define the functional (canonical) parameter

$$T(G) = 2\big(1 - D_\lambda(1/2)\big) \qquad (12.24)$$

which ranges between 0 and 1. Necessarily, $T(G) = 0$ and $T(G) = 1$, if independence and, respectively, complete dependence holds.

Suppose that there is a parametric family of bivariate EV dfs G_λ with a one–to–one relation between λ and $D_\lambda(1/2)$. One may introduce another representation of the family by

$$\vartheta = T(G_\lambda) = 2\big(1 - D_\lambda(1/2)\big). \qquad (12.25)$$

This is the canonical parameterization of a bivariate EV family[5] which will be studied more closely on the subsequent pages. In special cases, we obtain the following canonical parameters.

- Marshall–Olkin: $\vartheta = \lambda$;

- Hüsler–Reiss: $\vartheta = 2(1 - \Phi(1/\lambda))$.

Notice that the original and the canonical parameterization are identical for the Marshall–Olkin family (also see Section 12.2 for the Gumbel–McFadden model).

Canonical and Tail Dependence Parameters

Let again G be a bivariate EV df with Pickands dependence function D and canonical (functional) parameter $T(G) = 2(1 - D(1/2))$. An operational meaning of the canonical parameter can be expressed by means of a property of the survivor function $\bar{G}$ in the upper tail region. Applying (10.7) we get

$$
\begin{aligned}
\bar{G}(u, u) &= 1 - 2\exp(u) + \exp(2uD(1/2)) \\
&= T(G)|u| + O(u^2), \quad u \to 0.
\end{aligned}
\tag{12.26}
$$

Let X and Y be random variables with common EV df G. For the tail dependence parameter $\chi(q)$ at the level q, introduced in (2.60), one gets

$$
\begin{aligned}
\chi(q) &= P(Y > \log q \,|\, X > \log q) \\
&= \bar{G}(\log q, \log q)/(1 - q) \\
&= T(G) + O(1 - q), \quad q \to 1,
\end{aligned}
\tag{12.27}
$$

according to (12.26). Thus, the canonical parameter $T(G)$ of a bivariate EV df is the tail dependence parameter χ, cf. (2.61). This topic will be continued in Section 13.3.

Multivariate Minima, Limiting Distributions of Minima and the Min–Stability

Corresponding to the maximum of vectors $x_i = (x_{i,1}, \ldots, x_{i,d})$, one may take the componentwise minimum $\min_{i \le m} x_i = \left(\min_{i \le m} x_{i,1}, \ldots, \min_{i \le m} x_{i,d} \right)$. Let $X_1, \ldots, X_m$ be iid random vectors with common df F and survivor function $\bar{F}$. The survivor function of the minimum is given by

$$
P\left\{ \min_{i \le m} X_i > x \right\} = \bar{F}^m(x).
$$

---

[5] Falk, M. and Reiss, R.–D. (2001). Estimation of the canonical dependence parameter in a class of bivariate peaks–over–threshold models. Statist. Probab. Letters 52, 9–16

Moreover, the representation $\min_{i\leq m} X_i = -\max_{i\leq m}(-X_i)$ holds again. As in the univariate case, cf. (1.38) and $\overline{(1.38)}$, there is a one–to–one relationship between limiting dfs of maxima and minima. If

$$P\Big\{\max_{i\leq m}(-X_i) \leq b_m + a_m x\Big\} \to G(x), \qquad m \to \infty,$$

then for $c_m = a_m$ and $d_m = -b_m$ we have

$$P\Big\{\min_{i\leq m} X_i \leq d_m + c_m x\Big\} \to \bar{G}(-x), \qquad m \to \infty.$$

We see that the limiting dfs $\widetilde{G}$ of sample minima are necessarily of the form $\widetilde{G}(x) = \bar{G}(-x)$. Conversely, $G(x)$ can be written as the survivor function of $\widetilde{G}$ applied to $-x$.

Limiting dfs of minima can be characterized again by the min–stability. The survivor function $\bar{F}$ (and, thus, the df F) is min–stable, if $\bar{F}^m(d_m + c_m x) = \bar{F}(x)$ for certain vectors d_m and $c_m > 0$.

12.2 The Gumbel–McFadden Model

We first deal with certain bivariate dfs which are the Gumbel type II dfs as mentioned in [20], page 247, in the form of a min–stable df. D–variate versions are introduced in (12.32) and (12.33). The section is concluded with an application to utility–maximizing.

Bivariate Gumbel Type II Distributions

This is another model of EV dfs with exponential margins $G_{2,-1}(x) = \exp(x)$, $x < 0$,. The bivariate Gumbel type II dfs H_λ are parameterized by a dependence parameter λ which ranges from 1 to infinity. We have independence for $\lambda = 1$ and total dependence for $\lambda = \infty$. Let

$$H_\lambda(x,y) = \exp\Big(-\big((-x)^\lambda + (-y)^\lambda\big)^{1/\lambda}\Big), \qquad x,y < 0. \tag{12.28}$$

Notice that $H_1(x,y) = \exp(x)\exp(y)$ and $H_\lambda(x,y) \to \exp(\min\{x,y\}) =: H_\infty(x,y)$ for $x,y < 0$ as $\lambda \to \infty$. Taking partial derivatives one gets the density

$$h_\lambda(x,y) = H_\lambda(x,y)(xy)^{\lambda-1}\Big(\big((-x)^\lambda + (-y)^\lambda\big)^{2(1/\lambda-1)}$$
$$+ (\lambda-1)\big((-x)^\lambda + (-y)^\lambda\big)^{1/\lambda-2}\Big). \tag{12.29}$$

Several distributional properties of H_λ can be conveniently deduced from a representation in terms of independent random variables (due to L. Lee[6] and J.–C.

---

[6]Lee, L. (1979). Multivariate distributions having Weibull properties. J. Mult. Analysis 9, 267–277.

Lu and G.K. Bhattacharya[7]).

Let U, V_1, V_2 and Z_λ be independent random variables, where U is uniformly distributed on $[0,1]$, the V_j have the common exponential df $G_{2,-1}(x) = \exp(x)$, $x < 0$, and Z_λ is a discrete random variable with $P\{Z_\lambda = 0\} = 1 - 1/\lambda$ and $P\{Z_\lambda = 1\} = 1/\lambda$. Let $V = V_1 + Z_\lambda V_2$.

Random variables X_1 and X_2 with joint df H_λ have the representation

$$(X_1, X_2) = \left(U^{1/\lambda}V,\ (1-U)^{1/\lambda}V\right) \qquad \text{in distribution.} \tag{12.30}$$

Moments of Gumbel–McFadden distributions can be easily deduced from (12.30). In particular, there is a one–to–one relation between the dependence parameter λ and the correlation coefficient $\rho(\lambda)$ of L_λ. One gets

$$\rho(\lambda) = 2\Gamma^2(1 + 1/\lambda)/\Gamma(1 + 2/\lambda) - 1.$$

This function is plotted in Fig. 12.4.

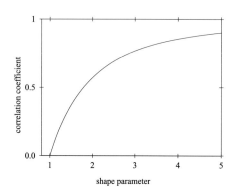

FIG. 12.4. Correlation coefficient ρ of H_λ plotted against the shape parameter λ.

The Pickands dependence function of the Gumbel df H_λ is

$$D_\lambda(z) = \left((1-z)^\lambda + z^\lambda\right)^{1/\lambda}. \tag{12.31}$$

The canonical parameter is given by $\vartheta = 2 - 2^{1/\lambda}$.

---

[7]Lu, J.–C. and Bhattacharya, G.K. (1991). Inference procedures for bivariate exponential model of Gumbel. Statist. Probab. Letters 12, 37–50.

Gumbel–McFadden Distributions

Certain d–variate extensions of the Gumbel distributions are due to Daniel Mc-Fadden[8]. We have

$$H_{\lambda,d}(\boldsymbol{x}) = \exp\left(-\left(\sum_{i\leq d}(-x_i)^\lambda\right)^{1/\lambda}\right), \qquad \boldsymbol{x} < 0, \qquad (12.32)$$

and as another extension

$$G(\boldsymbol{x}) = \exp\left(-\sum_{k\leq m} a_k\left(\sum_i(-x_i)^{\lambda(k)}\right)^{1/\lambda(k)}\right), \qquad \boldsymbol{x} < 0, \qquad (12.33)$$

where i varies over $B_k \subset \{1,\dots,d\}$, $\bigcup_{k\leq m} B_k = \{1,\dots,d\}$, $a_k > 0$, and $\lambda(k) \geq 1$.

To get standard exponential margins in the second case, the additional condition $\sum_{k\leq m} a_k I(i \in B_k) = 1$ must be satisfied for $i = 1,\dots,d$.

In the following lines it is shown that Gumbel–McFadden dfs $H_{\lambda,d}$ lead to a multinomial logit model in the utility–maximizing theory. In that application, EV dfs are convenient because one gets an elegant representation for certain choice probabilities.

An Application to Utility–Maximizing

It is the common situation that consumers can choose between different alternatives as, e.g., between different

- residential locations[8] described by certain configurations of attributes such as accessibility, quality of public services, neighborhood, etc.

- residential end–use energy configurations[9].

It is understood that a consumer chooses that configuration which provides a maximum utility.

The utility of a configuration, labeled by the index i, will be expressed by

$$U_i = v_i + X_i, \qquad i = 1,\dots,d,$$

where v_i is a function of the above mentioned attributes and of consumer's characteristics (family size, income), and X_i is an unobserved random variable.

---

[8] McFadden, D. (1978). Modelling the choice of residential location. In: Spatial Interaction Theory and Planning Models, pp. 75–96, A. Karlquist et al. (eds.), North Holland, Amsterdam. Reprinted in The Economics of Housing, Vol. I, pp. 531–552, J. Quigley (ed), Edward Elgar, London, 1997.

[9] Cowing, T.G. and McFadden, D.L. (1984). Microeconomic Modeling and Policy Analysis. Academic Press, Orlando.

The probability of the choice of the configuration with label i is

$$p_i = P\{U_i > U_j \text{ for } j = 1, \ldots, d \text{ with } j \neq i\}$$

which can be expressed by means of the ith partial derivative of the joint df F of the $X_1, \ldots, X_d$. We have

$$p_i = \int \frac{\partial}{\partial x_i} F\big((v_i + z - v_j)_{j \leq d}\big) \, dz. \tag{12.34}$$

This formula will be applied to dfs $F = G$, where G are EV dfs with univariate Gumbel margins.

Let $G(\boldsymbol{x}) = \exp(-\Psi(\boldsymbol{x}))$ be an EV df with univariate exponential margins. From (12.20) and (12.23) we know that the auxiliary function Ψ is homegeneous of order 1, that is

$$\Psi(a\boldsymbol{y}) = a\Psi(\boldsymbol{y}), \qquad \boldsymbol{y} < 0, \ a > 0, \tag{12.35}$$

which implies that the partial derivative is homogeneous of order 0, that is

$$\frac{\partial}{\partial x_i} \Psi(a\boldsymbol{y}) = \frac{\partial}{\partial x_i} \Psi(\boldsymbol{y}), \qquad \boldsymbol{y} < 0, \ a > 0.$$

In addition, EV dfs with univariate Gumbel margins are of the form

$$G(\boldsymbol{x}) = \exp\big(-\Psi\big((-e^{-x_j})_{j \leq d}\big)\big),$$

cf. (12.18).

Using the homogeneity properties one gets

$$
\begin{aligned}
p_i &= \int (-e^{-z}) \frac{\partial}{\partial x_i} \Psi\big((-e^{-(v_i+z-v_j)})_{j \leq d}\big) \exp\big(-\Psi\big((-e^{-(v_i+z-v_j)})_{j \leq d}\big)\big) \, dz \\
&= \int (-e^{-z}) \frac{\partial}{\partial x_i} \Psi\big((-e^{v_j})_{j \leq d}\big) \exp\big(-e^{-(v_i+z)}\Psi\big((-e^{v_j})_{j \leq d}\big)\big) \, dz \\
&= (-e^{v_i}) \frac{\partial}{\partial x_i} \Psi\big((-e^{v_j})_{j \leq d}\big) \big/ \Psi\big((-e^{v_j})_{j \leq d}\big). \tag{12.36}
\end{aligned}
$$

In the special case of a Gumbel–McFadden df $H_{\lambda,d}$ one gets the multinomial logit form

$$p_i = e^{v_i \lambda} \Big/ \sum_{j \leq d} e^{v_j \lambda}$$

for the choice probabilities p_i.

We have seen that the modeling of the random disturbances in utilities by means of EV distributions leads to attractive formulas for the choice probabilities.

12.3 Estimation in Extreme Value Models

A multivariate EV distribution is determined by the parameters (shape, location and scale parameter) in the single components and by the parameters of the copula or by the copula itself in the nonparametric approach.

The first class of parameters can be estimated by means of the methods presented in Part II; new estimation procedures must be developed for the second problem.

Estimation by Piecing–Together from Lower–Dimensional Margins

Let $\boldsymbol{x}_i = (x_{i,1}, \ldots, x_{i,d})$, $i = 1, \ldots, k$, be governed by the d–variate EV df G that is determined by the location, scale and shape parameter vectors $\boldsymbol{\mu}$, $\boldsymbol{\sigma}$ and $\boldsymbol{\gamma}$ and, in addition, by the copula C.

From (12.18) we know that G has the representation

$$G(\boldsymbol{x}) = C\Big(\big(G_{\gamma_j,\mu_j,\sigma_j}(x_j) \big)_{j \le d} \Big)$$

for vectors $\boldsymbol{x} = (x_1, \ldots, x_d)$, where $G_{\gamma_j,\mu_j,\sigma_j}$ is the jth marginal df of G. If $C \equiv C_\lambda$ is given in a parametric form, then we speak of a dependence parameter $\boldsymbol{\lambda}$.

To estimate G one must construct estimates of $\boldsymbol{\mu}, \boldsymbol{\sigma}, \boldsymbol{\gamma}$ and C (with C replaced by $\boldsymbol{\lambda}$ in the parametric case). One can easily find estimates of $\boldsymbol{\mu}, \boldsymbol{\sigma}$ and $\boldsymbol{\gamma}$ by taking estimates in the single components. Notice that the data $x_{1,j}, \ldots, x_{k,j}$ in the $j$the component are governed by the $j$the marginal df $G_{\gamma_j,\mu_j,\sigma_j}$ of G. Therefore, one may take estimates of the parameters γ_j, μ_j and σ_j as introduced in Section 4.1.

The definition of the copula C suggests to base the estimation of C or the unknown dependence parameter $\boldsymbol{\lambda}$ on the transformed vectors

$$\boldsymbol{z}_i = \big(G_{\hat{\gamma}_{j,k}, \hat{\mu}_{j,k}, \hat{\sigma}_{j,k}}(x_{i,j}) \big)_{j \le d}, \qquad i = 1, \ldots, k. \tag{12.37}$$

Notice that the $\boldsymbol{z}_i$ only depend on the $\boldsymbol{x}_i$. Subsequently, the $\boldsymbol{z}_i$ are regarded as vectors that are governed by C. Let $\widehat{C}_k$ be an estimate of C based on $\boldsymbol{z}_1, \ldots, \boldsymbol{z}_k$ within the copula model. If C is of the parametric form C_λ, then construct an estimate $\hat{\boldsymbol{\lambda}}_k$ of the parameter $\boldsymbol{\lambda}$.

Thus, the piecing–together method yields estimates $\hat{\boldsymbol{\gamma}}_k$, $\hat{\boldsymbol{\mu}}_k$, $\hat{\boldsymbol{\sigma}}_k$ and $\hat{\boldsymbol{\lambda}}_k$ of $\boldsymbol{\gamma}$, $\boldsymbol{\mu}, \boldsymbol{\sigma}$ and $\boldsymbol{\lambda}$ or, alternatively, the estimate

$$\widehat{G}_k(\boldsymbol{x}) = \widehat{C}_k\Big(\big(G_{\hat{\gamma}_{j,k}, \hat{\mu}_{j,k}, \hat{\sigma}_{j,k}}(x_j) \big)_{j \le d} \Big)$$

of the EV df G, where $\widehat{C}_k$ can be of the form $C_{\hat{\boldsymbol{\lambda}}_k}$ [10].

---

[10]For supplementary results concerning multivariate EV models see [16] and
  Tiago de Oliveira, J. (1989). Statistical decisions for bivariate extremes. In [14], pp. 246–261, or Smith, R.L., Tawn, J.A. and Yuen, H.K. (1990). Statistics of multivariate extremes. ISI Review 58, 47–58.

The Pickands Estimator in the Marshall–Olkin Model

We give a simple estimate of the dependence parameter λ in the model of standard Marshall–Olkin dfs, namely, for $(x_1, y_1), \ldots, (x_k, y_k)$,

$$\hat{\lambda}_k = 2 + k \Big/ \sum_{i \le k} \max\{x_i, y_i\}. \tag{12.38}$$

This estimate can be made plausible in the following way: from (12.11), deduce that $(2 - \lambda)\max\{X, Y\}$ is an exponential random variable with df $G_{2,-1}$, if (X, Y) has the Marshall–Olkin df M_λ. Therefore,

$$\frac{1}{k} \sum_{i \le k} \max\{x_i, y_i\} \approx E \max\{X, Y\} = -\frac{1}{2 - \lambda}$$

and $\hat{\lambda}_k \approx \lambda$. This yields estimates within an enlarged model.

EXAMPLE 12.3.1. (American Football (NFL) Data.) We partially repeat the analysis by Csörgő and Welsh[11]. The data were extracted from game summaries published in the Washington Post newspaper during three consecutive weekends in 1986. Consider the random game times to the first

- field goal (denoted by U);
- unconverted touchdown or safety (denoted by V),
- point–after touchdown (denoted by W).

Because the game time of the conversion attempt after a touchdown is zero, we have

$X = \min\{U, W\}$:   game time to the first kicking of the ball between the goalposts;

$Y = \min\{V, W\}$:   game time to the first moving of the ball into an endzone.

Notice that $X = Y$ if the first score is a point–after touchdown which happens with a positive probability. Assuming for a while that the random variables U, V, W are exponential it is reasonable to assume that (X, Y) has a Marshall–Olkin distribution. Table 12.1 contains the first and last three data vectors.

TABLE 12.1. Scoring times (minutes : seconds) from 42 American Football games.

| x: | 2:03 | 9:03 | 0:51 | $\cdots$ | 19:39 | 17:50 | 10:51 |
| y: | 3:59 | 9:03 | 0:51 | $\cdots$ | 10:42 | 17:50 | 38:04 |

The data (x_i, y_i), expressed in decimal minutes, are stored in the file football.dat. Because we introduced the Marshall–Olkin distributions in the version for maxima, we first change the signs of the data. Estimation in the EV model shows that a Weibull modeling for the single components is adequate, yet with $\alpha \ne -1$. The MLE(EV1) procedure provides the parameters $\alpha_1 = -1.39$, $\mu_1 = 0$, $\sigma_1 = 9.92$ and $\alpha_2 = -1.18$,

---

[11]Csörgő, S. and Welsh, A.H. (1989). Testing for exponential and Marshall–Olkin distributions. J. Statist. Plan. Inference 23, 287–300.

$\mu_2 = 0$, $\sigma_2 = 14.34$ in the first and second component. Thus, we have Weibull components, yet we still assume a bivariate Marshall–Olkin structure.

Check that $-(-X/\sigma)^{-\alpha}$ has the exponential df $G_{2,-1}$ if X has the df $G_{2,\alpha,0,\sigma}$. Therefore, a modeling by standard Marshall–Olkin dfs is more adequate for the transformed data

$$(x_i', y_i') = \left(-(-x_i/12)^{1.2}, -(-y_i/12)^{1.2} \right), \qquad i = 1, \ldots, 42. \tag{12.39}$$

Based on (x_i', y_i') one obtains the estimate $\hat{\lambda} = 0.63$ for the dependence parameter λ. Finally, the converse transformations provide an estimated df for the original data.

An application of the ML method is a bit more complicated, because one must deal with more sophisticated densities, namely densities with respect to the sum of the 2–dimensional Lebesgue measure on the plane and the 1–dimensional Lebesgue measure on the main diagonal.

Estimation in the Gumbel–McFadden and Hüsler–Reiss Models

First, specify a model for the univariate margins. We will primarily deal with PTEs (piecing–together estimates) so that first the parameters in the univariate margins are estimated by means of one of the estimation methods in Chapter 4. After a transformation, as explained in (12.37), we may assume that the data are governed by the df L_λ or H_λ.

Thus, an estimate of the dependence parameter λ or the canonical parameter ϑ must be constructed.

- Moment Method: because of the one–to–one relationship between the dependence parameter λ and the correlation coefficient ρ, one obtains an estimate of λ based on the sample correlation coefficient.

- Pickands Estimator: assume that the dfs G are given in the Pickands representation (12.20). Deduce that $((2 - T(G)) \max(X, Y)$ is an exponential random variable with exponential df $G_{2,-1}$, where $T(G)$ is the canonical parameter. Therefore, generalizing (12.38) we conclude that

$$\widehat{T}(G)_k = 2 + k \Big/ \sum_{i \leq k} \max\{x_i, y_i\} \tag{12.40}$$

is an estimator of the functional parameter $T(G)$, which also provides an estimator of the canonical parameter ϑ within the parametric framework.

- Maximum Likelihood Method: from the density in (12.29) or (12.16), deduce the likelihood equation and calculate the MLE numerically. The preceding moment or Pickands estimate may serve as the initial value of the iteration procedure.

Such PTEs may also serve as initial values of an iteration procedure to evaluate the MLE in the full bivariate model.

EXAMPLE 12.3.2. (Ozone Concentration in the San Francisco Bay Area.) We partially repeat the analysis of ozone concentration in the aforementioned article by Harry Joe[4]. The data set (stored in the file cm–ozon1.dat) consists of 120 weekly maxima of hourly averages of ozone concentrations measured in parts per hundred million for the years 1983–1987 for each of the five monitoring stations (Concord (cc), Pittsburg (pt), San Jose (sj), Vallejo (va), Santa Rosa (st)). Weeks within the months April to September are only taken, because there are smaller maxima in the winter months.

The first 21 maxima are listed in Table 12.2 to give a first impression of the data.

TABLE 12.2. The first 21 maxima at 5 stations.

No.	cc	pt	sj	va	st	No.	cc	pt	sj	va	st	No.	cc	pt	sj	va	st
1.	7	6	6	6	5	8.	10	11	11	8	5	15.	13	8	15	8	7
2.	6	7	5	5	4	9.	7	7	6	4	4	16.	8	8	8	5	4
3.	6	7	8	5	6	10.	11	14	13	13	6	17.	7	8	6	3	4
4.	6	6	5	4	4	11.	12	10	12	8	5	18.	9	8	8	4	5
5.	6	6	5	4	4	12.	7	7	7	4	4	19.	13	11	14	12	7
6.	5	6	5	4	4	13.	8	8	8	5	5	20.	15	13	13	10	7
7.	9	9	7	7	5	14.	8	7	7	5	4	21.	5	6	6	4	4

Next, MLEs for the Gumbel (EV0) and the full EV model are calculated for the single components. The estimates suggest to take a Gumbel modeling for the margins. This is supported by a nonparametric visualization of the data. We have not analyzed any other data in this book with a better fit to a parametric model (perhaps with the exception of Michelson's data concerning the velocity of light in the air). Of course the stronger discrete nature of the data becomes visible.

TABLE 12.3. MLE(EV0) and MLE(EV) for the single stations.

	Univariate Parameters								
	cc			pt			sj		
	γ	μ	σ	γ	μ	σ	γ	μ	σ
MLE(EV0)	0	7.21	2.02	0	6.81	1.62	0	7.23	2.13
MLE(EV)	-0.06	7.28	2.06	0.06	6.76	1.58	0.00	7.23	2.13

	va			st		
	γ	μ	σ	γ	μ	σ
MLE(EV0)	0	5.40	1.66	0	4.54	1.10
MLE(EV)	0.12	5.29	1.58	0.11	4.48	1.05

We also include in Table 12.4 the estimated pairwise dependence parameters λ at the 5 different stations.

Dependence Parameter λ			
	cc	pt	sj	va
pt	2.7			
sj	1.9	1.5		
va	1.7	1.6	1.5	
st	1.3	1.2	1.2	1.6

TABLE 12.4. MLEs of pairwise dependence parameters λ.

As it was already reported in the article by H. Joe[4] that the Hüsler–Reiss modeling is adequate for the present data set. This judgement is supported by 3–D plots in Fig. 12.5 of a bivariate kernel density for the cc–pt data (38 distinct pairs with certain numbers of multiplicities) and the estimated parametric density.

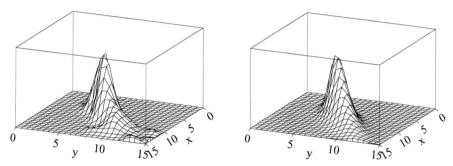

FIG. 12.5. Bivariate kernel density (left) and parametric density (right) for cc–pt data.

What has been achieved in the multivariate framework is promising yet this approach is still in a research stadium.

12.4 A Spectral Decomposition Methodology

We merely state some basic properties and results about a certain spectral decomposition in the special case of bivariate distributions. For a continuation we refer to Section 13.3. A detailed account of this methodology, including the d–dimensional case, is given in Falk et al. [17].

A Spectral Decomposition

Any vector $(x, y) \in (-\infty, 0]^2$ with $(x, y) \neq (0, 0)$ can be uniquely represented by means of the angular and radial components z and c with $z = x/(x + y) \in [0, 1]$ and $c = x + y \leq 0$. We have

$$(x, y) = (cz, c(1 - z)). \tag{12.41}$$

Any df H with support $(-\infty, 0]^2$ may be written in the form $H(cz, c(1-z))$. Putting

$$H_z(c) = H(cz, c(1-z)), \quad z \in [0,1], \ c \le 0, \tag{12.42}$$

one gets a one–to–one representation of H by means of the family of univariate dfs $\{H_z : z \in [0,1]\}$ which we call spectral decomposition of H. Check that H_0 and H_1 are the marginal dfs of H in the 1st and 2nd component.

First, we provide two examples where H is the df of independent exponentially and, respectively, uniformly distributed random variables:

(a) if $H(x, y) = \exp(x + y)$, $x, y \le 0$, then

$$H_z(c) = \exp(c), \quad c \le 0,$$

(b) if $H(x, y) = (1+x)(1+y)$, $-1 \le x, y \le 0$, then

$$H_z(c) = 1 + c + c^2 z(1 - z), \quad c \le 0, \ 1 + c + c^2 z(1 - z) \le 0.$$

Next we extend (a) to bivariate EV dfs in general.

(c) A bivariate EV df G with Pickands dependence function D, cf. (12.20), has the spectral dfs

$$G_z(c) = \exp(cD(z)), \quad c \le 0.$$

Notice that G_z is an exponential df with reciprocal scale parameter $D(z)$. In particular, the spectral dfs are univariate EV dfs.

We also make use of the partial derivative

$$h_z(c) = \frac{\partial}{\partial c} H_z(c). \tag{12.43}$$

In the preceding examples (a)–(c) we have

(a') $h_z(c) = \exp(c)$, $c \le 0$;

(b') $h_z(c) = 1 + 2cz(1 - z)$, $c \le 0$, $1 + c + c^2 z(1 - z) \le 0$,

(c') $g_z(c) = D(z) \exp(cD(z))$, $c \le 0$.

A Spectral Condition

Assume that H is a bivariate df with support $(-\infty, 0]$ such that

$$H_z(c) = 1 + cg(z)(1 + o(1)), \quad c \uparrow 0, \ z \in [-1, 0], \tag{12.44}$$

where $g(0) = g(1) = 1$. Then, the following assertions hold, cf. [17], Theorem 5.3.2, which is formulated for the d–variate case:

- $g(z) = D(z)$ is a Pickands dependence function,

- H is in the domain of attraction of the EV df G with Pickands dependence function; more precisely,

$$H^n \left(\frac{x}{n}, \frac{y}{n} \right) \to G(x, y), \quad n \to \infty.$$

It is apparent that condition (12.44) is satisfied with $D(z) = 1$ in the cases (a) and (b), and for $D(z)$ in the general case (c) of EV dfs. Because of the standardization $D(0) = D(1) = 1$ we get for the marginal dfs

$$H_0(c) = 1 + c(1 + o(1)) \quad \text{and} \quad H_1(c) = 1 + c(1 + o(1)).$$

Therefore, it suggests itself to transform original data to $(-1, 0)$–uniform or exponential data before applying results based on this condition.

Condition (12.44) can be verified by the following "differentiable" version: if $h_z(c) > 0$ for c close to zero, $z \in [0, 1]$, and

$$h_z(c) = g(z)(1 + o(1)), \quad c \uparrow 0, \ z \in [0, 1], \tag{12.45}$$

where $g(0) = g(1) = 1$, then condition (12.44) holds.

It is evident that condition (12.45) is satisfied for the examples in (a')–(c') at a certain rate. In Section 13.3 we formulate an expansion for $h_z(c)$ which specifies the rate at which (12.45) is attained. This expansion is applied to the testing of tail dependence against tail independence. Section 13.3 also provides further examples for which condition (12.45) is satisfied.

The Spectral Decomposition of a GP Distribution Function

The spectral decomposition of a bivariate GP df $W = 1 + \log G$, which will be introduced in Section 13.1, consists of uniform dfs, we have

$$W_z(c) = 1 + cD(z), \quad -1/D(z) \leq c \leq 0,$$

for $0 \leq z \leq 1$, where D is again the pertaining Pickands dependence function.

Thus, the univariate margins of bivariate GP dfs are again GP dfs and, more general, the spectral dfs of bivariate GP dfs are GP dfs.

The Random Angular and Radial Components

Let (X, Y) be a random vector with joint df H which has the support in $(-\infty, 0]^2$. Then, the vector

$$(X/(X + Y), X + Y)$$

may be called Pickands transform with the angular and radial component in the
1st and 2nd component (cf. Falk et al. [17], pages 150–153).

It turns out that the angular and radial components have remarkable, stochastic properties. They are conditionally independent conditioned on $X + Y > c$
for GP dfs W under mild conditions. We have unconditional independence if
one takes a different version of the GP df, namely that truncated outside of
$\{(x, y) \in (-\infty, 0]^2 : x + y > c\}$ with $c > -1$. Moreover, the asymptotic conditional independence holds for dfs in a neighborhood of W.

This result for GP distributions is closely related to that for spherical distributions, see Section 11.2. In that section, the radial component is the L_2–norm of
the spherical random vector, whereas the radial component $X + Y$ in the present
section is related to the L_1–norm.

The testing of tail dependence against tail independence in Section 13.3 will
be based on the radial component.

Chapter 13

Multivariate Peaks Over Threshold

co–authored by M. Falk[1]

We already realized in the univariate case that, from the conceptual viewpoint, the peaks–over–threshold method is a bit more complicated than the annual maxima method. One cannot expect that the questions are getting simpler in the multivariate setting. Subsequently, our attention is primarily restricted to the bivariate case, this topic is fully worked out in [16], 2nd ed., for any dimension. A new result about the testing of tail dependence is added in Section 13.3.

13.1 Nonparametric and Parametric Generalized Pareto Models

We introduce bivariate generalized Pareto (GP) distributions and deal with the concepts of canonical and tail–dependence parameters within this framework.

Introducing Bivariate GP Distributions

In analogy to the univariate case, define a bivariate GP distribution[2] pertaining to an EV df G by

$$W(x, y) = 1 + \log G(x, y), \qquad \text{if } \log G(x, y) \geq -1. \tag{13.1}$$

---

[1]Katholische Universität Eichstätt; now at the University of Würzburg.

[2]Kaufmann, E. and Reiss, R.–D. (1995). Approximation rates for multivariate exceedances. J. Statist. Plan. Inf. 45, 235–245; see also the DMV Seminar Volume [16].

The univariate margins of the bivariate GP df W are just the univariate GP dfs pertaining to the margins of the EV df G.

If G is an EV df such that the Pickands representation (12.20) holds, then

$$W(x,y) = 1 + (x+y)D\left(\frac{y}{x+y}\right), \qquad \text{if } (x+y)D\left(\frac{y}{x+y}\right) > -1, \qquad (13.2)$$

where D is the Pickands dependence function of G. In this case, the univariate margins of W are both the uniform df $W_{2,-1} = 1+x$ on $[-1,0]$.

Because $D \le 1$, we know that the condition $\log G(x,y) \ge -1$ is satisfied if $x+y \ge -1$. In the subsequent calculations, we assume that the latter condition is satisfied to avoid technical complications.

If $\widetilde{W}$ is any bivariate GP df with margins $\widetilde{W}_1$ and $\widetilde{W}_2$, then

$$W(x,y) = \widetilde{W}\big(\widetilde{W}_1^{-1}(1+x), \widetilde{W}_2^{-1}(1+y)\big), \qquad -1 < x, y < 0, \qquad (13.3)$$

is of the standard form (13.2).

Tajvidi's Definition of Bivariate GP Distributions

The definition of a bivariate GP df is not as self–evident as in the univariate case. We mention the relation of the preceding definition of a GP df to another one.

Consider the map

$$m_{(u,v)}(x,y) = \big(\max(x,u), \max(y,v)\big) \qquad (13.4)$$

which replaces marginal values $x < u$ and, respectively, $y < v$ by the marginal thresholds u and v as illustrated in Fig. 13.1.

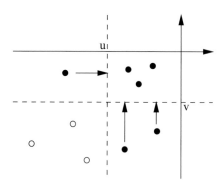

Fig. 13.1. Putting points (x,y) to $m_{(u,v)}(x,y)$.

We refer to Section 13.3, where the map $m_{u,v}$ is once again utilized in the context of a point process approximation.

Let (X, Y) be a random vector with GP df W as introduced in (13.1) and let u, v be thresholds such that $\log G(u, v) = -1$. Then, the random vector $m_{(u,v)}(X, Y)$ has the df

$$\widetilde{W}(x, y) = \begin{cases} W(x, y) & (x, y) \geq (u, v) \\ & \text{if} \\ 0 & \text{otherwise} \end{cases} \tag{13.5}$$

which is a GP df as defined by Tajvidi[3].

Thus, both definitions are closely related to each other. We remark that

- both distributions are identical in the region $\{(x, y) : x > u, y > v\}$,

- in contrast to the GP df W, the modification $\widetilde{W}$ does not possess the property that the margins are again GP dfs.

In view of the latter remark, we prefer to deal with GP dfs as introduced in (13.1). Using such dfs also facilitates transformations in the univariate margins.

Canonical and Tail Dependence Parameters

Now, let (X, Y) be a random vector with df bivariate GP df W as given in (13.2). Applying (10.7) one gets for $-1/2 \leq u < 0$,

$$\begin{aligned} P(Y > u | X > u) &= \frac{2|u| - 2|u| D_W(1/2)}{|u|} \\ &= 2\left(1 - D_W(1/2)\right) \\ &=: T(W), \end{aligned} \tag{13.6}$$

where D_W is the Pickands dependence function pertaining to W. We see that the conditional probabilities do not depend on the threshold u.

Notice that $T(W)$ is just the canonical parameter introduced in (12.25) for bivariate EV dfs. Thus, for a parametric family of GP dfs W_ϑ, for which there is a one–to–one relation between the original parameter λ and $T(W_\lambda)$, we may introduce another representation by taking the canonical parameter

$$\vartheta = T(W_\lambda).$$

It is apparent from (13.6) that the survivor function $\overline{W}_\vartheta$ satisfies

$$\overline{W}(u, u)/|u| = T(W), \qquad -1/2 \leq u \leq 0. \tag{13.7}$$

---

[3]Tajvidi, N. (1996). Characterization and some statistical aspects of univariate and multivariate generalised Pareto distributions. PhD Thesis, Dept. of Mathematics, University of Göteborg.

Recall from (12.26) that such a relation approximately holds for the pertaining EV df.

In addition, we have

$$T(W) = \chi(q), \qquad q \geq 1/2, \tag{13.8}$$

where $\chi(q)$ is the tail dependence parameter at the level q introduced in (2.60).

Parametric Models of Bivariate GP Distributions

Explicit representations of W_ϑ, where ϑ is the canonical parameter, are given for the bivariate Marshall–Olkin, Gumbel–McFadden and Hüsler–Reiss dfs.

- (Marshall–Olkin–GP dfs.) Because $\vartheta = \lambda$, we obtain from (13.1) and (12.11),

$$W_\vartheta(x,y) = 1 + (1-\vartheta)(x+y) + \vartheta \min\{x,y\}, \tag{13.9}$$

  whenever $x, y \leq 0$ and the right–hand side exceeds zero.

- (Gumbel–McFadden–GP dfs.) Because $\vartheta = 2 - 2^{1/\lambda}$, we have

$$\lambda = (\log 2)/\log(2 - \vartheta).$$

  (13.1) and (12.28) yield

$$W_\vartheta(x,y) = 1 - \left((-x)^{\frac{\log 2}{\log(2-\vartheta)}} + (-y)^{\frac{\log 2}{\log(2-\vartheta)}} \right)^{\frac{\log(2-\vartheta)}{\log 2}}, \tag{13.10}$$

  whenever $x, y \leq 0$ and the right–hand side exceeds zero.

- (Hüsler–Reiss–GP dfs.) Because $\vartheta = 2(1 - \Phi(1/\lambda))$, we have $\lambda = 1/\Phi^{-1}(1 - \vartheta/2)$. Deduce from (13.1) and (12.19) that

$$W_\vartheta(x,y) = 1 + \psi_\vartheta\left(\frac{y}{x}\right) y + \psi_\vartheta\left(\frac{x}{y}\right) x, \tag{13.11}$$

  where the auxiliary function ψ_ϑ is defined by

$$\psi_\vartheta(z) = \Phi\left(\Phi^{-1}\left(1 - \frac{\vartheta}{2}\right) + \frac{1}{2\Phi^{-1}\left(1 - \frac{\vartheta}{2}\right)} \log(z) \right), \qquad 0 < z < \infty,$$

  whenever $x, y < 0$ and the right–hand side in (13.11) exceeds zero.

In all three cases we have $W_0(x,y) = 1 + (x+y)$ and $W_1(x,y) = 1 + \min\{x,y\}$. It is remarkable that W_0 is the uniform distribution on the line $\{(x,y) : x, y \leq 0, x + y = -1\}$.

The Canonical Dependence Function

We define the canonical dependence function of a bivariate GP df W by

$$T_W(z) = 2(1 - D_W(z)), \qquad (13.12)$$

where D_W is again the Pickands dependence function pertaining to W.

In conjunction with EV distributions, Huang[4] studied the stable tail dependence function

$$l(x, y) = (x + y)D\left(\frac{y}{x + y}\right)$$

which is the function $-\Psi(x, y)$ on page 304. For obvious reasons we prefer to work with a real–valued function which varies between zero and one.

The value of the canonical dependence function $T_W(1/2)$ at $z = 1/2$ is the canonical parameter $T(W)$. We also write T_ϑ etc., if there is a parametric family of GP dfs reresented in the canonical parameterization. Given the canonical parameterization we have $T_0(z) = 0$ and $T_1(z) = \min(2z, 2(1 - z))$, if tail independence and, respectively, total tail dependence hold.

In Fig. 13.2 we plot several canonical dependence functions T_ϑ for Gumbel–McFadden and Marshall–Olkin dfs ranging from the case of tail independence to the one of total dependence.

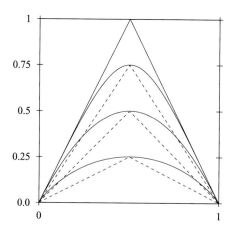

FIG. 13.2. Canonical dependence functions T_ϑ for Gumbel–McFadden (solid) and Marshall–Olkin (dashed) distribution with canonical parameters $\vartheta = 0, 0.25, 0.5, 0.75, 1$.

We provide a representation of the canonical dependence function T_W by means of the survivor function $\overline{W}$ which extends that for the canonical parameter, see (13.7).

---

[4] Huang, X. (1992). Statistics of bivariate extreme values. PhD Thesis, Erasmus University Rotterdam.

Let W be a standard bivariate GP df as given in (13.2). From the representation (10.7) of a survivor function deduce that

$$2\overline{W}(-d(1-z), -dz)/d = T_W(z) \qquad (13.13)$$

for $0 < d \le 1$ and $0 \le z \le 1$. Notice that the formula (13.7) for the canonical parameter $T(W) = T_W(1/2)$ is a special case. In addition, one must take $|u| = d/2$. Here, z and d may be regarded as direction and distance measures.

Therefore, by estimating the survivor function $\overline{W}$ in (13.13)—for certain selected values d which may depend on z—one gets estimates of the canonical dependence function T_W and the canonical parameter $T(W)$.

13.2 Estimation of the Canonical Dependence Function

In this section we estimate the canonical dependence function T_W by means of a sample version of the survivor function. Consequently, the actual df F can be estimated in the upper tail region by the piecing–together method. The results presented here should be regarded as a preliminary introduction[5].

Estimation Within the Standard GP Model

Let (x_i, y_i) be governed by the standard GP df W as given in (13.2). In view of (13.13) one may take

$$\widehat{T}_n(z) = \frac{2}{nd} \sum_{i \le n} I(x_i > -d(1-z),\ y_i > -dz) \qquad (13.14)$$

as an estimate of the canonical dependence function $T_W(z)$, where $0 < d \le 1$ can be selected by the statistician. Spezializing this to $z = 1/2$ one gets by $\widehat{T}_n(1/2)$ an estimate of the canonical parameter $T(W)$.

This approach to estimating the canonical dependence function includes the case where the modeling of an actual survivor function $\overline{F}$ by means of $\overline{W}$ is sufficiently accurate for $x, y > u \ge -1$. In that case, one must take $d = |u|$.

Transforming Univariate GP Margins

In contrast to the preceding lines we consider the more general situation of bivariate GP dfs $\widetilde{W}$ having the GP margins $\widetilde{W}_1$ and $\widetilde{W}_2$ which are not necessarily equal to the uniform df on the interval $[-1, 0]$.

---

[5]For further results see Falk, M. and Reiss, R.–D. (2003). Efficient estimators and LAN in canonical bivariate pot models. J. Mult. Analysis 84, 190–207.

1. Use estimated univariate GP distributions to transform the original data to the standard form of $[-1, 0]$–uniform data;

2. estimate the canonical dependence function (the canonical parameter) as in (13.14) based on the transformed data,

3. use the estimated canonical dependence function and univariate GP dfs to construct an estimate of the GP df $\widetilde{W}$.

This approach can be also applied to the situation, where univariate GP dfs $\widetilde{W}_1$ and $\widetilde{W}_2$ are fitted to the upper tails of actual univariate margins F_1 and F_2 above the thresholds $v(1)$ and $v(2)$. Assume that the bivariate GP modeling of the actual bivariate distribution is valid for $x, y > u = \max(u(1), u(2))$, where the $u(i)$ are the transformed thresholds $v(i)$.

Arbitrary Margins

Let X and Y be random variables having continuous dfs F_X and F_Y. Put

$$p(d, z) := P\{X > F_X^{-1}(1 - d(1 - z)), Y > F_Y^{-1}(1 - dz)\}. \tag{13.15}$$

A natural estimator of the probability $p(d, z)$ is

$$\hat{p}_n(d, z) = \frac{1}{n} \sum_{i \leq n} I\left(x_i > x_{[n(1-d(1-z))]:n}, y_i > y_{[n(1-dz)]:n}\right),$$

where (x_i, y_i) are realizations of (X, Y).

Applying the quantile transformation one gets

$$
\begin{aligned}
p(d, z) &= P\{U > -d(1 - z), V > -dz\} \\
&= \bar{F}\left(-d(1 - z), -dz\right)
\end{aligned}
$$

for $0 < d \leq 1$ and $0 \leq z \leq 1$, where U and V are $[-1, 0]$–uniform random variables with common df F. In view of (13.13) it is plausible to assume that for all $0 \leq z \leq 1$,

$$\left|p(d, z) - \overline{W}\left(-d(1 - z), -dz\right)\right|/d \to 0, \qquad d \to 0, \tag{13.16}$$

where W is a standard GP df. For the probabilities in (13.15) one gets

$$2p(d, z)/d \to T_W(z), \qquad d \to 0,$$

Thus, T_W may be regarded as a limiting canonical dependence function. Putting $d = k/n$, one arrives at the estimator

$$\widehat{T}_{n,k}(z) = \frac{2}{k} \sum_{i \leq n} I\left(x_i > x_{n-[k(1-z)]:n}, y_i > y_{n-[kz]:n}\right) \tag{13.17}$$

of the limiting canonical dependence function. For the direction $z = 1/2$ one gets again an estimator of the canonical parameter.

In Fig. 13.3 we plot the estimate $\widehat{T}_{n,k}(z)$ of the canonical dependence function based on $n = 500$ Gumbel–McFadden data generated under the parameter $\vartheta = 0.5$ with $k = 60$. For $z = 0.5$ one gets an exceptional accuarate estimate of the canonical parameter.

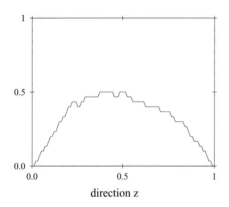

FIG. 13.3. Estimated canonical dependence function generated under the canonical parameter $\vartheta = 0.5$.

A somewhat related estimator was examined by Deheuvels[6] in the context of "tail probabilities". We refer to Huang[4] and Drees and Huang[7] for recent asymptotic results and a survey of the literature.

Estimating by Piecing–Together

Piecing together

- the estimated univariate GP dfs,

- the estimated parametric bivariate standard GP df

one gets an approximation to the original bivariate df in the upper tail region. This approximation can be extrapolated to higher tail regions outside of the range of the data

If this is done within the nonparametric framework—using the estimated canonical dependence function—then one should be aware that the resulting df is

---

[6]Deheuvels, P. (1980). Some applications to the dependence functions in statistical inference: nonparametric estimates of extreme value distributions, and a Kiefer type universal bound for the uniform test of independence. In: Nonparametric Statistical Inference, pp. 183–201, B.V. Gnedenko et al. (eds), North Holland, Amsterdam.

[7]Drees, H. and Huang, X. (1998). Best attainable rates of convergence for estimators of the stable tail dependence function. J. Mult. Analysis 64, 25–47.

not a bivariate GP df. This can be achieved by using a concave majorant of the sample canonical dependence function.

We present once again the scatterplot and contour plot in Fig. 10.3 and add an extrapolated contour plot.

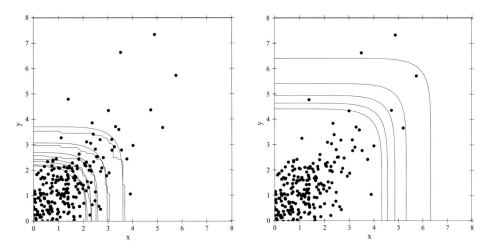

FIG. 13.4. (left.) Contour plots of theoretical and sample survivor function. (right.) Estimated parametric extrapolation to the upper tail region.

13.3 About Tail Independence
co–authored by M. Frick[8]

In this section we discuss some well–known measures of independence which characterize different degrees of tail independence and indicate the relationship of these parameters to each other. Moreover, we study a test of tail dependence against tail independence based on the radial component.

Recall from (12.10) that

$$\bar{\Phi}_\rho(\boldsymbol{b}_n + \boldsymbol{b}_n^{-1}\boldsymbol{x}) = O\big(n^{-2/(1+\rho)}(\log n)^{-\rho/(1+\rho)}\big),$$

where $\bar{\Phi}_\rho(x, y) = P\{X > x, Y > y\}$ is the survivor function of a standard normal vector (X, Y) with correlation coefficient ρ such that $-1 < \rho < 1$. Thus, according to (12.9), the right–hand side determines the rate at which the random maxima in each component become independent as the sample size goes to infinity. The latter property was addressed as the upper tail independence of the bivariate normal df.

---

[8]University of Siegen.

The Coefficient of Tail Dependence for Normal Vectors

In an article by Ledford and Tawn[9], relation (12.10) is reformulated and strengthened with (X, Y) replaced by the pertaining copula random vector $(U, V) = (\Phi(X), \Phi(Y))$ where Φ is the univariate standard normal df.

These authors prove that

$$P\{U > u, V > u\} \sim c(\rho)(1 - u)^{2/(1+\rho)}(-\log(1 - u))^{-\rho/(1+\rho)}, \quad u \to 1, \quad (13.18)$$

where $c(\rho) = (1 + \rho)^{3/2}(1 - \rho)^{-1/2}(4\pi)^{-\rho/(1+\rho)}$.

In that context, Coles et al.[10] introduced the coefficient of tail dependence at the level u, namely,

$$\bar{\chi}(u) = \frac{2\log P\{U > u\}}{\log P\{U > u, V > u\}} - 1, \quad (13.19)$$

and the coefficient of tail dependence

$$\bar{\chi} = \lim_{u \to 1} \bar{\chi}(u). \quad (13.20)$$

It is easy to verify that

$$\bar{\chi} = \rho \quad (13.21)$$

for the particular case of $(0, 1)$–uniformly distributed random variables U and V with normal dependence structure as in (13.18). As pointed out by Coles et al., the relation (13.21) "provides a useful benchmark for interpreting the magnitude of $\bar{\chi}$ in general models."

The reason for introducing the coefficient of tail dependence as another tail dependence parameter is the desire to distinguish between pairs of random variables which are both tail independent, that is

$$\chi = \lim_{u \to 1} \chi(u) := \lim_{u \to 1} \frac{P\{U > u, V > u\}}{P\{U > u\}} = 0,$$

cf. (2.61), but have different degrees of independence at an asymptotic level of higher order.

The Coefficient of Tail Dependence in General Models

For non–normal vectors, the coefficients of tail dependence $\bar{\chi}(u)$ and $\bar{\chi}$ are defined in analogy to (13.19) and (13.20). We list some properties of $\bar{\chi}(u)$ and $\bar{\chi}$:

---

[9]Ledford, A.W. und Tawn, J.A. (1996). Statistics for near independence in multivariate extreme values. Biometrika 83, 169–187.

[10]Coles, S., Heffernan, J.E. and Tawn, J.A. (1999). Dependence measures for extreme value analyses. Extremes 2, 339–365.

- $\bar{\chi}(u)$ and $\bar{\chi}$ are symmetric in U and V;

- $\bar{\chi}(u)$ and $\bar{\chi}$ range between -1 and 1;

- if $U = V$, then $\bar{\chi} = 1$.

The pair $(\chi, \bar{\chi})$ of tail dependence parameters may be employed to describe the tail dependence structure of two random variables:

$\chi > 0, \bar{\chi} = 1$ tail dependence with χ determining the degree of dependence.

$\chi = 0, \bar{\chi} < 1$ tail independence with $\bar{\chi}$ determining the degree of dependence.

Another Coefficient of Tail Dependence

We start again with a normal copula random vector (U, V) pertaining to a standard normal vector with correlation coefficient ρ with $-1 < \rho < 1$.
Relation (13.18) can be written

$$P\{U > u, V > u\} \sim L(1 - u)(1 - u)^{1/\eta}, \quad u \to 1, \tag{13.22}$$

where $\eta = (1 + \rho)/2$ and $L(1 - u) = c(\rho)(-\log(1 - u))^{-\rho/(1+\rho)}$ with $c(\rho)$ as in (13.18). The term η is again called coefficient of tail dependence.
Generally, if for a copula random vector (U, V), a relation

$$P\{U > u, V > u\} \sim L(1 - u)(1 - u)^{1/\eta}, \quad u \to 1, \tag{13.23}$$

holds, where L is a slowly varying function at 0 (i.e. as $u \to 1$), then η is called coefficient of tail dependence. It is evident that (13.22) is a special case. Notice that $0 < \eta \le 1$.

EXAMPLE 13.3.1. (Morgenstern distributions.) The copula of the Morgenstern df with parameter $-1 < \alpha \le 1$ is given by

$$C_\alpha(u, v) = uv[1 + \alpha(1 - u)(1 - v)],$$

see, e.g., Heffernan[11]. The bivariate survivor function satisfies the relation

$$P\{U > u, V > u\} \sim (1 + \alpha)(1 - u)^2, \quad u \to 1,$$

so that $\eta = 1/2$ and $L(1 - u) = 1 + \alpha$.

---

[11] Heffernan, J.E. (2000). A directory of coefficients of tail independence. Extremes 3, 279–290.

For further examples we refer to the article by Heffernan. We indicate the relationship between the different dependence parameters χ, $\bar{\chi}$ and η.

$$\bar{\chi} = 2\eta - 1;$$

$$\chi = \begin{cases} c & \text{if } \bar{\chi} = 1 \text{ and } L(u) \xrightarrow{u \to 1} c \geq 0, \\ 0 & \text{if } \bar{\chi} < 1. \end{cases} \qquad (13.24)$$

In Ledford and Tawn[12] one may find related formulas for the case of unequal thresholds u and v.

Estimating the Coefficient of Tail Dependence

In analogy to (2.62), a sample version pertaining to $\bar{\chi}(u)$ based on data (x_i, y_i), $i = 1, \ldots, n$, is

$$\bar{\chi}_n(u) = \frac{2 \log(1 - u)}{\log\left(\frac{1}{n} \sum_{i \leq n} I(x_i > x_{[nu]:n}, y_i > y_{[nu]:n})\right)} - 1, \qquad (13.25)$$

which is an estimator of $\bar{\chi}(u)$ as well as $\bar{\chi}$.

EXAMPLE 13.3.2. The illustration in Fig. 13.5 concerns a data set consisting of $n = 2,894$ three–hourly–measurements of the surge and wave heights taken at Newlyn, a coastal town in England, see Example 13.3.6. The sample version $\bar{\chi}_n(u)$ is plotted against the level u. The plot was generated by using the source code written by Coles, Heffernan and Tawn[13]. This plot suggests an estimate of $\bar{\chi}_n = 0.5$ of $\bar{\chi}$. Near $u = 1$ there is a larger variation of $\bar{\chi}_n(u)$ due to the fact that the estimate is merely based on a smaller number of extremes.

It would be desirable to get some theoretical results for this estimator (as well as for the serial version which is mentioned in the following lines).

Another Auto–Tail–Dependence Function

In analogy to the auto–tail–dependence function $\rho(u, h)$ for the parameter $\chi(u)$, see (2.63), one may introduce a auto–tail–dependence function $\bar{\rho}(u, h)$ for the tail–independence parameter $\bar{\chi}(u)$ at the level u.

---

[12]Ledford, A.W. und Tawn, J.A. (1997). Modelling dependence within joint tail regions. J.R. Statist. Soc. B 59, 475–499.

[13]www.maths.lancs.ac.uk/˜currie/Code/DependenceMeasuresForExtremes.S and -.txt

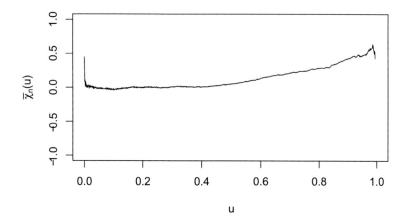

FIG. 13.5. The sample coefficient of tail dependence $\bar{\chi}_n(u)$ at the level u plotted against u for the wave and surge data.

Let again $X_1, \ldots, X_n$ be a series of identically distributed random variables with common df F. Then, for $i \leq n - h$, put

$$
\begin{aligned}
\bar{\rho}(u, h) &= \frac{2 \log P\{X_i > F^{-1}(u)\}}{\log P\{X_i > F^{-1}(u), X_{i+h} > F^{-1}(u)\}} - 1 \\
&= \frac{2 \log P\{X_1 > F^{-1}(u)\}}{\log P\{X_1 > F^{-1}(u), X_{1+h} > F^{-1}(u)\}} - 1,
\end{aligned} \quad (13.26)
$$

where we implicitly assume stationarity in the dependencies.

Likewise define an auto–tail–dependence function $\bar{\rho}(h)$ by

$$
\bar{\rho}(h) = \lim_{u \to 1} \bar{\rho}(u, h). \quad (13.27)
$$

Again $\bar{\rho}(u, h)$ and $\bar{\rho}(h)$ can be estimated by the sample versions

$$
\bar{\rho}_n(u, h) = \frac{2 \log(1 - u)}{\log \left(\frac{1}{n-h} \sum_{i \leq n-h} I(\min(x_i, x_{i+h}) > x_{[nu]:n}) \right)} - 1 \quad (13.28)
$$

based on the data $x_1, \ldots, x_n$.

EXAMPLE 13.3.3. (Gaussian AR(1) series.) We study once more a Gaussian AR(1) series as in Example 2.5.3 with autocorrelation function $\rho(h) = d^h$. It is evident that the autocorrelation function $\rho(h)$ and the auto–tail–dependence function $\bar{\rho}(h)$ are identical. Therefore, the sample versions estimate the same functions and should coincide to some extent.

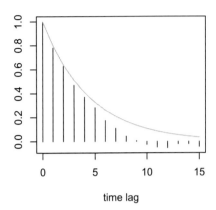
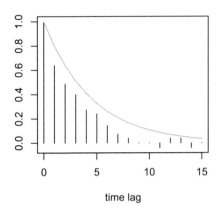

FIG. 13.6. Autocorrelation functions (left) and auto–tail–dependence functions (right).

In Fig. 13.6 we plot the sample autocorrelation function and sample auto–tail–independence function for $u = 0.8$ based on 200 Gaussian AR(1) data under the parameter $d = 0.8$. The theoretical correlation function $\rho(h) = d^h$ is included.

The sample autocorrelation function, which is based on the whole data set, provides a more accurate estimate of the theoretical curve than the sample auto-tail-dependence function because the latter sample function is merely based on the extremes.

A Spectral Expansion of Length 2

Next we make use of the spectral decomposition methodology, outlined in Section 12.4, to characterize the degree of tail independence.

First we state a refinement of condition (12.45) given in Section 12.4. Let X and Y be random variables having the joint bivariate df H with support $(-\infty, 0]^2$. Assume that H satisfies a differentiable spectral expansion of length 2, i.e.,

$$h_z(c) := \frac{\partial}{\partial c} H_z(c) = D(z) + B(c)A(z) + o(B(c)), \quad \text{as } c \uparrow 0, \qquad (13.29)$$

where B is regularly varying with exponent of variation $\beta > 0$. We say that H satisfies a differentiable spectral expansion of length 2 with exponent of variation β. Notice that B satisfies $\lim_{c\uparrow 0} B(c) = 0$.

An important example is given by $B(c) = |c|^\beta$. For this special case with $\beta = 1$, condition (13.29) was introduced in Falk et al. [16], 2nd ed. The present form is due to Frick et al.[14]

---

[14]Frick, M., Kaufmann, E. and Reiss, R.–D. (2006). Testing the tail–dependence based on the radial component, submitted.

Because (13.29) implies (12.45) we know that the function D in the expansion is necessarily a Pickands dependence function, cf. (12.20).

We provide two examples of distributions for which (13.29) is satisfied with $D(z) = 1$ which is the Pickands dependence function for the case of independent marginal maxima.

EXAMPLE 13.3.4. (Bivariate Standard Normal Distribution) We consider again the bivariate standard normal distribution with correlation $\rho \in (0, 1)$. Let H be its df after transformation to $(-1, 0)$–uniformly distributed margins. It satisfies the expansion

$$h_z(c) = 1 + B(c)A(z) + o(B(c)), \qquad \text{as } c \uparrow 0, \qquad (13.30)$$

with

$$B(c) = |c|^{2/(1+\rho)-1}\tilde{L}(c),$$

where $\tilde{L}(c) := c(\rho)(-\log|c|)^{-\rho/(1+\rho)}$ with $c(\rho)$ as in (13.18), and

$$A(z) = \frac{2}{1+\rho}(z(1-z))^{1/(1+\rho)}.$$

The function $\tilde{L}$ is slowly varying so that the function B is regularly varying with exponent of variation $\beta = 2/(1+\rho) - 1$. Substituting this result in equation (13.33) we receive again the relationship $\bar{\chi} = \rho$ (see (13.21)).

Additionally,

$$L(1-u) = (1-u)^{-\beta}B(-(1-u)) \quad \text{and} \quad \eta = (1+\beta)^{-1} \qquad (13.31)$$

in (13.22).

In the following example, we extend the representation of $h_z(c)$ in the case of independent, $(-1, 0)$–uniformly distributed random variables, cf. (c′) on page 310, to Morgenstern random vectors.

EXAMPLE 13.3.5. (Morgenstern distributions.) A transformation of the Morgenstern df, whose copula form was given in Example 13.3.1, to $(-1, 0)$–uniformly distributed margins leads to the df

$$H(u, v) = (1+u)(1+v)(1+\alpha uv)$$

which satisfies the expansion

$$h_z(c) = 1 + c(1+\alpha)2z(1-z) + o(c), \quad c \uparrow 0.$$

Therefore, $D(z) = 1$, $A(z) = -2z(1-z)$, and $B(c) = -(1+\alpha)c$, the latter being a regularly varying function with exponent of variation $\beta = 1$. Recalling from Example 13.3.1 that $\eta = 1/2$, both (13.24) and the subsequent formula (13.33) imply that $\bar{\chi} = 0$.

Relationships Between Measures of Dependence

Let again H be the df of a random vector (X, Y) satisfying a differentiable spectral expansion of length 2. According to (13.6) we have

$$\chi = \lim_{c \uparrow 0} P(Y > c | X > c) = 2(1 - D(1/2)), \tag{13.32}$$

where D is again the Pickands dependence function, and the expression $2(1 - D(1/2))$ is the canonical parameter.

This implies the following characterization of tail dependence :

If $D \neq 1$, we have tail dependence.

If $D = 1$, we have tail independence.

In addition, if $D = 1$ one obtains a relationship between the exponent of variation β (in the spectral expansion of length 2) and the coefficient of tail dependence $\bar{\chi}$. For that purpose let, in addition, B be absolutely continuous with a monotone density. This is satisfied, e.g., for the standard case of a regularly varying function $B(c) = |c|^{\beta}$.

Under these conditions one can prove that

$$\bar{\chi} = \frac{1 - \beta}{1 + \beta}. \tag{13.33}$$

Notice that $\bar{\chi} \to 1$ if $\beta \to 0$, and $\bar{\chi} \to -1$ if $\beta \to \infty$.

Asymptotic Distributions of the Radial Component

We assume that the spectral expansion (13.29) is valid and the partial derivatives $h_z(c) = \partial/\partial c\, H_z(c)$ and $\tilde{h}_c(z) = \partial/\partial z\, H_z(c)$ are continuous.

Under these conditions, the conditional asymptotic distribution of the radial component $X + Y$ for increasing thresholds c is provided (a result due to Frick et al. in the article cited on page 326, who extended a result in Falk et al. [16], 2nd ed., from the special case of $\beta = 1$ to $\beta > 0$).

(i) $D \neq 1$ implies

$$P(X + Y > ct | X + Y > c) \longrightarrow t =: F_0(t), \quad c \uparrow 0,$$

uniformly for $t \in [0, 1]$,

(ii) $D = 1$ and $\beta > 0$ implies

$$P(X + Y > ct | X + Y > c) \to t^{1+\beta} =: F_\beta(t), \quad c \uparrow 0,$$

uniformly for $t \in [0, 1]$ provided that

$$(2 + \beta) \int A(z)\, dz - A(0) - A(1) \neq 0. \tag{13.34}$$

Condition (13.34) is satisfied, e.g., for normal and Morgenstern dfs. This condition is not satisfied for the bivariate generalized Pareto df, special considerations are required in this case (for a discussion see Falk et al. [16] and the above mentioned article by Frick et al.). This result will be applied to the testing of tail dependence based on the radial component.

Selection of the Null–Hypothesis

We continue our discussion on page 293 about the selection of the model: "before using a more complex model, one should check whether the EV model for independent margins or a related generalized Pareto model is applicable." Therefore, we wish to check whether the more complex model with unknown Pickands dependence function D can be replaced by the simpler one where $D = 1$.

About the selection of the hypotheses, there is a well–known general advice (e.g., in the booklet by J. Pfanzagl[15], page 95, translated from German): "As null–hypothesis select the opposite of that you want to prove and try to reject the null–hypothesis." In our special case we want to prove the tail independence and, therefore, take tail dependence as the null–hypothesis.

We are well aware that statisticians often do not follow this advice primarily due to technical reasons.

Testing Tail Dependence Against Tail Independence

We are testing a simple null–hypothesis H_0, representing dependence, against a composite alternative H_1, representing the various degrees of independence.

For that purpose we merely have to deal with the asymptotic conditional distributions of the test statistic $X + Y$. We are testing

$$H_0 : \; F_0(t) = t \quad \text{against} \quad H_1 : \; F_\beta(t) = t^{1+\beta}, \; \beta > 0,$$

based on the the radial components

$$C_i = (X_i + Y_i)/c, \quad i = 1, \ldots, n,$$

with $C_i < 1$, $i = 1, \ldots, n$. The testing procedure is carried out conditioned on the random sample size of exceedances, see page 234. Denote by $\tilde{C}_i$, $i = 1, \ldots, m$, the iid random variables in the conditional set–up.

The Neyman–Pearson test at the level α for testing F_0 against the fixed alternative F_β is independent of the parameter $\beta > 0$. Notice that this test is based on the densities $f_\beta(t) = (1 + \beta)t^\beta$, $0 \le t \le 1$. Therefore, one gets a uniformly most powerful test for the testing against the composite alternative.

---

[15]Pfanzagl, J. (1974). *Allgemeine Methodenlehre der Statistik II.* Walter de Gruyter, Berlin.

For iid random variables $\tilde{C}_i$, $i = 1, \ldots, m$, with common df F_0, the Neyman–Pearson test statistic $\sum_{i=1}^{m} \log \tilde{C}_i$ is distributed according to the gamma df

$$H_m(t) = \exp(t) \sum_{i=0}^{m-1} \frac{(-t)^i}{i!}, \quad t \leq 0,$$

on the negative half–line with parameter m. Therefore, the Neyman–Pearson test at the level α is given by the critical region

$$C_{m,\alpha} = \left\{ \sum_{i=1}^{m} \log \tilde{C}_i > H_m^{-1}(1 - \alpha) \right\}.$$

In the aforementioned article by Frick et al. one may also find a bias–corrected MLE for β.

Power Functions and P–Values

Evaluating $P(C_{m,\alpha})$ under iid random variables Z_i, which are distributed according to F_β, one gets the power function

$$\zeta_{m,\alpha}(\beta) = 1 - H_m\left((1 + \beta)H_m^{-1}(1 - \alpha)\right), \quad \beta \geq 0,$$

for the level–α–test.

The p–value of the optimal test, finally, is given by

$$p(\tilde{c}) = 1 - H_m\left(\sum_{i=1}^{m} \log \tilde{c}_i\right) \approx \Phi\left(-\frac{\sum_{i=1}^{m} \log \tilde{c}_i + m}{m^{1/2}}\right), \tag{13.35}$$

with $\tilde{c} = (\tilde{c}_1, \cdots, \tilde{c}_m)$, according to the central limit theorem.

One may also derive an approximate representation of the power function by

$$\zeta_{m,\alpha}(\beta) \approx 1 - \Phi((1 + \beta)\Phi^{-1}(1 - \alpha) - \beta m^{1/2}).$$

Particularly, for $m = 361$ and $\alpha = 0.01$ we have

$$\zeta_{361,0.01}(\beta) = 1 - \Phi((1 + \beta)\Phi^{-1}(0.99) - 19\beta).$$

This function together with some other power functions is displayed in Figure 13.7.

Data Transformation

Real data $(v_1, w_1), \ldots, (v_n, w_n)$ are independent realizations of a random vector (V, W) with common unknown df. These pairs have to be transformed to the left

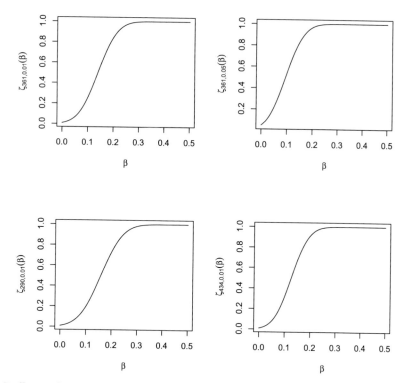

FIG. 13.7. Power functions for $m = 361$ and $\alpha = 0.01$, 0.05 (above), and $m = 290$, 434 and $\alpha = 0.01$ (below).

lower quadrant to make the preceding test procedure applicable. Such a transformation can be achieved as explained in Section 10.3. Subsequently, we make use of transformations by means of the sample dfs $\widehat{F}_n(\boldsymbol{v};\cdot)$ and $\widehat{F}_n(\boldsymbol{w};\cdot)$ in the single components. We have

$$x_i = \widehat{F}_n(\boldsymbol{v};v_i) \quad \text{and} \quad y_i = \widehat{F}_n(\boldsymbol{w};w_i), \quad i = 1,\ldots,n.$$

Recall that the resulting data are relative ranks with values in the interval $[0, 1]$ or $(0, 1)$ if the ranks are divided by $n + 1$ in place of n. Then the data are shifted to $(-1, 0)$. This procedure is illustrated in the following example.

EXAMPLE 13.3.6. (Wave and Surge Heights at Newlyn, England.) The wave and surge data set was originally recorded by Coles and Tawn. It consists of $2,894$ three–hourly– measurements of the surge and wave heights taken at Newlyn, a coastal town in England. Ledford and Tawn as well as Coles et al. have already analyzed the dependence structure of these data with the motivation that flooding is likely if both surge and wave heights reach extreme levels.

Now we are going to test the extremal dependence of the two components by applying the uniformly most powerful test. In a first step the marginal data are transformed by means of the marginal univariate empirical dfs. Then they are shifted from $(0,1) \times (0,1)$ to $(-1,0) \times (-1,0)$. The data set in its original form and after transformation to $(0,1)$–uniformly distributed margins is displayed in Figure 13.8.

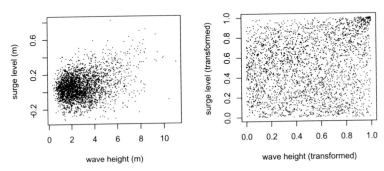

FIG. 13.8. Wave heights and surge levels at Newlyn: original data (left) and transformed data set with $[0,1]$–uniformly distributed margins (right)

Testing the Tail Dependence for the Wave and Surge Heights

Next we proceed as in Falk et al. [17], page 189: fix $c < 0$ and consider those observations $x_i + y_i$ exceeding the threshold c. These data are subsequently denoted by $c_1, \ldots, c_m$. The threshold c is chosen in such a way that the number m of exceedances is about 10% to 15% of the total number $n = 2,894$. Therefore, we take values c from -0.46 to -0.35. It may be worthwhile to investigate more closely the impact of the threshold c on the performance of the test. Figure 13.9 shows the transformed full data set and the part near zero together with threshold lines.

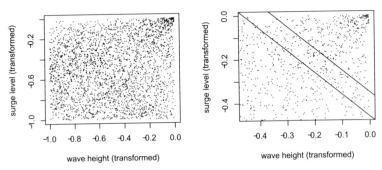

FIG. 13.9. The transformed full data set, shifted to the negative quadrant (left) and data above the threshold lines corresponding to $c = -0.46$ and $c = -0.35$ (right).

We assume that our results are applicable to the c_i/c, $i = 1, \ldots, m$. Substituting the x_i in (13.35) by c_i/c, $i = 1, \ldots, m$, one gets approximate p–values. Table 1 provides some examples of p–values for different thresholds c together with the number m of exceedances.

TABLE 13.1. p–values for different thresholds c.

c	-0.46	-0.45	-0.44	-0.43	-0.42	-0.41
m	431	414	400	390	376	361
p–value	0.00028	0.00085	0.00158	0.00162	0.00305	0.00665
c	-0.4	-0.39	-0.38	-0.37	-0.36	-0.35
m	353	338	325	313	300	294
p–value	0.00542	0.01179	0.01939	0.02816	0.04601	0.03291

Having in mind a significance level of $\alpha = 0.01$ the null–hypothesis is accepted for the bigger values of c, up from -0.39. For lower thresholds the p–values are rather low, suggesting a rejection of the null–hypothesis. Remembering that we are considering limiting distributions as $c \uparrow 0$ we may nevertheless decide to accept the null–hypothesis entailing the acception of an extremal tail–dependence. Yet we have to be aware that this conclusion is not straightforward and we must not ignore that the null–hypothesis is always rejected if we choose a significance level of $\alpha = 0.05$, for example.

13.4 The Point Process Approach to the Multivariate POT Method

We shortly introduce the point process models related to the multivariate GP models and possible estimation procedures within such models.

The Models

We shortly describe the relationship between

- the original statistical modeling by means of iid random vectors (X_i, Y_i), $i = 1, \ldots, n$, with common bivariate df F,

- the model determined by bivariate GP dfs $\widetilde{W}$ as introduced in (13.5).

Let (x_i, y_i), $i = 1, \ldots, n$, be generated according to F. Let (x_i', y_i'), $i = 1, \ldots, k$, be those (x_i, y_i) for which either $x_i > u$ or $y_i > v$ (taken in the original order of the outcome). Let

$$(\tilde{x}_i, \tilde{y}_i) := m_{(u,v)}(x_i', y_i') = \big(\max(x_i', u), \max(y_i', v) \big)$$

be the vectors introduced in (13.4). Within a certain error bound (see [16], Section 5.1), the $(\tilde{x}_i, \tilde{y}_i)$ can be regarded as realizations of random vectors

$$(\widetilde{X}_i, \widetilde{Y}_i), \qquad i = 1, \ldots, K(n), \tag{13.36}$$

where

- $(\widetilde{X}_1, \widetilde{Y}_1)$, $(\widetilde{X}_2, \widetilde{Y}_2)$, $(\widetilde{X}_3, \widetilde{Y}_3)$, ... are iid random vectors with common GP df $\widetilde{W}(x, y)$ in the form described in (13.5),

- $K(n)$ is a binomial or a Poisson random variable which is independent of the $(\widetilde{X}_i, \widetilde{Y}_i)$.

Thus, the series of random vectors in (13.36) constitute a binomial or a Poisson process depending on the choice of $K(n)$.

Estimation in Point Process Models

Of course, first of all one may also use the estimators presented before within the point process framework.

The ML approach was applied, for example, by Davison and Smith in the article mentioned on page 121. To compute the likelihood function specify densities of point processes as it was done in Section 9.3 (also see Section 9.5) with the help of results in [43], Section 3.1.

Part IV

Topics in Hydrology and Environmental Sciences

Chapter 14

Flood Frequency Analysis

co–authored by J.R.M. Hosking[1]

We first summarize and supplement in Section 14.1 the at–site analysis done before in conjunction with annual flood series.

Section 14.2 deals with the partial duration series of daily discharges. Our program was already outlined in Section 2.5, namely we want to handle the serial correlation and seasonal variation in the data. Our primary interest still concerns the calculation of the T–year threshold, yet also briefly discuss the question of seasonal variation in its own right. To catch the seasonal variation of discharges over a higher level, we introduce inhomogeneous Poisson processes with marks which distributionally depend on the time of their occurrence (Section 14.2). The handling of such partial duration series is the main topic of this chapter. A trend is included in Section 14.3.

In Sections 14.3–14.5, we also deal with the regional flood frequency analysis Our primary references are the paper by Hosking et al., cf. page 120, and the book [29].

14.1 Analyzing Annual Flood Series

The traditional approach of dealing with floods is to use annual maxima. On the one hand, one is avoiding the problems of serial correlation and seasonal variation; on the other hand, one is losing information contained in the data. A first modification of this approach is to base the inference on seasonal or monthly maxima.

---

[1]IBM Research Division, Thomas J. Watson Research Center; co–authored the 2nd edition.

In Section 4.3 we introduced several distributions, such as Wakeby, two–component and gamma distributions, which are used in the hydrological literature besides EV distributions for the modeling of annual or seasonal floods. In this chapter, we only employ EV distributions.

Estimation of the T–Year Flood Level Based on Annual Maxima

We repeat the analysis of annual floods as it was already done for the Feather River discharges, see Examples 4.1.1 and 5.1.1. We computed 50 and 100–year discharge levels based on the estimated Gumbel (EV0) and exponential (GP0) distributions.

To make the results in this and the subsequent section comparable, we repeat the aforementioned analysis with respect to the Moselle River data.

EXAMPLE 14.1.1. (Continuation of Example 2.5.2 about the Moselle River Data.) The MLE(EV) will be applied to the annual maximum levels for water years running from Nov. to Oct. of consecutive calendar years. Based on 31 annual maxima one obtains the parameters $\gamma = -0.40$, $\mu = 8.00$ and $\sigma = 1.74$. Moreover, the right endpoint of the estimated Weibull distribution is 12.37. The 50 and 100–year levels are $u(50) = 11.45$ and $u(100) = 11.67$. According to these estimates, there was a 100–year flood, with a level of 11.73 meters, around Christmas time in the year 1993.

In view of the small number of data, we do not carry out a tail estimation for the Moselle River within a GP model based on exceedances over a certain level.

Modifications of the Annual Maxima Approach

If the number of years is small, there is a greater need to extract more information out of a sample of daily recorded discharges or water levels. The number of extreme data can be increased by extracting monthly or seasonal maxima for the statistical analysis, where independence may still be assumed, yet one is exposed to the seasonal variation. We do not go into details but concentrate our attention to the partial duration approach which will be outlined in the next section.

14.2   Analyzing Partial Duration Series

Series of flood peaks are investigated under the fairly general conditions that the frequency of occurrence and the magnitude of floods exhibit a seasonal dependency.

Modeling by an Inhomogeneous Poisson Process

After a declustering of the data, the modeling

- of frequencies by means of an inhomogeneous Poisson process, and

- of the magnitudes by stochastically independent, time–dependent marks

is adequate.

Subsequently, we use a nonparametric approach with the exception that the marks are assumed to be distributed according to generalized Pareto (GP) dfs. A parametric modeling for the marginal distributions of the marks is indispensable when the usual extrapolation to extraordinary high flood levels must be carried out.

We model the frequencies and the marks by means of an inhomogeneous Poisson process in the time scale with mean value function Λ, and marks with dfs $F(\cdot|t_i)$ at time t_i. This constitutes a Poisson(Λ, F), cf. Section 9.5.

The flood peaks are those marks exceeding a threshold v. According to (9.31) and (9.32) this situation is described by a Poisson$(\Lambda_v, F^{[v]})$ with mean value function

$$\Lambda_v(t) = \int_0^t (1 - F(v|s)) \, d\Lambda(s) \tag{14.1}$$

in the time scale, and marks distributed according to

$$F^{[v]}(w|t) = \big(F(w|t) - F(v|t)\big)/\big(1 - F(v|t)\big), \qquad w \geq v. \tag{14.2}$$

The T–Year Flood Level

We extend the computations on page 252 concerning the T–year return level to the new framework. (14.1) and (14.2) yield that the first exceedance time with respect to the threshold v is the random variable $\tau_{1,v} = \Lambda_v^{-1}(X)$, where X is a standard exponential random variable. Hence, the T–year return level is the solution to the equation

$$E(\tau_{1,v}) = T. \tag{14.3}$$

An estimate of the T–year return level is obtained by plugging in estimates of the mean value function Λ and of the conditional df $F = F(\cdot|\cdot)$.

Subsequently, we assume that there is only a seasonal dependence of $F(\cdot|s)$ and $\Lambda(s)$. Then, according to (14.1),

$$\Lambda_v(T) = T \int_0^1 (1 - F(v|s)) \, d\Lambda(s) =: T\psi(v), \qquad T = 1, 2, 3, \ldots \ . \tag{14.4}$$

When the inhomogeneous mean value functions $\Lambda_v(t)$ are replaced by the homogeneous mean value functions $\widetilde{\Lambda}_v(t) = t\psi(v)$, then $\psi(v)$ is the intensity in the time domain, and the first exceedance time is an exponential random variable with expectation $1/\psi(v)$, see Section 9.2 for details. This expectation is equal to T, if

$$T\psi(v) = 1. \tag{14.5}$$

Therefore, one obtains a T–year return level $\tilde{v}(T)$ as a solution to this equation. This T–year return level may serve as an approximation to the one in the original setting.

We also give an alternative interpretation of $\tilde{v}(T)$: let N_v be the inhomogeneous Poisson counting process with mean value function Λ_v. Because

$$1 = T\psi(v) = \tilde{\Lambda}_{\tilde{v}(T)}(T) = E\big(N_{\tilde{v}(T)}(T)\big) \tag{14.6}$$

we know that $\tilde{v}(T)$ is the threshold so that the mean number of exceedances up to time T is equal to 1, also see page 12.

Estimation of the T–Year Level

Let l be the number of years for which the flood measurements are available. Let again $N(t)$, $t \le l$, be the counting process pertaining to the exceedance times τ_i, $i \le N(l)$, of exceedances over u. It is apparent that

$$N_l(s) = \frac{1}{l} \sum_{j=0}^{l-1} \big(N(j+s) - N(j)\big), \qquad 0 \le s \le 1,$$

is an unbiased estimator of $\Lambda(s)$ for $0 \le s \le 1$. Plugging in N_l for Λ in $\psi(v) = \int_0^1 (1 - F(v|s))\, d\Lambda(s)$ one obtains by

$$
\begin{aligned}
\int_0^1 (1 - F(v|s))\, dN_l(s) &= \frac{1}{l} \sum_{i \le N(l)} \big(1 - F(v|\, j(i) + \eta_i)\big) \\
&= \frac{1}{l} \sum_{i \le N(l)} \big(1 - F(v|\eta_i)\big)
\end{aligned}
$$

an estimator of $\psi(v)$, where $j(i) + \eta_i = \tau_i$. Thus, we have $\eta_i = \tau_i$ modulo 1.

Next, $F(\cdot|\eta_i)$ will be replaced by a generalized Pareto (GP) df W. The final estimator of $\psi(v)$ is given by

$$\widehat{\psi}(v) = \frac{1}{l} \sum_{i \le N(l)} \big(1 - W_{\hat{\gamma}(\eta_i), u, \hat{\sigma}(\eta_i)}(v)\big), \tag{14.7}$$

where the estimators $\hat{\gamma}(\eta_i)$ and $\hat{\sigma}(\eta_i)$ of the shape and scale parameters are constructed in the following manner: let X_j be the mark pertaining to the time τ_j (and, thus, pertaining to η_j). Then, these estimators are based on those X_j for which $|\eta_j - \eta_i| \le b$, where $b > 0$ is a predetermined bandwidth. Alternatively, one may employ a nearest neighbor method, that is, take the marks X_j pertaining to the k values η_j closest to η_i, where k is predetermined (see Section 8.2).

Then, an estimator $\hat{v}(T)$ of the T–year level is obtained as the solution to

$$T\widehat{\psi}(v) = 1. \tag{14.8}$$

One gets again the estimator in the homogeneous case (see page 252) if the parameters are independent of the time scale.

EXAMPLE 14.2.1. (Continuation of Example 13.1.1 about the Moselle River Data.) After taking the exceedances over a base level $u = 5$ and the selection of cluster maxima (with run length $r = 7$) there are 131 observations within the 31 water years (starting in November). In the hydrological literature, it is suggested to take a base level and clusters such that there are three cluster maxima for each year on the average. Thus, we took a slightly greater number of exceedances.

In Fig. 14.1 (left), the exceedances are plotted against the day of occurrence within the water year. As one could expect the exceedances primarily occur in winter and spring time. It is remarkable that higher flood levels also occur in times with a lower flood frequency.

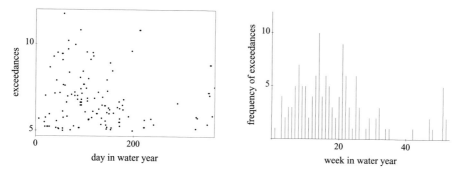

FIG. 14.1. (left.) Scatterplot of cluster maxima over $u = 5$ at days modulo 365. (right.) Frequency of exceedances within weeks modulo 52.

We also included a histogram of the frequencies plotted against the week of occurrence within the water year (Fig. 14.1 (right)).

The GP parameters $\gamma(i)$ and $\sigma(i)$ of the flood magnitudes are estimated by means of the MLE(GP) using the nearest neighbor approach with $k = 30$. The use of the Moment(GP) estimator would lead to a similar result. In Fig. 14.2, plots of the estimates are provided for all days within the water year where an exceedance occurred.

This is our interpretation: the estimates reflect to some extent what can also be seen in the scatterplot (Fig. 14.1 (left)). In winter time the cluster maxima are relatively homogeneously scattered within the range between 5 and 12, whereas in spring time very high floods are rare events. In the first case, the wide range is captured by the larger scale parameter, whereas the latter phenomenon is captured by the positive shape parameter. Based on Monte Carlo simulations we obtain 12.1 meter as an estimate of the 100–year threshold. Applying the Moment(GP) estimator one obtains higher T–year thresholds than for the MLE(GP).

Statistical inference for such series (particularly, the estimation of the T–year level) was carried out in the hydrological literature

- within the reduced setting of stationarity within certain seasons[2],

[2]Ashkar, F. and Rousselle, J. (1981). Design discharge as a random variable: a risk

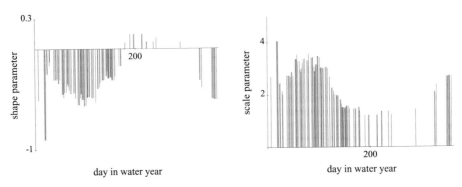

FIG. 14.2. Estimated shape parameters $\gamma(i)$ (left–hand side) and scale parameters $\sigma(i)$ (right–hand side) at days i.

- for a certain continuous modeling with trigonometric functions[3].

Moreover, it was assumed that the marks (the flood magnitudes) are exponentially distributed. For further explanations and references see Davison and Smith (cited on page 121), where also a decision is made for the exponential model according to a likelihood ratio test. We believe that for our example of Moselle data the GP modeling is adequate.

Flood Series with a Trend

The estimated T–year level may drastically increase in the presence of a slight trend. If a linear trend is included in the modeling, then much higher T–year levels were estimated for the Moselle river data[4]. We also refer to articles by R.L. Smith[5] and by H. Rootzén and N. Tajvidi[6].

study. Water Resour. Res. 17, 577–591.

Rasmussen, P.F. and Rosbjerg, D. (1991). Prediction uncertainty in seasonal partial duration series. Water Resour. Res. 27, 2875–2883.

[3]North, M. (1980). Time–dependent stochastic model of floods. J. Hydraul. Div. Am. Soc. Civ. Eng. 106, 649–665.

Nachtnebel, H.P. and Konecny, F. (1987). Risk analysis and time–dependent flood models. J. Hydrol. 91, 295–318.

[4]Reiss, R.–D. and Thomas, M. (2000). Extreme environmental data with a trend and a seasonal component. In: Environmental Engineering and Health Sciences (ed. J.A. Raynal et al.), 41–49, Water Resources Publications, Englewood.

[5]Smith, L.R. (1989). Extreme value analysis of environmental time series: an application to trend detection in round–level ozone. Statistical Science 4, 367–381.

[6]Rootzén, H. and Tajvidi, N. (1997). Extreme value statistics and wind storm losses: a case study. Scand. Actuarial. J., 70–94.

14.3   Regional Flood Frequency Analysis

We argued that the use of a partial duration series in place of an annual flood series is necessary to extract more information out of the available data. Another possibility is to include information from data recorded at nearby sites, respectively, sites having similar characteristics.

Regional Estimation of Parameters

In the regional analysis it is assumed that data are available at different sites with data $x_{j,i}$ at site j. We deal with data $y_{j,i}$ which are

- annual maxima, or

- excesses (exceedances) over a threshold u_j.

Subsequently, we primarily address the case of excesses.

Assume that the random excesses at site j are distributed according to the GP df $W_{\gamma,0,\sigma_j}$. Apparently this is the situation, where the accesses have a common shape parameter γ, yet the scale parameters are different from each other.

If $\hat{\gamma}_j$ is an estimate of γ at site j, then

$$\hat{\gamma}^R = \sum_j n_j \hat{\gamma}_j \Big/ \sum_j n_j \qquad (14.9)$$

is a regional estimate of γ, where the n_j are the sample sizes (number of excesses over the threshold u_j) at site j (cf. [29], page 7, where further references to the literature are provided). If the n_j are of a similar size, then a pooling of the data is appropriate.

An important question is the selection

- of the homogeneous region, that is, those sites which can be modeled by the same shape parameter,

- of an appropriate submodel of GP distributions.

The Index Flood Procedure

We introduce a special at–site estimation procedure which is also useful in conjunction with regional considerations.

Suppose that the excesses Y_i of discharges over the threshold u are distributed according to a GP df $W_{\gamma,0,\sigma}$ with shape and scale parameters γ and σ. Recall from (1.57) that Y_i has the expectation

$$m = \sigma/(1 - \gamma),$$

which is the index flood, under the condition $\gamma < 1$. Under this condition, the statistical inference is done within a certain submodel of GP dfs, where dfs with a very heavy upper tail are excluded.

The rescaled excesses $Y_i/m = Y_i(1 - \gamma)/\sigma$ have means equal to 1 and are distributed according to the GP df

$$
W_{\gamma,0,1-\gamma}(z) = 1 - \left(1 + \frac{\gamma}{1-\gamma} z\right)^{-1/\gamma} \quad \text{for} \quad
\begin{cases}
0 < z, & 0 \leq \gamma < 1, \\
& \text{if} \\
0 < z < \frac{1-\gamma}{|\gamma|}, & \gamma < 0;
\end{cases}
$$

$$(14.10)$$

with density

$$
w_{\gamma,0,1-\gamma}(z) = \frac{1}{1-\gamma}\left(1 + \frac{\gamma}{1-\gamma} z\right)^{-(1+1/\gamma)} \quad \text{for} \quad
\begin{cases}
0 < z, & 0 \leq \gamma < 1, \\
& \text{if} \\
0 < z < \frac{1-\gamma}{|\gamma|}, & \gamma < 0.
\end{cases}
$$

$$(14.11)$$

The unknown parameter $\gamma < 1$ may be estimated, for example, by the ML estimator $\hat{\gamma}$ in this submodel of GP dfs (using the Newton–Raphson iteration procedure).

Within the original model of GP dfs with shape parameter $\gamma < 1$ and scale parameter $\sigma > 0$ one gets the following two–step estimation procedure: let y_i be a realization of Y_i and denote again by $\bar{y}$ the sample mean of the y_i. Then, the excess degrees

$$
z_i = \frac{y_i}{\bar{y}}
$$

may be regarded as realizations under the GP df $W_{\gamma,0,1-\gamma}$ as given in (14.10). Let $\hat{\gamma}$ be the ML estimate (or any other estimate) of γ within the model of GP dfs

$$
\left\{ W_{\gamma,0,1-\gamma} : \ \gamma < 1 \right\}
$$

based on the excess degrees z_i. Because $\bar{y}$ is an estimate of the mean $m = \sigma/(1-\gamma)$, it is apparent that

$$
\hat{\sigma} = (1 - \hat{\gamma})\bar{y}
$$

is an estimate of the original scale parameter $\sigma > 0$.

The Sample Median Taken as Index Flood

If one takes the median in place of the mean, then the rescaled excesses have medians equal to 1 and are distributed according to a GP df

$$
\begin{array}{ll}
W_{0,0,1/\log 2} & \gamma = 0, \\
& \text{if} \\
W_{\gamma,0,\gamma/(2^\gamma - 1)} & \gamma \neq 0.
\end{array}
$$

For estimating the parameters γ and $\sigma > 0$ in the GP model of dfs $W_{\gamma,0,\sigma}$—without any restriction on the shape parameter γ—we use excess degrees with the sample mean replaced by the sample median.

Regional Estimation Using the Index Flood Procedure

Let again $y_{j,i}$ be the discharge excesses over a threshold u_j at site j which are governed by a GP df $W_{\gamma,0,\sigma_j}$.

Rescaling the GP dfs by the means $m_j = \sigma_j/(1-\gamma)$ as in the preceding lines, we obtain the common regional frequency distribution $F_\gamma = W_{\gamma,0,1-\gamma}$. Recall that m_j is the index flood for the site j. Notice that

$$F_{\gamma,m_j}(y) = F_\gamma(y/m_j) = W_{\gamma,0,\sigma_j}(y)$$

and, therefore, the index flood m_j is a scale parameter for the model of regional frequency distributions.

Let again

$$z_{j,i} = \frac{y_{j,i}}{\bar{y}_j}$$

be the excess degrees at site j, where $\bar{y}_j$ is the sample mean of the excesses $y_{j,i}$ at site j. Now, use the at–site estimates $\hat\gamma_j$ of γ to define the regional estimate $\hat\gamma^R$ in (14.9). In addition,

$$\hat\sigma_j = (1 - \hat\gamma)\bar{y}_j$$

is an estimate of the original scale parameter $\sigma_j = (1 - \gamma)m_j$.

Estimation of the T–Year Flood Level

At a given site, the T–year flood level may be estimated as in (14.7) and (14.8) with $\hat\gamma$ and $\hat\gamma(\eta_i)$ replaced by the corresponding regional estimates $\hat\gamma^R$ and $\hat\gamma^R(\eta_i)$.

14.4 The *L*–Moment Estimation Method

We give a short introduction to L–moments and the L–moment estimation method following closely the explanations in the book by Hosking and Wallis [29] about regional frequency analysis. L–moment estimators are obtained by equating sample L–moments with the pertaining L–moments corresponding to the procedure for ordinary moments, see page 86. As an application, L–moment estimators are dealt with under the Paretian modeling.

L–moment estimators are extensively applied in flood frequency in conjunction with the index flood procedure analysis due to their appealing small sample performance.

L–Moments

Recall from (1.33) that the ordinary jth moment is defined by

$$m_j = \int x^j \, dF(x) = \int_0^1 \left(F^{-1}(u) \right)^j \, du, \tag{14.12}$$

where the latter equality is a consequence of the quantile transformation, cf. page 38, and the transformation theorem for integrals.

The right–hand expressions will be replaced by certain probability weighted moments

$$\beta_r = \int_0^1 F^{-1}(u) u^r \, du \tag{14.13}$$

for $r = 0, 1, 2, \ldots$. Notice that β_0 is the mean m_1. Likewise, one could start with β_r replaced by $\alpha_r = \int_0^1 F^{-1}(u)(1 - u)^r \, du$.

The next step corresponds to the step from moments to centered moments to some extent. The first L–moment λ_1 is the mean, the second L–moment λ_2 corresponds to the standard deviation, and for $j \geq 3$ the jth L–moment is related to the jth central moment. More precisely, let

$$
\begin{aligned}
\lambda_1 &= \beta_0 \\
\lambda_2 &= 2\beta_1 - \beta_0 \\
\lambda_3 &= 6\beta_2 - 6\beta_1 + \beta_0 \\
\lambda_4 &= 20\beta_3 - 30\beta_2 + 12\beta_1 - \beta_0
\end{aligned}
$$

and, in general,

$$\lambda_j = \int_0^1 F^{-1}(u) P_{j-1}(u) \, du = \sum_{k=0}^{j-1} p_{j-1,k} \beta_k, \tag{14.14}$$

where

$$P_r(u) = \sum_{k=0}^r p_{r,k} u^k$$

are Legendre polynomials of degree r shifted to the interval $[0, 1]$ with the coefficients $p_{r,k} = (-1)^{r-k}(r + k)!/\left((k!)^2(r - k)!\right)$ for $r = 0, 1, 2, \ldots$. It is evident that all L–moments exist if the mean exists.

We have $P_0(x) = 1$, $P_r(1) = 1$ and $\int_0^1 P_r(u) P_s(u) \, du = 0$ for $r \neq s$. In particular, $\int_0^1 P_r(u) \, du = 0$ for $r \geq 1$ which yields that L–moments are independent of a location parameter for $j \geq 1$.

The representations

$$
\begin{aligned}
\lambda_2 &= E\left(X_{2:2} - X_{1:2}\right)/2 \\
\lambda_3 &= E\left(X_{3:3} - X_{2:3} - \left(X_{2:3} - X_{1:3}\right)\right)/3 \\
\lambda_4 &= E\left(X_{4:4} - X_{1:4} - 3\left(X_{3:4} - X_{2:4}\right)\right)/4
\end{aligned}
$$

reveal why L–moments exhibit characteristics of the distribution just as the standard deviation and the 3rd and 4th centered moments.

Also define the L–CV, the L–moment analogue of the coefficient of variation, by

$$\tau = \lambda_2/\lambda_1 \tag{14.15}$$

and the L–moment ratios

$$\tau_j = \lambda_j/\lambda_2 \tag{14.16}$$

with the L–skewness and the L–kurtosis as special cases for $j = 3, 4$. Notice that the L–CV is independent of a scale parameter, and the L–moment ratios are independent of location and scale parameters.

Sample L–Moments

One gets estimates of β_r for $r = 0, 1, 2, \ldots$ by replacing the qf F^{-1} in (14.13) by the sample qf F_n^{-1} as defined in (2.8). A slight modification leads to estimates

$$\tilde{\beta}_r = \frac{1}{n} \sum_{k \le n} \left(\frac{k}{n+1} \right)^r x_{k:n} .$$

Yet, unbiased estimators are obtained by using another modification, namely

$$\hat{\beta}_0 = \frac{1}{n} \sum_{k=1}^{n} x_{k:n} ,$$

$$\hat{\beta}_1 = \frac{1}{n} \sum_{k=2}^{n} \frac{k-1}{n-1} x_{k:n} ,$$

and, in general,

$$\hat{\beta}_r = \frac{1}{n} \sum_{k=r+1}^{n} \frac{(k-1)\cdots(k-r)}{(n-1)\cdots(n-r)} x_{k:n}.$$

Unbiased estimates $\hat{\lambda}_j$ of the L–moment λ_j are obtained by replacing the probability weighted moments β_k in (14.14) by the sample versions $\hat{\beta}_k$. We have

$$\hat{\lambda}_j = \sum_{k=0}^{j-1} p_{j-1,k} \hat{\beta}_k , \tag{14.17}$$

where, particularly, the first sample L–moment $\hat{\lambda}_1$ is the sample mean $\bar{x}$.

In this manner, on may also define the sample L–moment ratios $\hat{\tau}_j = \hat{\lambda}_j/\hat{\lambda}_2$ and the sample L–coefficient of variation $\hat{\tau} = \hat{\lambda}_2/\hat{\lambda}_1$.

L–Moment Estimation in the Pareto Model

We apply the L–moment estimation method

- to the Pareto model for excesses with unknown scale and shape parameters σ and γ,

- to the reduced Pareto model in (14.10) for excess degrees in conjunction with the index flood procedure.

It turns out that both approaches are equivalent.

The first two L–moments of a Pareto df $W_{\gamma,u,\sigma}$ are

$$\begin{aligned}
\lambda_1 &= u + \sigma/(1 - \gamma), \\
\lambda_2 &= \sigma/((1 - \gamma)(2 - \gamma)),
\end{aligned}$$

for $\gamma < 1$. Notice that $u = 0$, if we deal with excesses y_i.

L–moments estimates of the parameters γ and σ are solutions to

$$\begin{aligned}
\hat{\lambda}_1(\boldsymbol{y}) &= u + \sigma/(1 - \gamma), \\
\hat{\lambda}_2(\boldsymbol{y}) &= \sigma/((1 - \gamma)(2 - \gamma)),
\end{aligned}$$

where $\hat{\lambda}_j(\boldsymbol{y})$ are the sample L–moments based on the excesses y_i. The solutions are

$$\begin{aligned}
\hat{\gamma}(\boldsymbol{y}) &= 2 - \bar{y}/\hat{\lambda}_2(\boldsymbol{y}), \\
\hat{\sigma}(\boldsymbol{y}) &= (1 - \hat{\gamma}(\boldsymbol{y}))\bar{y}.
\end{aligned} \tag{14.18}$$

Within the model (14.10) for excess degrees we base the estimation of γ on the 2nd L–moment λ_2. One gets the estimate

$$\tilde{\gamma}(\boldsymbol{z}) = 2 - 1/\hat{\lambda}_2(\boldsymbol{z}),$$

where $\boldsymbol{z}$ is the sample of excess degrees $z_i = y_i/\bar{y}$. It is apparent that

$$\tilde{\gamma}(\boldsymbol{z}) = \hat{\gamma}(\boldsymbol{y}) \tag{14.19}$$

and, therefore, the L–moment method applied to the excesses y_i and, respectively, the index flood procedure in conjunction with the L–moment method applied to the excess degrees z_i leads to the same estimates of σ and γ within the Pareto model for excesses.

Simulations show that the L–moment estimator of γ has a better performance than the MLE in the model (14.10) for small and moderate sample sizes, if γ is around zero.

L–Moment Estimation in the Extreme Value Model

The first two *L*–moments and the *L*–skewness parameter of the EV df $G_{\gamma,\mu,\sigma}$ are

$$
\begin{aligned}
\lambda_1 &= \mu - \sigma(1 - \Gamma(1-\gamma))/\gamma, \\
\lambda_2 &= -\sigma(1 - 2^\gamma)\Gamma(1-\gamma)/\gamma, \\
\tau_3 &= 2(1 - 3^\gamma)/(1 - 2^\gamma) - 3 .
\end{aligned}
\tag{14.20}
$$

If γ is known, then

$$
\sigma = \frac{-\lambda_2 \gamma}{(1 - 2^\gamma)\Gamma(1-\gamma)},
\tag{14.21}
$$

and

$$
\mu = \lambda_1 + \sigma(1 - \Gamma(1-\gamma))/\gamma .
\tag{14.22}
$$

Thus, to get the *L*–moment estimates of γ, μ, σ, replace $\lambda_1, \lambda_2, \tau_3$ by the sample versions and compute a numerical solution to the third equation in (14.20).

14.5 A Bayesian Approach to Regional Estimation

In this section we give an outline of an approach to regional flood frequency analysis which originates in a paper by G. Kuczera[7].

A Bayesian At–Site Analysis

Assume that the annual floods at a given site are distributed with common log–normal df $F_{(\mu,\sigma)}(x) = \Phi_{\mu,\sigma}(\log(x))$ as defined in (1.60), where $\Phi_{\mu,\sigma}$ is the normal df with location and scale parameters μ and $\sigma > 0$. For such a distribution, the T–year flood (T–year level) is

$$
\begin{aligned}
u(T) &= F_{(\mu,\sigma)}^{-1}(1 - 1/T) \\
&= \exp\left(\mu + \sigma\Phi^{-1}(1 - 1/T)\right).
\end{aligned}
\tag{14.23}
$$

The estimation of $u(T)$ will be carried out in the normal model. If $z_1, \ldots, z_n$ are the annual floods then—according to the definition of the log–normal df—the transformed data $y_i = \log(z_i)$ are governed by the normal df $\Phi_{\mu,\sigma}$. The T–year flood $u(T)$ will be estimated by replacing in (14.23) the parameters μ and σ by the sample mean $\bar{y}$ and a Bayesian estimate of σ^2 based on $\boldsymbol{y} = (y_1, \ldots, y_n)$.

In the sequel, the normal dfs are represented by the variance $\vartheta = \sigma^2$ and the unknown mean μ is replaced by the sample mean $\bar{y}$. The latter corresponds to

---

[7]Kuczera, G. (1982). Combining site–specific and regional information: an empirical Bayes approach. Water Resour. Res. 18, 306–314.

taking a non–informative prior for μ. For the Bayesian estimation of ϑ we take a reciprocal gamma density

$$p_{a,b}(\vartheta) = \frac{1}{b\Gamma(a)}(\vartheta/b)^{-(1+a)}\exp(-b/\vartheta), \qquad \vartheta > 0,$$

as a prior for ϑ, see (3.43). Such densities are conjugate priors for ϑ. The posterior density is

$$p_{a,b}(\vartheta|\boldsymbol{y}) = p_{a',b'}(\vartheta)$$

with

$$a' = a + n/2$$

and

$$b' = b + ns^2/2,$$

where $s^2 = (1/n)\sum_{i\leq n}(y_i - \bar{y})^2$.

The Bayesian estimate of $\vartheta = \sigma^2$—as the mean of the posterior distribution—is

$$\hat{\sigma}^2 = \frac{b + ns^2/2}{a + n/2 - 1}. \qquad (14.24)$$

Therefore, one gets

$$\hat{u}(T) = \exp\left(\bar{y} + \hat{\sigma}\Phi^{-1}(1 - 1/T)\right)$$

as an estimate of the T–year flood $u(T)$.

Regional Moment Estimation of the Superparameters

Next, we assume that data sets of size n_j are recorded at k different sites. We want to use the regional information to estimate the superparameters a and b.

Let $\hat{\vartheta}_j$ be estimates of the unknown parameter ϑ_j at site j. From the Bayesian viewpoint (see page 243), the parameter ϑ_j was generated under a prior density $p_{a,b}$ which is the density of a random parameter θ. One may also regard the estimates $\hat{\vartheta}_j$ as realizations under $p_{a,b}$. It remains to estimate the superparameters a and b from $\hat{\vartheta}_j$, $j = 1,\ldots,k$.

This will be exemplified by applying the moment estimation method (see page 86) to the model of reciprocal gamma distributions. Because the mean and the variance of the reciprocal gamma distribution are given by $m(a,b) = b/(a-1)$ and $\mathrm{var}(a,b) = b^2/((a-1)^2(a-2))$ (see page 104) we have

$$b = (a - 1)m(a,b)$$

and

$$a = 2 + m(a,b)^2/\mathrm{var}(a,b)$$

for $a > 2$. Replacing $m(a,b)$ and $\mathrm{var}(a,b)$ by the sample versions based on the $\hat{\vartheta}_j$, $j = 1,\ldots,k$, one gets estimates $\hat{a}_k$ and $\hat{b}_k$ of the superparameters a and b. This leads to a new estimator in (14.24) and a regional estimate of the T–year flood.

Refined Regional Approach

It is more realistic to assume that the random parameters θ_j at sites j are not identically distributed. Suppose that

$$\theta_j = \boldsymbol{x}_j \boldsymbol{\beta}' + \varepsilon_j, \tag{14.25}$$

where $\boldsymbol{x}_j$ represents the known characteristics at site j and the ε_j are residuals. Thus, the prior distribution at site j has the parameters $a(\boldsymbol{x}_j)$ and $b(\boldsymbol{x}_j)$. In addition, let the estimates at sites j have the representations

$$\hat{\vartheta}_j = \vartheta_j + \eta_j \tag{14.26}$$

which yields that

$$\hat{\vartheta}_j = \boldsymbol{x}_j \boldsymbol{\beta}' + \varepsilon_j + \eta_j \ . \tag{14.27}$$

Within this multiple, linear regression problem one may get estimates of the mean and the variance of the prior distribution at site j and, therefore, of the superparameters $a(\boldsymbol{x}_j)$ and $b(\boldsymbol{x}_j)$ by using again the moment estimation method.

Further Applications of Empirical Bayes Estimation in Regional Analysis

The Bayesian argument in Kuczera's article was taken up by H.D. Fill and J.R. Stedinger[8] to justify forming a linear combination of two estimators with weights inversely proportional to the estimators variance. This is done within the annual maxima framework.

In this context, we also mention an article by H. Madsen and D. Rosbjerg[9] who apply the regression technique in conjunction with empirical Bayes estimation to regional partial duration series. Both articles contain exhaustive lists of relevant articles.

---

[8]Fill, H.D. and Stedinger, J.R. (1998). Using regional regression within index flood procedures and an empirical Bayesian estimator. J. Hydrology 210, 128–14.

[9]Madsen, D. and Rosbjerg, D. (1997). Generalized least squares and empirical Bayes estimation in regional partial duration series index–flood modeling. Water Resour. Res. 33, 771–781.

Chapter 15

Environmental Sciences

co–authored by R.W. Katz[1]

This chapter deals with the application of the statistics of extremes in the environmental sciences. Related chapters include those dealing with flood frequency analysis in hydrology (Chapter 14) and large claims in the insurance industry (Chapter 16). Consideration of extreme events commonly arises in the regulatory process related to the environment, particularly the selection of thresholds which aid in determining compliance and effectiveness. Statistical characteristics typical of environmental extremes include annual and diurnal cycles, as well as trends possibly attributable to anthropogenic activities.

One commonly unappreciated aspect in extreme value analysis for environmental variables is the potential of making use of covariates, particularly geophysical variables. Their incorporation into the analysis makes the resultant models both more efficient, e.g., in terms of quantile estimation, and more physically realistic.

Another feature, beyond the scope of this treatment, is the spatial dependence of extremes typically exhibited by fields of environmental data.

15.1   Environmental Extremes

In much of the environmental sciences, particularly impact assessment, extreme events play an important role. For example, earthquakes (and related tsunamis), fires, floods, or hurricanes can have devastating impacts, ranging from disturbances

---

[1]Institute for Study of Society and Environment, National Center for Atmospheric Research, Boulder, Colorado, USA. NCAR is sponsored by the National Science Foundation.

in ecosystems to economic impacts on society as well as loss of life. Such features indicate that there should be plentiful applications of the statistics of extremes to the environmental sciences.

Environmental Policy and Regulation

The implementation of environmental policy requires the development of regulations, such as the setting of standards for environmental variables by government agencies. In particular, these standards involve the selection of high (or low) thresholds, defining an extreme event. These circumstances imply that extreme value analysis ought to be an integral part of this process. Such standards could involve statistics as simple as whether the annual maxima exceeds a high threshold, e.g., in the context of air pollution. Yet they sometimes involve much more intricate quantities, whose motivation from a purely statistical perspective is lacking; e.g., the ozone standard set by the U.S. Environmental Protection Agency is defined in terms of the statistic, the daily average of the maximum of 8–hour running means, and is based on whether the fourth highest value of this quantity over a given year exceeds a threshold.

The monitoring of environmental variables to detect any trend in extremes, possibly attributable to human activity, provides important information for policy makers. Specifically, possible shifts in the frequency and intensity of extreme weather and climate events are one of the primary concerns under global climate change as part of the enhanced greenhouse effect.

Cycles

That environmental variables often possess marked diurnal and annual cycles is well established. Consequently, it is commonplace for statistical models of environmental time series to include an annual (and/or diurnal) component for the mean (as well as sometimes for the standard deviation). Yet in the statistics of environmental extremes such cyclical behavior is usually neglected. Although not necessarily identical in form to that for the mean, cyclical behavior in extremes ought to be anticipated. It is an inherent aspect of the process by which environmental extremes arise, so that taking into account the cyclical modulation of extremal characteristics would be physically appealing.

Trends

The monitoring and detection of trends in environmental extremes is important for a number of reasons. For one thing, any trends in extremes might serve as an early indicator of broader change, a catalyst for public policy intervention. In particular, it can be argued that the response of ecosystems is most sensitive to extreme events, not average conditions. On the other hand, it may be that regulations already have been imposed to "control" the environmental variable, with the hope of observing a diminished trend, i.e., as a measure of the effectiveness of environmental regulation. Typically, trend analysis of environmental variables focuses on the mean (as well as occasionally on the standard deviation), not the extremes per se. Yet there is no inherent reason why the form of trend in extremes need be identical to that for the mean for an example in which there is a trend in the extremes, but apparently none in the mean, see the article by R.L. Smith cited on page 342). So, for completeness, any trend in extremes ought to be directly modeled.

Covariates

Such cycles and trends can be incorporated into extreme value analysis as co-variates, in the form of simple deterministic functions of time, e.g., a sum of sine waves. Still a noteworthy feature of environmental variables is the influence of covariates, especially those geophysical in nature, which are not simply determin-istic functions, but rather random variables themselves. For example, air pollution concentration is effected by the prevailing meteorological conditions. This feature is well appreciated for statistics like the mean level of an environmental time se-ries. So it would be natural to anticipate that environmental extremes are likewise influenced by geophysical covariates. Yet it is less common to incorporate such covariates into environmental extreme value analysis, at least partly because of a lack of awareness by practitioners that this extension of extreme value analysis is feasible (a notable exception is the book by Coles [9], pages 107–114). In Chapters 8 and 9 we made some technical preparations.

Not only would the introduction of covariates result in increased precision in estimating high quantiles, but it would serve to make the statistics of extremes more realistic from an environmental science perspective.

Tail Behavior

The distribution of the impacts, in monetary terms, associated with extreme en-vironmental events—e.g., a hurricane or a flood—has a marked tendency to be heavy–tailed. Yet the distribution of the underlying geophysical phenomenon may not necessarily be heavy–tailed. Some geophysical variables, such as streamflow or precipitation, are clearly heavy–tailed; whereas others, such as wind speed or temperature, appear to have a light or bounded upper tail. Certain environmental variables, such as air pollution concentration, also do not appear to be heavy–tailed.

Because of the complex coupling among these processes (i.e., geophysical, environmental, and economic impact), the determination of the origin of extremal behavior in environmental impacts is non–trivial. In particular, the extent to which this heavy–tailed behavior in environmental impacts is "inherited" from the tail behavior of underlying geophysical and/or environmental variables is unclear. Re-calling the origin of the Pareto distribution as a model for the distribution of income or wealth, the heavy tail behavior of environmental impacts could well reflect primarily the aggregative nature of variables like income or wealth.

15.2 Inclusion of Covariates

In principle, the inclusion of covariates is feasible in any of the various approaches to extreme value analysis covered in this book, including both block maxima and peaks–over–threshold (pot). If the technique of maximum likelihood is adopted, then parameter estimation remains straightforward. Nevertheless, although the situation closely resembles that of generalized linear models [41], the well developed theory of estimation in that situation is not directly applicable. The extremal–based approach used here would not necessarily differ much from ordinary least squares in terms of point estimates. But substantial discrepancies would be possible for standard errors (or confidence intervals) for upper quantile estimators, particularly when dealing with a variable possessing a heavy–tailed distribution.

Block Maxima

Consider the EV df $G_{\gamma,\mu,\sigma}$ as the approximate distribution for block maxima, say corresponding to a random variable Y with observed data $y_1,\ldots,y_n$ (e.g., annual maximum of time series such as daily precipitation amount). Suppose that a covariate X is also available, say with observed data, $x_1,\ldots,x_n$, on the same time scale $t=1,\ldots,n$ as the block maxima. Given a value of covariate, say $X=x$, the conditional distribution of the block maxima is assumed to remain the EV, but now with parameters that possibly depend on x.

As an example, the location parameter μ and the logarithm of the scale parameter σ could be linear functions of x (applying the logarithmic transformation to preserve non–negativity of scale), whereas for simplicity the shape parameter γ might well be taken independent of the value x:

$$\gamma(x)=\gamma, \ \mu(x)=\mu_0+\mu_1 x, \ \log\sigma(x)=\sigma_0+\sigma_1 x. \tag{15.1}$$

Here $\mu_0,\mu_1,\sigma_0,\sigma_1$ and γ are unknown parameters to be estimated by maximum likelihood. Now the parameters of the EV df in (15.1) depend on a covariate which varies with time, writing $\gamma_t=\gamma(x_t)$, $\mu_t=\mu(x_t)$, and $\sigma_t=\sigma(x_t)$. For technical details we refer to Section 8.4.

This approach amounts to fitting non–identical EV dfs, whereas the conventional diagnostic displays (e.g., Q–Q plots) are predicated upon identical distributions. This non–stationarity can be removed by using the relationship between an arbitrary EV df and the standard Gumbel, i.e., EV0 df. Namely, if Y_t has an EV df with parameters γ_t, μ_t, and σ_t, then

$$\varepsilon_t=(1/\gamma_t)\log(1+\gamma_t(Y_t-\mu_t)/\sigma_t) \tag{15.2}$$

has a $G_{0,0,1}$ df (see Chapter 1). In practice, "residuals" can be obtained by substituting the time–dependent parameter estimates $\hat{\gamma}_t$, $\hat{\mu}_t$, and $\hat{\sigma}_t$, e.g., as in (15.1), along with the corresponding data (i.e., x_t's), into (15.2). Then a Q–Q plot, for instance, can be constructed for a standard Gumbel distribution.

Peaks–Over–Threshold Approach

The potential advantages of the pot approach over the block maxima approach become more readily apparent in the presence of covariates (for an early treatment of that question see the article by Davison and Smith, mentioned on page 121, in Section 3 about regression). For instance, allowing for an annual cycle in the block maxima approach would not even be feasible, except in an ad hoc and non–parsimonius manner such as taking block maxima over months or seasons. More generally, many physically–based covariates would naturally be measured on a time scale similar, if not identical, to that of the variables whose extremal behavior is being modeled (e.g., covariates such as pressure readings on a daily time scale in the case of modeling extreme high daily precipitation amounts). On the other hand, additional complications arise in the pot approach with covariates, including the need in some circumstances to permit the threshold to be time varying.

Poisson–GP (Two–Component) Model. By "two-component model" we refer to the situation in which the two individual components, the Poisson process—with rate parameter λ_t—for the occurrence of exceedances of the high threshold u and the GP df (with shape parameter γ_t and scale parameter $\sigma_t[u]$, which depends on the threshold u unlike the scale parameter for the EV df) for the excess over the threshold, are modeled separately. This approach was originally introduced in hydrology, for instance, to incorporate annual cycles in one or both of the components. This is the description of an exceedance process with an homogeneous Poisson process, see Section 9.1, or an inhomogeneous Poisson process, see Section 9.5, in the time scale and GP marks for the exceedances themselves. Analogous to the case of block maxima (15.1), it would be natural to express $\log \lambda_t$, $\log \sigma_t[u]$, and perhaps γ_t as functions of a covariate.

For the statistical validation of the model, the non–stationarity in the GP df can be removed by using the relationship between an arbitrary GP distribution and the standard exponential (i.e., GP0 df), corresponding to (15.2) for EV dfs. Namely, if Y_t has a GP df with parameters λ_t and $\sigma_t[u]$, then

$$\varepsilon_t = (1/\gamma_t) \log(1 + \gamma_t Y_t / \sigma_t[u]) \tag{15.3}$$

has a $W_{0,0,1}$ df (see Chapter 1). With marked dependence of the extremal characteristics on a covariate (e.g., in the case of a substantial trend or cycle), it may be that the threshold itself ought to vary with the covariate, i.e., with time t, writing u_t instead of u. For simplicity, we ignore this complication in the notation used here.

Point Process Approach. By "point process approach," we refer here to the treatment of the problem as a full–fledged two–dimensional, non–homogeneous point process, rather than modeling the two components separately[2]. Here the one–dimensional Poisson process for the occurrence of exceedances and the GP df for the excesses are modeled simultaneously, see Section 9.5, page 260. This approach has the advantage treating all of the uncertainty in parameter estimation in a unified fashion, e.g., making it more straightforward to determine confidence intervals for return levels.

Interpretation can be enhanced if the point process is parameterized in terms of the EV df. To do this, it is convenient to specify a time scaling parameter, say $h > 0$, e.g., $h \approx 1/365.25$ for annual maxima of daily data. Then the parameters of the point process, λ_t (i.e., rate parameter per unit time), γ_t, and $\sigma_t[u]$, are related to the parameters of the EV df, γ_t, μ_t, and σ_t, by

$$\log \lambda_t = -(1/\gamma_t) \log[1 + \gamma_t(u - \mu_t)/\sigma_t)], \quad \sigma_t[u] = \sigma_t + \gamma_t(u - \mu_t), \quad (15.4)$$

with the shape parameter γ_t being equivalent[3]. Now the model has at least a limited theoretical interpretation as arising as the limiting distribution which would be obtained if many observations were available at a fixed time t (instead of just one as in practice) and the largest value were taken. The nonlinearity and interaction in the relationships in (15.4) imply that expanding some of the parameters of the point process components as simple functions of a covariate would not necessarily correspond to nearly such a simple model in terms of the EV df parameters, see Section 9.5 for technical details.

Another question concerns how to extend the concepts of return period and level which originated in the context of stationarity, i.e., no covariates. One could always think in terms of an "effective" return period and level, consistent with the theoretical interpretation just discussed. We defer the discussion of possible extensions of these concepts until they arise in the context of specific examples later in this chapter.

Finally, the choice of the time scaling constant h should be viewed as arbitrary. Specifically, $\delta = h/h^*$ denotes the ratio between two different time scaling parameters (e.g., annual versus monthly maxima), then the corresponding parameters of the EV df (say γ, μ, and σ for time scale h versus γ^*, μ^*, and σ^* for time scale h^*) are related by[4]

$$\gamma^* = \gamma, \quad \mu^* = \mu + [\sigma^*(1 - \delta^{-\gamma})]/\gamma, \quad \sigma^* = \sigma\delta^\gamma. \quad (15.5)$$

[2]Smith, R.L. (2001). Extreme value statistics in meteorology and the environment. In *Environmental Statistics*, Chapter 8, 300–357 (NSF–CBMS conference notes). Available at www.unc.edu/depts/statistics/postscript/rs/envnotes.pdf.

[3]Katz, R.W., Parlange, M.B. and Naveau, P. (2002). Statistics of extremes in hydrology. *Advances in Water Resources* 25, 1287–1304.

[4]Katz, R.W., Brush, G.S. and Parlange, M.B. (2005). Statistics of extremes: Modeling ecological disturbances. *Ecology* 86, 1124–1134.

15.3 Example of Trend

Fitting a trend constitutes one of the simplest forms of covariate analysis. By means of an example, we contrast the traditional approach of trend analysis to that more in harmony with extreme value theory.

Traditional Approach

The traditional approach to trend analysis of environmental extremes would involve fitting a trend, ordinarily by the method of least squares, in the mean (and possibly in the standard deviation) of the original time series. Then the extreme value analysis is applied to residuals, presumed stationary, from this trend analysis (see Chapter 4). There are several possible drawbacks to such an approach. For one thing, it is in effect assumed that removing a trend in the overall mean would necessarily eliminate any trend in extremes as well. For another, such a two–stage estimation approach makes it difficult to account for all of the uncertainty in parameter estimation, with the error involved in constructing the residuals being usually neglected.

Extreme Value Approach

In the extreme value approach, we simply expand the parameters of the EV df as functions of time, e.g., as in (15.1). Of course, determining the appropriate functional form of trend is not necessarily a routine matter. It should further be recognized that such an analysis is not based on any statistical theory of extreme values for non–stationary time series per se. But recall the general interpretation for an EV df with time–dependent parameters as discussed earlier.

Example

In contrast to global warming, the "urban heat island" refers to a warming effect on the local climate attributable to urbanization. This phenomenon is well documented in terms of its effects on mean daily minimum and maximum temperature. Among other things, the warming trend is more substantial for minimum, rather than maximum, temperature.

So we analyze the extremes for a time series of daily minimum temperature during summer (i.e., July–August) at Phoenix, Arizona, USA, for the 43–year time period 1948–1990[5]. This area experience rapid urbanization during this time period, with a substantial trend in the mean of the distribution of daily minimum temperature being apparent. For simplicity, we adopt a block minima approach; strictly speaking, fitting the EV df for block maxima to negated block minima and

---

[5]Tarleton, L.F. and Katz, R.W. (1995). Statistical explanation for trends in extreme temperatures at Phoenix, Arizona. Journal of Climate 8, 1704–1708.

then converting the results back into the appropriate form for block minima, see Chapter 1, with the time series of summer minima being shown in Fig. 15.1.

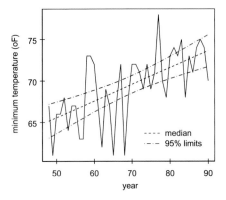

FIG. 15.1. Time series of summer (July–August) minimum of daily minimum temperature at Phoenix, Arizona, USA, 1948–1990. Also includes trend in conditional median of fitted EV df (i.e., effective two–year return level) along with point-wise approximate 95% confidence limits.

Table 15.1 summarizes the results of fitting the EV df to block minima by the method of maximum likelihood, with trends in the parameters of the form:

$$\gamma_t = \gamma, \quad \mu_t = \mu_0 + \mu_1 t, \quad \log \sigma_t = \sigma_0 + \sigma_1 t, \qquad \text{for } t = 1, \dots, 43. \qquad (15.6)$$

Three nested candidate models are compared:

(i) $\mu_1 = \sigma_1 = 0$, i.e., no trend in either μ or σ;

(ii) $\mu_1 \neq 0$, $\sigma_1 = 0$, i.e., trend in μ but not in σ;

(iii) $\mu_1 \neq 0$, $\sigma_1 \neq 0$, i.e., trends in both μ and σ.

The negative log–likelihoods (denoted by $-\log L$ in the table) indicate that model (ii) with a trend in only the location parameter is the superior fitting model.

TABLE 15.1. Maximum likelihood estimates of parameters (with standard errors of trend parameters given in parentheses) of EV df fitted by block minima approach for trend analysis of extreme low summer (July–August) temperatures (°F) at Phoenix, Arizona, USA, 1948–1990.

model	$\hat{\gamma}$	$\hat{\mu}_0$	$\hat{\mu}_1$	$\hat{\sigma}_0$	$\hat{\sigma}_1$	$-\log L$
$\mu_1 = \sigma_1 = 0$	-0.184	70.88	0	1.346	0	121.52
$\mu_1 \neq 0$, $\sigma_1 = 0$	-0.204	66.05	0.202 (0.041)	1.135	0	111.52
$\mu_1 \neq 0$, $\sigma_1 \neq 0$	-0.211	66.17	0.196 (0.041)	1.338	-0.009 (0.010)	111.11

The estimated slope of the trend in the location parameter is $\hat{\mu}_1 = 0.202$ °F per year for model (ii), or an estimate of the urban warming effect in terms of extreme low temperatures. Fig. 15.1 shows the corresponding trend in the median, i.e., effective two–year return level, of the conditional EV df for the best fitting model (ii). Pointwise 95% local confidence intervals for the conditional median are also included in Fig. 15.1. Using the expression for the return level for the lowest temperature during the tth summer as a function of the corresponding parameters of the EV df (see Chapter 1), these intervals can be derived from the large–sample standard errors for the maximum likelihood estimators.

A Q–Q plot for model (ii) based on (15.2) appears reasonably satisfactory (not shown). Nevertheless, the pattern in Fig. 15.1 suggests that an alternative form of trend in the location parameter, such as an abrupt shift, might be more appropriate. The same conclusion concerning the existence of a trend in extreme low summer temperatures is reached if the point process approach were applied instead.

15.4   Example of Cycle

Fitting cycles constitutes another simple form of covariate analysis. By means of an example, we contrast the traditional approach of seasonal analysis to that more in harmony with extreme value theory, as well as indicating an inherent advantage of the point process approach over block maxima.

Traditional Approach

The traditional approach to cyclic analysis of environmental extremes would involve fitting a cyclic function (e.g., sine wave), ordinarily by the method of least squares, to the mean (and possibly to the standard deviation) of the original time series. Then the extreme value analysis is applied to the residuals from the cyclic analysis. The drawbacks to this approach are identical to those for traditional trend analysis; namely, the possibility of a different cyclic form for extremes than that for the mean and standard deviation, as well as the lack of an integrated treatment of uncertainty.

Extreme Value Approach

Like the extreme value approach for modeling trends, we simply expand the parameters of the EV df as functions, in this case periodic, of time. For instance, in the following example, the location parameter and logarithm of the scale parameter of the EV df are both modeled as sine waves, whereas the shape parameter is assumed independent of time.

Example

In many regions, daily precipitation exhibits marked seasonality. Such cycles are well documented for both the probability of occurrence of precipitation and for the conditional mean (or median) amount of precipitation given its occurrence. Yet the seasonality in precipitation extremes is not typically explicitly taken into account in extreme value analysis of precipitation (e.g., to estimate upper quantiles of precipitation amount associated with flooding).

So we analyze the extremes for a time series of daily precipitation amount at Fort Collins, Colorado, USA, for the 100–year time period 1900–1999 (also see the article by Katz et al. (2002) mentioned on page 358). For such a long time series, it is convenient to summarize the data in terms of block maxima, with Fig. 15.2 showing the derived time series of annual maxima. Nevertheless, we will actually apply the point process approach to model the extremal behavior with seasonality being permitted. Although much year–to–year variation is present, no long–term trend is evident in Fig. 15.2. Because of a flood which occurred on 28 July 1997 (with an observed value of 4.63 inches), this data set is of special interest.

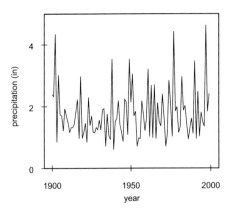

FIG. 15.2. Time series of annual maximum of daily precipitation amount at Fort Collins, Colorado, USA, 1900–1999.

Table 15.2 summarizes the results of fitting the EV df indirectly through the point process approach by the method of maximum likelihood. Annual cycles in the form of sine waves are incorporated for the location parameter and logarithm of the scale parameter:

$$
\begin{aligned}
\gamma_t &= \gamma, \\
\mu_t &= \mu_0 + \mu_1 \sin(2\pi t/T) + \mu_2 \cos(2\pi t/T), \\
\log \sigma_t &= \sigma_0 + \sigma_1 \sin(2\pi t/T) + \sigma_2 \cos(2\pi t/T),
\end{aligned}
\tag{15.7}
$$

for $t = 1, \ldots, 36524$ with time scaling constant $h = 1/T$, $T \approx 365.25$.

The threshold of $u = 0.395$ inches was chosen by trial and error, perhaps a bit lower than optimal for wetter times of the year, a compromise to avoid the complication of a time varying threshold. The parameter estimates are given for the case of no declustering (with little evidence, in general, of the need for declustering daily precipitation extremes), but quite similar results are obtained when runs declustering with $r = 1$ is applied (see Chapter 2).

Three nested candidate models are compared in Table 15.2:

(i) $\mu_1 = \mu_2 = \sigma_1 = \sigma_2 = 0$, i.e., no cycle in either μ or σ;

(ii) $\mu_1 \neq 0$, $\mu_2 \neq 0$, $\sigma_1 = \sigma_2 = 0$, i.e., cycle in μ but not in σ;

(iii) $\mu_1 \neq 0$, $\mu_2 \neq 0$, $\sigma_1 \neq 0$, $\sigma_2 \neq 0$, i.e., cycles in both μ and σ.

The negative log–likelihoods (denoted by $-\log L$ in Table 15.2) indicate that model (iii) with annual cycles in both the location parameter and the logarithm of the scale parameter of the EV df is the superior fitting model. A Q–Q plot for model (iii) based on (15.2), comparing the observed annual maxima to the EV df fitted indirectly by the point process approach, appears reasonably satisfactory (not shown). A more complex form of extremal model might include a sine wave for the shape parameter as well, or a sum of more than one sine wave for the location parameter and/or logarithm of the scale parameter.

TABLE 15.2. Maximum likelihood estimates of parameters of EV df (parameterized in terms of annual maxima, time scaling constant $h = 1/365.25$) fitted by point process approach (threshold $u = 0.395$ inches) for annual cycle in extreme high daily precipitation amount (inches) at Fort Collins, Colorado, USA, 1900-1999.

model	$\hat{\gamma}$	$\hat{\mu}_0$	$\hat{\mu}_1$	$\hat{\mu}_2$	$\hat{\sigma}_0$	$\hat{\sigma}_1$	$\hat{\sigma}_2$	$-\log L$
$\mu_1 = \mu_2 = \sigma_1 = \sigma_2 = 0$	0.212	1.383	0	0	−0.631	0	0	−1359.82
$\sigma_1 = \sigma_2 = 0$		0.103	1.306	0.082	−0.297	−0.762	0	0	−1521.45
no contraints		0.182	1.281	−0.085	−0.805	−0.847	−0.124	−0.602	−1604.29

For the best fitting model (iii), Fig. 15.3 shows the effective 100–year return level, with the parameters of the EV df being rescaled for sake of comparison using (15.5) to reflect the maximum of daily precipitation amount over a month instead of a year (i.e., time scaling constant $h^* = 12/365.25$). These estimated return levels range from a low in mid January of about 1.1 inches to a high in mid July of about 4.3 inches. To give a rough feeling for the actual annual cycle in extreme precipitation, the observed monthly maximum of daily precipitation is also included in Fig. 15.3. Consistent with the effective return levels for the fitted EV df, a marked tendency is evident toward higher precipitation extremes in summer than in winter.

It is also of interest to estimate the return period for the high precipitation in July 1997. With annual cycles in the parameters of the EV df, the determination

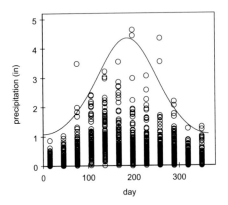

FIG. 15.3. Annual cycle in effective 100–year return level for fitted EV distribution for monthly maximum of Fort Collins daily precipitation. Observed values of monthly maximum of daily precipitation indicated by circles.

of a return period involves a product of probabilities that differ depending on the day of year (i.e., of the form $p_1 \times \cdots \times p_T$ instead of simply p^T, where p_t denotes the probability on the tth day of the year); for simplicity, treating the daily precipitation amounts as independent. Refitting only the data for the time period before the flood, i.e., 1900–1996, using model (iii), the estimated return period for the observed daily amount of 4.63 inches is roughly 50.8 years.

15.5   Example of Covariate

Fitting covariates which are random variables themselves, such as geophysical quantities, is no more involved than the examples of trends and cycles just presented. Yet such examples are more compelling, both in terms of the potential predictability of extremes (e.g., reflected in terms of confidence intervals for return levels whose length varies conditional on the covariate) and in terms of its appeal from an environmental science perspective.

Traditional Approach

Even in fields such as hydrology with a rich tradition of reliance on extreme value analysis, an inconsistency arises with extreme value theory typically being abandoned in favor of ordinary regression analysis, either for the entire data set or for a subset consisting only of extremes, if covariates are treated. As an ad hoc approach, sometimes an ordinary extremal analysis is conducted separately depending on a few discretized categories of the covariate (e.g., "above average" or "below average"). The disadvantages of such traditional approaches are somewhat subtle, with the central issue being the lack of a robust treatment of extremes.

Extreme Value Approach

Just like the extreme value approach for modeling trends or cycles, we expand the parameters of the EV df as functions of one or more physically based covariates. But now the issue of how to specify the functional form of such relationships is particularly challenging, ripe for input from environmental or geophysical scientists.

Example

It is well understood that heavy precipitation is associated with particularly meteorological conditions, involving other variables such as pressure. To demonstrate the viability of incorporating physically based covariates into extremal analysis, a prototypical example is considered. This example is not intended to constitute a realistic treatment of the various meteorological factors known to have an influence on heavy precipitation.

So we analyze the extremes for a time series of daily precipitation amount at Chico, California, USA, during the month of January over the time period 1907–1988, with fours years have been eliminated from the data set because of missing values (also see the paper by Katz et al. mentioned on page 362). As a covariate, the January average sea level pressure at a single grid point (actually derived from observations within a grid box) over the Pacific Ocean adjacent to the California coast is used, say a random variable denoted by Y. Fig. 15.4 shows a scatterplot of the January maxima of daily precipitation amount versus the pressure covariate, with at least a weak tendency for lower maximum precipitation when the average pressure is higher being apparent.

Rather than using monthly block maxima (as, for simplicity, in Fig. 15.4), we actually fit the EV df indirectly through the point process approach by the method of maximum likelihood, with the results being summarized in Table 15.3. Given average pressure $X = x$, the location parameter and the logarithm of the scale parameter of the conditional EV df are assumed linear functions of y:

$$\gamma(x) = x, \quad \mu(x) = \mu_0 + \mu_1 x, \quad \log \sigma(x) = \sigma_0 + \sigma_1 x. \tag{15.8}$$

These forms of functional relationship are intended solely for illustrative purposes, not necessarily being the most physically plausible. Because daily time series for the single month of January are being modeled, a time scaling constant of $h = 1/31$ is used. The threshold of $u = 40$ mm was selected by trial and error, with no declustering being applied.

Three nested candidate models are compared in Table 15.3:

(i) $\mu_1 = \sigma_1 = 0$, i.e., no variation with y;

(ii) $\mu_1 \neq 0$, $\sigma_1 = 0$, i.e., μ varies with y, but σ does not;

(iii) $\mu_1 \neq 0$, $\sigma_1 \neq 0$, i.e., both μ and σ vary with y.

The negative log–likelihoods (denoted by $-\log L$ in Table 15.3) suggest that model (ii), with only location parameter of the conditional EV df depending on pressure, is the superior fitting model. In particular, a likelihood ratio test for model (ii) versus model (iii) indicates only weak evidence that the scale parameter ought to be varied as well (p–value ≈ 0.209). A Q–Q plot for model (ii) based on (15.2), comparing the observed monthly maxima to the EV df fitted indirectly by the point process approach, appears reasonably satisfactory (not shown).

TABLE 15.3. Maximum likelihood estimates of parameters of EV df (parameterized in terms of annual maximum, time scaling constant $h = 1/31$) fitted by point process approach (threshold $u = 40$ mm) to daily precipitation amount (mm) at Chico, California, USA, conditional on pressure covariate (mb, with 1000 mb being subtracted), 1907–1988.

model	$\hat{\gamma}$	$\hat{\mu}_0$	$\hat{\mu}_1$	$\hat{\sigma}_0$	$\hat{\sigma}_1$	$-\log L$
$\mu_1 = \sigma_1 = 0$	0.198	35.49	0	2.226	0	244.10
$\mu_1 \neq 0,\ \sigma_1 = 0$	0.151	58.13	-1.361	2.315	0	235.27
$\mu_1 \neq 0,\ \sigma_1 \neq 0$	0.199	58.15	-1.284	2.979	-0.045	234.49

The estimated slope parameter in model (ii) is $\hat{\mu}_1 = -1.361$ mm per mb, or higher precipitation extremes being associated with lower pressure. For the best fitting model (ii), the conditional median of the fitted EV df (i.e., effective two–year return level) is also included in Fig. 15.4. Another way of interpreting the fitted model is in terms of effective return periods. For model (i) (i.e., no dependence on pressure), the estimate of the conventional 20–year return level is 73.0 mm. But this return level would correspond to effective return periods, based on fitted model (ii), ranging from 8.1 years for the lowest observed pressure to 44.9 years for the highest.

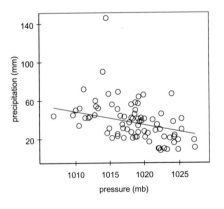

FIG. 15.4. Scatterplot of January maximum of daily precipitation amount at Chico, California, USA, versus January mean pressure. Also includes conditional median of fitted EV df (i.e., effective 2–year return level).

One natural extension of this example would be to allow the pressure covariate to vary from one day to the next, in meteorological terminology considering high frequency variations instead of just low frequency ones. The relationship between precipitation extremes and pressure could be simultaneously modeled over the entire year, but such a model might well entail the complication of a different relationship depending on the time of year to be physically realistic.

15.6 Numerical Methods and Software

Although not difficult in principle to program, most existing software routines for extreme value analysis do not make any provision for the incorporation of covariates, perhaps another explanation for alternatives to extreme value theory still being prevalent in applications. So a brief description focused on techniques for maximum likelihood estimation and on the available software is now provided.

Maximum Likelihood Estimation

Because the expansion of the parameters of extremal models in terms of one or more covariates can be readily incorporated in maximum likelihood estimation, the expressions for the likelihood function of the EV df, whether for the block maxima or point process approaches, are not repeated here for this specific situation (but see Chapter 5). It should be acknowledged that the presence of covariates makes the routine reliance on iterative numerical methods more problematic, with the possibility of multiple local maxima, etc.

Statistical Software

Software for extreme value analysis, which does make provision for covariates, includes the suite of S functions provided as a companion to the text by Stuart Coles [9]. These functions allow for covariates in fitting the EV df to block maxima, the GP df to excesses over a high threshold, and the EV df indirectly through the point process approach. Based on these same S functions (ported into R), the Extremes Toolkit provides a graphical interface so that users, particularly in the environmental and geophysical sciences, unfamiliar with R or S can still make use of the Coles software (gateway: www.isse.ucar.edu/extremevalues/extreme.html).

Part V

Topics in
Finance and Insurance

Chapter 16

Extreme Returns in Asset Prices

co–authored by

C.G. de Vries[1] **and S. Caserta**[2]

Throughout this chapter, we assume that speculative prices s_t like those per-taining to stocks, foreign currencies, futures etc. are evaluated at discrete times $t = 0, 1, 2, \ldots$, where the periods can be days or weeks. Thus, if s_0 is the price of an investment at time $t = 0$, then the return—the difference of prices taken relatively to the initial price—at time T is $(s_T - s_0)/s_0$. Our primary interest concerns daily returns under discrete compounding (arithmetic returns)

$$\tilde{r}_t = \frac{s_t - s_{t-1}}{s_{t-1}}$$

or the daily returns under continuous compounding (log–returns)

$$r_t = \log(s_t) - \log(s_{t-1}). \tag{16.1}$$

These quantities are close to each other if the ratio s_t/s_{t-1} is close to 1. We will focus on the latter concept. Log–returns are also called geometric returns in the financial literature.

The speculative return series generally exhibits two empirical properties. The first property is that the variability of the r_t shows a certain clustering which makes

---

[1]Tinbergen Institute and Erasmus University Rotterdam; co–authored the 1st and 2nd edition.

[2]Tinbergen Institute and Erasmus University Rotterdam; co–authored the 2nd edition.

the variance somewhat predictable. The other feature is that the random log–prices $\log S_t$ satisfy the martingale property. Both of these properties will be discussed in greater detail. Keep in mind that we write R_t and S_t when stochastic properties of random returns and prices are dealt with.

In Section 16.1, we collect stylized facts about financial data and give references to the history of statistical modeling of returns. In Section 16.2, daily stock market and exchange rate returns are visualized to get a first insight into financial data. The pot–methodology is applied in Section 16.3 to estimate the upper and lower tails of log–returns and cumulated return dfs. The loss/profit variable is represented by means of return variables in Section 16.4. Such representations make the statistical results of Section 16.3 applicable to the loss/profit variable.

As an application, we estimate the q–quantile of the loss/profit distribution in Section 16.5 under different statistical models for the returns. This is the so–called Value–at–Risk (VaR) problem faced by financial intermediaries like commercial banks. A related question is to evaluate the amount of capital which can be invested such that at time T the possible loss exceeds a certain limit l with probability q. Section 16.6 deals with the VaR methodology for a single derivative contract. Finally, we focus our attention on the modeling of returns by ARCH/GARCH and stochastic volatility (SV) series in Section 16.7 and predict the conditional VaR in Section 16.8.

16.1 Stylized Facts and Historical Remarks

In this section, we collect some basic facts about returns series of financial data. We compare the relationship between arithmetic and log–returns, deal with the weekend and calendar effects and discuss the concept of market efficiency in conjunction with the martingale property.

Arithmetic Returns and Log–Returns

The subsequent explanations about the advantage of using log–returns r_t compared to arithmetic returns $\tilde{r}_t$ heavily relies on a discussion given by Philippe Jorion [33].

- For the statistical modeling, e.g., by means of normal dfs, it is desirable that the range of the returns is unbounded. This property is merely satisfied by log–returns because the arithmetic returns are bounded from below by -1.

- Some economic facts can be related to each other in a simpler manner by using log–returns. For example, let s_t be the exchange rate of the U.S. dollar against the British pound and let $r_t = \log(s_t/s_{t-1})$ be the pertaining log–return. Then, the exchange rate of the British pound against the U.S. dollar is $1/s_t$. This yields that the log–return, from the viewpoint of a British investor, is just $-r_t = \log((1/s_t)/(1/s_{t-1}))$.

- It is apparent that the price s_T at time T can be regained from the daily returns $r_1, \ldots, r_T$ and the initial price s_0 by

$$s_T = s_0 \exp \left(\sum_{t \leq T} r_t \right). \tag{16.2}$$

Conversely, the T–day log–return $\log s_T - \log s_0$ of the period from 0 to T is the sum of the daily log–returns which is a useful property in the statistical context.

Weekend and Calendar Effects

Before we can analyze any speculative return series, we must deal with the fact that returns are computed at equally–spaced moments in time, but that business time does not coincide with physical time. For example, stock markets are closed at night and during weekends. This poses a problem for the definition of returns at higher frequencies. The following ad hoc procedures can be employed to deal with the weekend effect.

1. (Returns with Respect to Trading Days.) Just take the prices for the given trading days and compute the returns.

2. (Omitting Monday Returns.) Omit the days for which prices are not recorded including the consecutive day. Thus, after a weekend, the Monday returns are also omitted.

3. (Distributing Monday Returns.) The return registered after a gap (e.g., the return recorded on Monday) is equally distributed over the relevant days (e.g., if r is the return on Monday, then $r/3$ is taken as the return on Saturday, Sunday and Monday).

Distributing or omitting the Monday returns is a very crude method; a thorough analysis and a refined treatment of the weekend effect may be desirable for certain questions. Of course, one may also take just the original returns with the potential disadvantage that the Monday may exhibit the behavior of returns accumulated over three days. In the book [54] by S. Taylor there is a detailed discussion of further calendar effects such as the dependence of the mean and standard deviation of returns on the day in the week or the month in the year. Such effects will be neglected in this chapter.

The economic background of financial time series is also well described in the books [1] and [12].

Market Efficiency and the Martingale Property

The martingale property of the random return series $\{R_t\}$ derives from economic insight. The returns $\{R_t\}$ are regarded as martingale innovations, a property which is characterized by the condition that the conditional expectation of the future return R_t given the past returns value $r_{t-1}, \ldots, r_0$ with $r_0 = \log s_0$ (or, equivalently, given the prices $s_{t-1}, \ldots, s_0$) is equal to zero.

On the contrary, suppose that

$$E(R_t|s_{t-1}, \ldots, s_0) \neq 0$$

holds. Let as assume that the conditional expectation is positive so that the price is expected to rise. In speculative markets, lots of arbitrageurs quickly eliminate unfair gambles: buying at todays low price s_{t-1} is more than a fair gamble. If all agents have this information, they will all want to buy at s_{t-1} in order to resell at the higher expected price tomorrow. Yet if many try to buy, this already drives up the price today and the expected gain disappears.

If the martingale property of returns is accepted, then the returns are necessarily uncorrelated.

Some Historical Remarks

The hypothesis that logarithmic speculative prices form a martingale series was first raised by Bachelier[3]. Specifically, he assumed normally distributed returns.

Writing

$$R_t = \mu + \sigma W_t$$

for the random log–returns, where the W_t are iid standard normal random variables, μ is the drift parameter, and $\sigma > 0$ is the so–called volatility, one arrives in (16.2) at a discrete version of the famous Black–Scholes[4] model, namely,

$$S_T = S_0 \exp\left(\mu T + \sigma \sum_{t \leq T} W_t\right). \tag{16.3}$$

Later, Benoit Mandelbrot[5] discovered that the speculative return series, i.e., the innovations of the martingale, are fat–tailed distributed. He suggested the modeling of speculative returns by means of non–normal, sum–stable random variables (see also [40], [1], and the book by F.X. Diebold[6]). But this latter model conflicts with the fact that return series generally exhibit bounded second moments, see, e.g., Akgiray and Booth[7] and the literature cited therein.

---

[3]Bachelier, L.J.B.A. (1900). Théorie de la Speculation. Gauthier–Villars, Paris.

[4]Black, F. and Scholes, M. (1973). The pricing of options and corporate liabilities. J. Political Economy 81, 637–659.

[5]Mandelbrot, B.B. (1963). The variation of certain speculative prices. J. Business 36, 394–419.

[6]Diebold, F.X. (1989). Empirical Modeling of Exchange Rate Dynamics. Lect. Notes in Economics and Math. Systems 303, Springer, Berlin.

[7]Akgiray, V. and Booth, G.G. (1988). The stable-law model of stock returns. J. Business & Economic Statistics 6, 51–57.

16.2 Empirical Evidence in Returns Series

Certain properties of financial series of stock market returns and exchange rate returns will be illustrated by means of scatterplots. One recognizes a clustering of higher volatility of returns which will also be captured by plots of sample variances over moving windows in the time scale (related to that what is done by means of moving averages).

Stock Market and Exchange Rate Data

Next, we partially repeat and extend the analysis made by Loretan and Phillips[8] concerning

- the Standard & Poors 500 stock market index from July 1962 to Dec. 1987 (stored in fm–poors.dat), and

- exchange rates of the British pound, Swiss franc, French franc, Deutsche mark, and Japanese yen measured relatively to the U.S. dollar from Dec. 1978 to Jan. 1991 (stored in fm–exchr.dat[9]).

We visualize the Standard & Poors 500 stock market data and the exchange rates of the British pound measured relatively to the U.S. dollar. Stock market and exchange rate data exhibit a similar behavior so that a joint presentation is justified.

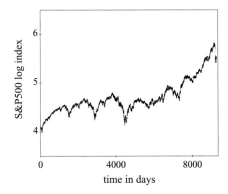 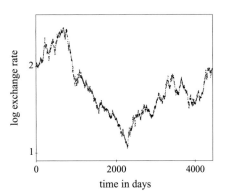

FIG. 16.1. (left.) Log–index $\log s_t$ of Standard & Poors 500 stock market. (right.) Log–exchange rate $\log s_t$ of British pound relative to U.S. dollar.

---

[8]Loretan, M. and Phillips, P.C.B. (1994). Testing the covariance stationarity of heavy–tailed time series. J. Empirical Finance 1, 211–248.

[9]Extended data sets are stored in fm–poor1.dat (from June 1952 to Dec. 1994) and in fm–exch1.dat (from Jan. 1971 to Feb. 1994).

Next, the equally (over the weekend) distributed returns are displayed in a scatterplot. In Oct. 1987, there are extraordinarily high and low returns to the stock market index which are not displayed in Fig. 16.2 (left).

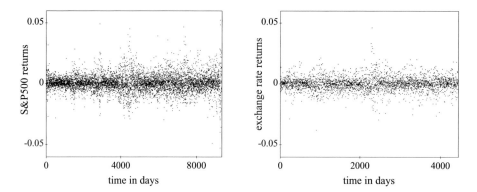

FIG. 16.2. (left.) Daily returns r_t to the Standard & Poors 500 stock market index. (right.) Daily exchange rate returns r_t to the British pound relative to U.S. dollar.

The scatterplot of the Standard & Poors log–returns (with changed signs) was partly displayed in Fig. 1.1 together with a plot of a moving sample quantiles.

The Sample Autocorrelations of Log–Returns

The serial autocorrelations for the returns (based on trading days or with the returns omitted after a gap) are practically equal to zero which supports the martingale hypothesis. In particular, the expectation of returns conditioned on the past is equal to zero. This is illustrated for the Standard & Poors market indices in Fig. 16.3.

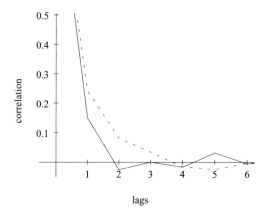

FIG. 16.3. Autocorrelation functions for the Standard & Poors log–returns w.r.t. the trading days (solid) and with the Monday returns distributed (dotted).

There is a slightly larger positive correlation for the lag $h = 1$, if the Monday returns are distributed over the weekend.

The Tranquility and Volatility of Returns

Series of daily squared returns r_t^2 from stock and foreign exchange markets are visualized by a scatterplot.

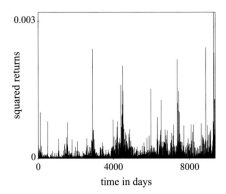
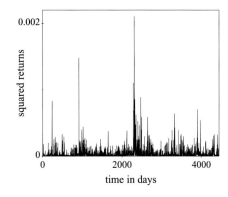

FIG. 16.4. Squared daily returns r_t^2 to (left) Standard & Poors 500 stock market index, and (right) exchange rates of the British pound measured relatively to the U.S. dollar.

The illustrations in Fig. 16.4 exhibit that there are periods of tranquility and volatility of the return series in a particularly impressive manner.

The periods of volatility can also be expressed by moving sample variances $(1/m) \sum_{i=t-m}^{t} (r_i - \bar{r}_t)^2$ based on the log–returns of the preceding m days.

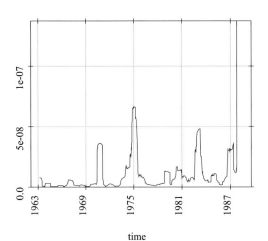

FIG. 16.5. Moving sample variance over a time horizon of $m = 250$ days for the Standard & Poors data.

This shows that, while the return process is a fair game, risk is spread unevenly. Hence, the risk of an investment is clustered through time and somewhat predictable. Supposedly, investors trade off return against risk, and hence portfolio management has to take this predictability into account. In technical terms, we may assume that the returns are conditionally heteroskedastic (the variance is varying conditioned on the past).

Trend and Symmetry

For currencies no trend has to be eliminated. Also, currencies are naturally symmetric due to the fact that the return, e.g., for the Deutsche mark on the U.S. dollar is the negative of the return of a Deutsche mark investment for U.S. dollars (see also page 372). Therefore, for exchange rate returns, the tails will also be estimated under the condition of symmetry. Tail estimators are also applied to returns reflected in 0.

For stock prices, one must first eliminate the positive trend, due to growth of the economy, to get to the martingale model.

16.3   Parametric Estimation
of the Tails of Returns

Again, we may visualize returns by means of the usual nonparametric tools like kernel densities, sample mean excess functions, etc., see Section 8.1 for a justification of classical extreme value statistics. The application of standard estimation procedures may also be relied on ergodic theory, see page 209 for references to the literature.

The overall impression of stock market and exchange rate data is that normal distributions or similar symmetric distributions can be well fitted to the central data, yet there seem to be fat tails. So even in those cases where we estimate light tails (exponential or beta tails) within the GP model, there seems to be a kurtosis larger than that of the normal distribution.

We estimate the upper and lower tail indices of stock market and exchange rate returns and compare our estimates with those of Loretan and Phillips (L&P) who always employed the Hill estimator (presumably to the returns with respect to the trading days).

Stock Market Returns

We start our investigations with the famous Standard & Poors stock market index which has been frequently analyzed in the literature.

There is a certain trend in the returns that can be captured by a quadratic least squares line to some extent. The trend is negligible, when GP parameters are estimated, as long as there is a time horizon of less than about 10 years.

EXAMPLE 16.3.1. (Estimating the Tail Indices of Returns to the Standard & Poors Index.) In Table 16.1, Hill(GP1), Moment(GP) and MLE(GP) estimates of the lower and upper tail index α, based on the daily returns to the Standard & Poors index, are displayed. The number k of lower or upper extremes was chosen according to the insight gained from the diagrams of the estimators.

We are also interested in the effect of handling the daily returns in different manners (taking the returns for each trading day or distributing or omitting returns after gaps).

TABLE 16.1. Returns with respect to trading days and returns after gaps distributed or omitted.

	Tail Indices α of Standard & Poors Index Returns							
	Lower ($k = 150$)				Upper ($k = 100$)			
	trad.	distr.	omit.	L&P	trad.	distr.	omit.	L&P
Hill(GP1)	3.39	3.79	4.33	3.59	3.81	3.70	3.70	3.86
Moment(GP)	2.77	2.59	4.40	–	4.26	3.67	3.67	–
MLE(GP)	3.25	2.20	4.26	–	4.32	3.52	3.52	–

The remarkable difference between the Moment(GP) or MLE(GP) estimates for the lower tail index—with returns after a gap distributed or omitted—is due to the drop from 282.70 to 224.84 of the Standard & Poors index during the well–known stock market crash on Monday, Oct. 19, 1987.

Omitting this extraordinary event from the data yields an underestimation of the risk entailed in negative returns.

The most important message from the preceding example is that for these stock market returns, one must take heavy upper and lower tails with a tail index α around 3 into account.

Related results are obtained for other stock market indices (as, e.g., for the Lufthansa and Allianz data stored in fm-lh.dat and fm-allia.dat).

Global Modeling of a Distribution, Local Modeling of the Tails

As already mentioned on page 61: one major goal of this book is to select distributions which simultaneously fit to the central as well as to the extreme data. The extreme part is analyzed by means of the tools of extreme value analysis.

In Example 2.3.2, we successfully fitted a mixture of two Gaussian distributions to the S&P500 data which were also dealt with in Example 16.3.1.

Applying the MLEs in the Student model as well as in the model of sum–stable distributions, see Sections 6.3 and 6.4, we get the estimates as listed in Table 16.2.

	Parameters	
	α	σ
Student	2.29	0.0034
sum–stable	1.34	0.0027

TABLE 16.2. MLEs of α and σ based on S&P data.

If the pertaining densities are additionally plotted in Fig. 2.11, they are hardly distinguishable from the Gaussian mixture and the kernel densities. Therefore, there are three candidates for the modeling of the Standard & Poors, namely, a Gaussian mixture, Student and sum-stable distribution. The insight gained from Example 16.3.1 speaks in favor of the Student modeling.

Exchange Rate Returns

For exchange rate returns we find light as well as heavy tails. The tail index is also estimated under the condition of symmetry. Under that condition one may base the estimation on a larger number of upper extremes.

EXAMPLE 16.3.2. (Estimating the Tail Indices of Daily Exchange Rate Returns: British Pound.) Our approach corresponds to that in Example 16.3.1.

TABLE 16.3. Estimating the tail indices.

	Tail Indices α of British Pound Returns								
	Lower ($k = 75$)				Upper ($k = 75$)				Symm. ($k = 150$)
	trad.	distr.	omit.	L&P	trad.	distr.	omit.	L&P	distr.
Hill(GP1)	3.85	4.03	4.03	2.89	3.65	4.25	3.75	3.44	3.95
Moment(GP)	12.88	4.51	4.51	–	5.48	3.62	5.05	–	4.72
MLE(GP)	238.0	4.28	4.28	–	5.98	3.33	5.40	–	4.85

There is a greater variability in the Moment and MLE estimates based on returns with respect to trading days. The diagrams of estimates provide more insight.

We remark that the Moment and ML estimates based on the symmetrized trading day data are around $\alpha = 6$. The overall picture is that a Paretian modeling of the lower and upper tails is justified for the British daily exchange rate returns.

The insight gained from Example 16.3.2 suggest to intensify an exploratory analysis of the data and to apply additionally the L–moment and related estimators, see Chapter 14. The latter has not been done yet.

A completely different modeling must be taken for the Swiss franc. It seems to us that an exponential modeling for the tails of the distribution is justified.

If the estimation is carried out in the unified generalized Pareto (GP) model, then the estimates of the shape parameter γ are close to zero in the present case. Therefore, $\alpha = 1/\gamma$ attains large positive as well as small negative values. In such cases, it is preferable to carry out the estimation in the γ–parameterization to avoid bigger differences in the estimates of α for varying k.

It is remarkable that the Hill estimator is incorrect, a fact which becomes apparent by comparing parametric and nonparametric plots, see also Fig. 5.2 (left) and the expansion in (6.38).

EXAMPLE 16.3.3. (Estimating the Tail Indices of Daily Exchange Rate Returns: Swiss Franc.) For daily exchange rate returns of the Swiss franc relative to the U.S. dollar, we obtain negative estimates of the lower tail index α. The estimates are based on equally distributed returns. When omitting Monday returns, one finds similar results. The chosen number k of extremes is 100.

TABLE 16.4. Estimating the tail indices α and γ of daily exchange rate returns of the Swiss franc relative to the U.S. dollar with equally distributed returns (including Hill estimates taken from Loretan and Phillips (L&P) in the α–parameterization).

| | Tail Indices α and γ of Swiss Franc Returns | | | | | | | |
| | Lower ($k = 100$) | | | Upper ($k = 100$) | | | Symmetric ($k = 200$) | |
	α	γ	L&P	α	γ	L&P	α	γ
Hill(GP1)	3.61	0.28	3.10	4.16	0.24	2.77	3.66	0.27
Moment(GP)	−9.09	−0.11	–	9.67	0.10	–	−9.35	−0.11
MLE(GP)	−5.88	−0.17	–	18.19	0.05	–	−31.81	−0.03

Moment(GP) and MLE(GP) estimates γ for the lower tail index are negative yet close to zero. When employing the armory of nonparametric tools such as sample excess functions, Q–Q plots, sample qfs, etc. and then comparing the nonparametric and parametric curves, there is doubt that an application of the Hill estimator is correct. On the basis of these results, we suggest a beta or an exponential modeling for the lower tail.

Moment(GP) and MLE(GP) estimates for the upper tail index can be negative for other choices of the number k of extremes. We suggest an exponential modeling of the upper tail in the case of Swiss exchange rate returns. The preceding remarks about the Hill estimator are applicable again. The estimates for the symmetrized sample confirm the preceding conclusions.

There is a greater variability in the estimates of the shape parameter for varying numbers k of extremes which could be reduced, to some extent, by smoothing these estimates over a moving window.

We shortly summarize results for the exchange rates of the French franc, Deutsche mark and yen against the U.S. dollar. We give a list of Moment(GP) and MLE(GP) estimates in the γ–parameterization.

TABLE 16.5. Estimating the lower and upper tail indices of distributed daily returns based on $k = 75$ extremes ($k = 150$ extremes for the symmetrized sample).

| | Estimating the Tail Index γ | | | | | |
| | Moment(GP) | | | MLE(GP) | | |
	Lower	Upper	Symmetric	Lower	Upper	Symmetric
French franc	−0.11	−0.05	−0.05	−0.09	−0.04	−0.05
Deutsche mark	−0.19	0.02	−0.05	−0.23	−0.04	−0.14
yen	0.06	0.08	0.007	−0.06	0.06	−0.03

We see that for all three cases an exponential modeling for the tails seems to be adequate.

A detailed exploration of extreme returns of further exchange rates, stock market indices and other speculative asset prices is desirable.

16.4   The Profit/Loss Variable and Risk Parameters

At the time of writing financial institutions like commercial banks have to meet new Capital Adequacy rules. A landmark is the Basle Committee accord of 1988 for credit risks which were supplemented by proposals for market risks starting 1993 (for details see the book by Jorion [33] and further supplements of the Basle Committee accord[10]). These rules require that a bank must have sufficient capital to meet losses on their exposures. In this context, we compute

- the Value–at–Risk (VaR) as the limit which is exceeded by the loss of a given speculative asset or a portfolio with a specified low probability,

- the Capital–at–Risk (CaR) as the amount which may be invested so that the loss exceeds a given limit with a specified low probability, see Section 16.5.

Thus, we fix either the invested capital (market value) or the limit and, then, compute the other variable. Further useful risk parameters of the loss/profit distribution will be introduced at the end of this section.

The VaR is a quantile of the loss (more precisely, profit/loss) distribution which must be estimated from the data. This goal is achieved in the following manner.

1. The loss variable is represented by means of the returns so that the results from the preceding section become applicable. This is done in the subsequent lines.

---

[10]Basle Committee on Banking Supervision (1996). Supplement to the Capital Accord to Incorporate Market Risk. Basle Committee on Banking Supervision, Basle.

2. We estimate the VaR based on different statistical models for the returns, particularly making use of the GP modeling as outlined in Section 16.3. This step will be carried out in the next section.

In both steps one must distinguish between the case of a single asset and the one of a portfolio.

Representing the Profit/Loss Variable by Means of Returns

Let V_0 and V_T be the market values of a single speculative asset or a portfolio at the times $t = 0$ and $t = T$. Usually, the time horizon is a day or a month (the latter corresponds to 20 trading days or, alternatively, to 30 days if the Monday returns are distributed over the weekend).

Losses and profits within the given period of T–days will be expressed by the loss (profit/loss) variable as the difference of the market values V_T and V_0. We have

$$L_T = -(V_T - V_0). \tag{16.4}$$

Notice that losses are measured as positive values. Conversely, there is a profit if L_T is negative. Under this convention, we may say that a loss is, for example, smaller than the 99% Value–at–Risk, see (16.11), with a probability of 99%.

The T–day loss L_T will be expressed by means of T–day returns

$$R_{(T)} = \sum_{t \leq T} (-R_t) \tag{16.5}$$

taken with a changed sign. Next, we distinguish between representations of the loss variable for a single asset and for a portfolio.

- (The Loss for a Single Asset.) The total market value V_t of an asset at time t can be expressed by $V_t = hS_t$, where h is the number of shares, which are held within the given period, and S_t is the price at time t. From (16.2), where prices are represented by means of the daily log–returns, we conclude that the loss at time T is

$$\begin{aligned} L_T &= V_0\big(1 - \exp(-R_{(T)})\big) \\ &\approx V_0 R_{(T)}. \end{aligned} \tag{16.6}$$

- (The Portfolio Loss.) Let $V_{t,j} = h_j S_{t,j}$ be the market value of the jth asset in the given portfolio at time t, where h_j are the numbers of shares which are held within the given period, and $S_{t,j}$ are the prices at time t. Notice that

$$V_t = \sum_j V_{t,j} \tag{16.7}$$

is the market value of the portfolio at time t.

We also introduce the vector of weights $\boldsymbol{w} = (w_1, \ldots, w_d)$, where $w_j = V_{0,j}/V_0$. Notice that the w_j sum to unity. The vector of weights determines the market strategy of the investor at the beginning of the period. We have

$$V_t = V_0 \sum_j w_j S_{t,j} = V_0 \boldsymbol{w} \boldsymbol{S}_t' \tag{16.8}$$

where $\boldsymbol{S}_t'$ is the transposed vector of prices $\boldsymbol{S}_t = (S_{t,1}, \ldots, S_{t,d})$.

From (16.2) and (16.7) deduce that the loss $L_T = -(V_T - V_0)$ of the portfolio at time T is

$$
\begin{aligned}
L_T &= V_0 \sum_j w_j \big(1 - \exp(-R_{(T,j)})\big) \\
&\approx V_0 \sum_j w_j R_{(T,j)} = V_0 \boldsymbol{w} \boldsymbol{R}_{(T)}',
\end{aligned} \tag{16.9}
$$

where $\boldsymbol{R}_{(T)} = (R_{(T,1)}, \ldots, R_{(T,d)})$ is the vector of random T–day log–returns (with changed sign) for the different asset.

If the portfolio consists of a single asset, then the formulas in (16.9) reduce to those in (16.6). The portfolio may also be treated like a single asset. The log–return of the portfolio is given by

$$R_t^* = \log V_t - \log V_{t-1} \tag{16.10}$$

with V_t as in (16.7). Now (16.6) is applicable with R_t^* in place of R_t.

The Value–at–Risk (VaR)

The Value–at–Risk parameter $\text{VaR}(T, q)$ is the q–quantile of the loss distribution. The VaR at the probability q satisfies the equation

$$P\{L_T \leq \text{VaR}(T, q)\} = q, \tag{16.11}$$

where L_T is the loss variable in (16.4).

We also speak of a VaR at the 99% or 95% level, if $q = 0.99$ or $q = 0.95$. Thus, for example, the loss is smaller than the VaR at the 99% level with a probability of 99%.

If the VaR is computed to fulfill capital adequacy requirements, then it is advisable to choose a higher percentage level, say 99% or 99.9%. To compare risks across different markets, a smaller level such as 95% can be appropriate.

The VaR depends on the distribution of the returns.

- (VaR of a Single Asset.) Let F_T denote the df of the T–day log–return $R_{(T)} = \sum_{t \leq T}(-R_t)$ with changed sign. Thus,

$$F_T(x) = P\{R_{(T)} \leq x\}. \tag{16.12}$$

Applying (16.6) we see that the VaR can be written

$$\text{VaR}(T,q) \quad = \quad V_0 \left(1 - \exp\left(-F_T^{-1}(q)\right)\right) \qquad (16.13)$$
$$\approx \quad V_0 F_T^{-1}(q), \qquad (16.14)$$

where V_0 is the market value at time $t = 0$.

The VaR is overestimated in (16.14) in view of the inequality $1 - \exp(-x) \leq x$, yet the error term is negligible if the q–quantile $F_T^{-1}(q)$ is not too large.

- (VaR of a Portfolio.) The Value–at–Risk $\text{VaR}(T,q)$ is the q–quantile and, respectively, the approximate q–quantile of the random variables in (16.9).

In Section 16.5, the VaR is computed in a more explicit manner based on different statistical models such as, e.g., the GP modeling for returns in the case of a single asset.

Further Risk Parameters

Further risk parameters may be defined as functional parameters of exceedance dfs pertaining to the loss L. Let F denote the df of L, and denote by $F^{[u]}$ the exceedance df at the threshold u (cf. (1.11)). An example of such a functional parameter is the expected shortfall which is the mean of $F^{[u]}$ with $u = \text{VaR}(q)$ which has the representation

$$E\big(L\big|L > \text{VaR}(q)\big) \quad = \quad \int_{\text{VaR}(q)}^{\infty} x \, dF^{[\text{VaR}(q)]}(x)$$
$$= \quad \frac{1}{1-q} \int_q^1 \text{VaR}(x) \, dx, \qquad (16.15)$$

where the second equation may be deduced with the help of the well–known formula $\int x \, dF(x) = \int_0^1 F^{-1}(q) \, dq$. We see that the expected shortfall at the level q is an "average" of the Value–at–Risks with $x \geq q$.

Perhaps, it would be more reasonable to call

$$E\big(L - \text{VaR}(q)\big|L > \text{VaR}(q)\big)$$

expected shortfall as the expected loss which is not covered by the allocated capital of the bank.

In Section 16.3 we estimated the upper tail of the distribution of returns and, thus, the exceedance df. Using a representation of the loss variable in Section 16.4 in terms of returns one can estimate the indicated risk parameters. Details will be given for the VaR in the subsequent section.

16.5   Evaluating the Value–at–Risk (VaR)

Next, we discuss the estimation of the VaR under certain models for the return
distribution. We start with the estimation of the VaR in the case of a single asset
based on the GP–modeling. The estimation by means of the sample q–quantile is
the second method. In the financial literature this method runs under the label
Historical Simulation (HS).

Estimating the VaR for a Single Asset: the GP Modeling

We take representations of the VaR, given in (16.13) and (16.14). For simplicity,
let the initial market value V_0 be equal to 1.

It remains to estimate the q–quantile $F_T^{-1}(q)$, where F_T is the df of the T–day
log–return $R_{(T)}$. Moreover, we assume that a GP df could be accurately fitted to
the upper tail of F_T as it was done in Section 15.3.

In Example 16.5.1, the VaR will be computed for $T = 1$ and $T = 30$ days
and several large probabilities q in the case of the Standard & Poors index with
the Monday returns distributed over the weekend.

EXAMPLE 16.5.1. (VaR for the Standard & Poors Index.) The Value–at–Risk VaR(T, q)
is estimated for $T = 1$ and $T = 30$ based on the 9314 equally distributed, daily returns
to the Standard & Poors Index.

For $T = 1$ the q–quantile of the daily returns with changed sign is estimated in a
parametric manner by means of the Moment(GP) estimator and our standard estimators
of the location and scale parameters. The estimation is based on $k = 150$ lower extreme
daily returns as in Example 16.3.1.

To estimate VaR$(30, q)$ the Moment(GP) estimator is applied to $k = 30$ extreme
30–days returns. The number of 30–days returns is 310. The estimated shape parameter
is $\alpha = 2.4$.

In both cases, the upper tail of the qf F_T^{-1} is estimated by the Pareto qf belonging
to the estimated parameters. The VaR, according to (16.14), is listed in Table 16.6 for
several values of q.

TABLE 16.6. Estimating the Value–at–Risk VaR(T, q) for the Standard & Poors Index.

	Value–at–Risk VaR(T, q)			
probability q	0.99	0.995	0.999	0.9995
$T = 1$	0.017	0.020	0.033	0.041
$T = 30$	0.128	0.169	0.333	0.435

For example, the loss within $T = 30$ days is smaller than VaR$(30, 0.99) = 0.128$
with a 99%–probability, if $V_0 = 1$ is invested at time $t = 0$.

We also give the exact values VaR$^* = 1-\exp(-\text{VaR})$ of the Value–at–Risk according to (16.13). As mentioned before, the values in Table 16.6 overestimate the exact values VaR*.

TABLE 16.7. Values VaR$^*(T, q) = 1 - \exp(-\text{VaR}(T, q))$ for the Standard & Poors Index.

	Value–at–Risk VaR$^*(T, q)$			
probability q	0.99	0.995	0.999	0.9995
$T = 1$	0.017	0.020	0.032	0.040
$T = 30$	0.120	0.155	0.283	0.353

For $T = 1$ the differences are negligible, yet there are remarkable differences for $T = 30$.

There is a remarkable coincidence between estimated parametric q–quantiles and sample q–quantiles when the parametric q–quantile lies inside the range of the sample.

Estimating the VaR: Historical Simulation

In Historical Simulation (HS), the loss/profit df of a given asset or a portfolio over a prescribed holding period of length T is simply given by the sample df of past losses and gains. We then translate this into a loss/profit qf and read of the VaR.

The advantage of HS relies on its simplicity and low implementation costs. But this very simplicity is also the cause of problems. The main problem is that extreme quantiles cannot be estimated, because extrapolation beyond past observation is impossible, and when estimation is possible the estimates are not reliable due to the lack of sufficient observations in the tail area. Moreover, the quantile estimators tend to be very volatile if a large observation enters the sample.

Estimating the VaR for a Portfolio:
the Variance–Covariance Method

This method is useful when we have to compute the VaR for a portfolio containing many different assets. In fact, this task becomes particularly easy when all asset returns are assumed to be normally distributed and the loss variable of the portfolio is a linear function of these. In this case, the VaR is a multiple of the portfolio standard deviation and the latter is a linear function of individual variances and covariances.

We assume that the T–day log–return (with changed sign)

$$R_{(T,j)} = \sum_{t \leq T}(-R_{t,j})$$

of the jth asset at time t is a Gaussian random variable with mean zero and unknown variance $\sigma_{T,j}^2 > 0$. In addition, assume that the $R_{(T,j)}$, $j \leq d$, are jointly Gaussian with a covariance matrix $\Sigma_T = (\sigma_{T,i,j})$ as introduced on page 280. Necessarily, $\sigma_{T,j,j} = \sigma_{T,j}^2$.

Let $V_{0,j}$ be the market value of the jth asset in the portfolio at time $t = 0$. According to (16.9), the loss L_T of the portfolio at time $t = T$ can be approximately represented by

$$L_T \approx V_0 \boldsymbol{w} \boldsymbol{R}'_{(T)},\tag{16.16}$$

where $\boldsymbol{R}_{(T)}$ is the vector of the T–day log–returns $R_{(t,j)}$. According to the well–known results for Gaussian random vectors (cf. page 280), the term $\boldsymbol{w}\boldsymbol{R}'_{(T)}$ in (16.16) is a Gaussian random variable with mean zero and standard deviation

$$\sigma_T = \sqrt{\boldsymbol{w}\Sigma_T\boldsymbol{w}'}.\tag{16.17}$$

Combining (16.16) and (16.17) we get

$$\text{VaR}(T,q) = V_0\sigma_T\Phi^{-1}(q),$$

where Φ^{-1} is the univariate standard Gaussian qf.

An estimate of the VaR is obtained by replacing the variances and covariances $\sigma_{T,i,j}$ in (16.17) by their sample versions $\hat{s}_{T,i,j}$, cf. (2.6), based on historical T–day log–returns.

If there is a single asset, or the portfolio is dealt with like a single asset, and the daily log–returns are uncorrelated, then $\sigma_T = V_0 T^{1/2}\sigma_1$ with σ_1 denoting the standard deviation of a daily log–return.

Estimating the VaR for a Portfolio: the Copula Method

A considerable improvement, compared to the simple variance–covariance method, is achieved when the univariate Gaussian margins are replaced by more realistic dfs. This is the copula method (cf. page 275) as already successfully applied in the Chapters 12 and 13 to estimate unknown multivariate EV and GP dfs.

We assume that the dependence structure of a multivariate Gaussian df $\Phi_{\widetilde{\Sigma}}$ (cf. (11.4)), with mean vector zero and covariance matrix $\widetilde{\Sigma}$, is still valid, however the univariate margins are unknown dfs F_j, say. Thus, we assume that the T–day log-returns $R_{(T,j)}$ have the joint df

$$F(\boldsymbol{x}) = \Phi_\Sigma\Big(\Phi^{-1}(F_1(x_1)),\dots,\Phi^{-1}(F_d(x_d))\Big),\tag{16.18}$$

where Σ is the correlation matrix pertaining to $\widetilde{\Sigma}$ and Φ^{-1} is the univariate standard Gaussian qf.

Notice that the preceding variance–covariance model is a special case with F_j being a Gaussian df with scale parameters σ_j. We discuss further possibilities.

- De Raaij and Raunig[11] report a successful implementation of mixtures of two Gaussian dfs (also see pages 31 and 380).

- If one goes one step further and is continuously mixing Gaussian distributions with respect to a gamma distribution, one arrives at a Student distribution (cf. (6.16) and page 380). The careful univariate extreme value analysis in Section 16.3 speaks in favor of such dfs which have Pareto–like tails.

- Further possibilities include sum–stable dfs yet, as mentioned before, this conflicts with the empirical evidence that log–returns exhibit bounded second moments.

The df in (16.18) can be estimated by applying the piecing–together method (used to estimate multivariate EV and GP dfs).

1. Estimate the df F_j by means of $\widehat{F}_j$ based on the univariate data $x_{i,j}$ in the jth component;

2. transform the data $x_{i,j}$ to

$$y_{i,j} = \Phi^{-1}\big(\widehat{F}_j(x_{i,j})\big)$$

which may be regarded as data governed by Φ_Σ;

3. estimate Σ by the sample correlation matrix $\widehat{\Sigma}$ based on the transformed data $y_{i,j}$,

4. take

$$\widehat{F}(\boldsymbol{x}) = \Phi_{\widehat{\Sigma}}\Big(\Phi^{-1}(\widehat{F}_1(x_1)), \ldots, \Phi^{-1}(\widehat{F}_d(x_d))\Big) \qquad (16.19)$$

as an estimate of $F(\boldsymbol{x})$.

For an application we refer to Example 10.3.1 where the copula method led to a curious modeling.

Using the tools from multivariate extreme value analysis, one should also analyze the validity of the chosen copula function. Further candidates of copula functions are, e.g., those pertaining to multivariate sum–stable or Student distributions.

---

[11] Raaij, de G. and Raunig, B. (1999). Value at risk approaches in the case of fat–tailed distributions of risk factors. Manuscript, Central Bank of Austria.

Estimating the VaR for a Portfolio:
a Multivariate Student Modeling

We extend the variance–covariance method to multivariate Student distributions. The loss L_T of a portfolio at time T is again represented by $V_0 \boldsymbol{w} \boldsymbol{R}'_{(T)}$ as in (16.16). Now we assume that the T–day log–returns $R_{T,j}$, $j \leq d$, are jointly Student distributed with shape parameter $\alpha > 0$ and parameter matrix Σ, see (11.11) and (11.12). Therefore, the log–returns have the representations

$$R_{T,j} = \frac{X_j}{(2Y/\alpha)^{1/2}},$$

where the X_j are jointly Gaussian with covariance matrix Σ. Consequently,

$$\boldsymbol{w}\boldsymbol{R}'_{(T)} = \frac{X}{(2Y/\alpha)^{1/2}},$$

where X is Gaussian with standard deviation $\sigma_T = \sqrt{\boldsymbol{w}\Sigma_T \boldsymbol{w}'}$ corresponding to (16.17), and L_T is a Student variable with shape parameter $\alpha > 0$ and scale parameter $V_0\sigma_T$. Therefore, one gets the Value–at–Risk

$$\mathrm{VaR}(T, q) = V_0 \sigma_T F_\alpha^{-1}(q),$$

where F_α denotes the standard Student df with shape parameter α.

One gets the variance–covariance result in the limit as $\alpha \to \infty$. The extended model is more flexible compared to the Gaussian model and includes the case of heavy–tailed distributions. There is still the restriction of equal shape parameters in the single components.

Capital–at–Risk (CaR) for a Single Asset

In the preceding lines the VaR was computed as the limit l such that $P\{L_T \leq l\} = q$, where L_T is the loss/profit variable for the period $t = 0$ and $t = T$. Conversely, one may fix the limit l and compute the Capital–at–Risk $\mathrm{CaR}(T, q, l)$ as the amount which can be invested such that the pertaining loss L_T does not exceed the limit l with a given probability q.

Using the representation (16.13) of the loss/profit variable L_T with V_0 replaced by $\mathrm{CaR}(T, q, l)$, one gets

$$P\{L_T \leq l\} = P\left\{\mathrm{CaR}(T, q, l)\left(1 - \exp\left(\sum_{t \leq T} R_t\right)\right) \leq l\right\}$$
$$= q,$$

if

$$\mathrm{CaR}(T, q, l) = l / \left(1 - \exp\left(-F_T^{-1}(q)\right)\right)$$
$$\approx l / F_T^{-1}(q).$$

Here, F_T is again the df of the T–day return $\sum_{t \leq T}(-R_t)$ with changed sign (see (16.12)). The ratio

$$c(T, q) = 1/F_T^{-1}(q) \tag{16.20}$$

is the capital/loss coefficient.

EXAMPLE 16.5.2. (Capital/Loss Coefficient for the Standard & Poors Index.) The capital/loss coefficient $c(T, q)$ is estimated for $T = 1$ and $T = 30$ based on the 9314 equally distributed daily returns to the Standard & Poors Index. Apparently, one must estimate the q–quantile of F_T.

It is evident from (16.20) that the capital/loss coefficients are just the reciprocals of the values computed for the VaR in Table 16.6.

TABLE 16.8. Estimating the Capital/Loss Coefficient for the Standard & Poors Index.

	Capital/Loss Coefficient $c(T, q)$			
probability q	0.99	0.995	0.999	0.9995
$T = 1$	58.1	50.0	30.3	24.4
$T = 30$	7.8	5.9	3.0	2.3

It is difficult, due to medium sample sizes, to extend the empirical approach to estimating the capital/loss coefficient beyond a period of a month.

Capital–at–Risk for a Portfolio

Let H_0 be an initial market strategy which determines the proportions of the different assets to each other. Let again $V_{t,j} = H_{0,j} P_{t,j}$ and $V_t = \sum_j V_{t,j}$ be the market values of the single assets and of the portfolio with respect to H_0.

We compute a constant b such that the loss/profit variable L_T for the period from $t = 0$ to $t = T$, belonging to the multiple bH_0 of the initial market strategy, fulfills the equation

$$P\{L_T \leq l\} = P\left\{b\sum_j V_{0,j}\left(1 - \exp\left(\sum_{t \leq T} R_{t,j}\right)\right) \leq l\right\}$$
$$= q.$$

This holds with $b = l \Big/ F_{T,H_0}^{-1}(q)$, where F_{T,H_0}^{-1} is the df of

$$\sum_j V_{0,j}\left(1 - \exp\left(\sum_{t \leq T} R_{t,j}\right)\right).$$

Therefore,

$$\mathrm{CaR}(T, q, l) = bV_0 = lV_0 \Big/ F_{T,H_0}^{-1}(q)$$

is the Capital–at–Risk at the probability q.

16.6   The VaR for a Single Derivative Contract

We start with a short introduction to derivative contracts with special emphasis laid on European call options, deduce the VaR for the call option under the Black–Scholes pricing and conclude the section with remarks concerning the general situation.

An Introduction to Derivative Contracts

A derivative contract is a contract, the value of which is dependent on the value of another asset, called the underlying. Options are a special kind of derivative contracts, because these depend non–linearly on the value of the underlying.

Options are traded on many underlyings such as single stocks, stock indices, exchange rates, etc. We concentrate on an option for which the underlying is a non–dividend paying stock. There are two basic types of options. A European call option (respectively, a put option) gives the owner the right to buy (to sell) the underlying asset by a certain future date T for a prefixed price X, where

- T is the expiration date or maturity,

- X is the strike price.

The prices and returns of the underlying are again denoted by S_t and R_t.

At maturity T the value of a European call option (called payoff) will be the amount

$$\overline{\mathrm{PF}} = \max(S_T - X, 0), \tag{16.21}$$

by which the stock price S_T exceeds the strike price X. The value for the put option is $\overline{\mathrm{PF}} = \max(X - S_T, 0)$.

The Black–Scholes Prices for European Call Options

A call option becomes more valuable as the stock price increases. The opposite holds for put options. The present value (at time t) of the option could be determined through discounting the payoff. However, S_T in (16.21) is unknown at time $t < T$, and one has to use a different market price.

Within the continuous–time version of the model presented in (16.3), in case the underlying is not dividend paying, this is the Black–Scholes price

$$\begin{aligned} &C(S_t, T - t) \\ &\quad := \quad S_t \Phi\big(d_1(S_t, T - t)\big) - X \exp\big(-r(T - t)\big) \Phi\big(d_2(S_t(T - t)\big), \end{aligned} \tag{16.22}$$

where r is the annual risk–free rate of interest, $T - t$ is the number of calendar days until expiration,

$$d_1(S_t, T - t) = \frac{\log \frac{S_t}{X} + \left(r + \frac{\sigma^2}{2}\right)(T - t)}{\sigma\sqrt{T - t}} \quad \text{and} \quad d_2(S_t, T - t) = d_1 - \sigma\sqrt{T - t}.$$

We have $S_{t+1} = S_t \exp(\mu + \sigma W)$, where W is a standard Gaussian random variable. For simplicity, we assume that $\mu = 0$. Therefore, σW is the log–return for the given time period. More precisely, one must deal with the conditional distribution of S_{t+1} conditioned on s_t because the price s_t is known at time t, yet this leads to the same result.

The profit/loss (again with a conversed sign) for the period from t to $t+1$ is

$$L = -\big(C(S_{t+1}, T - t - 1) - C(S_t, T - t)\big). \tag{16.23}$$

Therefore, the Value–at–Risk VaR(q) at the level q is the q-quantile of the distribution of L. We have $P\{L \leq \mathrm{VaR}(q)\} = q$ corresponding to (16.11).

We compute the VaR by means of linearization and also show that a direct computation is possible for the simple case of a call option.

Computation of the VaR by Linearization (Delta Method)

We deduce an approximate representation of the loss variable in (16.23) which is linear in the log–return variable σW of the underlying.

From the fact that

$$s\varphi\big(d_1(s, T - t)\big) - X \exp(-r(T_t))\varphi\big(d_2(s, T - t)\big) = 0$$

one gets

$$\frac{\partial}{\partial s} C(s, T - t) = \Phi\big(d_1(s, T - t)\big) =: \Delta(s, T - t)$$

which is called the Δ of the option.

Therefore, a Taylor expansion about S_t yields that

$$\begin{aligned}
L &\approx \Delta(S_t, T - t)(S_t - S_{t+1}) \\
&= S_t \Delta(S_t, T - t)(1 - \exp(\sigma W)) \\
&\approx S_t \Delta(S_t, T - t)\sigma W. \tag{16.24}
\end{aligned}$$

Using the last expression for the loss variable one gets

$$\mathrm{VaR}(q) = S_t \Delta(S_t, T - t)\sigma \Phi^{-1}(q) \tag{16.25}$$

for the one–day Value–at–Risk at the level q.

A more accurate linear approximation can be achieved when the linear term depending on the partial derivative $(\partial/\partial t)C(s, t)$ is added.

This kind of approach will underperform when more complicated kinds of derivatives are considered or when the underlying is highly non–normal.

Direct Computation of the VaR

The Black–Scholes formula for a call option can be also used to compute the VaR directly, because the price of the option is strictly increasing in the price of the underlying. The exact solution is

$$\text{VaR}(q) = C\big(S_t, T - t\big) - C\big(S_t \exp(-\sigma \Phi^{-1}(q)), T - t - 1\big). \tag{16.26}$$

We remark that (16.25) can also be deduced from this formula.

In the case of more complex derivatives, when the function C has no simple analytic expression, such an approach can still be used, and the VaR can be found by numerical approximation.

Computing the Black–Scholes Price

The preceding arguments heavily rely on the Black–Scholes pricing formula as specified in (16.22). We indicate in which manner this price for the European call option and other derivative contracts can be determined. This enables us to compute the VaR as well for other derivative contracts.

Under the conditions of the Black–Scholes model (16.3) we have

$$S_T = S_t \exp\left(\mu(T - t) + \sigma \sum_{k=t+1}^{T} W_k\right)$$

with S_t being observed at time t.

Let r be the annual risk–free rate of interest. In addition, assume that

$$E(S_T / S_t) = \exp(r(T - t)). \tag{16.27}$$

Thus, the expected return of the speculative asset corresponds to the risk–free rate of interest. Under this additional condition one obtains the equality

$$C(S_t, T - t) = E\big(\max(S_T - X, 0)\big) \exp(-r(T - t))$$

between the discounted expected option payoff and the Black–Scholes price.

Without condition (16.27), the Black–Scholes price can still be justified as a fair price by using the concept of hedge portfolios. Moreover, under certain regularity conditions, the fair price of any derivative contract can be computed by calculating the discounted expected value of the option payoff $\overline{\text{PF}}$ with respect to the risk-neutral probability, also called the equivalent martingale measure. Thereby, the original probability measure is replaced by another one under which the process of payoffs becomes a martingale. Yet, both probability measures coincide, if condition (16.27) is valid.

Value–at–Risk for General Derivative Contracts

For general derivative contracts we also obtain by

$$C(S_t, T - t) = E(\overline{\mathrm{PF}}) \exp(-r(T - t)) \tag{16.28}$$

the fair price, if the expectation is taken under the risk–neutral probability. The profit/loss with conversed sign is given in analogy to (16.23). One may try to apply the delta method or to carry out a direct computation of the VaR.

Another possibility is to use intensive simulation procedures, such as Monte Carlo or bootstraps, to compute the expected value of the payoff. For details we refer to Caserta et al.[12].

16.7 GARCH and Stochastic Volatility Structures

The aim of this section is to provide some theoretical insight in the tail–behavior of return distributions and to start a discussion about a semiparametric estimation of the VaR and the capital/loss coefficient. We primarily deal with time series $\{X_k\}$ that possess an ARCH (autoregressive conditional heterosketastic)[13] structure or, by generalizing this concept, a GARCH[14] structure.

We also mention stochastic volatility (SV) models which provide alternative models for the volatility in returns series.

First, we review some basic facts of the conditioning concept as outlined in Section 8.1. Recall that the mean and the variance of the conditional distribution are the conditional expectation $E(Y|x)$ and the conditional variance $V(Y|x)$ of Y given $X = x$.

Modeling the Conditional Heteroskedasticity and Distributions of Returns

Assume that the random return R_t at time t can be expressed as

$$R_t = \tilde{\sigma}_t \varepsilon_t, \tag{16.29}$$

where ε_t is a random innovation with $E(\varepsilon_t) = 0$ and $V(\varepsilon_t) = 1$, and $\tilde{\sigma}_t > 0$ is a random variable with finite expectation being independent of ε_t (and depending on the past). Values of $\tilde{\sigma}_t$ are denoted by σ_t.

---

[12]Caserta, S., Daníelsson, J. and de Vries, C.G. (1998). Abnormal returns, risk, and options in large data sets. Statistica Neerlandica 52, 324–335.

[13] Engle, R.F. (1982). Autoregressive conditional heteroscedasticity with estimates of the variance of United Kingdom inflation. Econometrica 50, 987–1007.

[14] Bollerslev, T. (1986). Generalized autoregressive conditional heteroskedasticity. J. Econometrics 31, 307–327.

Because of the independence of $\tilde{\sigma}_t$ and ε_t, the conditional distribution of R_t given $\tilde{\sigma}_t = \sigma_t$ is the distribution of $\sigma_t \varepsilon_t$, see (8.13). In the present context, $\tilde{\sigma}_t^2$ is also addressed as random variance or volatility.

Under the conditions above we have

(i) $E(R_t | \sigma_t) = 0$;

(ii) $V(R_t | \sigma_t) = \sigma_t^2$.

Thus, the conditional variance is not a fixed constant, a property addressed as conditional heteroskedasticity. In addition,

(iii) $E(R_t) = 0$;

(iv) $V(R_t) = E(\tilde{\sigma}_t^2)$.

Later on, further conditions will be imposed on $\tilde{\sigma}_t$ and ε_t so that $\{R_t\}$ is a martingale innovation scheme and, hence, a series of uncorrelated random variables.

Notice that the distribution of $\tilde{\sigma}_t \varepsilon_t$ is the mixture of the distributions $\sigma_t \varepsilon_t$ with respect to the distribution of $\tilde{\sigma}_t$. We mention three examples of innovations ε_t and random scale parameters $\tilde{\sigma}_t$:

- (Normal Innovations.) If ε_t is standard normal and $1/\tilde{\sigma}_t^2$ is a gamma random variable with shape parameter $r > 0$ (see (4.6)), then $R_t = \tilde{\sigma}_t \varepsilon_t$ has the density g_r which is the Student density as deduced in (6.16).

- (Laplace Innovations.) If ε_t is a Laplace (double–exponential) random variable with density $f(x) = \exp(-|x|)/2$ and $1/\tilde{\sigma}_t$ is a gamma random variable with shape parameter r, then $R_t = \tilde{\sigma}_t \varepsilon_t$ is a double–Pareto random variable with density (cf. also (5.28))

$$
\begin{aligned}
g_r(x) &= \int_0^\infty (\vartheta/2) \exp(-|x|) h_r(\vartheta) \, d\vartheta \\
&= r(1 + |x|)^{1+r}/2. \tag{16.30}
\end{aligned}
$$

- (Log–Normal Innovations.) If $\tilde{\sigma}_t$ and ε_t are log–normal, then R_t is log–normal.

ARCH(1) Series

The following scheme captures both the martingale feature and the observed clusters of volatility in speculative return series. These clusters are well described analytically, but not well understood from the economic point of view[15].

---

[15]See, e.g., Vries, de C.G. (1994). Stylized facts of nominal exchange rate returns. In: The Handbook of International Economics, R. van der Ploeg (ed.), Blackwell, Oxford, pp. 335–389.

Let $\{\varepsilon_t\}$ be a white–noise process of iid random variables satisfying the conditions $E(\varepsilon_t) = 0$ and $E(\varepsilon_t^2) = 1$. Let R_0 be an initial random variable which is independent of the innovations ε_t. Then, $R_t = \tilde{\sigma}_t \varepsilon_t$ with

$$\tilde{\sigma}_t^2 = \alpha_0 + \alpha_1 R_{t-1}^2, \qquad t \geq 1, \tag{16.31}$$

and $\alpha_0, \alpha_1 > 0$, is an ARCH(1) series. Notice that the conditions, specified in (16.29), are valid.

We discuss some properties of the series $\{R_t\}$. Because the ε_t and R_{t-i} are independent for each $i \geq 1$, the R_t are uncorrelated with expectations $E(R_t) = 0$ and variances

$$
\begin{aligned}
V(R_t) &= \alpha_0 \left(\sum_{j=0}^{t-1} \alpha_1^j \right) + \alpha_1^t V(R_0) \\
&= \alpha_0 \frac{1 - \alpha_1^t}{1 - \alpha_1} + \alpha_1^t V(R_0), \tag{16.32}
\end{aligned}
$$

where one must assume that $\alpha_1 \neq 1$ in the second representation.

If $\alpha_1 < 1$, then

$$V(R_t) \to \alpha_0/(1 - \alpha_1), \qquad t \to \infty,$$

and the ARCH series approximately satisfies the condition of weak stationarity. If, in addition, $V(R_0) = \alpha_0/(1 - \alpha_1)$, then $V(R_t) = V(R_0)$ and $\{R_t\}$ is a white–noise process. We remark that white–noise processes of this type are taken to model innovations in certain economic time series.

The special properties of the ARCH process are due to the fact that the R_t are uncorrelated yet not independent. As a consequence of (8.13), the conditional distribution of R_t given $R_{t-1} = r_{t-1}$ is the distribution of

$$\sigma(r_{t-1}) \varepsilon_t, \tag{16.33}$$

where

$$\sigma(r) = (\alpha_0 + \alpha_1 r^2)^{1/2}.$$

This is also the conditional distribution of R_t given the past $R_{t-1} = r_{t-1}, \ldots, R_0 = r_0$ up to time zero.

Especially, the property

$$E(R_t | r_{t-1}, \ldots, r_0) = 0$$

of martingale innovations holds. In addition, it is evident that $E(R_t | r_{t-1}) = 0$ and $V(R_t | r_{t-1}) = \sigma^2(r_{t-1})$.

Extremal Properties of ARCH Series

The distributional properties of the ARCH series are best analyzed by viewing the squared ARCH series R_t^2 as a stochastic difference equation. From (16.31) and (16.29) we obtain

$$
\begin{aligned}
R_t^2 &= \alpha_0 \varepsilon_t^2 + \alpha_1 \varepsilon_t^2 R_{t-1}^2 \\
&= B_t + A_t R_{t-1}^2,
\end{aligned}
\tag{16.34}
$$

say. This stochastic difference equation with iid pairs (B_t, A_t) is equivalent to the ARCH series up to a coin flip process for the sign. If $E(\log A_1) < 0$, and if there is a κ such that $E(A_1^\kappa) = 1$, $E(A_1^\kappa \log A_1) < \infty$, $0 < E(B_1^\kappa) < \infty$ and $B_1/(1 - A_1)$ is nondegenerate, then

$$
R_t^2 \to R_\infty^2 = \sum_{j<\infty} B_j \prod_{i \leq j-1} A_i, \qquad t \to \infty
\tag{16.35}
$$

in distribution.

Furthermore, R_∞^2 has a Pareto like tail with tail index κ, that is,

$$
P\{R_\infty^2 > x\} = (1 + o(1))cx^{-\kappa}.
$$

Note that this latter conclusion still follows if the innovations ε_t have a light tail. For example, if the ε_t are iid standard normal, the tail index κ for R_∞^2 can be computed from the condition that $E(A_1^\kappa) = 1$. We have

$$
\Gamma(\kappa + 1/2) = \pi^{1/2}(2\alpha_1)^{-\kappa}.
$$

For more details and references to the literature see de Haan et al.[16] and Basrak et al.[17]

Conditional Densities,
Quasi Maximum Likelihood Estimation

The question of estimating the tails of the unconditional distribution of R_t was already dealt with in the preceding section. Presently, this distribution (more precisely, the sequence of distributions) is regarded as a nuisance parameter. The aim is to estimate the parameters α_0 and α_1. We briefly indicate that the maximum likelihood principle is applicable by computing the joint density of the returns.

---

[16] Haan, de L., Resnick, S.I., Rootzén, H. and Vries, de C.G. (1990). Extremal behavior of solutions to a stochastic difference equation with applications to ARCH–processes. Stoch. Proc. Appl. 32, 214–224.

[17] Basrak, B., Davis, R.A. and Mikosch, T. (2002). Regular variation of GARCH processes. Stoch. Proc. Appl. 99, 95–115.

Let f be the density of the innovation ε_t, and let f_0 be the density of the initial random variable R_0. According to (16.33), the conditional density of R_t given $R_{t-1} = r_{t-1}, \ldots, R_0 = r_0$ (which is also the conditional density of R_t given $R_{t-1} = r_{t-1}$) is

$$f_t(r_t|r_{t-1}, \ldots, r_0) = f_t(r_t|r_{t-1}) = \frac{1}{\sigma(r_{t-1})} f\left(\frac{r_t}{\sigma(r_{t-1})}\right), \tag{16.36}$$

where $\sigma(r) = (\alpha_0 + \alpha_1 r^2)^{1/2}$. Therefore, the joint density of $R_0, \ldots, R_t$ is

$$f(r_0, \ldots, r_t) = f_0(r_0) \prod_{s \leq t} f_s(r_s|r_{s-1}) \tag{16.37}$$

according to (8.11).

It is well–known that consistent estimators of α_0 and α_1 are obtained by maximizing the likelihood function based on (16.37), whereby the term $f_0(r_0)$ can be omitted; that is, one is maximizing a conditional likelihood function. Moreover, the unknown density f of the innovations is replaced by the normal one.

The Extension to ARCH(p) Series

A first extension of an ARCH(1) series is achieved, if the stochastic volatility is of the form

$$\tilde{\sigma}_t^2 = \alpha_0 + \sum_{i \leq p} \alpha_i R_{t-i}^2 \tag{16.38}$$

with innovations ε_t being independent of the past random variables R_{1-i}, $i = 1, \ldots, p$.

Repeating the arguments in (16.33) one may verify that the conditional distribution of R_t given $R_{t-1} = r_{t-1}, \ldots, R_{t-p} = r_{t-p}$ is the distribution of

$$\sigma(r_{t-1}, \ldots, r_{t-p})\varepsilon_t, \tag{16.39}$$

where

$$\sigma(r_{t-1}, \ldots, r_{t-p}) = (\alpha_0 + \sum_{i \leq p} \alpha_i r_{t-i}^2)^{1/2}.$$

In (16.39) one also gets the conditional distribution of R_t given the past $R_{t-1} = r_{t-1}, \ldots, R_{1-p} = r_{1-p}$ up to time $1 - p$.

Extending (16.36) and (16.37), one gets the conditional densities

$$f_t(r_t|r_{t-1}, \ldots, r_{1-p}) = f_t(r_t|r_{t-1}, \ldots, r_{t-p}) \tag{16.40}$$

$$= \frac{1}{\sigma(r_{t-1}, \ldots, r_{t-p})} f\left(\frac{r_t}{\sigma(r_{t-1}, \ldots, r_{t-p})}\right),$$

and the joint density

$$f(r_{1-p}, \ldots, r_t) = f_0(r_{1-p}, \ldots, r_0) \prod_{s \leq t} f_s(r_s|r_{s-1}, \ldots, r_{s-p}) \tag{16.41}$$

of the returns R_t.

Stochastic Volatility (VS) Models

Let again $R_t = \tilde{\sigma}_t \varepsilon_t$ as in (16.29). In contrast to ARCH models we assume that the series $\{\tilde{\sigma}_t\}$ and $\{\varepsilon_t\}$ are independent. For example, let $\{\tilde{\sigma}_t\}$ be an AR series of the form

$$\tilde{\sigma}_t^2 = \beta_0 + \sum_{j \leq q} \beta_j \tilde{\sigma}_{t-j}^2 + \eta_t, \tag{16.42}$$

where $\{\eta_t\}$ is another innovation series. This is the product series model as dealt with in [54]. Another example is obtained if $\{\tilde{\sigma}_t\}$ is defined by means of an MA(∞) series. Let

$$\tilde{\sigma}_t^2 = c \exp \left(\sum_{j=0}^{\infty} \psi_j \eta_{t-j} \right). \tag{16.43}$$

Asymptotic results of extremes of such processes are obtained by F.J. Breidt and R.A. Davis[18].

The Extension to GARCH(p, q) Series

An extension of the concept of an ARCH series is achieved, if

$$\tilde{\sigma}_t^2 = \alpha_0 + \sum_{i \leq p} \alpha_i R_{t-i}^2 + \sum_{j \leq q} \beta_j \tilde{\sigma}_{t-j}^2, \qquad t \geq 1. \tag{16.44}$$

Then, $R_t = \tilde{\sigma}_t \varepsilon_t$, $t \geq 1$, is a GARCH(p, q) (a generalized ARCH) series. Notice that a GARCH($p, 0$) series is an ARCH(p) series. Thus, the extension is related to the step from an AR to an ARMA series.

The RiskMetrics^{TM} (RM) method deals with GARCH($1, 1$) series. RM is concerned with the calculation of the Value–at–Risk for a portfolio consisting of up to more than 450 assets. Market position, for example, can be entered through the RM interface and, then, the VaR will be provided.

The basic idea used by RM is that daily log–returns of an asset have zero mean and are generated according to a non–stationary GARCH($1, 1$) series

$$R_t = \tilde{\sigma}_t \varepsilon_t,$$

where the ε_t are iid standard normal innovations and

$$\tilde{\sigma}_t^2 = (1 - \lambda) R_{t-1}^2 + \lambda \tilde{\sigma}_{t-1}^2.$$

Using past log–returns on the asset it is possible to forecast the volatility of the daily return for the next day which in turn can be used to compute the one–day VaR.

---

[18]Breidt, F.J. and Davis, R.A. (1998). Extremes of stochastic volatility models. Ann. Appl. Probab. 8, 664–675.

The RiskMetrics^{TM} specification for λ, called the decay factor, is 0.94 for daily returns. This determination is based on minimization of the mean square variance forecast error with respect to λ for a number of asset returns. The decay factor is then a weighted average of individual optimal decay factors, see J.P. Morgan Bank[19] for details.

16.8 Predicting the Serial Conditional VaR
co–authored by A. Kozek[20] and C.S. Wehn[21]

In the preceding sections, we primarily studied the estimation of the Value–at–Risk (VaR) as a quantile of a stationary distribution of the loss variable L_t. In the conditional set–up, the VaR is a conditional quantile, cf. Section 8.1, which should be predicted based on observable quantities.

In the time series framework, we speak of a serial conditional quantile, cf. Section 8.3, and, therefore, also of a serial conditional VaR. At the end of Section 9.5, we already mentioned a covariate conditional VaR. The present section focuses on the serial conditioning within GARCH models. A more general formulation in terms of both serial and covariate information is possible, yet not considered in this book.

McNeil et al.[22] argue that both approaches, the unconditional as well as the conditional one, are relevant for risk management purposes. However, in view of the fact that most empirical time series used in market risk modeling experience non–stationarity and certain heteroskedastic properties, the conditional approach is particularily relevant in market risk modeling.

The Serial Conditional VaR

Let again
$$L_t = -(V_t - V_{t-1})$$
be the loss (profit/loss) variable at time t pertaining to the market values $V_{t-1} = hS_{t-1}$ and $V_t = hS_t$, where h is the fixed number of shares (the portfolio position), and S_t is the price at time t. Denote again by $R_t = \log S_t - \log S_{t-1} = \log V_t - \log V_{t-1}$ the log–return at time t.

The serial conditional VaR at the level q is the conditional q–quantile

$$\mathrm{VaR}(q; r_{t-1}, r_{t-2}, \ldots) := q(L_t | r_{t-1}, r_{t-2}, \ldots), \tag{16.45}$$

---

[19]J.P. Morgan (1996). RiskMetrics Technical Document (4th ed.). J.P. Morgan Bank, New York.

[20]Macquarie University, Sydney

[21]DekaBank, Frankfurt am Main

[22]McNeil, A.J., Frey, R., Embrechts, P. (2005). Quantitative Risk Management— Concepts, Techniques, Tools. Princeton University Press, Princeton.

as mentioned in (8.21), of the loss variable L_t given the past returns $r_{t-1}, r_{t-2}, \ldots$; thus, $q(L_t|r_{t-1}, r_{t-2}, \ldots)$ is the q–quantile of the conditional distribution of L_t given $R_{t-1} = r_{t-1}, R_{t-2} = r_{t-2}, \ldots$.

As in (16.6) one gets the representation

$$
\begin{aligned}
L_t &= V_{t-1}(1 - \exp(R_t)) \\
&\approx -V_{t-1}R_t.
\end{aligned}
\tag{16.46}
$$

In view of (16.46), one may replace in (16.45) the loss variable L_t by the approximate value $-V_{t-1}R_t$.

Notice that the past market value $V_{t-1} \equiv V_{t-1}(R_{t-1}, R_{t-2}, \ldots)$ is a non–random function of the past returns R_{t-i}, and V_{t-1} itself is known at time $t-1$. To some larger extent, we suppress the dependency on some unknown initial values or random variables in our notation when time series are studied.

In practical situations, the modeling of loss variables becomes even more complicated. We mention two important extensions:

1. A trading portfolio usually consists of a larger number of assets. Therefore, one has to consider the corresponding multivariate question.

2. The portfolio positions $h = h_t$ vary from one day to another, where h_t is the position in the asset V_t at time t. The assumption that the process h_t merely depends on the past returns $R_{t-1}, R_{t-2}, \ldots$ ensures that our analysis can be easily extended to time varying portfolio weights.

GARCH Models for the Returns, Again

The statistical modeling of the loss distribution is again formulated in terms of the log–returns R_t. As in (16.29) let

$$
R_t = \tilde{\sigma}_t \varepsilon_t, \qquad t \geq 1,
$$

where the ε_t are iid random innovations, and ε_t is independent of the random scale parameter $\tilde{\sigma}_t$. Because we are merely interested in q–quantiles of conditional distributions we do not necessarily impose any conditions on the moments of ε_t. Therefore, we also speak of a random scale parameter instead of a stochastic volatility. Specifically, we assume that $\tilde{\sigma}_t = \sigma(R_{t-1}, R_{t-2}, \ldots)$ and, hence,

$$
R_t = \sigma(R_{t-1}, R_{t-2}, \ldots)\varepsilon_t, \qquad t \geq 1,
\tag{16.47}
$$

where ε_t is independent of the past returns $R_{t-1}, R_{t-2}, \ldots$, and, consequently, the innovation ε_t is independent of the random scale parameter.

For the ARCH(p) series in (16.38) we have

$$
\sigma^2(R_{t-1}, R_{t-2}, \ldots) = \alpha_0 + \sum_{i \leq p} \alpha_i R_{t-i}^2.
\tag{16.48}
$$

For the GARCH(p,q) series in (16.44) the random scale parameter is recursively defined by

$$\sigma^2(R_{t-1}, R_{t-2}, \ldots) = \alpha_0 + \sum_{i \leq p} \alpha_i R_{t-i}^2 + \sum_{j \leq q} \beta_j \sigma^2(R_{t-j-1}, R_{t-j-2}, \ldots). \quad (16.49)$$

Within the ARCH(p) and GARCH(p,q) series, the distributions of the returns depend on the parameters α_i and, respectively, the parameters α_i and β_j, and the distribution of ε_t.

Representing GARCH as ARCH Models

By induction one gets a representation of $\tilde{\sigma}_t^2$ in terms of the returns R_t, see, e.g. Fan and Yao [18], (4.35). We have

$$
\begin{aligned}
\tilde{\sigma}_t^2 &= \frac{\alpha_0}{1 - \sum_{j=1}^q \beta_j} + \sum_{i=1}^p \alpha_i R_{t-i}^2 \\
&= + \sum_{i=1}^p \alpha_i \sum_{k \geq 1} \sum_{j_1=1}^q \cdots \sum_{j_k=1}^q \beta_{j_1} \times \cdots \times \beta_{j_k} R_{t-i-j_1-\cdots-j_k}^2 \quad (16.50)
\end{aligned}
$$

with respect to an infinite past. In practice, one has to use some initial values and a truncation, e.g., by truncating all terms having a non–positive index.

As a special case we deduce the representation in the RiskMetrics^{TM} (RM) model of a GARCH(1,1) series, where

$$\tilde{\sigma}_t^2 = (1 - \lambda) R_{t-1}^2 + \lambda \tilde{\sigma}_{t-1}^2,$$

and the innovations ε_t are standard normal, cf. Section 16.7, page 400.

In that case, we have

$$\tilde{\sigma}_t^2 = (1 - \lambda) \sum_{i \geq 1} \lambda^{i-1} R_{t-i}^2, \quad (16.51)$$

which can be addressed as a random scale parameter in an ARCH(∞) series. If $\lambda < 1$ it makes sense to take the first p terms in the sum, thus, getting an ARCH(p) series, because the deleted returns $R_{t-p-1}, R_{t-p-2}, \ldots$ enter only with very small cumulated weights $\sum_{i \geq p+1} \lambda^{i-1} = \lambda^p/(1-\lambda)$.

The general strategy is to estimate a smaller number of parameters in a GARCH series, yet to carry out further computations in a related (approximating) ARCH(p) series.

Conditional Distributions in GARCH Series

It is evident from (16.47) that the conditional distribution of R_t given $R_{t-1} = r_{t-1}, R_{t-2} = r_{t-2}, \ldots$ is the distribution of

$$\sigma(r_{t-1}, r_{t-2}, \ldots) \varepsilon_t. \quad (16.52)$$

Alternatively, the conditioning may simultaneously be based on certain past returns and past scale parameters: the conditional distribution of R_t given $R_{t-1} = r_{t-1}, \ldots, R_{t-p} = r_{t-p}, \tilde{\sigma}_{t-1} = \sigma_{t-1}, \ldots, \tilde{\sigma}_{t-q} = \sigma_{t-q}$ is the distribution of

$$\sigma_t \varepsilon_t \qquad (16.53)$$

with σ_t recursively defined by

$$\sigma_t^2 = \alpha_0 + \sum_{i \leq p} \alpha_i r_{t-i}^2 + \sum_{j \leq q} \beta_j \sigma_{t-j}^2.$$

The results in (16.52) and (16.53) are identical with $\sigma_t = \sigma(r_{t-1}, r_{t-2}, \ldots)$.

Based on these formulas one may write down conditional and joint densities and a likelihood function. In analogy to (16.40) and (16.41), one gets the conditional densities

$$f_t(r_t | r_{t-1}, r_{t-2}, \ldots) = \frac{1}{\sigma_t} f\left(\frac{r_t}{\sigma_t}\right), \qquad (16.54)$$

and the joint density

$$f(\ldots, r_{t-1}, r_t) = f_0(\ldots, r_{-1}, r_0) \prod_{s=1}^{t} \frac{1}{\sigma_s} f\left(\frac{r_s}{\sigma_s}\right) \qquad (16.55)$$

of the returns R_t, $t \geq 1$, including some initial values.

Likewise the conditional density of $R_1, \ldots R_t$ given the initial values is

$$f(r_1, \ldots, r_t | r_0, r_{-1}, \ldots) = \prod_{s=1}^{t} \frac{1}{\sigma_s} f\left(\frac{r_s}{\sigma_s}\right). \qquad (16.56)$$

Now we may build a likelihood function based on the joint density or the conditional likelihood function based on the conditional density (which leads to the same expression if the term depending on f_0 is omitted).

We provide two examples, namely, the cases where the innovation density f is the standard normal density φ or a Student density with shape parameter α.

- (Normal innovations.) In Section 16.7 we mentioned the result that returns have Pareto–like tails if the innovations ε_t are standard normal. Thus, normal innovations do not contradict our findings in the previous sections which spoke in favor for return distributions with heavy tails such as Student distributions. From (16.56) one easily gets the conditional likelihood function. For the conditional log–likelihood function one gets,

$$
\begin{aligned}
l(\boldsymbol{\alpha}, \boldsymbol{\beta}) &= -\sum_{s=1}^{t} \log \sigma_s + \sum_{s=1}^{t} \log \varphi(r_s/\sigma_s) \\
&= -\frac{t}{2} \log(2\pi) - \sum_{s=1}^{t} \log \sigma_s - \frac{1}{2} \sum_{s=1}^{t} \frac{r_s^2}{\sigma_s^2}, \qquad (16.57)
\end{aligned}
$$

where the right–hand side depends on the α_i and β_j via the σ_s; next one has to compute the likelihood equations.

- (Student innovations.) In the recent financial literature one observes a trend to take Student innovations in the GARCH modeling. Simulation studies indicate that this type of modeling for the innovations is compatible to Pareto–distributed returns. More precisely, if the innovations are Student-distributed, then one may conjecture that the returns are again Student–distributed with a shape parameter smaller than the initial one (personal communication by Petra Schupp).

We note the conditional log–likelihood function

$$l(\boldsymbol{\alpha}, \boldsymbol{\beta}) \;\; = \;\; t \log \left(\frac{\Gamma((\alpha + 1)/2)}{\sqrt{\pi(\alpha - 2)}\,\Gamma(\alpha/2)} \right)$$

$$- \sum_{s=1}^{t} \log \sigma_s - \frac{\alpha + 1}{2} \sum_{s=1}^{t} \log \left(1 + \frac{r_s^2}{\sigma_s^2(\alpha - 2)} \right), \quad (16.58)$$

where we take standardized Student innovations with shape parameter $\alpha > 2$ to make the result comparable to [18]. Recall from (1.62) that the standard deviation is equal to $\sqrt{\alpha/(\alpha - 2)}$. We do not know any result for the case of $\alpha \leq 2$.

The Serial Predictive Conditional VaR

We merely provide details in the special case of the RM model, see (16.51), where the innovation df F is the standard normal df Φ.

Within the RM modeling let again $\hat{\lambda} = 0.94$ be the specification of the parameter λ in the GARCH model as provided by RM (or some other estimate of λ). The innovations ε_t are assumed to be distributed according to Φ. Then, one gets the predictor

$$\widehat{\text{VaR}}(q; r_{t-1}, r_{t-2}, \ldots) = v_{t-1} \hat{\sigma}_t \Phi^{-1}(q), \qquad (16.59)$$

with

$$\hat{\sigma}_t = \sqrt{(1 - \hat{\lambda}) \sum_{i \geq 1} \hat{\lambda}^{i-1} r_{t-i}^2}, \qquad (16.60)$$

of the conditional Value–at–Risk $\text{VaR}(q; r_{t-1}, r_{t-2}, \ldots)$ based on the observed market value v_{t-1}, and the past observed returns r_{t-i}, where one only takes the first p terms in the sum.

It is worthwhile to reconsider all the steps which led to the prediction of the serial conditional VaR (as the conditional q–quantile of the loss variable L_t conditioned on the past returns):

1. replace the loss variable L_t by $-V_{t-1}R_t$

2. take a model $R_t = \tilde{\sigma}_t \varepsilon_t$ with iid innovations ε_t for the returns;

3. specialize this model further to a GARCH model like the RM model;

4. replace unknown parameters like the λ by estimates based on observable quantities (or by a value provided, e.g., by J.P. Morgan);

5. fix a df for the innovation ε_t;

6. take an appropriate smaller number p of terms in the sum of the random scale parameter which is the step from the GARCH model to a ARCH(p) model,

7. take the conditional q–quantile within this final model as a predictor of the serial conditional VaR.

The conditional distribution in step 6, with the unknown paramters replaced by estimates, may be addressed as a predictive distribution (as introduced in Section 8.1, page 237). One may also speak of a serial predictive VaR as a predictor of the serial conditional VaR.

In view of the longer list of conditions and approximations, one may ask after the accuracy of the prediction by means of this predictive VaR. In this context, the steps 2 to 5, which concern the statistical modeling and the estimation of unknown parameters, are of particular interest. Later on, we indicate a validation of the GARCH modeling by using the Rosenblatt transformation and, respectively, certain residuals.

Predicting the Serial Conditional Expected Shortfall

The predictive df
$$F(l|r_{t-1}, r_{t-2}, \dots) = \Phi(l/(v_{t-1}\hat{\sigma}_t)) \tag{16.61}$$

in (16.59) for the loss variable L_t provides more than a predictor of the serial conditional VaR.

For example, in place of a q–quantile we may use a different functional parameter such as the expected shortfall (which itself is frequently called conditional VaR in the literature; we hope that no confusion will arise due to this ambiguity). The serial conditional expected shortfall may be predicted by plugging in the predictive df. Other possible applications concern predictive intervals.

Validation for the GARCH Modeling
by Using the Rosenblatt Transformation

In the preceding sections, we employed exploratory tools to analyze financial data according to their postulated stationary distribution or their martingal (correlation) structure. To make tests applicable (if dependencies cannot be neclected), one needs complex theoretical results as, e.g., those developed in Chapter 7. Alternatively, standard tests can be applied to transformed data.

If the data come from independent, not necessarily identically distributed (innid) random variables, then the simple probability transformation helps to produce iid (0,1)–uniform data. Otherwise, as, e.g., in the GARCH case, one can use the Rosenblatt transformation, see (8.29).

In the present context, the Rosenblatt transformation was applied by Diebold et al. (in the article cited on page 238). We want to know whether the GARCH modeling for the returns series R_t and the estimation procedures can be justified. Let

$$F_t(\cdot|r_{t-1}, r_{t-2}, \ldots) = F(\cdot/\sigma(r_{t-1}, r_{t-2}, \ldots))$$

be the conditional df of R_t given $R_{t-1} = r_{t-1}$, $R_{t-2} = r_{t-2}, \ldots$, where F is the common df of the innovations ε_t, see (16.52). It is understood that unknown parameters in the conditional dfs are replaced by estimates based on the returns. Thus, one is using the predictive dfs (the "estimated conditional dfs") instead of the actual conditional dfs in the Rosenblatt transformation. Diebold et al. also compute the joint density of the transformed random variables when predictive dfs in place of actual conditional dfs are applied in the transformation.

According to the Rosenblatt transformation, the

$$y_t = F_t(r_t|r_{t-1}, r_{t-2}, \ldots)$$

may be regarded as iid $(0,1)$–uniform data if the predictive dfs are sufficiently accurate.

Now, the distributional properties as well as the serial independence can be analyzed. If, e.g., a test procedure rejects one of these properties then it is likely that one of our basic conditions is violated (or our estimation procedures in the GARCH model are not sufficiently accurate).

Diebold et al. argue in favor of exploratory analysis (just in the spirit of larger parts of the present book):

- ... when rejection occurs, the tests generally provide no guidance as to why.

- ... even if we know that rejection comes from violation of uniformity, we'd like to know more: What, precisely, is the nature of violation ...

- Is the dependence strong and important, or is iid an adequate approximation, even if strictly false?

They come to the conclusion: "The nonconstructive nature of tests of iid U(0,1) behavior, and the nonconstructive nature of related separate tests of iid and U(0,1), make us eager to adopt more revealing methods of exploratory data analysis."

The authors analyze the transformed data by means of histograms and sample autocorrelation functions. The latter tool is also applied to powers of centered data.

In simulation studies, they generate the data according to a GARCH(1,1) series (the data generating process) given by

$$\tilde{\sigma}_t^2 = 0.01 + 0.13 \times R_{t-1}^2 + 0.86 \times \tilde{\sigma}_{t-1}^2 \qquad (16.62)$$

which is close to the RM model. Yet, in place of standard normal innovations, these authors take standardized Student variables with shape parameter $\alpha = 6$ (six degrees of freedom). For the standardization one has to use the standard deviation which is equal to $\sqrt{3/2}$ for $\alpha = 6$, see (1.62) and (16.58).

The Rosenblatt transformation is carried out under the following sequences of predictive distributions:

- a sequence of iid standard normal or non–normal random variables;

- a GARCH(1,1) series with normal innovations,

- the correct model (with estimated parameters).

Diebold et al. conclude that "our density forecast evaluation procedures clearly and correctly revealed the strength and weakness of the various density forecasts."

A real data set of daily S&P returns is also analyzed within the framework of MA(1)–GARCH(1,1) model (again with Student innovations).

Despite the arguments above, we mention some test procedures which are employed for testing the simple null–hypothesis of iid $(0,1)$–uniform random variables. One may apply goodness–of–fit tests such as χ^2 or Kolmogorov–Smirnov tests. Likelihood–ratio tests may be applied to a binomially distributed number of exceedances, see Kupiec[23]. The Kupiec test can be regarded as a two–sided extension of the traffic light approach.

The traffic light approach is a binomial test based on the exceedances of the serial conditional VaR, i.e., marks in time, where the observed losses exceed the respective predicted VaR. By the Rosenblatt transformation it is ensured that these exceedances are binomially distributed if the serial conditional df is appropriately selected. This binomial test is especially relevant for regulatory purposes[24].

Another useful reference in the context of testing the conditional modeling is Berkowitz[25].

By statistical tests or explorative means like QQ–plot, PP–plot, histograms or autocorrelation functions, it is possible to iterate the different steps 1–7 above that

---

[23]Kupiec, P.H. (1995). Techniques for verifying the accuracy of risk measurement models. J. Derivatives 2, 73–84.

[24]Basel Committee on Banking Supervision: Supervisory Framework for 'Backtesting' in conjunction with the Internal Models Approach to Market Risk Capital Requirements, 1996.

[25]Berkowitz, J. (2001). Testing density forecasts, with applications to risk management, J. Business & Economic Statistics 19, 465–474.

led to the respective dfs[26]. If there is a significant serial correlation identified (or a significant autocorrelation of the squared returns), the specified GARCH/ARCH–model can be improved and, in addition, the estimated parameters should be reviewed (steps 3, 4 and 6 of the loss variable specification steps). If the distributional properties do not fit well (which again is explored by statistical or graphical means), the fixed df for the innovations is questionable (step 5). This procedure is called the "backtesting" of the risk model and should be conducted regularly to improve stepwise and iteratively the chosen model and its assumptions.

These pragmatic procedures can as well be employed in the case of multivariate asset returns and for time varying portfolio weights h_t (mentioned at the beginning of this section).

Validation for the GARCH Modeling by Using Residuals

Fan and Yao [18] analyze daily S&P returns (different from the one used by Diebold et al.) with a modeling of the innovations by Student distributions. Applying the conditional likelihood method these authors get the following GARCH(1,1) series (with estimated parameters)

$$\hat{\sigma}_t^2 = 0.007 + 0.047 \times r_{t-1}^2 + 0.945 \times \hat{\sigma}_{t-1}^2$$

with an additional estimated shape parameter $\alpha = 7.41$ of the Student distribution.

The validation for the GARCH modeling is based on the residuals

$$\hat{\varepsilon}_s = r_s/\hat{\sigma}_s, \qquad s \geq 1. \tag{16.63}$$

One can expect that the residuals $\hat{\varepsilon}_s$ have properties as the innovations ε_s to some extent.

Fan and Yao apply tests as well as exploratory tools to the residuals and conclude that Student modeling is more agreeable than the normal one. It is reported that there is no significant autocorrelation in the residual series and its squared series. These authors also discuss the question of predicting the conditional VaR based on the distribution of $\hat{R}_t = \hat{\sigma}_t \hat{\varepsilon}_{t,\alpha}$, where $\hat{\varepsilon}_{t,\alpha}$ corresponds to a Student–variable with estimated shape parameter α, and $\hat{\sigma}_t$ is the predictive scale parameter (with parameters estimated within the GARCH model).

Semiparametric Evaluation of the Conditional Df of Returns

We add some remarks about a predictive df of the return R_t within the GARCH setup, where the parameters α_i and β_j are estimated in a correct manner, yet the validity of the overall parametric modeling of the common innovation df F is questionable. Then, the residuals $\hat{\varepsilon}_s$ may still be regarded as observations under

---

[26]Wehn, C.S. (2005). Ansätze zur Validierung von Marktrisikomodellen—Systematisierung, Anwendungsmöglichkeiten und Grenzen der Verfahren. Shaker, Aachen.

F. Therefore, we may estimate F, based on the residuals, by means of the sample df $\hat{F}_{t-1}$ or a kernel df $\hat{F}_{t-1,b}$ with bandwidth $b > 0$ as defined in (2.17). One may as well apply the extreme value technique to estimate the upper tail of F.

By piecing together the predictive scale parameter $\hat{\sigma}_t$ and the estimates $\hat{F}_{t-1}$ or $\hat{F}_{t-1,b}$ of F, one gets the predictive dfs $\hat{F}_{t-1}(\cdot/\hat{\sigma}_t)$ or $\hat{F}_{t-1,b}(\cdot/\hat{\sigma}_t)$ of the return R_t. In order to get a predictive VaR, one has to use quantiles of the predictive dfs. The approach, using the kernel qf introduced below (2.17), would also be an option. In conjunction with the kernel df approach, special refined methods of selecting the quantile are available[27] [28].

An Empirical Evaluation of the Conditional Df of Returns

Within the ARCH(p) setting we predicted the conditional df $F_t(\cdot|r_{t-1}, \ldots, r_{t-p}) = F(\cdot/\sigma(r_{t-1}, \ldots, r_{t-p}))$ of the return R_t given $R_{t-1} = r_{t-1}, \ldots, R_{t-p} = r_{t-p}$, where F is the common df of the innovations ε_t. A predictive version was provided by replacing the unknown parameters α_i by estimates $\hat{\alpha}_i$ based on the past returns. One may ask whether an empirical approach is possible by estimating the conditional df in a direct manner under certain weak assumptions imposed on the time series of returns. There is a positive answer if p is small, yet our answer is negative due to the "curse of dimensionality" for the more interesting case of larger p. For simplicity, we merely give details for $p = 1$.

In view of the corresponding property of an ARCH(1)–series, one may assume that the returns R_t satisfy the technical condition of a Markov chain with stationary transition df $F(\cdot|r)$. In the special case of an ARCH(1)–series we have $F(\cdot|r) = F(\cdot/\sigma(r))$. Now we proceed as in Section 8.2 with the iid condition replaced by the Markov condition. Let $r_{t(1)}, \ldots, r_{t(k)}$, $1 < t(j) < t$, be the returns in the past for which $r_{t(j)-1}$ is close to the fixed value r. Then, the $r_{t(j)}$ may be regarded as observations under $F(\cdot|r)$, and the sample df $F_k(\cdot|r)$, based on the $r_{t(j)}$, as an estimator of the conditional df $F(\cdot|r)$.

---

[27] Kozek, A. (2003). On M–estimators and normal quantiles. Ann. Statist. 31, 1170–1185.

[28] Jaschke, S., Stahl, G. and Stehle, R. (2006). Value–at–risk forecasts under scrutiny—the German experience. To appear in Quantitative Finance.

Chapter 17

The Impact of Large Claims on Actuarial Decisions

co–authored by M. Radtke[1]

In this chapter, we elaborate on and develop some ideas which were already pre-sented in Section 1.1. Recall that the expectation of the total claim amount de-termines the net premium. Based on the net premium, the insurer determines the total premium that must be paid by the policy holder. We start in Section 17.1 with the calculation of the df, expectation and variance of the total claim amount.

From our viewpoint, reinsurance is of particular interest, because the excesses of large or catastrophic claim sizes over a certain higher threshold are covered by the reinsurer. Special names for threshold are limit, priority or retentation level. One may distinguish between the following reinsurance treaties:

- excess–of–loss (XL) reinsurance, when the reinsurer pays the excess of a certain fixed limit for individual claim sizes;

- stop–loss reinsurance or total loss insurance, when the reinsurer covers the excess of a certain limit for the total claim amount of an insurer's portfolio,

- ECOMOR reinsurance, which is a modification of the XL treaty with the kth largest individual claim size taken as a random limit (thus, the reinsurer only pays excesses of the kth largest claim size).

The net premium for the next period can be estimated in a nonparametric manner by means of the total claim amount of preceding periods. It is suggested to also employ a parametric approach in the reinsurance business.

---

[1] Kölnische Rückversicherung; co–authored the first edition.

We merely deal with risks in conjunction with the XL reinsurance treaty. The required modifications for the ECOMOR treaty are apparent from our viewpoint[2]. The restriction to the XL treaty is justified by the fact that the ECOMOR treaty seems to have less practical relevance.

In XL reinsurance, the estimation of the net premium can be done again within the generalized Pareto (GP) model (Section 17.2) or some other statistical model such as the Benktander II or truncated converse Weibull model.

The segmentation of a portfolio with respect to the probable maximum loss (PML) of single risks is dealt with in Section 17.3. The segmentation is necessary to adjust the tarification to the risks of individual policy holders. We pursue an empirical and a parametric statistical approach towards this important question. The parametric one is based on the estimation of the mean of GP distributions.

Another important question is the choice of an adequate initial reserve (capital) for a portfolio. We will introduce a concept based on finite ruin theory (see, e.g., the book by Gerber [22]) in which the initial reserve becomes a parameter which can be estimated by the insurer. For that purpose, one must formulate a certain criterion which determines the initial reserve in a unique manner: we suggest using a T–year initial reserve for a $q \times 100\%$ ruin probability. This is the initial reserve for a portfolio such that ruin occurs within the next T years with a probability of $q \times 100\%$. Reasonable quantities are a time horizon of $T = 10$ years and a predetermined ruin probability of 1% or 5%. These ideas are pursued in Section 17.4 within the framework of risk processes. This chapter is concluded with some remarks about asymptotic ruin theory (Section 17.5).

17.1   Numbers of Claims
and the Total Claim Amount

Let $S_n = \sum_{i \leq n} X_i$ denote the total (aggregate) claim amount of the first n of the random claims sizes $X_1, X_2, X_3, \ldots$. Then, the total claim amount for a given period can be written

$$S_N = \sum_{i \leq N} X_i, \tag{17.1}$$

where N is the random number of claims occurring within this period. Remember that the expectation $E(S_N)$ of the total claim amount is the net premium that must be estimated by the actuary.

The Total Claims DF

The df, expectation and variance of the total claim amount S_N will be computed under the conditions of the homogeneous risk model.

---

[2]For details and further references, see, e.g., Kremer, E. (1992). The total claims amount of largest claims reinsurance revisited. Blätter DGVM 22, 431–439.

(Homogeneous Risks Model): the claim sizes $X_1, X_2, X_3, \ldots$ are iid random variables with common df F. Additionally, the claim number N and the sequence $X_1, X_2, X_3, \ldots$ are independent.

With calculations similar to those in (1.54) we obtain for the total claims df

$$
\begin{aligned}
P\{S_N \le x\} &= \sum_{n=0}^{\infty} P\{S_N \le x, N = n\} \\
&= \sum_{n=0}^{\infty} P\{N = n\} F^{n*}(x),
\end{aligned}
\tag{17.2}
$$

where $S_0 = 0$, $F^{0*}(x) = I(0 \le x)$ and $F^{n*}(x) = P\{S_n \le x\}$ is the n–fold convolution, cf. page 30, of F again.

Next, the total claims df will be written in a more explicit form for binomial, Poisson, negative binomial and geometric claim numbers[3].

- (Compound Binomial.) If the number N of claims is a binomial random variable with parameters n and p, then the total claims df is

$$
P\{S_N \le x\} = \sum_{k=0}^{n} B_{n,p}\{k\} F^{k*}(x).
\tag{17.3}
$$

- (Compound Poisson.) If the number N of claims is a Poisson random variable with parameter λ, then the total claims df is

$$
P\{S_N \le x\} = \sum_{k=0}^{\infty} P_\lambda\{k\} F^{k*}(x).
\tag{17.4}
$$

- (Compound Negative Binomial.) If the number N of claims is a negative binomial random variable with parameters r and p, see (3.31), then the total claims df is

$$
P\{S_N \le x\} = \sum_{k=0}^{\infty} B_{r,p}^{-}\{k\} F^{k*}(x).
\tag{17.5}
$$

- (A Simplified Representation in the Geometric Case.) If $r = 1$ in the preceding example, then N is geometric. Thus,

$$
\begin{aligned}
P\{N = k\} &= B_{1,p}^{-}\{k\} \\
&= p(1 - p)^k, \qquad k = 0, 1, 2, 3, \ldots,
\end{aligned}
$$

---

[3] Also see Kuon, S., Radtke, M. and Reich, A. (1993). An appropriate way to switch from the individual risk model to the collective one. Astin Bulletin 23, 23–54.

for some p with $0 < p < 1$, and

$$P\{S_N \leq x\} = p \sum_{k=0}^{\infty} (1-p)^k F^{k*}(x). \qquad (17.6)$$

In addition, assume that the claim sizes are exponentially distributed with common df $F(x) = 1 - e^{-x/\sigma}$, $x > 0$. Recall from (4.7) that the convolution F^{k*} is a gamma df with density

$$f^{k*}(x) = \sigma^{-1}(x/\sigma)^{k-1} \exp(-x/\sigma)/(k-1)!, \qquad x > 0.$$

Check that

$$\frac{p}{1-p} \sum_{k=1}^{\infty} (1-p)^k f^{k*}(x) = \sigma^{-1} p e^{-px/\sigma}, \qquad x \geq 0,$$

and, hence, equality also holds for the pertaining dfs. By combining this with (17.6), one obtains the representation

$$P\{S_N \leq x\} = p + (1-p)(1 - e^{-px/\sigma}), \qquad x \geq 0, \qquad (17.7)$$

of the total claims df.

Next, we verify that the net premium[4]—the expectation of the total claim amount—within the given period is

$$E(S_N) = E(X)E(N), \qquad (17.8)$$

where X is a random variable with the same distribution as the X_i. Because the random variables S_n and $I(N = n)$ are independent, we have

$$
\begin{aligned}
E(S_N) &= \sum_{n=0}^{\infty} E\big(S_N I(N = n)\big) \\
&= \sum_{n=1}^{\infty} E(S_n) E(I(N = n)) \\
&= E(X) \sum_{n=1}^{\infty} n P\{N = n\} \\
&= E(X)E(N)
\end{aligned}
$$

and, thus, (17.8) holds.

The variance of S_N may be computed in a similar manner. One obtains

$$V(S_N) = V(X)E(N) + (EX)^2 V(N). \qquad (17.9)$$

---

[4]For an elementary introduction to premium principles see Straub, E. (1989). Non–Life Insurance Mathematics. Springer, Berlin.

17.2 Estimation of the Net Premium

The net premium will be estimated in a nonparametric and a parametric manner under various conditions.

Nonparametric Estimation of the Net Premium

Assume that the claim arrival times $T_1, T_2, T_3, \ldots$ constitute a claim number process $N(t) = \sum_{i=1}^{\infty} I(T_i \leq t)$, $t \geq 0$, which has a constant arrival rate, that is,

$$E\big(N(t_2) - N(t_1)\big) = (t_2 - t_1)E(N(1)), \qquad t_1 < t_2. \qquad (17.10)$$

The estimation of the net premium $E(S_N)$ for the next period will be based on the total claim amount $S_{N(T)}$ of the past T years. Under the conditions of the homogeneous risk model, one obtains according to (17.8),

$$E(S_N) = E(N(1))E(X),$$

and, under condition (17.10),

$$E(S_{N(T)}) = TE(N(1))E(X).$$

Therefore, $S_{N(T)}/T$ is an unbiased estimator of the net premium $E(S_N)$. The variance of this estimator can easily be deduced from (17.9).

EXAMPLE 17.2.1. (Large Norwegian Fire Claim Data.) A data set of large fire claim data (stored in the file it–fire2.dat) was extensively analyzed by R. Schnieper[5].

TABLE 17.1. Fire claims sizes over 22.0 million NKr from 1983 to 1992.

year	claim size (in millions)	year	claim size (in millions)
1983	42.719		23.208
1984	105.860	1990	37.772
1986	29.172		34.126
	22.654		27.990
1987	61.992	1992	53.472
	35.000		36.269
1988	26.891		31.088
1989	25.590		25.907
	24.130		

---

[5]Schnieper, R. (1993). Praktische Erfahrungen mit Grossschadenverteilungen. Mitteil. Schweiz. Verein. Versicherungsmath., 149–165.

The original fire claim data over a priority of 10.0 million NKr (Norwegian crowns) were corrected by means of a certain trend function. The original data are indexed "as if" they occurred in 1993. Table 17.1 contains the corrected data. In addition, the original data were made anonymous so that the portfolio cannot be identified (personal communication).

We estimate the net premium for the excess claim sizes over the given priority of $u = 22$ million NKr and another one, namely $u = 50$ million NKr. In the first case, the estimation is based on the given 17 claim sizes; in the second case, only 3 claim sizes are available and, thus, the estimate is less reliable. The nonparametric estimates of the net premium (XL) are 26.98 and, respectively, 7.13 million NKr.

If the estimation of the net premium must be based on such a small number of claim sizes as in the preceding example, then a parametric approach should be employed.

Parametric Estimation of the Net Premium for Excess Claims

Assume that the arrival times and the claim sizes satisfy the conditions of a Poisson(λ, F) process, cf. page 248. Notice that λ is the mean number of claims in a period of unit length and F is the df of the excesses over the priority u. We assume that $F = W_{\gamma,u,\sigma}$ is a GP df. The mean claim size $E(X)$ is the mean of the GP df (in the γ–parameterization). The net premium for the next period is

$$E(S_N) = \lambda m_{W_{\gamma,u,\sigma}},$$

where m_F again denotes the mean of a df F. Thus, by estimating the parameters λ, γ and σ one obtains an estimate of the net premium.

Notice that

$$\lambda_{N(T)} = N(T)/T$$

is an estimator of $\lambda = E(N)$. If $\gamma_{N(T)}$ and $\sigma_{N(T)}$ are estimators of the parameters γ and σ, then the mean

$$m_{N(T)} := u + \frac{\sigma_{N(T)}}{1 + \gamma_{N(T)}}$$

of the GP df $W_{\gamma_{N(T)},u,\sigma_{N(T)}}$ is an estimator of the mean claim size $E(X) = m_{W_{\gamma,u,\sigma}}$. Therefore, $\lambda_{N(T)} m_{N(T)}$ is an estimator of the net premium $E(S_N)$.

EXAMPLE 17.2.2. (Continuation of Example 17.2.1.) Apparently, the estimated parameter for the number of claims within a single year—the number of claims divided by the number of years—is $\lambda = 1.7$.

The parametric estimation of the claim size df was carried out in the GP0, GP1 and GP models with left endpoint $u = 22.0$. The estimated parameters are listed in Table 17.2. Necessarily, the nonparametric estimate and the MLE in the exponential model (GP0) lead exactly to the same net premium.

TABLE 17.2. Parametric estimation of the mean claim size (excesses), the net premium (XL) in millions of NKr and nonparametric estimate of net premium (XL).

	γ-Parameterization			Mean Claim Size	Net Premium (XL)
	γ	μ	σ	$E(X)$	$E(S_N)$
nonparametric	–	–	–		26.98
MLE(GP0)	0.0	22.0	15.873	15.87	26.98
Hill(GP1)	0.451	22.0	9.915	18.06	30.70
MLE(GP)	0.254	22.0	11.948	16.02	27.23
Moment(GP)	0.293	22.0	12.060	17.06	29.00

In Fig. 17.1, the sample mean excess functions and the mean excess functions of GP dfs pertaining to the parametric estimates are plotted. These functions are given within the range of 20 to 35 millions NKr. First of all we are interested in the question as to whether the sample mean excess function is sufficiently close to a straight line so that the GP modeling is acceptable.

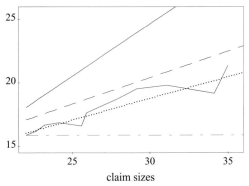

FIG. 17.1. Sample mean excess function and mean excess functions for Hill(GP1) (solid), Moment(GP) (dashed), MLE(GP) (dotted) and also MLE(GP0) (dashed–dotted).

A comparison of the mean excess functions leads to the conclusion that a GP modeling is acceptable. In addition, the plots speak in favor of the MLE(GP) which, consequently, will be employed in the following computations.

For small sample sizes as in the preceding example it is desirable to include further information in the statistical analysis. One possibility is to use Bayesian analysis as it is outlined in Section 17.6. Another one is to pool data over different portfolios (as it was also done by Schnieper in the paper cited on page 415). Such a pooling was also done in the regional frequency analysis (cf. Section 14.4).

Analyzing large Danish fire claim data, McNeil[6] justifies the GP modeling for the upper tail of the claim size distribution and obtains an ML estimate $\gamma = 0.497$.

---

[6]McNeil, A.J. (1997). Estimating the tails of loss severity distributions using extreme value theory. ASTIN Bulletin 27, 117–137.

Estimating the Net Premium for Extraordinarily High Priorities

We regard a priority $v > u$ as extraordinarily high when the number of observed excesses over v is very small (including the case that no claim size over v has been observed). This is the typical situation when v is the predetermined retentation level yet one wants to base the statistical inference on a larger number of claims. Within the parametric framework, the preceding parametric estimates $\gamma_{N(T)}$ and $\sigma_{N(T)}$—obtained for the priority u—can still be employed to estimate the net premium for claim sizes over $v > u$.

Recall from page 249 that the claim arrival times and claim sizes over v satisfy the conditions of a $\text{Poisson}\big(\tilde{\lambda}, W^{[v]}_{\gamma,u,\sigma}\big)$ process, where $\tilde{\lambda} = \lambda(1 - W_{\gamma,u,\sigma}(v))$. We obtain

$$\tilde{\lambda}_{N(T)} = \frac{N(T)}{T}\left(1 + \gamma_{N(T)}\frac{v-u}{\sigma_{N(T)}}\right)^{-1/\gamma_{N(T)}}$$

as an estimator of $\tilde{\lambda} = E(N)$. From (1.45), we found that

$$W^{[v]}_{\gamma,u,\sigma} = W_{\gamma,v,\sigma+\gamma(v-u)}.$$

Plugging in the estimators for γ and σ, one obtains by the mean

$$\widetilde{m}_{N(T)} = v + \frac{\sigma_{N(T)} + \gamma_{N(T)}(v-u)}{1 - \gamma_{N(T)}}$$

an estimator of the mean claim size $E(X)$. By combining these estimators, one obtains $\tilde{\lambda}_{N(T)}\widetilde{m}_{N(T)}$ as an estimator of the net premium $E(S_N) = E(N)E(X)$ for the excesses of the claim sizes over v.

EXAMPLE 17.2.3. (Continuation of Example 17.2.1.) The estimated parameters of the GP distribution (based on the 17 claim sizes) for the retentation level of 50 millions are given in the following table.

TABLE 17.3. Parametric estimation of the mean claim size (excesses) and the net premium (XL) in millions of NKr (and nonparametric estimate of net premium (XL)) for retentation level of 50 million.

| | γ-Parameterization | | | Mean Claim Size | Net Premium (XL) |
	γ	μ	σ	$E(X)$	$E(S_N)$
nonparametric	–	–	–		7.13
MLE(GP)	0.254	50.0	19.06	25.55	6.90

Notice that the estimate of the shape parameter γ is identical to that of the retentation level $u = 22$.

Finally, we remark that there could be a positive trend in the frequency of large claims. Yet, such phenomena will be only dealt with in conjunction with partial duration flood series (Section 14.2).

Combining the Nonparametric and Parametric Approach

Nonparametric and parametric calculations can be combined to estimate the claim size df

- over a fixed priority u in a parametric manner,

- below the priority in a nonparametric manner.

Let $\widehat{F}_n(\boldsymbol{x}; \cdot)$ be the sample df based on the claim sizes x_i. Let $W_{\gamma, \tilde{\mu}, \tilde{\sigma}}$ be the GP df fitted to the upper tail of $\widehat{F}_n(\boldsymbol{x}; \cdot)$, cf. (2.35). Both dfs can be pieced together smoothly because $W_{\gamma, \tilde{\mu}, \tilde{\sigma}}(u) = \widehat{F}_n(\boldsymbol{x}; u)$. The df

$$\widehat{F}_n(\boldsymbol{x}; x) I(x \leq u) + W_{\gamma, \tilde{\mu}, \tilde{\sigma}}(x) I(x > u)$$

is such a nonparametric–parametric estimate of the claim size df. Such a procedure must also be used when the parametric hypothesis is valid only for a threshold larger than that predetermined by the XL treaty.

Let us review the justification for such an approach. We want to utilize a nontrivial model for the tail of the distribution, even in regions where possibly no data are available. A parametric modeling seems to be the only reasonable approach to that question. On the other hand, there is a greater bulk of data available in the center of the distribution so that a nonparametric estimate can have a higher accuracy than a parametric estimate if the parametric modeling is incorrect.

17.3 Segmentation According to the Probable Maximum Loss

In this section, we deal with the tarification of policies with respect to the probable maximum loss (PML)[7], especially in the property business with losses caused, e.g., by fire, storm and earthquake. Besides a nonparametric estimation of the mean claim sizes in the different PML groups we pursue a parametric approaches within GP models.

---

[7]Gerathewohl, K. (1976). Rückversicherung: Grundlagen und Praxis. Verlag Versicherungswirtschaft.

Mean Claim Sizes in Dependence on the Probable Maximum Loss

It is evident that the loss potential is not homogeneous for all risks of a portfolio, but it depends on the underlying exposure of the individual risk of a policy. A particularly important figure is the PML of the individual risk that describes the maximum single claim size covered by the policy. This value is estimated by underwriting experts on the basis of additional information of the individual risk and can be thought of as the upper endpoint of the individual claim size distribution.

Thus, besides a claim size variable X for single policies, there is a covariate Z, namely the PML. One is interested in the conditional expectation $E(X|z)$ of X given $Z = z$, where z is the PML for a given policy. The aim is to estimate the conditional mean claim size $E(X|z)$ as a function of the PML z (also see Section 6.6). It is apparent that the estimation of the conditional mean claim size is a regression problem. The next steps are to estimate the mean claim number and the mean total claim amount (net premium) in dependence on the PML.

In practice, this is done for each sub–portfolio defined by a PML group of risks within specified priorities. We introduce a greater variety of methods to estimate the mean claim size within a PML group. We are particularly interested in PMLs and claim sizes over a higher priority so that it is tempting to use also parametric approaches besides empirical ones.

Estimating the Mean Claim Size for a PML Group by Using Claim Degrees

The ith PML group is determined by those risks with PMLs z between boundaries p_i and p_{i+1}, where

$$u = p_0 < p_1 < \cdots < p_{m-1} < p_m = \infty$$

is a predetermined partition. We write n_i for the number of claims and $x_{i,j}$ for the claim sizes belonging to the ith PML group. The PML pertaining to $x_{i,j}$ is denoted by $z_{i,j}$. Note that $n_i^{-1} \sum_{j \leq n_i} x_{i,j}$ is an estimate of the mean claim size within the ith PML group.

In order to make the results in the different PML groups comparable, we introduce

- the claim degrees $x_{i,j}/z_{i,j}$, and
- the empirical mean degrees $d_i = n_i^{-1} \sum_{j \leq n_i} x_{i,j}/z_{i,j}$.

The mean claim size in the ith group can be estimated by

$$m_i = q_i d_i$$

where $q_i = (p_i + p_{i+1})/2$ are the centers of the PML intervals. The advantage of using mean degrees d_i for estimating the mean claim sizes is that one may smooth the variation of these values by using, e.g., a polynomial least squares function.

EXAMPLE 17.3.1. (A Segmented Fire–Industrial Portfolio.) We deal with a middle–sized fire–industrial portfolio segmented into 25 PML groups. It is indexed and coded. Each recorded claim size exceeds $u = 100$ thousand currency units. A first impression of the data set (stored in im–pmlfi.dat) can be gained from Table 17.4.

TABLE 17.4. PMLs and claim sizes in fire–industrial portfolio in thousands.

No. i	From p_i	To p_{i+1}	PML $z_{i,j}$	Claim Size $x_{i,j}$	Claim Degree $x_{i,j}/z_{i,j}$
1	100	250	.	.	.
2	250	500	434	123	0.28341
2	250	500	324	254	0.78395
3	500	750	727	534	0.73427
.	.	.	.	.	.
25	150000	200000	183186	176	0.00096
25	150000	200000	169666	607	0.00358
25	150000	200000	165994	161	0.00097

The largest claim size of over 37 millions occurred in the 24th PML group. No claims are recorded in the first PML group. An overview of the data is obtained by the scatterplot in Fig. 17.2. One recognizes that the recorded claim sizes in the highest PML group with PMLs between 150 and 200 millions are relatively small.

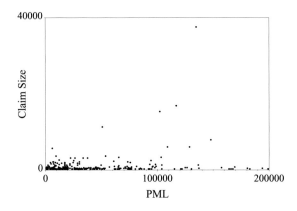

FIG. 17.2. Scatterplot of claim sizes x_j plotted against the PMLs z_j.

The special feature of the largest PML group may be caused by a systematic under-writing effect of these large scaled risks, i.e., the insurance company takes some individual risk management measures in order to avoid heavy losses.

Another aspect is that these large PML risks normally are composed of a number of differently located smaller risks—a greater industrial firm with different locations and buildings—which produce independently small losses and, therefore, a total loss is less probable.

In Fig. 17.3, the mean claim degrees d_i are plotted against the group number i. A least squares line is added (the mean degrees of the second and third group are omitted from that analysis).

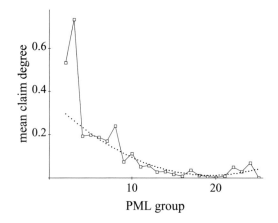

FIG. 17.3. Plotting the mean degrees d_i against the group number i including a least squares line (dotted).

It is typical that the mean degrees d_i are decreasing in i. This is partially due to the fact that for larger PMLs there is a smaller threshold for the claim degrees (therefore, this evaluation of the mean degree is a length–biased estimation).

Up to now, it is not clarified whether a plotting and smoothing should be done against the centers of the PML intervals instead of the group numbers. Using the smoothed mean degrees d_i' from the least squares line, one obtains the estimate

$$m_i' = q_i d_i' \qquad (17.11)$$

of the mean claim size in the ith PML group.

Relative Frequencies of Segmented Portfolios and Estimating the Net Premium in PML Groups

We introduce the relative claim frequencies

$$f_i = n_i / r_i$$

where n_i is again the number of claims and r_i is the number of risks belonging to the ith PML group.

EXAMPLE 17.3.2. (Continuation of Example 17.3.1.) In Table 17.5, we list the number of risks r_i and the claim numbers n_i for each of the PML groups.

TABLE 17.5. Number of risks r_i and number of claims n_i in ith PML group.

	Group Nr. i											
	2	3	4	5	6	7	8	9	10	11	12	13
r_i	2049	1658	1673	2297	1732	2536	1709	1186	1749	1669	1349	726
n_i	2	1	3	5	4	14	1	13	22	22	25	36

	14	15	16	17	18	19	20	21	22	23	24	25
r_i	719	254	194	123	76	61	34	36	24	26	15	28
n_i	36	24	22	9	13	19	8	11	5	7	5	9

We also plot the relative frequencies $f_i = n_i/r_i$ of claim numbers against the group number i and include a least squares line. The increase of the plot can also be explained as an effect of a length–biased estimation.

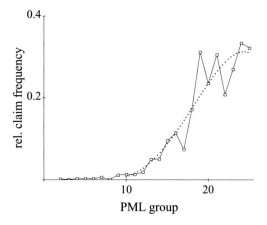

FIG. 17.4. Plotting the relative frequencies $f_i = n_i/r_i$ of claim numbers against the group number i (including a least squares (dotted) line).

Let m_i' be as in (17.11) and f_i' the smoothed relative frequencies. Then, besides $\sum_{j=1}^{n_i} x_{i,j}$ one obtains by

$$m_i' f_i' r_i$$

an estimate of the net premium for the ith PML group. Subsequently, m_i' will be replaced by parametric estimates of the mean claim size. These estimates are either based on claim degrees or on the original claim sizes. The smoothing by least squares lines is done for the parameters of GP distributions.

A Parametric Approach

The mean claim size in the ith PML group can be estimated in a parametric manner. Specify a GP model for the claim degrees or the original claim sizes in the ith PML group and use the mean of the estimated GP distribution for estimating the mean claim degree or the mean claim size. For example, the claim degree in the ith group are between $u_i = u/p_{i+1}$ and 1 and, thus, a beta (GP2) modeling is possible (with unknown shape parameter α_i). Note that the mean of this beta df is

$$(1 + u_i|\alpha_i|/(1 + |\alpha_i|).$$

Use, e.g., the Hill(GP2) estimator for evaluating α_i.

A Parametric Nearest Neighbor Approach

The disadvantage of the previous method is that, for some of the PML groups, the parametric estimation must be based on a very small number of claims. Recall that the mean claim size within a PML group is related to the conditional expectation $E(X|z)$ of the claim size X conditioned on the PML $Z = z$.

The estimation of the conditional expectation $E(X|z)$ can also be based on the claim sizes y_j, $j = 1, \ldots, k$, pertaining to the k PMLs z_j closest to z (for an introduction to the nearest neighbor method see Section 7.2). Then,

$$m_k = k^{-1} \sum_{j \leq k} y_j$$

is an empirical estimate of the conditional mean claim size $E(X|z)$.

Likewise, the estimation can be carried out within a GP model. For that purpose, compute estimates $\hat{\gamma}_k(z)$ and $\hat{\sigma}_k(z)$ of the shape and scale parameters $\gamma(z)$ and $\sigma(z)$ based on the y_j within a GP model and use the mean of the GP distribution $W_{\hat{\gamma}_k(z),u,\hat{\sigma}_k(z)}$ as an estimate of $E(X|z)$.

EXAMPLE 17.3.3. (Continuation of Example 17.3.1.) For the given fire–industrial claim sizes, the estimates $\hat{\gamma}_k(z)$ vary around the value 0.9 which corresponds to a Pareto modeling. In the further calculation, we take the constant estimate $\hat{\gamma}_k(z) = 0.88$, cf. (17.13). The scale parameters $\sigma(z)$ of the GP distribution $W_{0.88,u,\sigma(z)}$ are estimated by means of MLE's for unknown scale parameters.

Motivated by Example 17.3.3, we use the simplified modeling of claim size distributions conditioned on $Z = z$ by means of GP dfs $W_{\gamma_0,u,\sigma(z)}$, where γ_0 is a predetermined shape parameter which is independent of z. If $\hat{\sigma}_k(z)$ are estimates of $\sigma(z)$ (first, a smoothing may be carried out for the $\hat{\sigma}_k(z)$), then the mean

$$\widehat{m}_{i,k} = u + \hat{\sigma}_k(q_i)/(1 - \gamma_0) \tag{17.12}$$

of $W_{\gamma_0,u,\hat{\sigma}_k(q_i)}$ is an estimate of the mean claim size in the ith PML group.

EXAMPLE 17.3.4. (Continuation of Example 17.3.3.) Our computations only concern the 10th to 24th PML group (thus, we have PMLs from 5 to 150 millions). We tabulate the estimates m'_i and $\widehat{m}_{i,k}$ in (17.11) and (17.12) of the mean claim sizes in the different groups for $k = 30$. The smoothing was carried out with least squares lines of degree 2.

TABLE 17.6. Estimated mean claim sizes m'_i and $\widehat{m}_{i,k}$ in ith PML group.

	Group Nr. i														
	10	11	12	13	14	15	16	17	18	19	20	21	22	23	24
m'_i	435	569	730	879	982	951	807	635	521	550	808	1380	2352	4747	10139
$\widehat{m}_{i,k}$	1367	1334	1292	1248	1208	1203	1254	1361	1524	1743	2017	2348	2734	3418	4526

Notice the remarkable differences in the total amounts of the estimates as well as the different variations from group to group. Up to now we can not offer a final conclusion which of the estimation methods is preferable for the given PML data set.

Further modifications of the presented approaches are possible. For example, the smoothing of mean degrees, relative frequencies and GP parameters may be done by fitting least squares lines to the log–values or the estimated GP distributions may be truncated at the upper boundary of the PML segment.

A Collective Viewpoint

We clarify in which way we determined the mysterious shape parameter $\gamma_0 = 0.88$ in Example 17.3.3. The collective claim size df for the portfolio (the unconditional claim size df) is

$$F_{\gamma, Z}(x) = \int W_{\gamma, u, \sigma(z)}(x) \, dF_Z(z)$$

where F_Z is the df of Z.

For γ close to 1 and large x, the conditional of $W_{\gamma, u, \sigma(z)}$ can be replaced by the Pareto df $W_{1, 1/\gamma, 0, \sigma(z)/\gamma}$ in the α–parameterization (cf. (6.38)). Therefore,

$$F_{\gamma, Z}(x) \approx \int W_{1, 1/\gamma, 0, \sigma(z)/\gamma}(x) \, dF_Z(z) \; = \; W_{1, 1/\gamma, 0, \sigma_0}(x) \tag{17.13}$$

for larger x, where

$$\sigma_0 = \left(\int \sigma(z)^{1/\gamma} \, dF_Z(z) \right)^{\gamma} / \gamma.$$

Thus, a Pareto modeling is adequate for the portfolio if this is justified in the segments. The value $\gamma = 0.88$ in Example 17.3.3 was estimated within this collective approach.

The standard reference for collective risk theory is the book by Bühlmann which was mentioned on page 5. Also see the aforementioned article by Kuon, Radtke and Reich.

17.4 The Risk Process
and the T–Year Initial Reserve

In Sections 17.1–17.3, the total claim amount S_N was evaluated within a certain fixed period. When the time t varies, we write $N(t)$ and $S(t)$ in place of N and S_N, thereby obtaining the claim number and total claims processes.

Next, we deal with the risk process, which is primarily based on the initial reserve, denoted by s, and the total claims process. The initial reserve is a variable of the system which can be chosen by the insurer. Ruin occurs when the risk process becomes negative.

The primary aim in risk theory is the calculation of the ultimate ruin probability[8] or an upper bound (Lundberg inequality) of that quantity (also see Section 17.5). Knowledge about the ruin probability within a finite time horizon is preferred by practitioners.

For us the initial reserve is the central parameter that must be estimated by the actuary. We estimated that initial reserve s such that ruin within a time span of length T occurs with a probability of $q \times 100\%$.

Are Reserves for Single Portfolios of Practical Importance?

To measure the performance of a single portfolio from an economic viewpoint, the insurer must take two aspects into account, namely

- the profitability, i.e., the expected profit from running the portfolio over a certain period of time, and

- a certain fluctuation potential

of a portfolio. The profitability can be deduced from the estimated net premium and the related premium income.

Yet, for an economic evaluation of a portfolio, it is also necessary to quantify the fluctuations of the results of a portfolio over time. A commonly used approach is to consider the risk process and derive some initial reserve which is necessary to avoid a technical ruin in a fixed finite time horizon with a certain probability.

This reserve can be interpreted as the security capital the insurer needs to carry the collective risk of the portfolio. It is a kind of fluctuation reserve which ensures that the company does not become ruined over time by possible deviations from the expected results of the portfolio.

In this context, another important aspect is that risk processes allow comparison of different portfolios not only by their profitability, but also by the required initial reserves.

---

[8]See, e.g., Vylder, de F. (1997). Advanced Risk Theory. Editrans de l'Université Bruxelles.

For example, to compare a fire portfolio, a portfolio consisting of natural perils (like earthquakes and storms) and some type of motor insurance portfolio, it is obviously necessary to deal with substantially different reserves. By applying this approach to all business over all lines, the insurance company is able to conduct the technical solvency process. This also enables the supervisory authority to evaluate a solvency margin for the liabilities of the company.

Net and Total Premiums

The net premium (at time t) is the expectation $E(S(t))$ of the total claim amount $S(t)$. Assuming the independence of the claim size process

$$X_1, X_2, X_3, \ldots$$

and the claim arrival process

$$T_1, T_2, T_3, \ldots,$$

our basic condition in Section 17.1 is valid, namely the independence of the claim size process $X_1, X_2, X_3, \ldots$ and the claim numbers $N(t)$.

We also assume that the claim sizes X_i are identically distributed. Denote the mean claim size by $E(X)$ again. Recall from (17.8) that the net premium can be written

$$E(S(t)) = E(X)E(N(t)).$$

If the claim number process $N(t)$, $t \geq 0$ is a homogeneous Poisson process with intensity λ, then

$$E(S(t)) = E(X)\lambda t. \tag{17.14}$$

If $N(t)$, $t \geq 0$ is a Pólya–Lundberg process, then a corresponding formula holds with λ replaced by $\alpha\sigma$.

Assuming that the total premium $c(t)$—the total amount to be paid by the policy holders to compensate the future losses—is a multiple of the net premium, one may write

$$c(t) = (1 + \rho)E(S(t)), \tag{17.15}$$

where ρ is a constant called the safety loading.

The Risk Process

We introduce the risk (surplus, reserve) process in the form

$$U(t) = s + E(S(t)) + b(t) - S(t), \qquad t \geq 0, \tag{17.16}$$

where the single terms are

- the initial insurer's reserve $s = U(0) \geq 0$ for a given portfolio;

- the net premium $E(S(t))$ up to time t;

- the safety function (mean surplus) $b(t)$ at time t which is

  - the difference $\rho E(S(t))$ between the total and net premiums (with ρ denoting the safety loading in (17.15))
  - plus the interest income for the accumulated reserve
  - minus expenses, taxes and dividends etc.,

- the total claim amount $S(t) = S_{N(t)} = \sum_{i \leq N(t)} X_i$ up to time t.

The mean of the reserve variable $U(t)$ is

$$E(U(t)) = s + b(t), \qquad (17.17)$$

and, therefore, $b(t)$ is the surplus that can be added (in the mean) to the reserve. It is clear that the safety function $b(t)$ must be nonnegative.

Special Safety Functions

Insurance mathematics primarily concerns safety functions $b(t) = c(t) - E(S(t))$ which are the differences between the total premium and the net premium. If $c(t)$ is given as in (17.15), then

$$b(t) = \rho E(X) E(N(t)) = \rho E(X) \lambda t$$

for a homogeneous Poisson process with intensity λ.

If $\rho > 0$, then the mean reserve is linearly increasing according to (17.17); on the other hand, the outstanding result of insurance mathematics is that ruin occurs with a positive probability (see the next Section 17.5).

Accumulating reserves which are rapidly increasing is perhaps not a desirable goal. We also suggest to consider safety functions of the form

$$b(t) = \rho E(X) \lambda t^{\beta} \qquad (17.18)$$

for Poisson processes with intensity λ (and related functions for other claim number processes).

Note that the safety exponent $\beta = 1$ is taken in the classical framework. If $\beta \leq 1/2$, then ruin occurs with probability one, if the claim sizes satisfy the conditions of the law of the iterated logarithm.

EXAMPLE 17.4.1. (Continuation of Example 17.2.1.) We again consider the claim sizes over the retentation level of 22 million NKr. For the estimated parameters, several paths of the risk process—with initial reserve $s = 250$, safety exponent $\beta = 0.3, 1$ and safety loading $\rho = 0.1$—are generated and displayed in Fig. 17.5.

Let us summarize what has been achieved up to now:

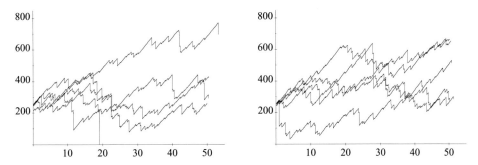

FIG. 17.5. Paths of risk processes with $\beta = 0.3$ (left) and $\beta = 1$ (right).

- we estimated the unknown parameters γ, σ and λ, thereby getting an estimate of the mean claim size $E(X) = u + \sigma/(1 + \gamma)$ and of the net premium $\lambda E(X)$,

- we fixed a safety function $\rho E(X)\lambda t^{\beta}$, where the choice of the safety coefficients ρ and β is presumably dictated by the market and other external conditions.

The knowledge of these constants enables us to simulate risk process paths. This also yields that any functional parameter of the risk process can be estimated, and the estimate can be computed by means of the simulation technique. This will be exemplified in conjunction with the T–year initial reserve for a given $q \times 100\%$ ruin probability.

Ruin Times and Ruin Probabilities within an Infinite und Finite Time Horizon

Ruin occurs at the time when the reserve variable $U(t)$ becomes negative. Let τ_s be the ruin time (with the convention that $\tau_s = \infty$ if no ruin occurs) of the risk process $U(t)$, $t \geq 0$, starting with an initial reserve $s = U(0)$. The ruin time can be written

$$\tau_s = \inf\{t : U(t) < 0\}.$$

Consider the ruin time df

$$H_s(x) = P\{\tau_s \leq x\}$$

which can be a defected df, since non–ruin may occur with a positive probability. The ultimate ruin probability

$$\psi(s) = P\{\tau_s < \infty\} \tag{17.19}$$

as a function of the initial reserve s will be dealt with in Section 17.5. This is the quantity that is studied in ruin theory.

Subsequently, we examine ruin within a certain finite time horizon T. The probability of ruin up to time T is

$$\psi_T(s) = H_s(T) = P\{\tau_s \leq T\}.$$

We will calculate such ruin probabilities by simulations[9]. Apparently, the ultimate ruin probability $\psi(s)$ is the limit of $\psi_T(s)$ as T goes to infinity.

One may also deal with early warning times

$$\tau_{s,w} = \inf\{t : U(t) < w\}, \tag{17.20}$$

where $w > 0$ is an early warning limit for the insurer. Because

$$U(t) = s + E(S(t)) + b(t) - S(t) < w$$

if, and only if,

$$(s - w) + E(S(t)) + b(t) - S(t) < 0,$$

we have $\tau_{s,w} = \tau_{(s-w)}$, so that this case can be treated within the previous framework.

The T–Year Initial Reserve for a $q \times 100\%$ Ruin Probability

Choose the initial reserve s such that ruin occurs with a probability q within a time span of length T, where, e.g., $q = 0.01$ or $q = 0.05$ and $T = 10$ or $T = 50$ years. The value $s(q, T)$ is is called a T–year initial reserve[10]. Apparently, it depends on the underlying risk process of the specific portfolio.

Note that $s(q, T)$ is a solution to the equation $\psi_T(s) = H_s(T) = q$ which can be written as

$$H_s^{-1}(q) = T. \tag{17.21}$$

Thus, find the initial reserve s such that the ruin time qf H_s^{-1} evaluated at q is equal to T. Check that $s(q, T)$ is a T–year threshold according to the q–quantile criterion (see page 251) for the process $S(t) - (E(S(t)) + b(t))$.

A closely related concept would be the mean T–year initial reserve as the solution to $E\tau_s = T$, yet the mean seems to be infinite even for finite ruin time rvs (for a safety exponent $\beta \leq 1/2$).

The Initial Risk Contour Plot

By employing the simulation technique, one can plot a sample qf as an estimate of the ruin time qf H_s^{-1} for several initial reserves s. This plot is the initial risk

---

[9]Also see Vylder, de F. and Goovaerts, M.J. (1988). Recursive calculation of finite–time ruin probabilities. Insurance: Mathematics and Economics 7, 1–7.

[10]Presented at the 35th ASTIN meeting (Cologne, 1996) of the DAV.

contour plot. Notice that one gets a sample version of a contour plot of the function $s(q, T)$.

Now, by employing an iteration procedure, one can calculate the desired estimate of the T–year initial reserve for a predetermined $q \times 100\%$ ruin probability. Knowledge about the accuracy of this procedure can be gained by constructing a parametric bootstrap confidence interval, see Section 3.2.

EXAMPLE 17.4.2. (Continuation of Example 17.2.1.) We again consider the claim sizes over the retentation level of 22 million NKr. For the estimated parameters, we compute the initial reserve contour lines for $s = 100, 200, 300$. The risk process is taken for the safety exponent $\beta = 1$ and the safety loading $\rho = 0.1$.

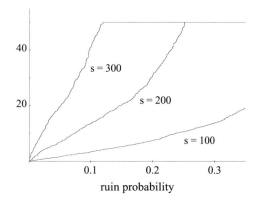

FIG. 17.6. Initial risk contour plot for the initial reserves $s = 100, 200, 300$.

From the contour plot, one may deduce that, e.g., the 10–year initial reserve for the 5% ruin probability is about 250. The estimate obtained by an iteration procedure is $\hat{s}(0.05, 10) = 225$.

We also compute an upper bootstrap confidence bound. First, let us recollect the bootstrap procedure for this special application. The bootstrapping is based on the $\mathrm{Poisson}(\lambda, G_{\gamma, \mu, \sigma})$ process as introduced on page 248, where $\lambda = 1.7$, $\gamma = 0.254$, $\mu = 0$ and $\sigma = 11.948$. Moreover, the sampling time is $T = 10$. To obtain a bootstrap sample for the initial reserve, generate the claim number and claim sizes according to the specified Poisson process and estimate the parameters by means of the MLE(GP). Compute the initial reserve according to the estimated parameters and store this value in a file. Repeat this procedure according to the bootstrap sample size. We obtained the 80% and 90% upper bootstrap confidence bounds 334 and 508. These larger bounds are not surprising in view of the small sample size of 17 claims (in the mean).

It is apparent that such contour lines can be utilized to simulate a T–year initial reserve. This can be done by using a StatPascal program[11].

---

[11]Supplementary details can be found in Reiss, R.–D., Radtke, M. and Thomas, M.

17.5   Elements of Ruin Theory

Assume that the claim number process is a homogeneous Poisson process with intensity $\lambda > 0$. Assume that the total premium is $c(t) = (1 + \rho)E(X)\lambda t$, where

$$\rho > 0 \qquad (17.22)$$

is the safety loading (see (17.15)). We deal with risk processes

$$U(t) = s + (1 + \rho)E(X)\lambda t - S(t).$$

Denote the ultimate ruin probability by $\psi(s)$ again, given an initial reserve s.

We refer to the review article by Embrechts and Klüppelberg[12] and to [11] for a detailed account of extremal ruin theory.

Exact Evaluation of Ruin Probabilities

Let F denote the claim size df again. Because $0 < E(X) < \infty$, we know from (2.22) that $\int_0^\infty (1 - F(y))\, dy = E(X)$, and, hence, $h(y) = (1 - F(y))/E(X)$, $y \ge 0$, is a density of a df, say H. Because

$$1 - H(x) = \frac{1}{EX} \int_x^\infty (1 - F(y))\, dy$$

we see that the survivor function of H can be expressed by tail probabilities (cf. (2.24)).

Under the present conditions, one obtains

$$1 - \psi(s) = p \sum_{k=0}^\infty (1 - p)^k H^{k*}(s), \qquad s \ge 0 \qquad (17.23)$$

for the probability of non–ruin[13] where

$$p = \rho/(1 + \rho).$$

Therefore, one must compute a compound geometric df (cf. (17.6)) in order to evaluate ruin probabilities.

(1997). The T–year initial reserve. Technical Report, Center for Stochastic Processes, Chapel Hill.

[12]Embrechts, P. and Klüppelberg, C. (1993). Some aspects of insurance mathematics. Theory Probab. Appl. 38, 262–295.

[13]See, e.g., Hipp, C. and Michel, R. (1990). Risikotheorie: Stochastische Modelle und Statistische Methoden. Verlag Versicherungswirtschaft, Karlsruhe.

Alternatively, the ruin probability function may be written in terms of the survivor functions of H^{k*}. We have

$$\psi(s) = p \sum_{k=1}^{\infty} (1-p)^k (1 - H^{k*}(s)), \qquad s \geq 0. \tag{17.24}$$

Deduce that $\psi(s)$ tends to zero as $s \to \infty$.

We compute the auxiliary df H for two special cases. In addition, an explicit representation of the ruin probability function ψ can be given for exponential claim sizes (Erlang model).

- (Exponential Claim Sizes.) Assume that the claim sizes are exponentially distributed. More precisely, let $F(x) = 1 - e^{-x/\sigma}$, $x \geq 0$. Hence, $EX = \sigma$. Then, $H(x) = F(x)$. From (17.7) and (17.23), one obtains

$$\psi(s) = \frac{1}{1+\rho} \exp\left(-\frac{\rho s}{(1+\rho)E(X)}\right), \qquad s \geq 0. \tag{17.25}$$

  Thus, the ruin probabilities decrease with an exponential rate to zero as the initial capital s goes to infinity.

- (Pareto Claim Sizes.) If X is a Pareto random variable with df $W_{1,\alpha,\mu,\sigma}$ with lower endpoint $\mu + \sigma > 0$ and $\alpha > 1$, then $E(X) = \mu + \sigma\alpha/(\alpha - 1)$ and

$$1 - H(x) = \frac{\sigma}{E(X)(\alpha - 1)} \left(\frac{x - \mu}{\sigma}\right)^{-(\alpha - 1)} \tag{17.26}$$

  for $x \geq \mu + \sigma$.

Lower Bounds of Ruin Probabilities

A simple lower bound of $\psi(s)$ can be easily deduced from (17.24). Because $H^{k*} \leq H(s)$, one gets

$$\psi(s) \geq (1 - H(s))/(1 + \rho). \tag{17.27}$$

- (Exponential Claim Sizes.) Again, let $F(x) = 1 - e^{-x/\sigma}$, $x \geq 0$. We have

$$\psi(s) \geq \exp(-E(X)/s)/(1 + \rho).$$

  We see that the simple lower bound is close to the exact one in (17.25).

- (Pareto Claim Sizes.) From (17.26), we obtain the lower bound

$$\psi(s) \geq \frac{\sigma}{E(X)(\alpha - 1)(1 + \rho)} \left(\frac{s - \mu}{\sigma}\right)^{-(\alpha - 1)} \tag{17.28}$$

  for $s \geq \mu + \sigma$. Thus, in contrast to the preceding examples, one gets an arithmetic rate of the lower bound as $s \to \infty$.

We refer to the aforementioned book by Hipp and Michel for more refined lower bounds.

Upper Bounds: The Lundberg Inequality

An upper bound for the ultimate ruin probability is provided by the Lundberg inequality. Assume that $R > 0$ satisfies the equation

$$\int_0^\infty e^{Ry}(1 - F(y))\, dy = (1 + \rho)E(X). \tag{17.29}$$

Such a constant R is called an adjustment coefficient. Then, the inequality

$$\psi(s) \le \exp(-Rs), \qquad s \ge 0, \tag{17.30}$$

holds.

- (Exponential Claim Sizes With Mean $\mu > 0$.) We have

$$R = \frac{\rho}{(1 + \rho)E(X)} \tag{17.31}$$

  and, hence, the upper bound exceeds the exact value in (17.25) by a factor $1 + \rho$.

- (Pareto Claim Sizes.) For Pareto claim size dfs the Lundberg inequality is not satisfied according to (17.28). Notice that the left–hand side of (17.29) is equal to infinity.

Ruin Probabilities in the Pareto Case Revisited

Direct calculations lead to an upper bound of the ultimate ruin probability $\psi(s)$ for Pareto claim size dfs $W_{1,\alpha,\mu,\sigma}$ with $\alpha > 1$. We have[14]

$$\psi(s) \approx \frac{\sigma}{E(X)(\alpha - 1)\rho}\left(\frac{s}{\sigma}\right)^{-(\alpha-1)}, \qquad s \to \infty. \tag{17.32}$$

We see that the lower bound in (17.28) is exact, except of a factor $(\rho + 1)/\rho$.

17.6 Credibility (Bayesian) Estimation of the Net Premium

As in Section 17.1 we deal with an excess–of–loss (XL) reinsurance treaty. The primary aim is to estimate the net premium which is the expectation of the total

---

[14] Bahr von, B. (1975). Asymptotic ruin probabilities when exponential moments do not exist. Scand. Actuarial J., 6–10.

In greater generality dealt with in Embrechts, P. and Veraverbeke, N. (1982). Estimates for the probability of ruin with special emphasis on the probability of large claims. Insurance: Mathematics and Economics 1, 55–72.

claim amount

$$S_N = \sum_{i \leq N} (X_i - u)$$

for the next period, where N is the random number of claims occurring in the period, and the X_i are the random claim exceedances over the priority u.

Throughout, we assume that the exceedance times and the exceedances can be jointly modeled by a Poisson(λ, F) process. Thus, the exceedance times are described by a homogeneous Poisson process with intensity $\lambda > 0$, and the exceedances have the common df F. The net premium is

$$E(S_N) = \lambda(m_F - u) \qquad (17.33)$$

with m_F denoting the mean of F. The estimate of the net premium is based on the exceedance times and exceedances of the past T years. In contrast to Section 17.1, we merely deal with the estimation of the net premium within certain Poisson–Pareto models.

Our attention is focused on the Bayes estimation—that is, the exact credibility estimation—of the net premium. A justification for a restriction to the exact credibility estimation was formulated by S.A. Klugman [35], page 64: "Most of the difficulties ... are due to actuaries having spent the past half–century seeking linear solutions to the estimation problems. At a time when the cost of computation was high this was a valuable endeavor. The logical solution is to drop the linear approximation and seek the true Bayesian solution to the problem. It will have the smallest *mse* and provides a variety of other beneficial properties".

Recall that the mean of a Pareto distribution is equal to infinity if the shape parameter α is smaller or equal to 1, cf. (1.44). This yields that the support of a prior distribution must be restricted to parameters $\alpha > 1$, when the Bayes estimation of the mean is dealt with. This is the reason that we introduce truncated gamma distributions in the subsequent lines.

Truncated Gamma Distributions

A gamma distribution, truncated left of $q \geq 0$, with shape and reciprocal scale parameters $s > 0$ and $d > 0$, cf. (3.42), serves as a prior. The pertaining df and density are denoted by $H_{s,d,q}$ and $h_{s,d,q}$. One gets gamma distributions if $q = 0$. The gamma density $h_{s,d,q}$, truncated left of $q \geq 0$, is given by

$$h_{s,d,q}(\alpha) = h_{s,d}(\alpha)/\big(1 - H_{s,d}(q)\big), \qquad \alpha > q. \qquad (17.34)$$

It turns out that truncated gamma densities are conjugate priors for our models. That is, the posterior densities are of the same type.

Straightforward calculations yield that such a truncated gamma distribution has the mean

$$M_{s,d,q} = \big(s(1 - H_{s+1,d}(q))\big)/\big(d(1 - H_{s,d}(q))\big) \qquad (17.35)$$

and the variance

$$V_{s,d,q} = \frac{s\Big((s+1)(1 - H_{s+2,d}(q))(1 - H_{s,d}(q)) - s(1 - H_{s+1,d}(q))^2\Big)}{d^2(1 - H_{s,d}(q))^2}. \quad (17.36)$$

These formulas are useful

- to specify a gamma prior with predetermined moments,
- to compute the numerical value of a Bayes estimate.

Estimation of the Net Premium for Poisson Processes with Pareto GP1$(u, \mu = 0)$ Marks

For the modeling of exceedances over the priority u we use dfs within the restricted Pareto GP1$(u, \mu = 0)$ model which are given by

$$W_{1,\alpha,0,u}(y) = 1 - (y/u)^{-\alpha}, \qquad y \geq u, \quad (17.37)$$

where $\alpha > 0$. These dfs are used for the modeling of exceedances over a priority u.

The pertaining excesses have the mean $u/(\alpha - 1)$ and, therefore, the net premium (cf. (17.33)) is given by

$$m(\lambda, \alpha) = \frac{\lambda u}{\alpha - 1}, \qquad \alpha > 1. \quad (17.38)$$

This functional can be written in a product form as given in (9.16). We further apply the concept as developed in Section 9.3. The prior is $p(\lambda, \alpha) = h_{r,c}(\lambda)h_{s,d,q}(\alpha)$ with $q > 1$. The posterior with respect to the parameter α is $p_2(\alpha|\boldsymbol{y}) = h_{s',d',q}(\alpha)$ with s' and d' as in (5.6).

One obtains the Bayes estimate

$$\widehat{m}(\{t_i, y_i\}) = \lambda_k^* m_{k,q}^*(\boldsymbol{y}) \quad (17.39)$$

of the net premium $m(\lambda, \alpha)$ with λ_k^* as in (3.53) and

$$m_{k,q}^*(\boldsymbol{y}) = \int_q^\infty \frac{u}{\alpha - 1} h_{s',d',q}(\alpha)\, d\alpha. \quad (17.40)$$

The integral must be numerically evaluated. Notice that $m_{k,q}^*(\boldsymbol{y})$ is a Bayes estimate of the mean of excesses over u.

Estimation of the Net Premium for Poisson Processes with Pareto GP1(u) Marks

In contrast to the preceding lines, we assume that the exceedances over the priority u are distributed according to a Pareto df

$$F_{\alpha,\eta}(y) = 1 - \Big(1 + \frac{y - u}{\eta u}\Big)^{-\alpha}, \qquad y \geq u, \quad (17.41)$$

where $\alpha, \eta > 0$, cf. (5.9). Thus, we deal with the full Pareto model in the (α, η)–parameterization. One obtains the restricted GP1$(u, \mu = 0)$ model if $\eta = 1$.

The excesses have the mean $u\eta/(\alpha - 1)$ and, therefore, the net premium is given by

$$m(\lambda, \alpha, \eta) = \frac{\lambda u \eta}{\alpha - 1}, \qquad \alpha > 1. \tag{17.42}$$

The prior is $p(\lambda, \alpha, \eta) = h_{r,c}(\lambda) h_{s,d,q}(\alpha) f(\eta)$ with $q > 1$, and f is another probability density on the positive half–line. The posterior with respect to the parameter vector (α, η) is $p_2(\alpha, \eta | \boldsymbol{y}) = h_{s', d'(\eta), q}(\alpha) \tilde{f}(\eta)$ with s', $d'(\eta)$ and $\tilde{f}(\eta)$ as in (5.13) and (5.14).

One obtains the Bayes estimate

$$\widetilde{m}(\{t_i, y_i\}) = \lambda_k^* m_{k,q}^{**}(\boldsymbol{y}) \tag{17.43}$$

of the net premium $m(\lambda, \alpha, \eta)$ with λ_k^* as in (3.53) and

$$m_{k,q}^{**}(\boldsymbol{y}) = \int_0^\infty \eta \left(\int_q^\infty \frac{u}{\alpha - 1} h_{s', d'(\eta), q}(\alpha) \, d\alpha \right) \tilde{f}(\eta) \, d\eta. \tag{17.44}$$

Estimation of Parameters

We also shortly note the estimates of the parameters α and η if a prior $h_{s,d,q}$ is taken for α.

- The restricted Pareto model GP1$(u, \mu = 0)$: from (17.35) deduce that the Bayes estimate of $\alpha > q \geq 0$ is

$$\alpha_{k,q}^*(\boldsymbol{y}) = \alpha_k^*(\boldsymbol{y}) \frac{1 - H_{s'+1, d'}(q)}{1 - H_{s', d'}(q)}, \tag{17.45}$$

  where $\alpha_k^*(\boldsymbol{y})$ is the Bayes estimate in the case of the gamma prior (with $q = 0$; see (5.7)). Notice that $\alpha_{k,0}^*(\boldsymbol{y}) = \alpha_k^*(\boldsymbol{y})$.

- The full Pareto model GP1$^*(u)$: extending the formulas in (5.15) and (5.16), one gets

$$\alpha_{k,q}^*(\boldsymbol{y}) = \int \frac{s'}{d'(\eta)} \frac{(1 - H_{s'+1, d'(\eta)}(q))}{(1 - H_{s', d'(\eta)}(q))} \tilde{f}(\eta) \, d\eta, \qquad q \geq 0, \tag{17.46}$$

  and, independently of q, $\eta_{k,q}^*(\boldsymbol{y}) = \eta_k^*(\boldsymbol{y})$ as Bayes estimates of $\alpha > q \geq 0$ and $\eta > 0$.

Part VI

Topics in
Material and Life Sciences

Chapter 18

Material Sciences

In Section 18.1 the blocks method (in other words, annual maxima or Gumbel method) is applied to corrosion engineering. We are particularly interested in the service life of items exposed to corrosion. Our primary sources are the book by Kowaka et al., [37] and a review article by T. Shibata[1].

Section 18.2 concerns a stereological problem in conjunction with extreme value analysis. We are interested in extremes of 3–dimensional quantities of which we merely observe a 2-dimensional image.

18.1   Extremal Corrosion Engineering

In corrosion analysis the role of time periods is played by several units such as steel tanks. One can also deal with a single area which is divided into T subareas of the same size. This corresponds to dividing a given period into T blocks.

For each of the units or subareas, which are exposed to corrosion, we observe the maximum pit depth (or, e.g., a measurement concerning stress corrosion cracking), and thus get a data set $y_1, \ldots, y_T$ of maxima. In the statistical modeling we get random variables $Y_1, \ldots, Y_T$ which are distributed according to a Gumbel df $G_{0,\mu,\sigma}$ or, in general, according to an extreme value df $G_{\gamma,\mu,\sigma}$.

The unified extreme value model was introduced by Laycock et al.[2] in the field of corrosion analysis.

Assume that there are T units or an area of size T exposed to corrosion. Our final goal is to determine the T–unit service life which is the period of $x(T, l)$ years such that the T–unit depth $u(T)$ for $x(T, l)$ years is equal to an allowable margin of thinning l.

---

[1]Shibata, T. Application of extreme value statistics to corrosion. In [15], Vol. II, 327–336.

[2]Laycock, P.J., Cottis, R.A. and Scarf, P.A. (1990). Extrapolation of extreme pit depths in space and time. J. Electrochem. Soc. 137, 64–69.

Recall from page 6 that the T–unit pit depth $u(T)$ for a given time period of length x is the level once exceeded on the average.

Estimation of a T–Unit Depth

Let Y_i be the maximum pit depth of the ith unit within the given period of x years. The T–unit depth $u(T)$ is the solution to the equation

$$E\left(\sum_{i \leq T} I(Y_i \geq u)\right) = 1$$

which are the quantiles

$$u(T) = G^{-1}_{0,\mu,\sigma}(1 - 1/T)$$

or

$$u(T) = G^{-1}_{\gamma,\mu,\sigma}(1 - 1/T)$$

corresponding to (1.13).

It is evident that the T–unit depth also depends on the service time x. Occasionally, we also write $u(T, x)$ in place of $u(T)$. Necessarily, $u(T, x)$ is an increasing function in x.

The value $u(T)$ is evaluated by estimating the parameters μ, σ or γ, μ and σ as it is done in Chapter 4 within the Gumbel EV0 model or the full extreme value model EV.

EXAMPLE 18.1.1. (Maximum Pit Depths of Steel Tanks.) The maximum pit depths (in mm) of steel tanks exposed to cyclic dry/wet (by sea water) conditions are measured after 4 years of service (stored in gu–pit04.dat). This data set is taken from Kowaka et al., [37], Table 18.1[3].

These authors estimate a Gumbel distribution with location and scale parameters $\mu = 1.03$ and $\sigma = 0.37$ by visual inspection of a Gumbel probability paper. The 100–unit depth $u(100)$ is estimated as 2.76mm.

TABLE 18.1. Maximum pit depth measured in mm after 4 years of service.

0.3	0.7	0.8	0.9	1.1	1.2	1.3	1.4	1.5	1.6
0.7	0.7	0.8	1.0	1.1	1.2	1.3	1.4	1.6	2.0
0.7	0.8	0.9	1.0	1.1	1.2	1.3	1.4	1.6	2.1
0.7	0.8	0.9	1.0	1.1	1.2	1.3	1.4	1.6	2.3
0.7	0.8	0.9	1.0	1.1	1.3	1.4	1.5	1.6	2.5

---

[3]With reference given to Tsuge, H. (1983). Archive of 51st Corrosion Symposium, JSCE, page 16.

This result is confirmed by the MLE(EV0). One gets a Gumbel distribution with location and scale parameters $\mu = 1.00$ and $\sigma = 0.36$. The adequacy is also confirmed by the MLE(EV) and other diagnostic tools. The estimate for the 100–unit depth $u(100)$ is 2.65mm.

Varying the Service Time

Subsequently, we compute the T–unit depths for different years of service. We write $u(T, x)$ in place of $u(T)$ to emphasize the dependence of the T–unit depth from the service time x.

In addition to the data analyzed in Example 18.1.1 for a service time of $x = 4$ years, we also deal with data for service times $x = 6, 12$.

EXAMPLE 18.1.2. (Continuation of Example 18.1.1 About Maximum Pit Depths of Steel Tanks.) We also analyze a sample of 32 maximum pit depths after a service time of 6 years (stored in gu–pit06.dat) and a sample of 30 maximum pit depths after a service time of 12 years (stored in gu–pit12.dat). These data sets are also taken from Kowaka et al. [37]. We compare the estimates obtained by the MLE(EV0) with those in [37].

In the subsequent table we compare the estimated Gumbel parameters μ, σ and the pertaining 100–unit depths $u(100)$ in [37] with the values obtained by the MLE(EV0). The corresponding parameters of the MLE(EV) are added. Recall that Gumbel distributions possess the shape parameter $\gamma = 0$.

TABLE 18.2. Estimated EV parameters and 100–unit pit depths.

	Estimated EV0/EV parameters and 100–unit depths								
	Kowaka et al.			MLE(EV0)			MLE(EV)		
x	4	6	12	4	6	12	4	6	12
$\gamma(x)$	0	0	0	0	0	0	−0.07	−0.07	0.38
$\mu(x)$	1.03	1.16	1.43	1.00	1.17	1.58	1.01	1.19	1.40
$\sigma(x)$	0.37	0.49	1.23	0.36	0.41	0.94	0.36	0.42	0.79
$u(100, x)$	2.76	3.4	7.1	2.65	3.1	6.0	2.43	2.82	11.2

One recognizes a greater difference of the estimated parameters in the Gumbel model for the service time of 12 years. Within the EV–model one gets a shape parameter $\gamma = 0.38$, which indicates a heavier upper tail of the distribution. Thus, the shape parameter γ significantly deviates from $\gamma = 0$ which is the shape parameter for Gumbel distributions.

Whereas the Gumbel distribution—with the larger scale parameter $\sigma = 1.23$ of

Kowaka et al.—catches the heavier upper tail it does not fit to the lower data. Thus, these authors locally fitted a Gumbel distribution to the upper extremes of the maxima $y_1, \ldots, y_T$.

The Service Life Criterion

Based on different T–unit depths $u(T, x_k)$ we select the service life $x(T, l)$ which is the number of years x such that the T–unit depth $u(T, x)$ is equal to an allowable margin of thinning l. Thus, given T units exposed to corrosion there is one maximum pit depths which will exceed the depth of the amount l within $x(T, l)$ years on the average.

EXAMPLE 18.1.3. (Continuation of the Preceding Examples.) The margin of thinning is taken equal to $l = 5\mathrm{mm}$ in the preceding example. By fitting a linear and a quadratic regression line to the 100–unit depths $u(100, x)$ based on the MLE(EV) estimates for the three different years $x = 4, 6, 12$, one obtains estimates of $\hat{x}(100, 5) = 6.9$ and $\hat{x}(100, 5) = 8.5$ years for the service life compared to $\hat{x}(100, 5) = 8.2$ years which was estimated by Kawata et al.

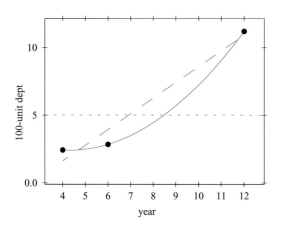

FIG. 18.1. 100–unit depths plotted against years, fitted linear (dashed) and quadratic (solid) least squares lines, constant (dotted) line with value equal to the margin of thinning $l = 5\mathrm{mm}$.

We merely dealt with the blocks method, yet one could also apply the peaks–over–threshold (pot) method. Then, one should take care of a possible clustering of larger pit depths. This question is discussed by Scarf and Laycock[4]. A global modeling of pit depth by means of log–normal distributions may be found in a paper by Komukai and Kasahara[5].

[4]Scarf, P.A. and Laycock, P.J. (1994). Applications of extreme value theory in corrosion engineering. In: [15], Vol. II, 313–320.

[5]Komukai, S. and Kasahara, K. (1994). On the requirements for a reasonable extreme value prediction of maximum pits on hot–water–supply copper tubing. In: [15], Vol. II, 321–326.

18.2   Stereology of Extremes
co–authored by E. Kaufmann[6]

This section is devoted to the well–known Wicksell[7] "corpuscle problem" which concerns the size and number of inclusions in a given material (such as particles of oxides in a piece of steel). If we assume that the particles are spheres, then the size is determined by the radius R which is regarded as a random variable. Besides the df F of R one is interested in the number of particles.

Information about the spheres can be gained by those which become visible on the surface area. On a plane surface one observes circles which are determined by the radius S with a df which is denoted by $W(F)$. This df is addressed as the Wicksell transformation of F. Drawing conclusions about the initial 3–dimensional objects from the observable 2–dimensional ones is a genuine problem in stereology. Our attention will be focused on the modeling of the upper tails of F and $W(F)$.

The Wicksell Transformation

At the beginning we compute the Wicksel transformation $W(F)$ as a conditional df of the radius variable S and provide different representations.

A sphere in the Euclidean space is determined by the coordinates x, y, z of the center and the radius r. We assume that $0 \le z \le T$. The sphere intersects the x, y–plane if $r > z$. In that case one observes a circle in the x, y–plane with radius $s = \sqrt{r^2 - z^2}$.

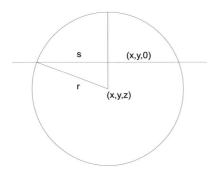

FIG. 18.2. Computation of circle radius s.

Within the stochastical formulation we assume that the random radius $R > 0$ and the random coordinate Z are independent, R has the df F, and Z is uniformly distributed on $[0, T]$. Thus, $S = \sqrt{R^2 - Z^2}$ is the random radius of the circle in the x, y–plane under the condition that $S^2 = R^2 - Z^2 > 0$; that is, there is an intersection with the x, y–plane.

---

[6]University of Siegen.

[7]Wicksell, S.D. (1925). The corpuscle problem I. Biometrika 17, 84–99, and Wicksell, S.D. (1926). The corpuscle problem II. Biometrika 18, 152–172.

We compute the survivor probability $P\{S > s\}$ for $s > 0$ and the conditional survivor function

$$\overline{W_T(F)}(s) = P(S > s | S > 0).$$

Applying results for conditional distributions, namely (8.4) and (8.13), one gets

$$
\begin{aligned}
P\{S > s\} &= \int P\{R > \sqrt{s^2 + z^2}\} \, d\mathcal{L}(Z)(z) \\
&= \frac{1}{T} \int_0^T \overline{F}\left(\sqrt{s^2 + z^2}\right) dz,
\end{aligned}
$$

where $\overline{F}$ is again the survivor function of F. Therefore,

$$\overline{W_T(F)}(s) = \frac{\int_0^T \overline{F}\left(\sqrt{s^2 + z^2}\right) dz}{\int_0^T \overline{F}(z) \, dz}. \qquad (18.1)$$

If $\int_0^\infty \overline{F}(z) \, dz < \infty$, then the "tail probability" formula (2.24) yields

$$
\begin{aligned}
\overline{W_T(F)}(s) \quad &\rightarrow_{T \to \infty} \quad \frac{\int_0^\infty \overline{F}\left(\sqrt{s^2 + z^2}\right) dz}{\int_0^\infty z \, dF(z)} \\
&=: \quad \frac{H(s)}{H(0)} = \overline{W(F)}(s), \qquad (18.2)
\end{aligned}
$$

where the df $W(F)$ is the Wicksell transformation of F.

Alternative Representations of the Wicksell Transformation

We add alternative representations of the numerator $H(s)$. Applying the substitution rule and, in a second step, writing $\overline{F}(z)$ as an integral with respect to F and interchanging the order of integration one gets

$$
\begin{aligned}
H(s) &= \int_s^\infty \left(z^2 - s^2\right)^{-1/2} z \overline{F}(z) \, dz & (18.3) \\
&= \int_s^\infty \sqrt{r^2 - s^2} \, dF(r). & (18.4)
\end{aligned}
$$

With the help of (18.4) verify that $H(s) = \int_s^\infty h(z) \, dz$ with

$$h(s) = s \int_s^\infty \left(r^2 - s^2\right)^{-1/2} dF(r). \qquad (18.5)$$

Hence, h is a density pertaining to H. This also yields that $w(F) := h/H(0)$ is a density of the Wicksell transformation $W(F)$.

Poisson Process Representations

In the preceding lines we merely considered the radii of the spheres and circles. Next, we include the centers and the number of occurances. These quantities are jointly described by a marked Poisson process. Thus, our 2–step approach resembles that employed for exceedances, where we first introduced the exceedance df and, later on, the Poisson process of exceedances.

The centers of the spheres are regarded as points of a homogeneous Poisson process with intensity $c > 0$. This is a Poisson process with intensity measure $c\lambda^3$ (see the end of Section 9.5), where λ^3 denotes the 3–dimensional Lebesgue measure. Secondly, let F be the df of the sphere radii which are regarded as marks. Then, a joint statistical model for the inclusions—for centers as well as radii—is given by a Poisson$(c\lambda^3, F)$ process with unknown intensity c and unknown df F.

The intersections with the x, y–plane, that is, the centers and radii of the circles, constitute a marked homogeneous Poisson$(2cH(0)\lambda^2, W(F))$ process, where $W(F)$ is the Wicksell transformation of F and $H(0)$ is the constant in (18.2). The new marked Poisson process is derived from the initial one by

- a truncation procedure, because only spheres contribute to the new process that intersect the plane,

- a transformation, due to the fact that circle coordinates instead of the original sphere coordinates are observed.

In practice, the surface area is a subset A in the x, y–plane with finite measure $\lambda^2(A) < \infty$. In addition, to analyze the size distribution of the circles, respectively, of the spheres, one merely has to consider the marginal Poisson process of circle radii. Under the postulated homogeneity, the circle centers provide no information about the radius df.

The intensity measure ν of the marginal Poisson process pertaining to the circle radii is given by the mean value representation

$$
\begin{aligned}
\nu(s, \infty) &= 2c\lambda^2(A)H(0)\overline{W(F)}(s) \\
&= 2c\lambda^2(A) \int_0^\infty \bar{F}(\sqrt{s^2 + z^2})\, dz, \qquad (18.6)
\end{aligned}
$$

where the latter equation follows from (18.2).

Hence, the circle radii may be treated as a random number N of independent and identically distributed random variables $S_1, S_2, \ldots$ with common df $W(F)$, where N is a Poisson random variable with parameter $2c\lambda^2(A)H(0)$ which is independent of the S_i.

The Art of Stereological Extreme Value Analysis

In the sequel, we study the relationship between the tails of radius distributions of spheres and circles. The modeling of the upper tail of the observable circle radii by means of GP dfs can regarded as business as usual. Our primary aim is to find a theoretical justification for a GP modeling for the sphere radii. This will be done by comparing

- max/pot–domains of attractions,

- the validity of certain von Mises conditions.

In addition, a counterexample shows that a GP upper tail for the circle radii does not necessarily justify such a modeling for the sphere radii.

Max–Domains of Attraction

We impose conditions on F which imply that the Wicksell transformation $W(F)$ belongs to the max–domain of attraction of an EV df G; in short $W(F) \in \mathcal{D}(G)$.

First we provide some details and a proof in the special case of Fréchet dfs $G_{1,\alpha}$. Assume that $F \in \mathcal{D}(G_{1,\alpha})$. From Section 1.3 we know that this condition holds if, and only if, the survivor function $\bar{F}$ of the sphere radius has a regularly varying tail with index $-\alpha < 0$; that is, for each $t > 0$ we have $\bar{F}(st)/\bar{F}(s) \to t^{-\alpha}$ as $s \to \infty$.

By substitution and employing the letter property, one gets

$$
\int_0^\infty \bar{F}(\sqrt{s^2 + z^2})\, dz = s \int_0^\infty \bar{F}(s\sqrt{1 + y^2})\, dy
$$

$$
\sim s\bar{F}(s) \int_0^\infty (1 + y^2)^{-\alpha/2}\, dy
$$

as $s \to \infty$, if $\alpha > 1$ (otherwise, the latter integral is infinite). Combining this with (18.2) one gets that $W(F)$ is a regularly varying function with index $1 - \alpha$. Hence, $W(F) \in \mathcal{D}(G_{1,\alpha-1})$, if $\alpha > 1$.

Such questions were treated in the article by Dress and Reiss, cited on page 33, in greater generality, with the preceding result as a special case. We have $W(F) \in \mathcal{D}(G_{i,\beta(\alpha)})$, if $F \in \mathcal{D}(G_{i,\alpha})$, for $i = 1, 2$, with the shape parameters $\beta(\alpha)$ given by

$$
\beta(\alpha) = \begin{cases} \alpha - 1 & i = 1,\ \alpha > 1, \\ & \text{if} \\ \alpha - 1/2 & i = 2,\ \alpha < 0. \end{cases} \tag{18.7}
$$

In the Gumbel case we have $W(F) \in \mathcal{D}(G_0)$, if $F \in \mathcal{D}(G_0)$. It is remarkable that we do not get any result for $W(F)$ with $-1/2 \le \beta(\alpha) < 0$.

Von Mises Conditions

If the df F has a density f which is positive in a left neighborhood of $\omega(F)$, then the following von Mises conditions are sufficient for F to belong to the domain of attraction of an EV df $G_{i,\alpha}$. We assume that

$$M(0): \qquad \int_t^{\omega(F)} (1 - F(u))\,du < \infty \text{ for some } t < \omega(F),$$

$$\lim_{t\uparrow\omega(F)} f(t) \int_t^{\omega(F)} (1 - F(u))\,du/(1 - F(t))^2 = 1;$$

$$M(1,\alpha): \quad \omega(F) = \infty, \quad \lim_{t\uparrow\infty} tf(t)/(1 - F(t)) = \alpha;$$

$$M(2,\alpha): \quad \omega(F) < \infty, \quad \lim_{t\uparrow\omega(F)} ((\omega(F) - t)f(t)/(1 - F(t)) = -\alpha.$$

We also write $M(0, \alpha)$ instead of $M(0)$. For a different von Mises condition, which involves the 1st derivative of the density f, we refer to (2.32).

If F satisfies condition $M(i, \alpha)$, then $F \in \mathcal{D}(G_{i,\alpha})$. The preceding conditions also entail the convergence in terms of the variational and Hellinger distances, see [42]. Under the restrictions on α given in (18.7), it was proved by Drees and Reiss that $W(F)$ fulfills $M(i, \beta(\alpha))$, if F fulfills $M(i, \alpha)$.

Most important, from our viewpoint, is the following converse result: if $\beta(\alpha) < -1/2$ or $\beta(\alpha) > 0$ and $W(F)$ fulfills $M(i, \beta(\alpha))$, then F is in the domain of attraction of $G_{i,\alpha}$. Therefore, if we come to the conclusion that the statistical modeling of circle radii by means of a GP df with the shape parameter $\beta(\alpha) < -1/2$ or $\beta(\alpha) > 0$ is appropriate, then we may assume that the upper tail of F is of GP type with shape parameter α.

The preceding results were extended to ellipsoidal particles by Hlubinka[8], where the particles (except the positions) are determined by a bivariate random vector.

A Counterexample

We focus our attention on the case $i = 2$ and $\beta(\alpha) = -1/2$. We provide an example of a Wicksell transformation $W(F)$, where $W(F)$ fulfills the von Mises condition $M(2, -1/2)$, yet F is not in the max–domain of attraction of an EV df. More precisely, we show that $W(F)$ satisfies a Weiss condition (belongs to a δ-neighborhood of a GP df), yet F is a df with the total mass concentrated in a single point (and, therefore, does not belong to the domain of attraction of an EV

---

[8]Hlubinka, D. (2003). Stereology of extremes; shape factor of spheroids. *Extremes* 6, 5–24.

df). Therefore, it is not self–evident to employ a GP modeling for the upper tail of F, if such a modeling is accepted for $W(F)$.

If the size of the inclusions is fix, say $s_0 > 0$, then one can deduce the density of the Wicksell transformation $W(F)$ from (18.5). Denote again by $w(F)$ the density of $W(F)$. We have

$$w(F)(s) \;=\; \frac{h(s)}{H(0)} \;=\; \frac{s}{s_0 \sqrt{s_0^2 - s^2}}$$

$$=\; w_{2,-1/2,\mu,\sigma}(s)\Big(1 + O\big(\overline{W}_{2,-1/2,\mu,\sigma}(s)\big)^2\Big) \qquad (18.8)$$

for $0 < s < s_0$, where $\mu = s_0$ and $\sigma = s_0/2$.

Wicksell Transformation and Exceedance Dfs

There is a nice relationship

$$W(F)^{[u]} = W(F^{[u]})^{[u]} \qquad (18.9)$$

between exceedance dfs of F and the pertaining Wicksell transformations $W(F)$, due to Anderson and Coles[9]. The proof is straightforward; we have

$$W(F)^{[u]}(s) \;=\; 1 - \frac{H(s)}{H(u)}$$

$$=\; 1 - \frac{\int_s^\infty \sqrt{r^2 - s^2}\, dF(r)}{\int_u^\infty \sqrt{r^2 - u^2}\, dF(r)}$$

$$=\; 1 - \frac{\int_s^\infty \sqrt{r^2 - s^2}\, dF^{[u]}(r)}{\int_u^\infty \sqrt{r^2 - u^2}\, dF^{[u]}(r)}$$

$$=\; W(F^{[u]})^{[u]}(s), \qquad s \geq u.$$

This result was applied by Anderson and Coles to estimate the upper tail of the initial df F based on exceedances of circle radii (see below).

Statistical Inference

In the article by Drees and Reiss, the estimation of the initial df F was treated in several steps. Based on the exceedances of circle radii over some threshold, a GP density with shape parameter $\hat{\beta}$ is fitted to the upper tail of the circle radius density $w(F)$. This tail estimator is combined with a nonparametric estimator of the central part of $w(F)$ by the piecing together method. In a second step, an estimate $\hat{f}$ of the sphere radius density f is computed by applying a smoothed

---

[9] Anderson, C.W. and Coles, S.G. (2002). The largest inclusions in a piece of steel. Extremes 5, 237–252.

expectation maximization (EMS) algorithm. Notice that the EMS algorithm is one of the common methods within the framework of ill–posed inverse problems such as the Wicksell corpuscle problem. Finally, the upper tail of the estimated density $\hat{f}$ is replaced by a GP density with shape parameter $\hat{\alpha}$ as established in (18.7). Likewise, one could generally investigate extremes in inverse problems.

An alternative estimation procedure is employed by Anderson and Coles. Again one assumes that a GP modeling for the upper tail of F is valid. In the subsequent lines, we make use of the γ–parametrization of GP dfs because then the left endpoint of the support is just the location parameter. Replacing $F^{[u]}$ by a GP df $W_{\gamma,u,\sigma}$ on the right–hand side of (18.9) one gets the truncated Wicksell transformation $W(W_{\gamma,u,\sigma})^{[u]}$. Notice that this df is the exceedance df for the pertaining circle radii above the threshold u. Moreover, this df has the density

$$w(W_{\gamma,u,\sigma})^{[u]}(s) = \frac{s \int_s^\infty (r^2 - s^2)^{-1/2} \, dW_{\gamma,u,\sigma}(r)}{\int_u^\infty \sqrt{r^2 - u^2} \, dW_{\gamma,u,\sigma}(r)}, \qquad s \geq u.$$

Now estimate the parameters γ and σ by applying the ML method to the circle radii exceedances.

Additional Literature on Maximum Inclusion Sizes

Applied work about the "maximum size of inclusions" started with articles by Murakami and Usuki[10] and Murakami et al.[11]. In that context, we also refer to a series of articles by Shi, Atkinson, Sellars and Anderson; see, e.g., the article by J.R. Yates et al.[12] and the references therein.

Further results about stereological extremes may be found in articles by Takahashi and Sibuya[13]. Notable is also a recent, theoretically oriented review about stereological particle analysis by Kötzer[14] which includes a longer list of references to the stereological literature.

---

[10]Murakami, Y. and Usuki, H. (1989). Quantitative evaluation of effects of non–metallic inclusions on fatigue strength of high strength steel II: fatigue limit evaluation based on statistics for extreme values of inclusion size. Int. J. Fatigue 11, 299–307.

[11]Murakami, Y., Uemura, Y. and Kawakami, K. (1989). Some problems in the application of statistics extreme values to the estimation of the maximum size of non–metallic inclusions in metals. Transactions Japan Soc. Mechan. Engineering 55, 58–62.

[12]Yates, J.R., Shi, G., Atkinson, H.V., Sellars, C.M. and Anderson, C.W. (2002). Fatigue tolerant design of steel components based on the size of large inclusions. Fatigue Fract. Engng. Mater. Struct. 25, 667–676.

[13]Takahashi, R. and Sibuya, M. (1996). The maximum size of the planar sections of random sheres and its application to metallurgy. Ann. Inst. Statist. Math. 48, 127–144, and Takahashi, R. and Sibuya, M. (1998). Prediction of the maximum size in Wicksell's corpuscle problem. Ann. Inst. Statist. Math. 50, 361–377.

[14]Kötzer, S. (2006). Geometric identities in stereological particle analysis. Image Anal. Stereol. 25, 63–74.

Chapter 19

Life Science

co–authored by E. Kaufmann[1]

In Section 19.1 we deal with the question whether human life spans are limited or unlimited. The celebrated Gompertz law will be central for our considerations. We particularly apply the results of Section 6.5, concerning penultimate distributions, to this question. Section 19.2 concerns the prediction of life tables by adopting a regression approach.

19.1 About the Longevity of Humans

We discuss the question whether the right endpoint of the life span df F of a population is infinite, in other words, that the life span is unlimited. This is primarily done within the framework of extreme value analysis, yet we also fit Gompertz and logistic distributions to extreme life spans.

Gompertz and Logistic Laws

The Gompertz distribution is the classical life span distribution. The Gompertz law (called Gompertz–Makeham formula by Gumbel) postulates a hazard (mortality) rate $h(x) = ae^{bx}$.

Recall from page 54 that the converse Gumbel df $\widetilde{G}_{0,\mu,\sigma}$, called Gompertz df, is the only df which satisfies the Gompertz law with $\mu = (\log(b/a)/b$ and $\sigma = 1/b$. A theoretical explanation for the adequacy of the Gompertz distribution, based on the cellular aging phenomena of limited replicability and deceleration of the

---

[1]University of Siegen.

process of mitosis, may be found in an article by Abernethy[2].

EXAMPLE 19.1.1. (Extreme Life Spans.) Our analysis concerns extreme life spans of women born before and around the year 1900 and later living in West Germany.

The given data are the ages of female persons at death in West–Germany in the year 1993[3]. None of the persons died at an age older than 111 years. The analysis is based on the ages at death of 90 or older.

TABLE 19.1. Frequencies of life spans (in years).

life span	90	91	92	93	94	95	96	97	98	99	100
frequency	12079	10273	8349	6449	5221	3871	2880	1987	1383	940	579
life span	101	102	103	104	105	106	107	108	109	110	111
frequency	340	207	95	63	36	16	9	4	1	1	1

Thus, our analysis is based on a subgroup of our target population described at the beginning. Those who died before 1993 and were alive after 1993 are not included in the study. Yet, the extreme life spans within the subgroup are governed by the same distribution as those of the entire population under the condition that the birth rate was nearly homogeneous during the relevant period. This idea can be made rigorous by a thinning argument within the point process framework, see [43], pages 68 and 69. For that purpose, consider uniformly distributed dates of birth V_i over the relevant time period and add life spans X_i. The thinning random variable is defined by $U_i = I(1993 \leq V_i + X_i < 1994)$.

A Gompertz density with location and scale parameters $\mu = 83.0$ and $\sigma = 9.8$ fits well to the histogram of the life span data above 95 years, see Fig. 19.1.

If the Gompertz df is accepted, then life span is unlimited. This is in agreement with recent experiments which showed (see, e.g., Carey et al.[4] about Mediterranean fruit flies and the article by Kolata[5]) that the average life span and the oldest ages strongly depend on the living conditions.

The insight gained from experiments and from a series of data sets may be summarized by the statement that "the studies are consistently failing to show any evidence for a pre–programmed limit to human life span," as nicely expressed

---

[2]Abernethy, J. (1998). Gompertzian mortality originates in the winding–down of the mitotic clock. J. Theoretical Biology 192, 419–435.

[3]Stored in the file um–lspge.dat (communicated by G.R. Heer, Federal Statistical Office).

[4]Carey, J.R., Liedo, P., Orozco, D. and Vaupel, J.W. (1992). Slowing of mortality rates at older ages in large medfly cohorts. Science 258, 457–461.

[5]Kolata, G., New views on life spans alter forecasts on elderly. The New York Times, Nov. 16, 1992.

by J.W. Vaupel (cited by Kolata).

Thatcher et al.[6] suggest to fit a logistic distribution to ages at death between 80 and 120. Such a modeling is very natural in view of the change in the population which can be observed around 95 years, and the fact that the logistic distribution is a certain mixture of Gompertz distributions with respect to a gamma distribution, cf. (4.11). The hazard function of a (converse) logistic df is given by

$$h(x) = ae^{bx}/(1 + \alpha e^{bx})$$

with $b = 1/\sigma$, $\alpha = \beta$ and $a = r\beta/\sigma$.

It was observed by Kannisto[7] and Himes et al.[8] that there is already a close fit to data in the special case where $a = \alpha$ which yields $r = \sigma =: s$. These distributions can be represented by the dfs

$$F_{s,\mu}(x) = 1 - (1 + \exp((x - \mu)/s))^{-s}$$

which have the densities

$$f_{s,\mu}(x) = \exp((x - \mu)/s)/(1 + \exp((x - \mu)/s))^{1+s},$$

where μ is a location parameter. In Fig. 19.1 (right) we give an illustration based on the same data as in Example 19.1.1.

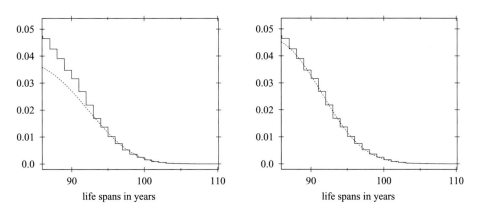

FIG. 19.1. Histogram of life span data and (left) Gompertz density with parameters $\mu = 83$, $\sigma = 9.8$, (right) logistic density $f_{s,\mu}$ with parameters $s = 7.25$, $\mu = 98.6$.

[6]Thatcher, A.R., Kannisto, V. and Vaupel, J.W.. (1998). The Force of Mortality at Ages 80 to 120. Odense University Press.

[7]Kannisto, V. (1992). Presentation at a workshop on old age mortality at Odense University.

[8]Himes, C.L., Preston, S.H. and Condran, G.A. (1994). A relational model of mortality at older ages in low mortality countries. Population Studies 48, 269–291.

For a logistic df there is again an infinite upper endpoint. In addition, these dfs—just as the Gompertz dfs—belong to the pot domain of attraction of the exponential df which is a special case of a generalized Pareto (GP) df. In the subsequent lines, the statistical inference is done in the GP model.

Analyzing the Upper Tail of the Life Span Distribution Within the Generalized Pareto Model

Aarssen and de Haan[9] applied the standard pot–method to Dutch life span data (stored in um–lspdu.dat) and estimated a negative shape parameter. This yields that the estimated generalized Pareto (GP) df has a finite upper limit. These authors go one step further and compute confidence intervals for the upper limit.

In the subsequent example we also make use of the standard pot–method.

EXAMPLE 19.1.2. (Continuation of Example 19.1.1.) The sample mean excess function is close to a straight line for life spans exceeding 95 years; there is a stronger deviation from this line below 95.

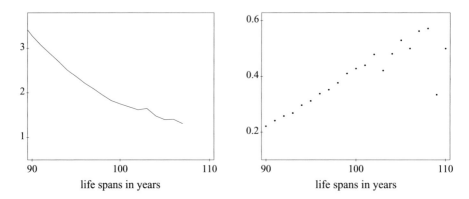

FIG. 19.2. (left.) Sample mean excess function. (right.) Sample hazard function.

The form of the mean excess function supports the conjecture that there is a long–living subpopulation which becomes visible beyond the age of 95 years (cf., e.g., the article of Perls (1995)). To a larger extent, our results merely concern this subpopulation.

It is clear that a beta df will be estimated within the GP model. Guided by simulations, we decided to base the estimation on 2500 extremes (just life spans ≥ 98 are taken). The estimated shape parameter γ within the generalized Pareto (GP) model

---

[9]Aarssen, K. and de Haan, L. (1994). On the maximal life span of humans. Mathematical Population Studies 4, 259–281.

is equal to -0.08. The right endpoint of the estimated beta distribution is equal to 122 years. This estimate fits well to the worldwide reported life span of Mrs Jeanne Calment (122 years and 164 days) who was born in Arles (France) on Feb. 21, 1875 and died in Arles on Aug. 4, 1997.

The estimation of a negative shape parameter is in accordance with the result by Aarssen and de Haan. In contrast to these authors we do not conclude that human life span has a finite upper limit. Therefore, we also make no attempt to estimate such an upper limit in accordance with Gumbel[10] who wisely refused to discuss this question. He suggested to estimate the "return value" $F^{-1}(1 - 1/n)$ or the expected value $E(X_{n:n})$ of the maximum $X_{n:n}$ of a sample of size n .

Using the concept of penultimate distributions one can show that an infinite upper limit is well compatible with extreme value theory although we estimated a beta distribution. Another argument which speaks in favor of the exponential df can be found in an article by Galambos and Macri[11].

Gompertz Law and Penultimate Approximation

In Fig. 19.1 a Gompertz df $\widetilde{G}_{0,\mu,\sigma}$ with location and scale parameters $\mu = 83$ and $\sigma = 9.8$ was fitted to the upper tail of the given data of life spans. The following considerations are made under the hypothesis that this Gompertz df is the actual life span df. We still utilize our standard modeling, that is, actual truncated dfs are replaced by GP dfs. For that purpose we compute the parameters of ultimate and penultimate GP dfs according to the theoretical results in Section 6.5.

The Gompertz df is truncated left of the threshold $u = 98$. According to (6.44) and (6.45), the approximating exponential df has the location parameter $u = 98$ and the scale parameter $\sigma(u) = (1 - \widetilde{G}_{0,\mu,\sigma}(u))/\tilde{g}_{0,\mu,\sigma}(u) \approx 2.12$.

The penultimate approach, see (6.47) and (6.48), leads to a beta df with the same location and scale parameters as in the ultimate case, and the shape parameter

$$\gamma(u) = \eta(1 - \widetilde{G}_{0,\mu,\sigma}(u)) = \big((1 - \widetilde{G}_{0,\mu,\sigma})/\tilde{g}_{0,\mu,\sigma}\big)'(u) \approx -0.22.$$

Plots of the densities show that the beta density is much closer to the truncated Gompertz density than the exponential density. Therefore, estimates based on exceedances of Gompertz data point to beta (GP2) dfs when the statistical inference is carried out in the GP model. Yet, estimating a beta df within the unified GP model does not necessarily yield that the actual upper endpoint (thus, the upper limit of life spans) is finite.

---

[10]Gumbel, E.J. (1933). Das Alter des Methusalem. Z. Schweizerische Statistik und Volkswirtschaft 69, 516–530.

[11]Galambos, J. and Macri, N. (2000). The life length of humans does not have a limit. J. Appl. Statist. Science.

In the case of a Gompertz df, the upper endpoint of the penultimate beta df is given by $u+\sigma$, which is equal to 107.8 for the parameters found above. This reveals that an approximation, measured in terms of the Hellinger distance, is not very sensitive to higher extremes. One may also fit a penultimate df to the truncation of the Gompertz df beyond the threshold $u = 111$, the observed maximum life span in our example. This procedure is related to the approximation of Gompertz maxima by a penultimate EV df. Applying (6.47) again one gets a penultimate beta df with shape parameter $\gamma(111) \approx -0.06$ and a right endpoint of 120.8.

The conclusion is that the penultimate approach leads to significantly higher accuracy than the approach by limiting dfs. In addition, the penultimate approximations reveal that a Gompertz hypothesis is compatible with estimated beta dfs within the framework of GP dfs.

19.2   Extrapolating Life Tables To Extreme Life Spans: A Regression Approach

Based on eleven German life tables from 1871 to 1986, we study the change of mortality rates of women over the time for fix ages and apply a regression approach to predict life tables. In a second step, we extrapolate the tail of the life span distribution by fitting a Gompertz distribution.

Introduction

Life tables are used by actuaries to calculate the premium of an insurance policy (life or health insurance, rental scheme etc.). The basic terms of interest are the mortality rates that will be separately estimated for both sexes and each age up to 100. In the 1986 life table, the estimator of the mortality rate of a x years old women is based on female babies born around the year $1986 - x$. Because people become older and older due to better standard of life, health care etc., the mortality rates of babies born today are overestimated.

In the following we study mortality rates of women over the time for fix ages and apply a regression approach to forecast life tables. In a second step, we extrapolate the tail of the life span distribution by fitting a Gompertz distribution. For a general overview concerning the theory of life spans we refer to L.A. Gavrilov and N.S. Gavrilova[12].

For a women of x years born within the year $t - x$ define the mortality rate $q_{x,t}$ as the (conditional) probability of death within one year. Denote by F_{t-x} the (continuous) life span distribution of a women born in the year $t - x$. Then, the

---

[12]Gavrilov, L.A. and Gavrilova, N.S. (1991). The Biology of Life Span: A Quantitative Approach. Harwood Academic Publishers, Chur (Russian edition by Nauka, Moscow (1986)).

theoretical mortality rate is

$$q_{x,t} := \frac{F_{t-x}(x+1) - F_{t-x}(x)}{1 - F_{t-x}(x)}, \qquad x = 0, 1, 2, \ldots \tag{19.1}$$

In the statistical literature, the term mortality rate is often used as a synonym for hazard rate. In our considerations these terms have a different meaning. The link between them will be given later.

A natural estimate of $q_{x,t}$ is given by the empirical estimate $\hat{q}_{x,t}$ which is defined by the number of women born in $t - x$ and dying at age between x and $x + 1$ divided by the total number of women born in $t - x$ exceeding an age of x years. In practice, one takes a small number of years to calculate the estimates.

The life span distribution of a women born within the year t can be derived from the mortality rates in a simple manner. We have

$$1 - F_t(x) = \prod_{j=0}^{x-1} \frac{1 - F_t(j+1)}{1 - F_t(j)} = \prod_{j=0}^{x-1}(1 - q_{j,t+j}), \qquad x = 0, 1, 2, \ldots \tag{19.2}$$

Because the calculation of an individual premium is based on the unknown mortality rates in the future one has to find a prediction procedure.

The Data

Our investigations are based on eleven German life tables[13] (consisting of estimated mortality rates $\hat{q}_{x,t}$) from the year 1871 to 1986 dealing with ages up to 100. In Table 19.2 we list the mortality rates $\hat{q}_{x,t}$ of women at ages $x = 90, 100$.

TABLE 19.2. Mortality rates $\hat{q}_{x,t}$ for women at ages $x = 90, 100$ taken from life tables at the years t.

t	1871	1881	1891	1901	1910	1924	1932	1949	1960	1970	1986
$\hat{q}_{90,t}$	.314	.306	.302	.296	.302	.263	.274	.259	.248	.234	.187
$\hat{q}_{100,t}$	.518	.447	.446	.420	.476	.402	.476	415	.380	.405	.382

It should be noted that the life tables differ from each other in methodology. The main differences concern the method of estimating mortality rates, smoothing techniques, the handling of special influences on mortality like the wave of influenza in 1969/70, and the number of years a life table is based on (e.g., 10 years (1871) and 3 years (1986)). In the following lines we ignore these differences and assume that the life tables consist of the empirical estimates as specified above.

---

[13]Stored in the files lm–lifem.dat for men and lm–lifew.dat for women (communicated by Winfried Hammes, Federal Statistical Office).

Regression over Time

Scatterplots of the mortality rates $\hat{q}_{x,t}$, for each fixed age x plotted against the years t, indicate a decrease of exponential or algebraic order (see also Table 1). We assume a polynomial regression model for the log–mortality rates $\log \hat{q}_{x,t}$ for fixed age x. Least squares polynomials of degree 2 fit particularly well to the data.

We have

$$\log \hat{q}_{x,t} \approx \hat{p}_{x,0} + \hat{p}_{x,1} t + \hat{p}_{x,2} t^2,$$

where $\hat{p}_{x,0}$, $\hat{p}_{x,1}$ and $\hat{p}_{x,2}$ are the estimated parameters.

The following illustrations show plots of the log–mortality rates $\log \hat{q}_{x,t}$ as functions in t and the fitted least squares polynomials of degree 2.

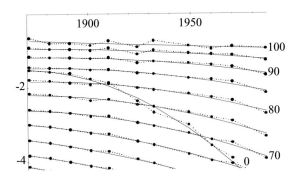

FIG. 19.3. Fitting least squares polynomials of degree 2 (solid lines) to log–transformed mortality rates (dotted).

It is obvious from Fig. 19.2 that linear regression curves—as used by Bomsdorf and Trimborn[14] and Lühr[15] do not adequately fit the transformed data. The larger reduction of mortality from the 1970 to the 1986 life table, which is also observed in newer abriged life tables[16] may serve as another argument for the non–linearity of the log–mortality rates. We remark that the mortality itself decreases with a slower rate for higher ages (in contrast to the log–mortality).

With the estimated polynomials p_x we estimate the theoretical mortality rates $q_{x,t}$ and the life span distribution F_t that is related to a cohort life table for women born in the year t.

---

[14]Bomsdorf, E. and Trimborn M. (1992). Sterbetafel 2000. Modellrechnungen der Sterbetafel. Zeitschrift für die gesamte Versicherungswissenschaft 81, 457–485.

[15]Lühr, K.–H. (1986). Neue Sterbetafeln für die Rentenversicherung. Blätter DGVM XVII, 485–513.

[16]Schmithals, B. and Schütz, E.U. (1995). Herleitung der DAV–Sterbetafel 1994 R für Rentenversicherungen (English title: Development of table DAV 1994 R for annuities). Blätter DGVM XXII, 29–69.

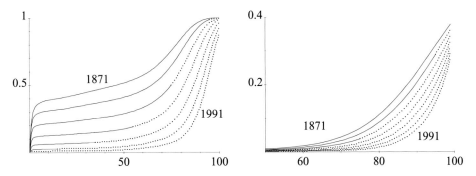

FIG. 19.4. Estimated life span distributions (left) and estimated mortality rates (right) by interpolation (solid) and extrapolation (dotted) for women born in 1871, 1891, ..., 1991.

Estimated life span distributions by interpolation means that the empirical estimators of the mortality rates are calculated within the period from 1871 to 1986 by interpolation.

Regression of Extreme Life Spans

In the following we are interested in extreme life spans. We apply again a regression approach to fit a parametric distribution to the mortality rates $\hat{q}_{x,t}$ of ages x exceeding 90 (also see Section 19.1). Note that within the life tables different smoothing techniques in the center as well as in the tail were applied. We assume that after the regression these manipulations are of no consequence for the general structure of the extreme life spans.

Recall the relation between the theoretical mortality rates $q_{x,t}$ as defined in (19.1) and the hazard rate. The latter one is defined by

$$h_t(x) := \frac{f_t(x)}{1 - F_t(x)}$$

where f_t denotes the density of F_t. Then

$$q_{x,t+x} = 1 - \frac{1 - F_t(x + 1)}{1 - F_t(x)} = 1 - \exp\left(-\int_x^{x+1} h_t(s)\, ds\right)$$

and, thus,

$$\int_x^{x+1} h_t(s)\, ds = -\log(1 - q_{x,t+x}).$$

To find a reasonable distribution that fits to the upper tail of the life span distribution, we apply again visualization techniques. For $90 \le x \le 100$, a scatter-plot of points $(x, \log(-\log(1 - q_{x,t+x})))$ is close to a straight line. Notice that for a Gompertz distribution one has the hazard rate $h_t(x) = \exp((t - \mu_t)/\sigma_t)/\sigma_t$ and, thus,

$$\log\left(\int_x^{x+1} h_t(s)\, ds\right) = \frac{x}{\sigma_t} - \frac{\mu_t}{\sigma_t} + \log\left(\exp\left(\frac{1}{\sigma_t}\right) - 1\right)$$

is a straight line. We fit a Gompertz distribution to the upper tail by the least square linear regression

$$\log\left(-\log(1 - \hat{q}_{x,t})\right) \approx \hat{a}_t x + \hat{b}_t, \qquad x = 90, \ldots, 100 \tag{19.3}$$

and obtain mortality rates $\hat{q}_{x,t}$ for women of 101 years and older by extrapolation in (19.3) and the formula

$$\hat{q}_{x,t} = 1 - \exp\left(-\exp(\hat{a}_t x + \hat{b}_t)\right), \qquad x = 101, 102, 103, \ldots$$

The estimated mortality rates and the corresponding life span distribution functions for high ages are plotted in Fig. 19.5.

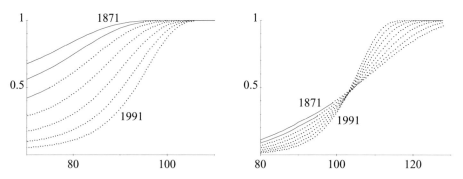

FIG. 19.5. Estimated life span distributions (left) and estimated mortality rates (right) with fitted Gompertz tail by interpolation (solid) and extrapolation (dotted) for women born in the years $1871, 1891, \ldots, 1991$.

We include tabulated values of three estimated survivor functions $\bar{F}_t = 1 - F_t$ for $t = 1901, 1921, 1991$ given in Fig. 19.5 (left). Notice that the women born in 1901 achieved the age 96 in the year of writing this article[17]. We see, for example, that the (estimated) probability of survival at the age of 96 years increases from .018 to .046 for the women born in the years 1901 and 1921.

We do not believe that a reliable forecast can be given for longer periods by using such a straightforward extrapolation technique. Forecasts for more than one or two decades merely indicate the theoretical consequences of our modeling.

---

[17]Reference is given to Kaufmann, E. and Hillgärtner, C. (1997). Extrapolating life tables to extreme life spans: a regression approach using XPL. In: 1st ed. of this book, Part V, pages 245–250.

TABLE 19.3. Estimated probability of survival $\bar{F}_t := 1 - F_t$ for women born in the years $t = 1901, 1921, 1991$.

age x	90	92	94	96	98	100	102	104	106	108
$\bar{F}_{1901}(x)$	.088	.057	.034	.018	.009	.004	.001	<.001	<.001	<.001
$\bar{F}_{1921}(x)$	.174	.121	.078	.046	.024	.011	.004	.001	<.001	<.001
$\bar{F}_{1991}(x)$	.664	.573	.466	.349	.232	.131	.059	.019	.004	<.001

The behavior of the mortality rates in Fig. 19.5 (right) is somewhat surprising. Primarily, it is not a result of the Gompertz (or converse Weibull) fit or the regression approach of degree 2 instead of degree 1. It rather depends on the weaker trend of reduction of mortality for elderly people. Fig. 19.5 (left) seems to confirm the *ecological crisis hypothesis* as reviewed (and rejected) in the above–mentioned book by Gavrilov and Gavrilova that states a biological limit for human life spans between 100 and 110 years. However, for a more detailed study it is essential, even in view of the existence of a long–living subpopulation visible behind an age of 95 years, cf. Perls[18] and Section 19.1, to take empirical mortality rates of women exceeding 100 years into account.

Notice that it is also possible to fit a converse Weibull distribution. In this case the hazard rate is of the form at^b. However, one obtains analogous illustrations for the converse Weibull distribution.

---

[18]Perls, T.T. (1995). Vitale Hochbetagte. Spektrum der Wissenschaft, März 3, 72–77.

Appendix:

First Steps towards Xtremes and StatPascal

Appendix A

The Menu System

A.1 Installation

Xtremes is a genuine Windows application so that any system running Windows 98 (and newer) or NT 4 (and newer) is also capable of executing Xtremes.

To facilitate the installation, the CD–ROM contains an installation program that copies all required files to your computer. The installation program starts automatically when you insert the CD–ROM. One may choose an installation directory and select certain optional parts of the system. If you are unsure, we recommend to accept the default options. Please make sure that you are logged in as an administrator when installing under Windows NT/2000/XP/2003 or Windows Vista 32–Bit.

Xtremes is deinstalled in the usual manner; i.e., by executing the option *Software* in the Windows Control Panel. Please consult the *Xtremes User Manual* for further information; a pdf version is accessible from its menu entry in the Xtremes section of the Windows start menu.

A.2 Overview and the Hierarchy

Xtremes is a statistical software system that possesses graphics facilities, a facility to generate and load data, an arsenal of diagnostic tools and statistical procedures, and a numerical part for Monte–Carlo simulations.

The development of Xtremes has been supported by Simon Budig (User-Formula facility), Martin Elsner (HTML help), Andreas Gaumann (help system), Sylvia Haßmann (censored data), Andreas Heimel (help system, ARMA estimators), Jens Olejak (minimum distance estimators), Wolfgang Merzenich (consultation on the StatPascal compiler), Reinhard Pfau (Linux port), Torsten Spillmann (XGPL plots), Karsten Tambor (early version of the multivatiate mode), and Arthur Böshans, Carsten Wehn, Lars Fischer, Ralf Pollnow (MS Excel frontend) whose help is gratefully acknowledged.

The illustration on the right–hand side shows the hierarchy of the Xtremes system:

- Univariate and Multivariate Modes: the system is partitioned into a univariate and a multivariate mode that can be selected in the toolbar.

- Domains D(ISCRETE), SUM, MAX and POT: select certain domains in the toolbar which correspond to different parametric models (discrete, Gaussian, extreme value (EV) or generalized Pareto (GP)) built for discrete data and data that are sums, maxima or exceedances (peaks–over–threshold).

- The Menu–Bar: select menus for handling and visualizing data, plotting distributions by analytical curves, estimating parameters etc. The *Visualize* menu and larger parts of the *Data* menu are independent of the different domains.

- Menus: commands of the system are available in different menus.

- Dialog Boxes: if a menu command requires further parameters, a dialog box is displayed.

- Graphics Windows: the result of a menu command is usually displayed in a graphics window.

- Local Menus: options and commands that are specific to a particular window or dialog box are provided by means of a local menu (available with a right–click somewhere in the window or the dialog box).

- Special Facilities: in the toolbar, select tools to manipulate a graphic. These tools change the action taking place when one drags the mouse in a graphics window. For example, to change the coordinate system, you may click on the coordinate tool ⊞ in the toolbar and, then, pull up a rectangle.

- Help System: a context–sensitive online help is available by pressing the F1–key at any time besides the general help facility (option *Index* in the *Help* menu).

- ACT, $, HYD Supplements: special options for insurance, finance and hydrology (Chapters 14–17) are available.

- UFO: select the UserFormula facility to enter your own formulas to plot curves, generate or to transform data (cf. Section A.5).

- STATPASCAL: the integrated programming language StatPascal is activated by means of the SP button (cf. Appendix B and StatPascal User Manual).

Keep in mind that the options in the *Visualize* and *Estimator* menus must be applied to an active data set (read from a file or generated in the *Data* menu). Instructions about the installation of the system are given in Section A.1.

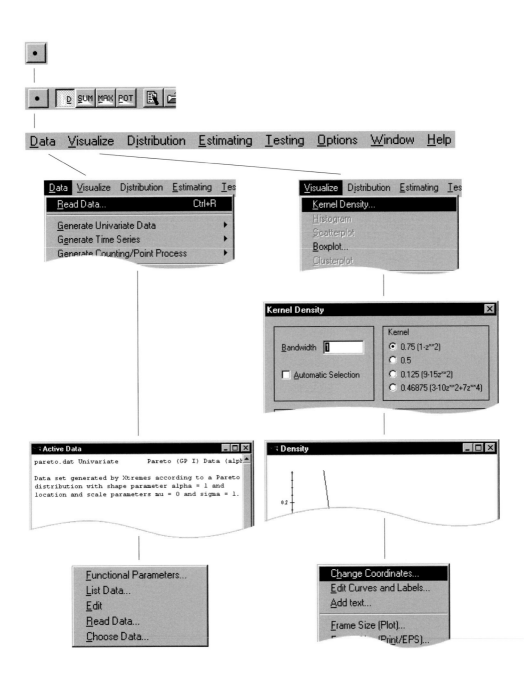

Professional Version

For details concerning the professional version of Xtremes we refer to the web–site *http://www.stat.math.uni-siegen.de/xtremes/.*

Under this web–site one is also informed about the latest news about Xtremes and possible updates.

A.3   Becoming Acquainted with the Menu System

A further possibility is to work closely with the online–help facility (press the F1–key or enter the *Help... Index*).

Plotting Histograms and Curves

Xtremes provides easy–to–use plotting facilities. The user can quickly plot a curve, adjust the coordinate system, get a list of the curves displayed in a plot window or export the picture via the clipboard. A special facility is available for producing EPS–files (see Documenting Illustrations on page 478).

We partly repeat the operations which are required to plot the histograms in Fig. 1.2. Make sure that the univariate D(iscrete) domain is active, that is, the first button in the toolbar shows a single bullet ▪ and the button labeled ℓ is pressed.

DEMO A.1. (Plotting Poisson Histograms.) (a) Select the option *Distribution... Poisson* in the univariate D(iscrete) domain and choose the parameter $\lambda = 10$. Execute the *Histogram* option.
(b) The tool (activating the option mouse mode) may be employed to adjust the positions of the plotted bars.

The user may change the parameters of a plotted histogram (or of a curve) by means of sliders which are an indispensable tool to work interactively with data. Select the parameter varying mouse mode from the toolbar and click onto a histogram or curve in a plot window to open a window with sliders for each parameter.

DEMO A.2. (Varying the Parameter of a Poisson Histogram.) First, plot a Poisson histogram with parameter $\lambda = 1$. Select the parameter varying mouse mode from the toolbar and click on the histogram of the Poisson distribution. Vary the parameter λ, and also change the boundaries for the possible parameter values. In Fig. A.1 there is a Poisson histogram with the pertaining slider and, in addition, a sample histogram.

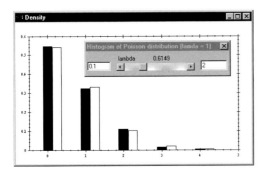

FIG. A.1. Interactive fitting of a Poisson histogram (right) to sample histogram (left).

Next, let us display a Pareto density on the screen. Now, activate the univariate POT domain, that is, the first button in the toolbar shows a single bullet and the button labeled POT is pressed.

DEMO A.3. (Plotting a Pareto Density.) Select the menu option *Distribution...* *Pareto* to open a dialog box asking for the parameters of the distribution The dialog box for the Pareto distribution is displayed in Fig. A.2. Click on the *Density* button to plot the Pareto density for the chosen parameters. A plot window opens showing the graph of the density.

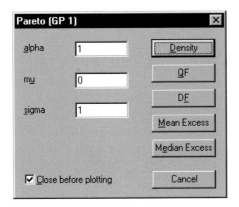

FIG. A.2. Dialog box *POT... Distribution... Pareto* in the univariate mode. Enter the parameters of the Pareto distribution and plot a curve using the buttons.

The coordinate system can be adjusted by using the local menu of the window. Click inside the window with the right mouse button and select the option *Change coordinates*. Xtremes stores the previous coordinates which may be restored by executing the option *Restore coordinates* in the toolbar (click the button) or by pressing the Backspace key.

A more direct, interactive facility to modify the coordinate system is available by selecting a certain mouse mode from the tool bar (explained in the next subsection).

One can get an overview of the curves plotted in a window utilizing the menu option *Edit Curves and Labels*. A list with all curves is displayed, and one can delete some curves or change their options like color, line style, brushes used to hatch histograms, etc. If one wants to know the parameters of a plotted curve, select the information mouse mode tool ⌗ and click onto the curve.

Pictures may be exported. Click the option *Print* in the local menu to send your window to the printer. You can also copy it to the Windows clipboard or create EPS–files. Advanced options to format a plot are described on page 478.

Selecting a Mouse Mode

The mouse mode determines what happens if you click into a window. The default mode just brings a window into the foreground.

Other mouse modes are employed to move or delete curves, change colors and plot options, add text, etc. Detailed explanations of all modes are given in Section A.4.

As an example, we describe the change of the coordinate system using a specific mouse mode. Activate the coordinate changing mouse mode by clicking the ⌗ tool in the toolbar. In this mode, a new coordinate system is selected by pulling up a rectangle: click into the graphics window (left mouse key), hold down the mouse key and move the cursor to the opposite corner. The rectangle may be pulled outside the window to enlarge the coordinate system.

Reading and Generating Data

At the beginning, the user should restrict himself to handling data included in the package or generated by Xtremes. To read a data set from the disk, execute the menu option *Data... Read Data*. The file dialog box of Windows appears.

Proceed to the `dat` folder and select any file. Xtremes loads this file and opens a window entitled `Active Sample` displaying information about the data set. Read another data set and notice that the description of the active data set changes.

Now, two data sets are kept in memory. One can choose the active data set from the ones already loaded by executing the menu option *Data... Choose Data*. A list of all data sets used in the current session is displayed and a new active data set can be selected. Keep in mind that all visualization and estimation procedures are based on the active data set.

If one needs to load multiple data sets from a different subdirectory, the *Drag and Drop* facility of Xtremes can be utilized. Just select the files in a directory listing of Windows and drop them anywhere on the Xtremes window.

Xtremes also enables the user generating data sets. Use the menu option *Data... Generate Univariate Data* and select a distribution from the menu (see Demo A.4). A dialog box opens asking for parameters, the sample size and a filename. Files are stored in the active directory of Xtremes.

After clicking OK the data set is generated and a short description appears in the `Active Sample` window.

Visualization of Data

A simple way to display data is in the form of a text. Load a data set or generate one using *Data... Generate Data* and select the menu option *Data... List Data*. Then, Xtremes opens a text window showing your data set. You can use the scroll bar to browse the data.

The *Visualize* menu contains options to display sample dfs, qfs, histograms, scatterplots, mean and median excess functions, among others. *Kernel Density* also provides options that reflect the data points at the right, left or both ends of the support. The bandwidth can be chosen by the user, an automatic selection (via cross–validation) is available.

The visualization options are also available in the local menu of a *List Data* window. They are applied to the displayed data set (rather than the active one) if selected from the local menu. An easy way to work with more than one data set is therefore to list them, minimize the windows and work with the local menus.

Time series (see Section A.4 for a description of the different types of data sets used in Xtremes) are visualized by means of the scatterplot option. Note that each scatterplot is displayed in a separate window. You can cut points from a scatterplot using the point selection (scissors) mouse mode tool ✂. The option *Least Squares Polynomial* in the local menu of the scatterplot window leads to a dialog box for polynomial regression.

The scatterplot option is also applicable to multivariate data. Depending on the active mode (univariate or multivariate), the user has to select two or three components. In the latter case, the points are displayed using a 3–D dynamic plot.

Applying Estimators to the Active Data

The three chapters in Part II of this book correspond to four different domains of Xtremes called *D(iscrete)*, *SUM*, *MAX* and *POT*. Each domain provides different distributions and estimators in the *Data... Generate Univariate Data, Distribution* and *Estimate* menus. One may switch between the different domains by means of the buttons D(ISCRETE), POT, MAX and SUM in the toolbar.

In the following example, we focus on estimators in the POT domain because it provides the richest facilities.

DEMO A.4. (Estimation Using the Hill Estimator.) To start, let us apply the Hill estimator to standard Pareto data.

(a) First, create a data set using *Data... Generate Univariate Data... Pareto(GP1)*.

(b) Next, execute the option *Estimating... Hill(GP1/GP2)* . Recall that generalized Pareto models are fitted to the upper tail of the distribution. Therefore, the estimator requires the number k of upper extremes to be used for the estimation. You can change the number of extremes by clicking the up or down arrows in the estimator dialog box.

(c) A plot of $\hat{\alpha}_{n,k}$ or $\hat{\sigma}_{n,k}$ as a function in k is obtained using the diagram option. Choose the desired parameters before clicking the button.

Various parametric curves (plotted with the estimated parameter values) can be selected from the estimator dialog box. Comparing these curves with the corresponding nonparametric ones, the user is able to judge visually the quality of the estimation.

Similar dialog boxes are provided within the MAX and SUM domains. One can work with the other parts of Xtremes while estimator dialogs are open. It is also possible to use two or more estimators at the same time to compare their results.

The Toolbar

The toolbar below the main menu provides a quick access to frequently used options of Xtremes. The tools enable the user to select different parametric distributions in the main menu. They are also used to select a mouse mode. We start with the tools already described in the Overview.

Switch menu bar from univariate to multivariate mode.

Switch menu bar from multivariate to univariate mode.

Activates pulldown menus for discrete models.

Activates pulldown menus for Gaussian models.

Activates pulldown menus for extreme value (EV) models.

Activates pulldown menus for generalized Pareto (GP) models.

Opens pulldown menu with options for hydrology data (Chapter 11).

Opens pulldown menu with options for insurances data (Chapter 12).

Opens pulldown menu with options for finance data (Chapter 13).

Opens pulldown menu providing UFO facilities.

Opens *StatPascal Editor Window* to enter and run StatPascal programs.

Next the tools are listed that are not described in the Overview.

Opens ASCII–editor window.

Opens the Windows file dialog box and loads a data set. The file dialog box provides options to delete or copy files.

The active data set is displayed in a text window.

Restores coordinate system in active window to the size before the last change.

The toolbar is also used to select a mouse mode. The mouse mode determines the action taking place when the user clicks into a window or onto a curve[1].

Standard mouse mode: no special actions occurs.

Option mouse mode*: changes display options of a curve (e.g., color, line styles, number of supporting points, etc.). The actual dialog box depends on the type of the curve.

Parameter varying mouse mode*: opens window with sliders for each parameter of a curve. Parameters are changed dynamically while sliders are dragged.

Clipboard mouse mode*:

- moves the curves to the Xtremes clipboard window. When this mode is applied in the Xtremes clipboard window, the systems asks for a destination window;

- a curve can be directly dragged to a different plot window (also to a scatterplot window), if the left mouse button is kept pressed until the cursor is located in the destination window.

Deleting mouse mode*: deletes curves from a plot window.

---

[1]Mouse modes, where one must click onto a curve, are marked with *.

Information mouse mode*: displays parameters of curve.

Coordinate changing mouse mode: adjust the coordinate system by pulling up a rectangle or, in the trivariate setup, rotate the coordinate system.

Point selection mouse mode: use this mode to cut off points in a bivariate scatterplot. Options of the local menu of a scatterplot do not use the cut points.

Line drawing mouse mode: adds straight lines to a plot window.

Label mouse mode: adds text labels to a plot. See page 478 for details.

Curve tabulating mouse mode*: the supporting points of a curve can be tabulated by storing them into a bivariate data set.

For that purpose, adjust the coordinate system and the number of supporting points (enter the *Change Coordinates* box to adjust the range of the supporting points and use the option tool 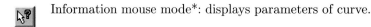 to select the number of supporting points).

A.4   Technical Aspects of Xtremes

This section discusses two technical aspects of Xtremes, namely the format of data sets and the mechanisms provided to export graphics.

Format of Data Sets

Data sets are stored as plain ASCII files. Certain specifications can be given at the top of the file, such as the type of the data set and the sample size. Moreover, one may include a shorter and a more detailed description.

Data sets can be entered by utilizing any text editor available under MS–DOS or WINDOWS. It is possible to use the integrated editor, yet one should be aware of the fact that Windows 95/98 and ME limits the text size of the editor to 64 KBytes. Under Windows NT/2000/XP and Vista 32–Bit, text files of arbitrary size can be handled.

We start with an example showing the data entry using the integrated editor. Suppose you want to create a univariate data set with the following values: 1, 3.5, 7, −4.

Start the editor by selecting the editor button in the toolbar and click on the *Header* button in the toolbar of the editor window. A dialog box asking for the type of the data set opens. Select *Univariate Data* to create a template of a univariate data set and fill in the following fields:

```
Xtremes Univariate Data
Type: Artificial example
\begin(description)
This is an artificial data set. It
was entered using the integrated editor.
\end(description)
Sample Size: 4
1
3.5
7
-4
17
```

The first line defines the type of the data set—in the present case Xtremes Univariate Data. A list of all types is given below. The second line starts with Type: and provides a short description which will be shown in the list of loaded data sets (*Data... Choose Data*). It is also added to the description of curves based on this data set. The description must be restricted to one line.

Between the lines \begin(description) and \end(description), a longer description may be added. It is displayed in the Active Sample window. The next line determines the size of the data set.

Then, the data are listed, one value for each line. After having typed the text, save it to a file (e.g., in the subdirectory \dat). Afterwards, your data set becomes the active one. One may also simulate a data set of the desired type using the option *Generate... .*

Xtremes particularly supports the following data types.

- **Xtremes Univariate Data.** Real data $x_1, \ldots, x_n$ in any order, as presented above. Execute *Data... Transform Data... Sort* to sort these data according to their magnitude.

- **Xtremes Time Series.** Pairs (i, x_i) of integers i and reals x_i as, e.g.,

1	17.5
2	-2
3	0.34
4	0.001

  The discrete time must be given in increasing order. Some of the pairs (i, x_i) can be omitted (see, e.g., ct–sulco.dat), so that the entry Sample Size is not necessarily the number of data points within the file. It may be larger than the time of the last point if values were omitted at the end of the file.

- **Xtremes Multivariate Data.** Multivariate data $(x_{i,1}, \ldots, x_{i,m})$ are stored using m entries on a line.

Moreover, the line after `Sample Size` contains an entry defining the dimension m of the data set. It is followed by m names surrounded by quotation marks. They define the headers for the corresponding column, e.g.,

Sample Size: 12

Dimension: 4

"Month"	"SO2"	"NO"	"O3"
1	75.2	13.4	17.2
2	83.1	17.9	15.4
3	.	12.8	11.3
4	43.9	15.3	11.3

Missing values are indicated as a dot. It is possible to combine related univariate data sets of different length to one multivariate data set. The rows containing a dot are ignored when the multivariate data set is transformed or converted.

- **Data Sets Without Header.** Xtremes can also load plain ASCII files containing just a matrix of data, without any headers. Such data sets are treated as multivariate. Moreover, one can use decimal points or decimal commas within a data set.

Discrete, grouped and censored data types are also available. Please consult the *Xtremes User Manual* for details.

Data can be converted from one type into another by the option *Data... Convert to*. All canonical conversions are available. There are also some special conversions.

One can apply the UserFormula facility to perform transformations not covered by the menu system. More sophisticated conversions are accomplished by means of StatPascal programs.

Documenting Illustrations

Xtremes provides various tools to change the outer appearance of a plot and to export it to other systems. We start with a description of advanced plot options (like different colors and line styles) that are used to prepare pictures for exporting. The following options are available:

- Coordinate System: the coordinate system is either displayed within the window or on a rectangle around the actual plot area. These options are controlled in the *Change Coordinates* box of the local menu. The portion of the plot area may be changed to provide space for the attachment of labels outside the frame using the *Frame Size* option.

- Line Styles and Colors: the option mouse mode tool 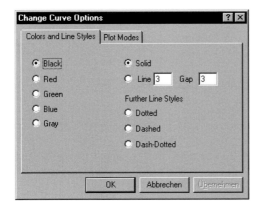 is used to change the plot options of a curve. A left–click onto the curves opens a dialog box (cf. Fig. A.3). The user may select

    - predefined line styles,
    - define his own line style by specifying the length of curve segments and gaps

  as well as the thickness of the curve. For example, choose the values

    - 1 and 1 to produce a dotted line,
    - 4 and 4 to produce a dashed line.

  These procedures lead to a better result on printed pages than the use of predefined line styles (except of the solid line).

  Different sizes and hatch styles are provided for histograms. The local menu of a scatterplot window provides the *Options* entry to change the point size.

- Adding Text: select the label mouse mode tool and click at the position where you want to put your label. The font and position of the text may be changed using the parameter varying and option mouse mode tools. It is possible to display vertical text or to move a label to the edge of the window. Labels are treated like curves, so they may be moved to another window or deleted in the same way.

The box for curve options is presented in the following Fig. A.3. We suggest to use the solid line option or the "line" and "gap" facility.

FIG. A.3. Specify your own line style in the input fields "line" and "gap".

The contents of an Xtremes plot window may be exported, either by printing a window, saving it as an EPS (Encapsulated Postscript) file or storing it in the Windows clipboard.

- Printing: first, select the option *Frame Size (Print/EPS)* (cf. Fig. A.4) from the local menu of the active window to define the size of the picture and provide space for the frame (see the next Demo A.1). Then, select *Print* to copy the contents to your printer. *Printer Setup* is utilized to change options of the printer.

- Saving an EPS file: first, set the size of the picture and frame, as in the previous case. Then, select *Save as EPS file* to store the contents in the EPS format. Xtremes asks for a filename.

- Copying to the clipboard (Option *Copy to Clipboard* in the local menu): the contents of the active window are copied to the clipboard in the standard bitmap format. It is possible to insert the contents of the clipboard in other applications like painting programs or word processors.

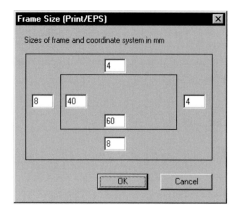

FIG. A.4. *Frame Size (Print/EPS)* dialog box. The user selects the size of the coordinate system and provides space for text displayed outside the actual picture.

In the following demo further explanations are provided about the dialog box in Fig. A.4.

DEMO A.5. (Printing a Graphics Window.) Select the *Frame Size (Print/EPS) option* from the local menu and enter the size of your picture in the dialog box. The values shown in Fig. A.4 entail a picture of the size 72mm × 52mm, the actual plot area comprises 60mm × 40mm. After that, proceed with the *Print option* to copy the active window to your printer or select *Save as EPS file* to create an EPS file that can be included in other documents.

LATEX–users may employ the *epsf* macro package, which provides commands to include postscript files into LATEX documents (e.g., the commands \epsfxsize=72mm \epsfbox{picture.eps} load picture.eps and scale it to a horizontal size of 72mm).

A.5 The UserFormula (UFO) Facilities

With UserFormula (UFO), the user can type in formulas that are used

- to evaluate expressions using a calculator;

- to plot univariate or bivariate curves;

- to generate data sets;

- to transform existing data sets.

The formulas are entered by using the notation of common programming languages.

We give an overview of the functions that are available in UserFormula expressions and describe the options of the UserFormula menu, which opens after clicking the UFO button UFO in the toolbar. Operations that are too complicated for UserFormula may be handled by using the integrated programming language StatPascal, introduced in Appendix B.

Overview

One can access all distributions implemented in Xtremes by calls to predefined functions. There are three different groups of predefined functions.

- Standard mathematical functions like `abs(x)` (absolute value), `exp(x)` (exponential function), `log(x)` (natural logarithm) or `sqrt(x)` (square root), among others.

- Function calls—partly including a shape parameter `a`—under which one may generate data, such as `betadata(a)` or `gumbeldata`. The returned values are independent for successive calls and governed by the respective distribution in its standard form. In addition, $[0, 1)$–uniform data may be called by the function `random`.

- Functions for densities, qfs and dfs (again partially including a shape parameter `a`) such as:

  - `betadensity(a,x)`, `betadf(a,x)`, `betaqf(a,x)`;
  - `gaussiandensity(x)`, `gaussiandf(x)`, `gaussianqf(x)`; etc.

The last curve plotted within an Xtremes plot window is available under the name `actualcurve`.

The chapter *Library Functions* within the *StatPascal User Manual* gives a detailed description of all predefined functions that are available within the UserFormula facility.

Calculator

The calculator allows the user to type in a formula and evaluate it. Fig. B.5 shows the calculator dialog box.

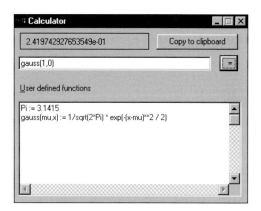

FIG. A.5. *Calculator* dialog box. Formulas typed in the upper edit field are evaluated. The lower edit field defines variables and functions also available in other parts of Xtremes.

In the lower part of the calculator window, you can define your own functions and variables. Write your definitions in the edit field *User Defined Functions* and click on the =−button. The definitions thus made are available within all dialog boxes providing a UserFormula facility. They can also be used in all edit fields where a real value is expected, e.g., in the dialog boxes used for plots of parametric curves. For example, a Gaussian density including a location parameter is defined in the following way:

```
Pi:=3.1415
gauss(mu,x):=1/sqrt(2*Pi)*exp(-(x-mu)**2/2)
```

The formulas are stored in the file `formula.txt` within the working directory. They are loaded again upon the next start of Xtremes.

Plotting Curves

The graph of a function $x \to f(x)$ or $x \to f(\boldsymbol{p}, x)$ may be plotted in every graphics window. The optional parameter vector $\boldsymbol{p} = (p1, p2, p3)$ is changed by using the parameter varying mouse mode tool ⟨⟩. Instead of $p1$ one may also use p. Within the multivariate mode, a surface plot of a function $(x, y) \to f(x, y)$ is performed.

DEMO A.6. (Plotting Gaussian Densities with Varying Location Parameter.) Click the UFO button ⟨UFO⟩ in the toolbar and select the option *Plot curve* from the popup menu. Now, type the formula `1/sqrt(2*3.1415) * exp(-(x-p1)**2/2)` in the edit field labeled *f(x) or f(p,x)* in the dialog box. If you have entered the definitions shown in the *Calculator* box, you can also write `gauss(p1, x)`.

Especially note the option for the destination window. Xtremes lists all open windows, and you can also enter the name of a new window. Select OK to plot the curve.

Generating Data

The UserFormula facility may be employed to generate univariate data sets. Click the UFO button and select the option *Generate Data...* A dialog box similar to the one used for plotting curves is utilized.

Now, the user must specify a quantile function (qf) `Q(x)` that is applied to $[0, 1)$–uniform data. For example, use `-log(x)` to generate standard exponential data.

Data distributed according to the distributions implemented in Xtremes is available by means of the predefined functions `*data` (where `*` is replaced by the name of the distribution). For example, one might also write `exponentialdata` in the above example.

Transforming Data

The UserFormula facility offers the transformation of univariate or multivariate data sets and time series. When you select the option *Transform Data...* in the UFO menu, Xtremes asks for a transformation depending on the type of the active data set.

- Univariate Data x_i: specify a transformation T to generate the data $T(x_i)$.

- Time Series (t_i, x_i): specify two functions $T_1(t, x)$ and $T_2(t, x)$ to obtain the time series values $(T_1(t_i, x_i), T_2(t_i, x_i))$. Note that real–valued times are allowed.

- Multivariate Data $(x_{i,1}, \ldots, x_{i,m})$: specify transformations T_j. The system generates

$$(T_1(x_{i,1}, \ldots, x_{i,m}), \ldots, T_k(x_{i,1}, \ldots, x_{i,m})).$$

In addition to the transformation, one must specify k names for the columns of the transformed data set. See Demo B.8.

DEMO A.7. (Smoothing a Data Set Using Polynomial Regression.) Convert the data set to a time series, display a scatterplot and add a regression polynomial. Now, the polynomial is available as `actualcurve`. Therefore, one can apply the transformation $T_1(t, x) = t$ and $T_2(t, x) = $ actualcurve(t) to store the values of the polynomial, evaluated at the times t of the original time series.

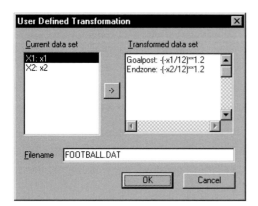

FIG. A.6. *Transform Data* dialog
box for multivariate data. The transformation in (12.39) is applied to
football.dat (with changed signs).

DEMO A.8. (UFO Transformation of Multivariate Data.) Read football.dat (cf.
Example 8.2.1) and change the signs (*Data... Transform Data... Change Signs*).
Choose UFO and apply *Transform Data*. The dialog box *Transform Data* lists the column
names X1 and X2 of the current data set on the left–hand side (see Fig. B.6) together
with the variable names x1 and x2 assigned to the values in the columns. In the edit
field on the right–hand side, the user must define the names of the ith column (using the
arrow button one may edit a template of the transformation $T(x_1, x_2) = (x_1, x_2)$). In our
example, we use the names Goalpost and Endzone and add the transformed variables
-(-x1/12)**1.2 and -(-x1/12)**1.2. Finally, press OK to execute the transformation.

Appendix B

The StatPascal Programming Language

To enhance the flexibility of the system beyond the possibilities of the UserFormula facility, it is supplemented by the integrated Pascal–like programming language StatPascal. StatPascal programs are handled in the StatPascal editor window which can be opened by selecting the SP button.

Exaggerating a bit, one can say that the pull–down menu system serves as a platform for learning the specific functions and procedures available in StatPascal. Thus, many options in the menus and dialog boxes have their counterparts as functions in StatPascal.

When a StatPascal program is executed, use the implemented

- dialog boxes,

- plot windows and a StatPascal window for the output.

One may also attach StatPascal programs to the menu bar which provides a facility to extend the menu system. We assume that the reader has a working knowledge of the Pascal language.

In contrast to other common statistical languages, StatPascal is a strongly typed language which is compiled and executed by an abstract stack machine. StatPascal is therefore usually faster than other systems.

The *StatPascal User Manual* contains a formal description of the syntax and an alphabetical list of all library functions. The first chapters of the *StatPascal User Manual* also include an introduction to basic (Pascal) programming techniques.

The installation program copies a pdf version of the *StatPascal User Manual and Reference* to the file `spmanual.pdf` in the `sp` subdirectory.

B.1   Programming with StatPascal: First Steps

This section enables the user to take the first steps into the StatPascal environment. We mention the StatPascal editor and introduce some simple programs.

The StatPascal Editor

A StatPascal editor window is opened by selecting the [SP] button within the toolbar. It provides the usual editing facilities of Windows. Text blocks can be exchanged by means of the clipboard utilizing the commands listed in Appendix A. Under Windows 95, the maximum file size of the editor window is limited to 64 KBytes.

    The toolbar enables the user to save and load text files. The *Run* option [⚡] is a short cut for the compilation and execution of a program. Fig. B.1 shows the StatPascal editor window, where the following tools are available.

New: erases the text within the StatPascal editor window.

Load: opens the file dialog box and loads a StatPascal program.

Save: writes the text in the StatPascal editor window to a disk file. If no filename has been provided, the *Save as* option is activated.

Save as: writes the text in the StatPascal editor window to a disk file after asking for a file name.

Run: compiles and executes the program in the StatPascal editor window.

Compile: compiles a program and stores the resulting binary under a filename, executes it or locates the position of a runtime error with the source file.

Compiler Options: opens the *Compiler Options* dialog box, controlling parameters of the compiler and runtime environment.

Help: opens the StatPascal online help.

The "Hello, World" Program

We start with the traditional "Hello, world" program to demonstrate the techniques of entering, compiling and running a StatPascal program. The user will immediately recognize that the program looks like its Pascal equivalent.

```
program hello;
begin
    writeln ('Hello, world!')
end.
```

Type in the program, save it to a file (*Save as*, 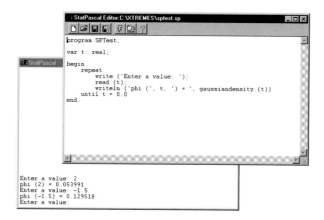 and click the *Run* button to execute the program. If the program contains no errors, Xtremes opens the StatPascal window displaying the output.

The StatPascal Window

The procedures *write*, *writeln* and *read* of the Pascal language are provided to perform input and output operations in the StatPascal window. In Fig. B.1, we see a program in the StatPascal editor window and its output in the StatPascal window.

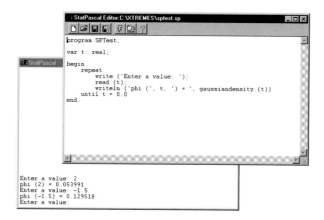

FIG. B.1. StatPascal editor window with example program (front) and its output in the StatPascal window (back).

The example program asks for real numbers t and displays the Gaussian density $\varphi(t)$. The predefined routines *MessageBox*, *DialogBox* and *MenuBox* (which are described in the StatPascal manual) provide an alternative using dialog boxes.

Data Types and Structures

StatPascal provides most of the data types and data structures of the Pascal language. The predefined types *boolean, char, integer, real* and *string*

are available, and the user can define new types using all data structures of Pascal (with the exception of variant records). Our examples use only the predefined data types in conjunction with the standard Pascal data structure array and the new data structure vector which is introduced below.

Elements of a StatPascal Program

A StatPascal program consists of up to eight different sections. In the following table, we give a short explanation of the different sections of a program. As one could see from the previous example programs, most of these sections are optional.

program *name*;	a StatPascal program starts with the reserved word **program**, followed by the name of the program
uses ... ;	the optional **uses** lists libraries which are used by the program
label ... ;	the optional **label** starts the declaration of the labels
const ... ;	the optional **const** defines constant values
type ... ;	the optional **type** is used to assign names to user defined data types (see page 489)
var ... ;	the optional **var** declares the variables used in a program
procedure ... ; **function** ... ;	an arbitrary number of functions and procedures may be declared (see page 489)
begin ... **end.**	the mandatory main program contains the instructions performed by the program

A tutorial on basic programming techniques with an introduction to standard Pascal can be found in the StatPascal manual.

Vector and Matrix Types

StatPascal implements a new data structure **vector** which is similar to the array structure. Yet, one does not have to specify the number of elements when declaring a vector. A vector type is defined using the declaration

vector of *type.*

We start with a simple example which shows the usage of a real vector. The following program generates a Gaussian data set of size 100 under the location and scale parameters 2 and 3, and stores the data in the vector x. Then the sample mean and sample variance of the simulated data set are displayed.

```
program example;
var x: vector of real;
begin
    x := 2 + 3 * GaussianData (100);
    writeln (mean (x), variance (x))
end.
```

Readers who are familiar with other statistical languages should note that the usual arithmetic and logical expressions (with componentwise operations) as well as index operations are supported.

Vectors can also be used as arguments and return types of functions. The language provides implicit looping over the components of a vector if a function operates on the base type of a vector structure.

The following program demonstrates further vector operations. We perform a numerical integration of a real–valued function f, defined on the interval $[a, b]$, using the approximation

$$\int_a^b f(x)\, dx \approx \sum_{i=1}^n \frac{f(c_i) + f(c_{i+1})}{2} \frac{b-a}{n}$$

where $c_i = a + (i-1)(b-a)/n$ for $i = 1, \ldots, n+1$. The function f as well as the parameters a, b and n are provided as arguments.

```
type realfunc = function (real): real;

function integrate (f: realfunc; a, b: real; n: integer): real;
    var fc: vector of real;
    begin
        fc := f (realvect (a, b, n + 1));
        return sum (fc [1..n] + fc [2..(n+1)]) * (b-a) / (2*n)
    end;
```

We start with a type declaration for the functional parameter. The first assignment within the function *integrate* calculates the values $f(c_i)$, $i = 1, \ldots, n+1$ and stores them in the variable *fc*.

Note that the call to the predefined function *realvect* returns a real vector with $n+1$ equally spaced points between a and b, which is given as an argument in the call of *f*.

In the second statement, we generate two integer vectors containing the values from 1 to n and from 2 to $n+1$, which serve as indices to *fc*. The index operation yields two real vectors with the values $(f(c_1), \ldots, f(c_n))$ and $(f(c_2), \ldots, f(c_{n+1}))$. The $+$ operator adds these vectors componentwise, and the predefined function *sum* calculates the sum of the components of the resulting vector. Finally, the value of the integral is returned.

Next, we define a function *square* and calculate its integral.

```
function square (x: real): real;
    begin
        return x * x
    end;

begin
    writeln (integrate (square, 0, 1, 100))   (* 0.33335 *)
end.
```

The data structure **matrix** represents two–dimensional arrays where the number of rows and columns are determined at run time. The language provides an implicit conversion from two–dimensional arrays to matrices. One can also construct a matrix using the predefined function *MakeMatrix*, which fills a matrix with the components of a vector. Matrices can be used in arithmetic operations. The multiplication of two matrices or of a matrix and a vector perform the usual mathematical matrix operations. As an example, we show a program that prints 100 random variables simulated under a bivariate Gaussian distribution with covariance matrix

$$\Sigma = \begin{pmatrix} 1 & 0.2 \\ 0.2 & 1.5 \end{pmatrix}.$$

Note that chol (S) returns a matrix C such that $S = CC^t$.

```
program bivgauss;
var S, C: matrix of real;
    i: integer;
begin
    S := MakeMatrix (combine (1.0, 0.2, 0.2, 1.5), 2, 2);
    C := chol (S);
    for i := 1 to 100 do
        writeln (C * GaussianData (2))
end.
```

Consult the StatPascal Reference Manual for further information about vectors and matrices.

B.2   Plotting Curves

StatPascal allows the user to open an Xtremes plot window and to display curves and scatterplots in it. These windows and curves exactly act like the ones available from the menu system.

In the following, we only discuss univariate curves and scatterplots. The **sp** subdirectory of the Xtremes directory contains various example programs that demonstrate the graphical facilities of StatPascal.

Univariate Curves

Xtremes provides a predefined function *plot* which is utilized to plot univariate curves. The function requires two vectors containing the points x_i and values y_i, the destination window and a description of the curve. A linear interpolation of the given points is displayed.

For example, the following program plots a Gaussian density in two Xtremes plot windows.

```
program gaussplot;
const n = 100;
var x, y: vector of real;
begin
    x := realvect (-3, 3, n);
    y := gaussiandensity (x);
    plot (x, y, 'Density 1', 'Gaussian density');
    plot (x, y, 'Density 2', 'Gaussian density')
end.
```

Fig. B.2 shows the output of the program. Two Xtremes plot windows (*Density 1* and *Density 2*) are opened by calls to *plot*. The curves are displayed as solid black lines. One can change the plot options using the procedures listed at the end of this section.

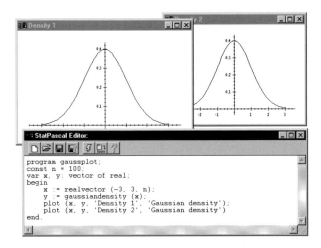

FIG. B.2. The program **gaussplot** and its output in two Xtremes plot windows.

Scatterplots

The *scatterplot* procedure, displaying scatterplots, is similar to the *plot* procedure. The routine requires three parameters: two arrays defining the points and the name of the scatterplot window. In the above example, the call to plot must be replaced by `scatterplot (x, y, 'Scatterplot')` to obtain a scatterplot of the points $(x_1, y_1), \ldots, (x_n, y_n)$.

B.3   Generating and Accessing Data

An important facility of StatPascal is the implementation of routines for data generation and data transformation not covered by the menu system or UserFormula. In this section, we introduce functions and procedures to exchange data between StatPascal and Xtremes. Data stored in a StatPascal program (e.g., in a vector) are not used directly within Xtremes, and data sets loaded in Xtremes are not used by StatPascal automatically. Instead, all data transfer is accomplished by calling predefined functions and procedures. They give the user access to the active data set from within a StatPascal program and allow to pass data collected in a StatPascal vector to Xtremes, thus creating a new active data set.

We start with an example for generating standard Pareto data under the shape parameter $\alpha = 1$.

```
program Pareto;
const n = 100;
      alpha = 1.0;
var x : vector of real;
begin
    x := paretodata (alpha, n);
    createunivariate (x, 'pareto.dat', 'Description')
end.
```

Here n Pareto data are generated independently by the function *paretodata* and stored in the vector x. The call to *createunivariate* passes the data to Xtremes, that is, the data set stored in x is saved to the file pareto.dat which is then the active one. In addition, a short comment is added. After having run the program, all options of the menu system can be applied to the new data set.

Passing Data from StatPascal to Xtremes

We now provide a more systematic description of the generation of data sets by StatPascal. Four different procedures are provided to pass data collected in a vector from StatPascal to Xtremes. In the following examples, the data are saved to filename.dat in the working directory. One may create data sets of the following types.

Xtremes Univariate Data: data $x_1, \ldots, x_n$ are collected in a real vector given as argument to the call of the predefined procedure *createunivariate*.

```
var x: vector of real;
    ...
```

```
    createunivariate (x, 'filename.dat', 'Description');
```

Instead of a vector, a one–dimensional real array may be given as well.

Xtremes Time Series: in addition to the previous case, a vector containing the times t_i of the observations must be provided.

```
    var   x : vector of real;
          t : vector of integer;
    ...
    createtimeseries (t, x, 'filename.dat', 'Description');
```

Xtremes Censored Data: besides a real vector containing the censored data, there is an integer vector with the censoring information.

```
    var   z : vector of real;
          delta : vector of integer;
    ...
    createcensored (z, delta, 'filename.dat', 'Description');
```

Xtremes Multivariate Data: the data $x_{i,j}$ are collected in a real matrix. In addition, a string with the column names, separated by '|', must be provided.

```
    var x: matrix of real;
        h: string;
    ...
    h := 'Day|Month|...';
    createmultivariate (x, h, 'filename.dat', 'Description');
```

Note that a two–dimensional array can be provided instead of a matrix type, because the language supports an implicit type conversion from two–dimensional arrays to matrix types.

As a result of such a procedure you will get an active data set of the type as specified by the command **create....** The Active Data window opens showing the name of your data set and the description provided in the last argument.

Passing Data from Xtremes to StatPascal

Next, let us consider the case where active data are dealt with by StatPascal. The active data set is accessed by means of the following functions:

samplesize	size of the active data set;
dimension	dimension of the active data of type Xtremes Multivariate Data. This function can also be applied to univariate data or a time series, yielding 1 or 2, respectively;

data(i)	$x_{i:n}$ if $x_1, \ldots, x_n$ are Xtremes Univariate Data. Use the function call *data(i,1)* to access the unsorted data;
data(i,j)	$x_{i,j}$ if $(x_{1,1}, x_{1,2}), \ldots, (x_{n,1}, x_{n,2})$ is the active time series. Multivariate data are dealt with in the same way. If a grouped data set is active, then *data(i,1)* returns the cell boundary t_i and *data(i,2)* the frequencies n_i in cell $[t_i, t_{i+1})$. Moreover, censored data are treated like multivariate data with the censored data in the first component, the censoring information in the second and the weights of the Kaplan–Meier estimate in the third one;
columndata(i)	vector with the (unsorted) data in the ith column of the active data set;
rowdata(i)	vector with the data in the ith row of the active data set;
columnname(i)	name of the ith column. This function yields an empty string if not applied to a multivariate data set.

DEMO B.1. (Translation of a Univariate Data Set.) We employ StatPascal to add the value 5 to univariate data. Note that the vector structure allows us to deal with data sets of any size.

```
program translation;
var x: vector of real;
begin
    x := columndata (1);
    createunivariate (x + 5, 'demo.dat', '')
end.
```

We used the function call `columndata (1)` to access the unsorted data set.

Author Index

Subject Index

501

Bibliography

[1] Baillie, R.T. and McMahon, P.C. (1989). *The Foreign Exchange Market: Theory and Econometric Evidence.* Cambridge University Press, Cambridge.

[2] Beirlant, J., Teugels, J.L. and Vynckier, P. (1996). Practical Analysis of Extreme Values. Leuven University Press, Leuven.

[3] Best, P. (1998). *Implementing Value at Risk.* Wiley, Chichester.

[4] Bobée, B. and Ashkar, F. (1991). *The Gamma Family and Derived Distributions Applied in Hydrology.* Water Resources Publications, Littleton.

[5] Brockwell, P.J. and Davis, R.A. (1987). *Time Series: Theory and Methods.* Springer, New York.

[6] Castillo, E. (1988). *Extreme Value Theory in Engineering.* Academic Press, Boston.

[7] Christoph, G. and Wolf, W. (1992). Convergence Theorems with a Stable Limit Law. Akademie Verlag. Berlin.

[8] Cleveland, W.S. (1993). Visualizing Data. Hobart Press, New Jersey.

[9] Coles, S. (2001). An Introduction to the Statistical Modeling of Extreme Values. Springer, London.

[10] Daykin, C.D., Pentikäinen, T. and Pesonen, M. (1994). *Practical Risk Theory for Actuaries.* Chapman & Hall, London.

[11] Embrechts, P., Klüppelberg, C. and Mikosch, T. (1997). *Modelling Extremal Events for Insurance and Finance.* Springer, New York.

[12] Enders, W. (1995). *Applied Econometric Time Series.* Wiley, New York.

[13] *Extreme Values: Floods and Droughts.* (1994). Proceedings of International Conference on Stochastic and Statistical Methods in Hydrology and Environmental Engineering, Vol. 1, 1993, K.W. Hipel (ed.), Kluwer, Dordrecht.

[14] *Extreme Value Theory.* (1989). Proceedings of Oberwolfach Meeting, 1987, J. Hüsler and R.–D. Reiss (eds.), Lect. Notes in Statistics 51, Springer, New York.

[15] *Extreme Value Theory and Applications.* (1994). Proceedings of Gaithersburg Conference, 1993, J. Galambos et al. (eds.), Vol. 1: Kluwer, Dortrecht, Vol. 2: Journal Research NIST, Washington, Vol. 3: NIST Special Publication 866, Washington.

[16] Falk, M., Hüsler, J. and Reiss, R.-D. (1994). *Laws of Small Numbers: Extremes and Rare Events.* DMV–Seminar Bd 23. Birkhäuser, Basel.

[17] Falk, M., Hüsler, J. and Reiss, R.-D. (2004). *Laws of Small Numbers: Extremes and Rare Events.* 2nd ed., Birkhäuser, Basel.

[18] Fan, J. and Yao, Q. (2003). *Nonlinear Time Series.* Springer, New York.

[19] Fang, K.-T., Kotz, S. and Ng, K.-W. (1990). *Symmetric Multivariate and Related Distributions.* Chapman and Hall, London.

[20] Galambos, J. (1987). *The Asymptotic Theory of Extreme Order Statistics.* 2nd ed. Krieger: Malabar, Florida (1st ed., Wiley, New York, 1978).

[21] Gentleman J.F. and Whitmore G.A. (1995). *Case Studies in Data Analysis.* Lect. Notes Statist. 94, Springer–Verlag, New York.

[22] Gerber, H.-U. (1979). *An Introduction to Mathematical Risk Theory.* Huebner Foundation Monograph 8, Philadelphia.

[23] Gray, H.L. and Schucany, W.R. (1972). *The Generalized Jackknife Statistic.* Marcel Dekker.

[24] Gumbel, E.J. (1958). *Statistics of Extremes.* Columbia Univ. Press, New York.

[25] Haan, L. de (1970). *On Regular Variation and its Application to the Weak Convergence of Sample Extremes.* Math. Centre Tracts 32, Amsterdam.

[26] Hamilton, J.D. (1994). *Time Series Analysis.* Princeton University Press, Princeton.

[27] Hampel, F.R., Ronchetti, E.M., Rousseeuw, P.J. and Stahel, A.W. (1986). *Robust Statistics.* Wiley, New York.

[28] Hogg, R.V. and Klugman, S.A. (1984). *Loss Distributions.* Wiley, New York.

[29] Hosking, J.R.M. and Wallis, J.R. (1997). *Regional Frequency Analysis.* Cambridge Univ. Press, Cambridge.

[30] Huber, P.J. (1981). *Robust Statistics.* Wiley, New York.

[31] Johnson, N.L. and Kotz, S. (1970). *Distributions in Statistics: Continuous Univariate Distributions–1, –2.* Houghton Mifflin, Boston.

[32] Johnson, N.L. and Kotz, S. (1972). *Distributions in Statistics: Continuous Multivariate Distributions.* Wiley, New York.

[33] Jorion, P. (2001). *Value at Risk: The New Benchmark for Controlling Market Risk.* (first edition 1997), McGraw–Hill, New York.

[34] Kinnison, R.P. (1985). *Applied Extreme Value Statistics.* Battelle Press, Columbus.

[35] Klugman, S.A. (1992). *Bayesian Statistics in Actuarial Sciences.* Kluwer, Dordrecht.

[36] Klugman, S.A., Panjer, H.H. and Willmot, G.E. (1998). *Loss Models, From Data to Decisions.* Wiley, New York.

[37] Kowaka, M., Tsuge, H., Akashi, M., Masamura, K. and Ishimoto, H. (1994). *Introduction to Life Prediction of Industrial Plant Materials, Application of the Extreme Value Statistical Method for Corrosion Analysis.* Allerton Press (Japanese version, 1984, Maruzen, Tokyo.)

[38] Kutoyants, Yu.A. (1998). *Statistical Inference for Spatial Poisson Processes.* Lect. Notes in Statistics 134, Springer, New York.

[39] Leadbetter, M.R., Lindgren, G. and Rootzén, H. (1983). *Extremes and Related Properties of Random Sequences and Processes.* Springer, New York.

[40] Mandelbrot, B.B. (1997). *Fractals and Scaling in Finance.* Springer, New York.

[41] McCullagh, P. and Nelder, J.A. (1989). Generalized Linear Models. Second Edition. Chapman and Hall, London.

[42] Reiss, R.–D. (1989). *Approximate Distributions of Order Statistics.* Springer, New York.

[43] Reiss, R.–D. (1993). *A Course on Point Processes.* Springer, New York.

[44] Resnick, S.I. (1987). *Extreme Values, Regular Variation, and Point Processes.* Springer, New York.

[45] Rosbjerg, D. (1993). *Partial Duration Series in Water Resources.* Institute of Hydrodynamics and Hydraulic Engineering. Technical University, Lyngby.

[46] Samorodnitsky, G. and Taqqu, M.S. (1994). *Stable Non–Gaussian Random Processes.* Chapman & Hall, New York.

[47] SCOR Notes: *Catastrophe Risks.* 1993, J. Lemaire (ed.)

[48] Serfling, R.J. (1980). *Approximation Theorems of Mathematical Statistics.* Wiley, New York.

[49] Shao, J. and Tu, D. (1995). *The Jackknife and the Bootstrap.* Springer, New York.

[50] Simonoff, J.S. (1996). *Smoothing Methods in Statistics.* Springer, New York.

[51] *Statistics for the Environment 2, Water Related Issues.* (1997). Proceedings for the SPRUCE Conference, Rothamsted, 1993, V. Barnett and K.F. Turkman (eds.), Wiley, Chichester.

[52] *Statistics for the Environment 3, Pollution Assessment and Control.* (1997). Proceedings for the SPRUCE Conference, Merida, 1995, V. Barnett and K.F. Turkman (eds.), Wiley, Chichester.

[53] *The Handbook of Risk Management and Analysis.* (1996). C. Alexander (ed.) Wiley, Chichester.

[54] Taylor, S. (1986). *Modeling Financial Time Series.* Wiley, Chichester.

[55] Tukey, J.W. (1977). *Exploratory Data Analysis.* Addison–Wesley, Reading.